TechOne: Automotive Heating and Air Conditioning

TechOne: Automotive Heating and Air Conditioning

Russell Carrigan

Texas State Technical College
Sweetwater, TX

John Eichelberger

St. Philip's College
San Antonio, TX

THOMSON
DELMAR LEARNING™

Australia Canada Mexico Singapore Spain United Kingdom United States

THOMSON
DELMAR LEARNING

TechOne: Automotive Heating and Air Conditioning
Russell Carrigan and John Eichelberger

Vice President, Technology and Trades SBU:
David Garza

Director of Learning Solutions:
Sandy Clark

Senior Acquisitions Editor:
David Boello

Product Manager:
Matthew Thouin

Channel Manager:
William Lawrenson

Marketing Coordinator:
Mark Pierro

Production Director:
Mary Ellen Black

Production Editor:
Barbara L. Diaz

Art/Design Specialist:
Cheri Plasse

Technology Project Manager:
Kevin Smith

Technology Project Specialist:
Linda Verde

Editorial Assistant:
Andrea Domkowski

Printed in the United States of America
1 2 3 4 5 XX 08 07 06

For more information contact
Thomson Delmar Learning
Executive Woods
5 Maxwell Drive, PO Box 8007,
Clifton Park, NY 12065-8007
Or find us on the World Wide Web at
www.delmarlearning.com

Library of Congress Cataloging-in-Publication Data
Carrigan, Russell.
TechOne. Automotive heating & air conditioning / Russell Carrigan, John Eichelberger.
p. cm.
Includes index.
ISBN 1-4018-3989-4 (pbk.)
1. Automobiles--Heating and ventilation. 2. Automobiles--Air conditioning. I. Title: Automotive heating & air conditioning. II. Eichelberger, John. III. Title.
TL271.C37 2006
629.2'772--dc22
2006000856

NOTICE TO THE READER

Contents

Preface

THE SERIES

Welcome to Thomson Delmar Learning's *TechOne*, a state-of-the-art series designed to respond to today's automotive instructor and student needs. *TechOne* offers current, concise information on ASE and other specific subject areas, combining classroom theory, diagnosis, and repair into one easy-to-use volume.

You'll notice several differences from a traditional textbook. First, a large number of short chapters divide complex material into chunks. Instructors can give tight, detailed reading assignments that students will find easier to digest. These shorter chapters can be taught in almost any order, allowing instructors to pick and choose the material that best reflects the depth, direction, and pace of their individual classes.

TechOne also features an art-intensive approach to suit today's visual learners—images drive the chapters. From drawings to photos, you will find more art to better understand the systems, parts, and procedures under discussion. Look also for helpful graphics that draw attention to key points in features such as You Should Know and Interesting Fact.

Just as importantly, each *TechOne* starts off with a section on Safety and Communication, which stresses safe work practices, tool competence, and familiarity with workplace "soft skills," such as customer communication and the roles necessary to succeed as an automotive technician. From there, learners are ready to tackle the technical material in successive sections, ultimately leading them to the real test—an ASE practice exam in the Appendix.

THE SUPPLEMENTS

TechOne comes with an **Instructor's Manual** that includes answers to all chapter-end review questions and a complete correlation of the text to NATEF standards. A **CD-ROM,** included with each Instructor's Manual, consists of **PowerPoint Slides** for classroom presentations and a **Computerized Testbank** with hundreds of questions to aid in creating tests and quizzes. Chapter-end review questions from the text have also been redesigned into adaptable **Electronic Worksheets,** so instructors can modify questions if desired to create in-class assignments or homework.

Flexibility is the key to *TechOne*. For those who would like to purchase jobsheets, Delmar Learning's NATEF Standards jobsheets are a good match. Topics cover the eight ASE-subjects areas, plus advanced topical areas, and include:

- Engine Repair
- Automatic Transmissions
- Manual Transmissions
- Suspension & Steering
- Brakes
- Electrical & Electronic Systems
- Heating & Air Conditioning
- Engine Performance
- Fuels and Automotive Emissions

Visit **www.autoed.com** for a complete catalog.

A NOTE TO THE STUDENT

There are now more computers on a car than aboard the first spacecraft, and even gifted backyard mechanics long ago turned their cars over to automotive professionals for diagnosis and repair. That's a statement about the nation's need for the knowledge and skills you'll develop as you continue your studies. Whether you eventually choose a career as a certified or licensed technician, service writer or manager, automotive engineer—or even decide to open your own shop—hard work will give you the opportunity to become one of the 840,000 automotive professionals providing and maintaining safe and efficient automobiles on our roads. As a member of a technically proficient, cutting-edge workforce, you'll fill a need, and, even better, you'll have a career to feel proud of.

Best of luck in your studies,
The Editors of Thomson Delmar Learning

About the Authors

Russell Carrigan is an automotive technology instructor at Texas State Technical College, located in Sweetwater, Texas. He is an ASE Master Technician with L1 certification and has been a part of the automotive industry since 1988. He has advanced to the position of automotive instructor, having started his career as an apprentice technician. During his service career, he achieved master technician status and was employed as a service advisor before entering the field of education as an instructor.

John Eichelberger, an ASE Master and Emission Certified automotive technician, has been an instructor at St. Philip's College in San Antonio, Texas since 1994. He has served the Automotive Technology Department as Department Chairman as well as teaching GM ASEP, Ford ASSET, ACDelco TSEP, and General Motors Training Center Courses. Prior to his teaching career, John gained 25 years of experience as an automotive technician and shop owner. In addition, he has served as president of the International Association of General Motors Automotive Service Educational Programs (IAGMASEP) and as vice-president of the Texas Independent Automotive Association (TIAA).

John has been married to his wife Shirley for 29 years. They have raised three amazing children together: JR, Shannon, and Luke. In addition to hunting and fishing, John fills his free time building stock cars that his son races at the local half-mile NASCAR race track.

Acknowledgments

The authors and publisher would like to thank the following reviewers, whose technical expertise was invaluable in creating this text:

John Christopherson
Wyoming Tech Institute
Laramie, WY

C. Neel Flannagan
Aiken Technical College
Aiken, SC

James W. Haun
Walla Walla Community College
Walla Walla, WA

Donald Lumsdon
Ivy Tech State College
Terre Haute, IN

Michael Malczewski
College of DuPage
Glen Ellyn, IL

Features of this Text

TechOne includes a variety of learning aids designed to encourage student comprehension of complex automotive concepts, diagnostics, and repair. Look for these helpful features:

Section Openers provide students with a **Section Table of Contents** and **Objectives** to focus the learner on the section's goals.

Interesting Facts spark student attention with industry trivia or history. Interesting facts appear on the section openers and are then scattered throughout the chapters to maintain reader interest.

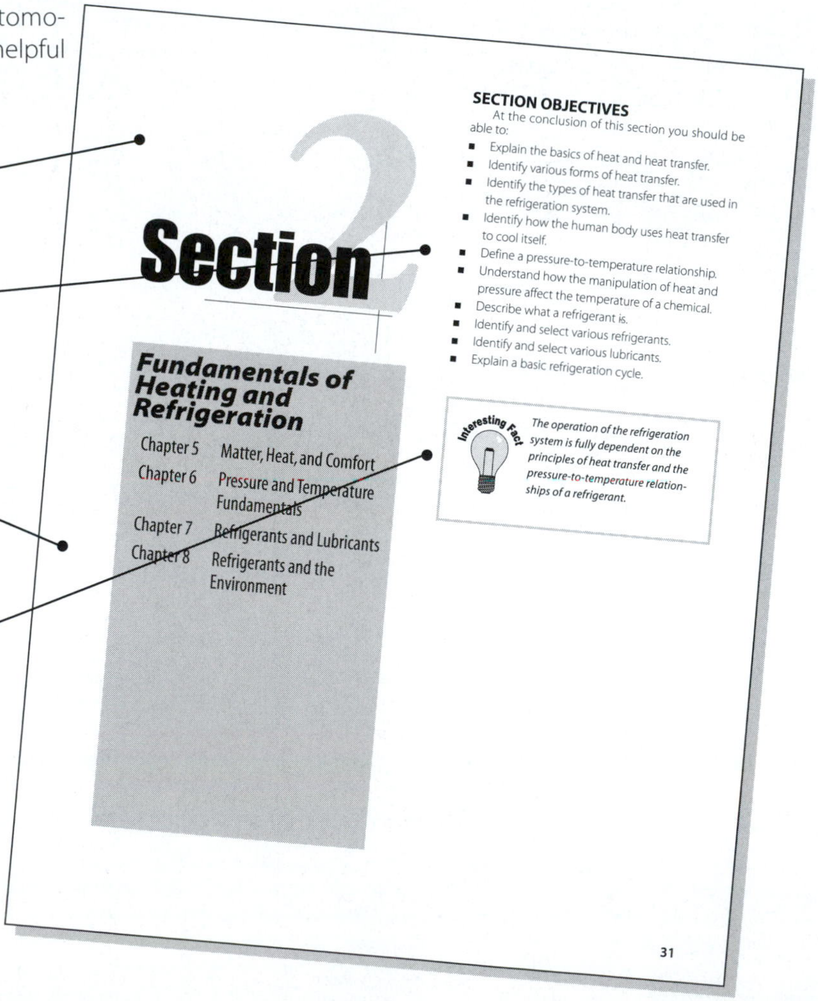

Chapter 5

Matter, Heat, and Comfort

Introduction

As a heating and air conditioning technician it is important that you understand the physical principles of matter, and heat, and their effects on personal comfort. These principles are the backbone of how the heating, air conditioning, and engine cooling systems operate. Although you may not consciously think about these principles when diagnosing or repairing one of these systems, understanding these principles will help you to develop a deeper understanding of what occurs within each of these systems. What must be considered during all automotive technical studies is that your main objective is not necessarily learning how to repair a system, but instead how to diagnose a system effectively. This cannot be accomplished without a thorough understanding of a system's operation.

MATTER

The first thing that we must identify is **matter**. Matter is defined as anything that occupies space and has mass. Matter is composed of single elements or a combination of elements that occur in nature. Some of the more common elements that we can easily identify with are also widely used in the construction and operation of the automobile heating, A/C, and cooling systems:

- Carbon: found in some quantity in all organic compounds. Carbon is found in the fuels, lubricants, and materials used in the construction of automobiles. Carbon is also used in the manufacture of air conditioning system refrigerants.
- Chlorine: a highly reactive chemical that easily combines with other elements to form compounds. Because it easily combines with other elements, it is not found in an uncombined state in nature. In order to obtain pure chlorine, chlorine-containing compounds are broken down to extract the chlorine. One of these common compounds is sodium chloride, commonly known as salt. Chlorine is used in the manufacture of some automotive refrigerants.
- Aluminum: the most common metal found within the crust of the earth. Aluminum is always found combined with another material, such as clay or mica, and it must be separated from its host compound to be used. Once refined, pure aluminum is a soft, lightweight, and ductile material that is also an excellent conductor of heat and electricity and provides excellent resistance to corrosion. The strength and rigidity of aluminum can be improved significantly by combining it with other materials such as copper, manganese, magnesium, or zinc to form an **alloy**. An alloy is a material that is manufactured from a combination of two or more materials. This is done to provide specific desirable characteristics from metals. Aluminum alloys are common in the manufacture of many automotive components.

Atoms

The atom is the smallest particle of an element. An atom still retains all of the physical characteristics of the element from which it was extracted. Atoms from different elements have different characteristics; for example, iron atoms are different from aluminum atoms. Atoms are formed of yet smaller particles called protons, electrons, and neutrons. Although we will not go into the specific construction of the atom, you should know that it is

33

An **Introduction** orients readers at the beginning of each new chapter. **Technical Terms** are bolded in the text on first reference and are defined.

Chapter 4 Diagnostic and Service Tools • 23

ods, including electronic detection, the use of fluorescent dye, liquid bubble testing, and ultrasonic tests, all of which are viable in the detection of specific leaks.

Electronic Leak Detectors

The electronic leak detector, sometimes called a "sniffer," uses a portable electronic module with a specially tipped wand to detect refrigerant leaks (see **Figure 2**). The wand is introduced into a suspected leak area. As the wand moves through the suspect area, the air is sampled and the amount of refrigerant in the sample is determined by the detector. When an excessive amount of refrigerant is detected, the operator may be notified by an audible signal, a visual signal, or both. Most modern electronic leak detectors can detect both R-12 and R-134A refrigerant leaks as slight as .5oz (14g) per year. However, some instruments may be able to detect leaks as small as .1oz (3g) per year. The electronic leak detector provides quick and easy test results for those refrigerant leaks that are consistently present. The disadvantages to this type of equipment is that the tip of wand must be in near contact with the test area. The farther away the tip, the less accurate the results. Second, because of the relative sensitivity of the electronic leak detector, small concentrations of moisture or traces from other substances may cause the detector to falsely indicate a leak. A great deal of experience is required to operate the electronic leak detector accurately. Always refer to the manufacturers' operators manual for proper procedures in the calibration and operation of your specific piece of equipment.

Interesting Fact: *Because electronic leak detectors "sniff" leaking refrigerant, they are often referred to as "sniffers."*

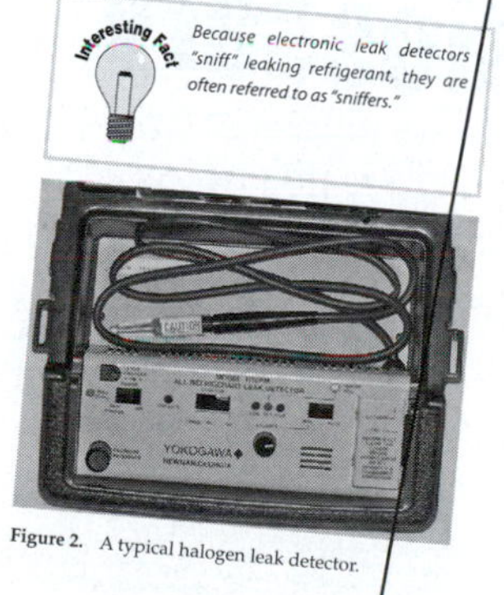

Figure 2. A typical halogen leak detector.

You Should Know: *Because some designs of electronic leak detectors can produce high voltage sparks, the detector should only be used in well-ventilated areas free from any highly flammable materials or vapors.*

Fluorescent Dye Systems

The fluorescent dye system is a visual method of identifying refrigerant leaks, and in recent years has gained wide acceptance as a highly dependable method of detecting hard-to-find refrigerant leaks (see **Figure 3**). The dye method is often used to locate leaks in areas that can be visually recognized but cannot be accessed with an electronic detector.

A specific amount of refrigerant dye is introduced into the refrigeration system. This is accomplished by mixing the dye with refrigerant oil or using special equipment to force the dye into a charged system. Once introduced, the system is then operated under normal conditions. As the refrigerant oil circulates throughout the system, the dye and oil become mixed and the mixture travels throughout the system. Because oil also mixes with the refrigerant, any time that refrigerant leaks from the system it will take a small amount of oil with it. This often leaves an oil film at the location of the leak. Leaks can then be spotted using a portable ultraviolet light. Under ultraviolet light, the dye will appear as a bright distinguishable fluorescent yellow or orange color. The main disadvantage of this method is that the system may have to be operated for several hours before a leak may appear, which often results in a return trip for the customer.

Interesting Fact: *Some refrigeration oils may contain fluorescent dye as part of the additive package.*

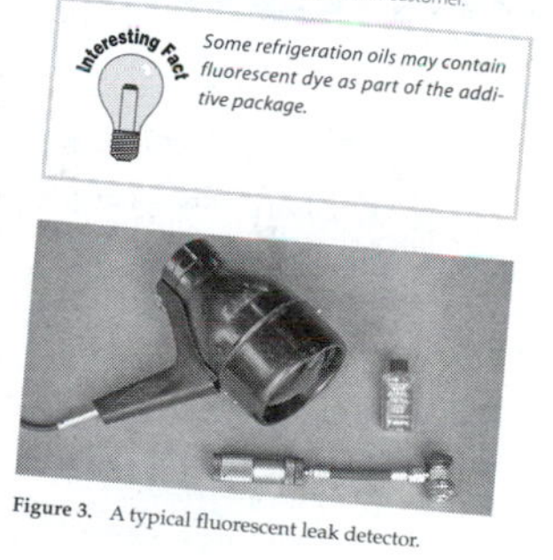
Figure 3. A typical fluorescent leak detector.

Chapter 4 Diagnostic and Service Tools • 29

grooves located on the pulley face. This tool is more compact and easier to use.

- Seal tools: these tool sets are made for a specific seal design and consist of seal removers that grip the body of the seal. Specially designed seal installers press against the body of the seal to press it into the compressor. Additionally, the set can include a tapered seal protector that slides over the threads of the compressor shaft to protect the seal against abrasion; the protector is usually slightly tapered to assist the seal in sliding over the shaft and prevent abrasion damage.

Summary

- Technicians may use multiple thermometers when diagnosing an A/C system problem. There are three common types of thermometers: dial, digital, and infrared.
- Leak detection is among the most common tasks performed by A/C repair technicians. Leak detection methods include halogen, fluorescent dyes, ultrasonic tests, and the use of a bubble solution.
- A scan tool provides data about the operation of electronic A/C system components.
- The refrigerant identifier determines the amount of specific chemicals present in a closed A/C system.
- The manifold gauge set allows the technician to monitor pressures within an operating A/C system. Additionally, the gauge set allows the technician to add refrigerant, remove refrigerant, or pull a vacuum on the system.
- The vacuum pump is used to remove air and moisture from the refrigeration system. The efficiency of the vacuum pump is directly related to the condition of the vacuum pump oil; therefore, oil must be serviced regularly.
- The scale is used to accurately measure the amount of refrigerant being introduced into the A/C system. Some scales have the ability to stop refrigerant flow when the proper amount has been introduced.
- A recovery machine is used to capture A/C system refrigerant and store it in a separate storage tank. Most recovery stations have the ability to recycle refrigerant.
- Recovery, recycling, and recharging stations have the ability to perform all related refrigerant storage tasks: recovery, recycling, evacuation, and recharging.
- Many special service tools are required to perform specific A/C system repair tasks. The most common tools include spring lock tools, orifice tube removal tools, and compressor service equipment.

Review Questions

1. List three different styles of thermometers and the advantages of each.
2. Technician A says that excessive moisture in a test area may cause an electronic leak detector to indicate false results. Technician B says that bubble solution will mix with the refrigerant oil to indicate a leak. Who is correct?
   A. Technician A
   B. Technician B
   C. Both Technician A and Technician B
   D. Neither Technician A nor Technician B
3. Which of the following substances can a refrigerant identifier detect?
   A. R-12
   B. R-22
   C. Hydrocarbons
   D. All of the above
4. All of the following statements about a manifold gauge set are true *except*:
   A. A compound gauge can read both pressure and vacuum.
   B. The hand valves isolate each respective pressure from the service hose.
   C. The yellow hose is used for service access.
   D. When the hand valves are closed, the gauges will not read pressure.
5. Which of the following statements concerning a vacuum pump are true?
   A. A vacuum pump uses specially developed oil.
   B. Moisture from the A/C system collects in the vacuum pump oil.
   C. Vacuum pump oil must be serviced only when the oil level is low.
   D. Both A and B
6. Which of the following is not the function of a recovery, recycling, and recharging station?
   A. To detect refrigerant leaks.
   B. To measure the amount of refrigerant recovered from the system.
   C. To evacuate the A/C system.
   D. To recharge the A/C system.
7. Which of the following tools should be used to install an A/C compressor clutch?
   A. An orifice tube remover
   B. A hammer
   C. Spring lock remover
   D. A clutch plate installer

You Should Know informs the reader whenever special safety cautions, warnings, or other important points deserve emphasis.

A **Summary** concludes each chapter in short, bulleted sentences. **Review Questions** are constructed in a variety of formats, including ASE-style, challenging students to prove they've mastered the material.

An **ASE Practice Exam** is found in the **Appendix** of every *TechOne* book, followed by a **Bilingual Glossary,** which offers Spanish translations of Technical Terms alongside their English counterparts.

ASE PRACTICE EXAM FOR HEATING & AIR CONDITIONING

1. Technician A says that a properly written repair order will contain information about the customer, specific information about his or her vehicle, and a detailed description of the customer's concern.
   Technician B says that the repair order is the communication link between the customer and the technician. Who is correct?
   A. Technician A
   B. Technician B
   C. Both Technician A and Technician B
   D. Neither Technician A nor Technician B
2. Technician A says that service bulletins are very general in nature and cover a wide variety of concerns on several different vehicles.
   Technician B says that service bulletins may be issued by many different sources. Who is correct?
   A. Technician A
   B. Technician B
   C. Both Technician A and Technician B
   D. Neither Technician A nor Technician B
3. Technician A says that at the very least the vacuum pump oil should be serviced after 10 hours of operation.
   Technician B says that the vacuum pump oil should never have to be changed unless it is connected to a contaminated system. Who is correct?
   A. Technician A
   B. Technician B
   C. Both Technician A and Technician B
   D. Neither Technician A nor Technician B
4. Technician A says that latent heat can be measured with an infrared thermometer.
   Technician B says that sensible heat can be considered "hidden" heat. Who is correct?
   A. Technician A
   B. Technician B
   C. Both Technician A and Technician B
   D. Neither Technician A nor Technician B
5. Technician A says that heat transfer occurs when heat energy moves from one body of matter to another.
   Technician B says that conduction, convection, and evaporation are forms of heat transfer. Who is correct?
   A. Technician A
   B. Technician B
   C. Both Technician A and Technician B
   D. Neither Technician A nor Technician B
6. Technician A says that when pressure is applied to a liquid in a sealed container, the temperature of the liquid will increase.
   Technician B says that when heat is applied to a liquid in a sealed container, the pressure will increase. Who is correct?
   A. Technician A
   B. Technician B
   C. Both Technician A and Technician B
   D. Neither Technician A nor Technician B
7. Technician A says that the use of a heat exchanger in an air conditioning system allows heat energy to be transferred from the chemical refrigerant to the atmosphere.
   Technician B says that the use of a heat exchanger in an air conditioning system allows heat energy to be

1

Bilingual Glossary

Absolute zero The temperature at which no energy is present in a material.
Cero absoluto *Temperatura en la cual no hay energía en un material.*

AC Mode HVAC operating mode where air is drawn in from outside the vehicle, cooled, and discharged through the dash vents.
Modo de corriente alterna *Modo de operación HVAC dónde el aire se obtiene de afuera del vehículo, se enfría y se desecha por las salidas de aire en el tablero.*

Actuators A device that changes electrical signals provided by the computer into mechanical actions.
Actuadores *Dispositivo que cambia las señales eléctricas que proporciona la computadora en acciones mecánicas.*

Aerated A condition that exists when air becomes mixed with a liquid, creating small bubbles within the liquid.
Aireado *Condición que existe cuando el aire se combina con un líquido, creando así pequeñas burbujas en el líquido.*

Air Distribution Ductwork and associated controls that directs the air through the HVAC system.
Distribución del aire *Ductos? y controles correspondientes que dirigen el aire a través del sistema HVAC.*

Alloy A material that is manufactured from two or more materials that provides specific desirable results.
Aleación *Material que se fabrica de dos o más materiales que proporcionan resultados deseados específicos.*

ASE See National Institute for Automotive Service Excellence

ASE *Vea Instituto Nacional para la Excelencia en el Servicio Automovilístico*

Automatic Temperature Control (ATC) HVAC operating mode that self regulates its parameters to maintain a cabin temperature set by the occupants.
Control automático de la temperatura (CAT) *Modo de operación HVAC que autorregula sus parámetros para mantener la temperatura en cabina que seleccionan sus ocupantes.*

Axial Plate The axial plate is an offset concentric plate attached to the driveshaft at an angle. As the plate rotates it forces the pistons to and fro.
Placa separadora *La placa separadora es una placa concéntrica desviada que está sujeta en ángulo al eje motor. Al dar vuelta la placa, fuerza los pistones hacia enfrente y hacia atrás.*

Azeotrope A new compound that assumes chemical properties and characteristics that are different from either of the parent chemicals.
Mezcla de temperatura de ebullición constante *Un nuevo material para pulimentar que asume propiedades químicas y características que son diferentes de cualquiera de las sustancias químicas originales.*

Balance pressure A pressure equal to that found in the evaporator, which assists the TXV in making smooth valve adjustments.
Presión de balance *Presión que es igual a la que se encuentra en el evaporador, la cual ayuda al TXV a hacer los ajustes de la válvula más suaves.*

Ballast resistor A large capacity resistor that is able to absorb a large amount of voltage, reducing the amount

303

A comprehensive **Index** will help instructors and students pinpoint information in the text.

309

Section 1

Working as an Air Conditioning Technician

Chapter 1	Safe Work Practices
Chapter 2	Working as a Heating and Air Conditioning Technician
Chapter 3	Diagnostic Resources
Chapter 4	Diagnostic and Service Tools

SECTION OBJECTIVES

At the conclusion of this section you should be able to:

- Understand your role in providing a safe work environment.
- Work safely in the automotive shop environment.
- Understand what is expected from you as an air conditioning technician.
- Understand the role of the technician in an automotive shop.
- Recognize the diagnostic resources available to you.
- Use service manuals.
- Apply service bulletins to a diagnostic procedure.
- Apply the diagnostic process.
- Identify air conditioning specialty equipment.
- Know how to select and use air conditioning specialty equipment.

A good technician possess good mechanical aptitude, strong reading and other academic skills, and a strong desire to tackle challenging problems, and consistently demonstrates good safe work practices.

Chapter 1 Safe Work Practices

Introduction

Safety is an important topic, important to you and those around you. Mishaps in the automotive shop will happen from to time to time; however, most major injuries from these mishaps are preventable. Although the safety issues presented in this chapter are categorized, true safe work practices are based on nothing else but common sense and knowledge of safety procedures. Here are some guidelines to better prepare you in the case of an accident:

- Keep a list of emergency phone numbers close to the phone and know the proper procedure for calling local emergency services (911 or otherwise).
- Make sure you know the location and contents of the shop's first aid kit.
- Always summon emergency medical services (EMS) immediately after stabilizing victim in the event of a severe emergency.
- Be familiar with the location and operation of all fire extinguishers. Know the fire evacuation routes for the entire building.
- Be familiar with the location and operation of the eye wash station and chemical shower **(Figure 1)**.
- If someone is overcome by carbon monoxide, get him or her fresh air immediately.
- Burns should be cooled immediately by rinsing them with water.
- Whenever there is severe bleeding from a wound, try to stop the bleeding by applying pressure with clean gauze on or around the wound and get medical help.
- Never move someone who may have broken bones unless the person's life is otherwise endangered. Moving that person may cause additional injury. Call for medical assistance.

Figure 1. You should know the exact location of the eye wash station and chemical shower and the procedure required to use them.

- Your supervisor should be immediately informed of all accidents that occur in the shop.

EYE PROTECTION

Eye protection should be worn whenever you are working in the shop. There are many types of eye protection available, as shown in **Figure 2**. To provide adequate eye protection, safety glasses have lenses made of polycarbonate plastic. They also offer some sort of side protection. For

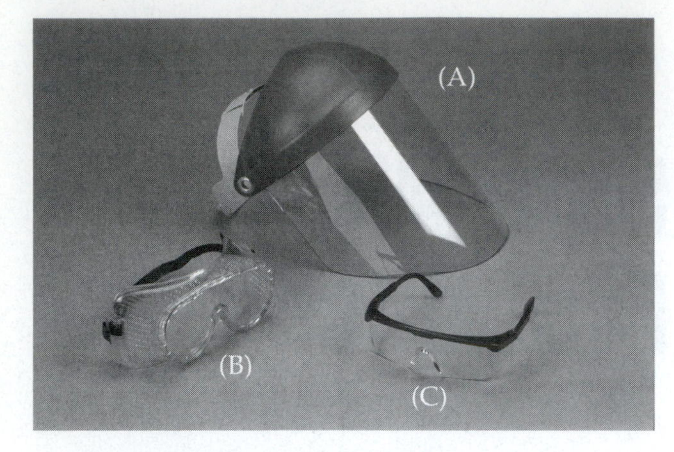

Figure 2. (A) Face shield (B) Safety goggles (C) Safety glasses.

nearly all services performed on the vehicle, eye protection should be worn. This is especially true when you are working under the vehicle.

Some procedures may require that you wear other eye protection in addition to safety glasses. A face shield not only gives added protection to your eyes but also protects the rest of your face. A face shield should be used when grinding, using pressurized cleaning equipment, or in any other situation when serious damage could occur to your face.

If chemicals such as battery acid, fuel, or solvents get into your eyes, flush them continuously with clean water. Have someone call a doctor and get medical help immediately. Many shops have eye wash stations or safety showers that should be used whenever you or someone else has been sprayed or splashed with a chemical. You must be familiar with their location and operation.

EAR PROTECTION

Ear protection should be worn any time that you are subject to loud or high-pitched noises. The practice of drying components off with compressed air is one of the most common performed in the automotive shop, but it can be one of the most damaging to your ears. Among other tasks that can damage your ears is the use of air chisels and air-powered drills. There are many comfortable types of protection available. Two common methods of ear protection are illustrated in **Figure 3**.

CLOTHING AND GLOVES

Your clothing should be well fitted and comfortable but made with strong material. If you have long hair, tie it back or tuck it under a cap. Never wear rings, watches, bracelets, and neck chains. These can easily get caught in moving parts and cause serious injury.

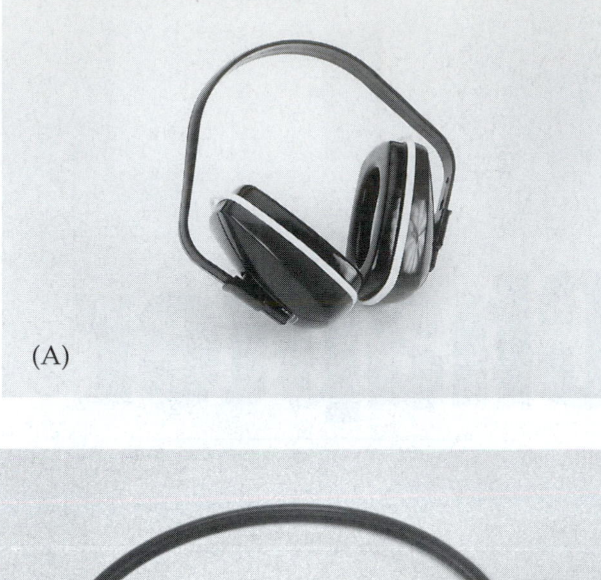

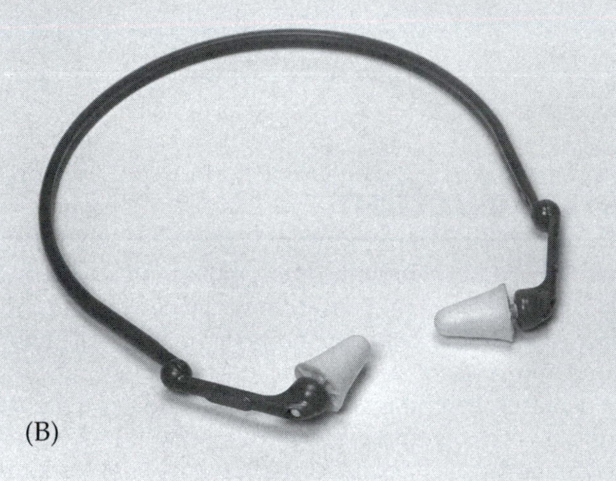

Figure 3. (A) Ear muffs (B) Ear plugs.

Automotive work involves the handling of many heavy objects, which can be accidentally dropped on your feet or toes. Always wear shoes or boots that are constructed of leather or similar material and that are equipped with no-slip soles. Steel-tipped safety shoes can give added protection to your feet.

Good hand protection is often overlooked. A scrape, cut, or burn can limit your effectiveness at work for many days. Typically, it is difficult to perform many automotive tasks with gloves on because of the lost sense of touch. However, in the last several years, several different styles of gloves made of materials such as leather and Kevlar have become available. These offer good protection but are thin enough to offer some sense of touch. Because chemical agents can be absorbed through the skin, thin latex gloves have become popular among technicians. These gloves allow the technician an incredible sense of touch and provide protection from most chemicals; however, they are ineffective against scrapes or cuts. The gloves come in a variety of different thicknesses **(Figure 4)**.

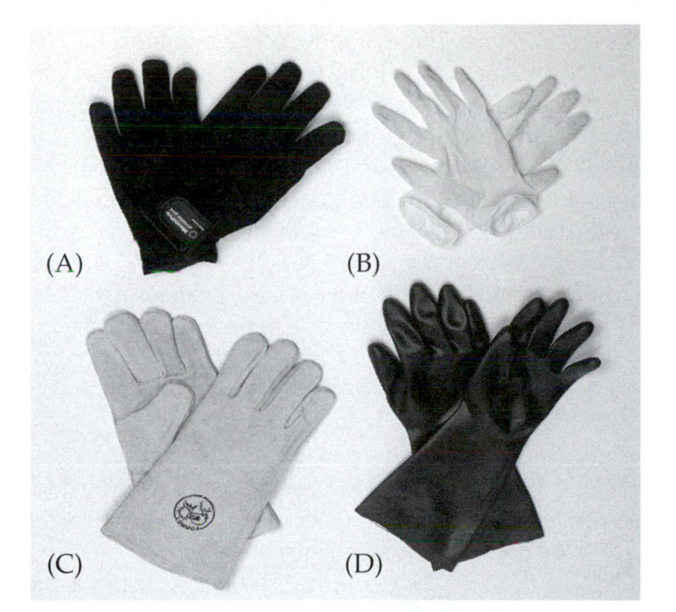

Figure 4. (A) Mechanics gloves (B) Latex gloves (C) Welding gloves (D) Chemical resistant gloves.

SAFE WORK AREAS

Your work area is your responsibility and should be kept clean and safe. Here are some simple steps to a safe work area:

1. The floor and bench tops should be kept clean, dry, and orderly. Any oil, coolant, or grease on the floor can make it slippery and should be cleaned up immediately. Slips can result in serious injuries.
2. Aisles and walkways should be kept clean and wide enough to easily move through.
3. Make sure the work areas around machines are large enough to safely operate the machine.
4. Make sure all drain covers are snugly in place. Open drains or covers that are not flush to the floor can cause toe, ankle, and leg injuries.
5. If you notice a dangerous situation or a piece of defective equipment, it is your responsibility to correct the situation or notify someone who can correct it. In the meantime, the electrical power to the equipment should disabled and an "out of order" sign should be placed in a visible location.

HAZARDOUS MATERIALS

As an auto technician, you will come in contact with hazardous materials on a daily basis. These may include cleaners, solvents, fuels, used rags, and sealers. In dealing with these materials, it is important that you know how these materials should be stored, handled, used, and disposed of. Most solvents and other chemicals used in an auto shop have warning and caution labels that should be read and understood by everyone that uses them. "Right-to-Know" laws concerning hazardous materials and wastes protect every employee in a shop. The general intent of these laws is for employers to provide a safe working place, as it relates to hazardous materials. All employees must be trained to understand their rights under the legislation, including the nature of the hazardous chemicals in their workplace, the labeling of chemicals, and the information about each chemical listed and described on Material Safety Data Sheets (MSDS). These sheets are available from the manufacturers and suppliers of the chemicals. They detail the chemical composition and precautionary information for all products that can present health or safety hazards. The Canadian version of the MSDS is the Workplace Hazardous Materials Information Systems (WHMIS).

You Should Know *When handling any hazardous material, always wear the appropriate safety protection. Always follow the correct procedures when using the material and be familiar with the information given on the MSDS for that material.*

A list of all hazardous materials used in the shop should be posted for employees to see. Shops must maintain documentation on the hazardous chemicals in the workplace, proof of training programs, records of accidents or spill incidents, satisfaction of employee requests for specific chemical information via the MSDS, and a general right-to-know compliance procedure manual utilized within the shop **(Figure 5)**.

Figure 5. MSDS information should be located in an accessible place in a "Right to Know" station like this one.

You Should Know *Whenever a hazardous material is moved from its original container into a different container, the new container must be marked with the applicable information concerning health, fire, and reactivity hazards that the material poses.*

FIRE HAZARDS AND PREVENTION

Many items around a typical shop are a potential fire hazard. These items include gasoline, diesel fuel, cleaning solvents, and dirty rags. Each of these items must be properly stored **(Figure 6)**. Each of these should be treated as potential firebombs and handled and stored properly.

It is imperative that before a fire occurs you know the location of the fire extinguishers and fire alarms in the shop and the procedures required to use them. You also should be aware of the different types of fires and the fire extinguishers used to put out these types of fires. There are basically four types of fires. **Figure 7** illustrates the symbols as they appear on the fire extinguisher and provides additional information on how each class is best extinguished, the fuels involved with each class, and the type of extinguisher material needed. Typically, fire extinguishers found in automotive repair shops are charged with chemicals to fight more than one class of fire.

USING A FIRE EXTINGUISHER

Remember, during a fire, never open doors or windows unless it is absolutely necessary; the extra draft will only make the fire worse. Make sure the fire department is contacted before or during your attempt to extinguish a fire. To extinguish a fire, stand 6 to 10 feet from the fire. Hold the extinguisher firmly in an upright position. Aim the nozzle at the base of the fire and use a side-to-side motion, sweeping the entire width of the fire. Stay low to avoid inhaling the smoke. If it gets too hot or too smoky, get out. Remember; never go back into a burning building for anything. To help remember how to use an extinguisher, remember the word "PASS":

Pull the pin from the handle of the extinguisher.
Aim the extinguisher's nozzle at the base of the fire.
Squeeze the handle.
Sweep the entire width of the fire with the contents of the extinguisher.

If there is not a fire extinguisher handy, a blanket or fender cover may be used to smother the flames. You must be careful when doing this because the heat of the fire may burn you and the blanket. If a fire is contained in a drain pan or container, smother the fire with the extinguisher's chemical or foam. If the fire is too great to smother, move everyone away from the fire and call the local fire department. A simple under-the-hood fire can cause the total destruction of the car and the building and can take some lives. You must be able to respond quickly and precisely to avoid a disaster.

LIFT SAFETY

Always be careful when raising a vehicle on a lift or a hoist. Adapters and hoist plates must be positioned correctly on twin post and rail-type lifts to prevent damage to the underbody of the vehicle. There are specific lift points. These points allow the weight of the vehicle to be evenly supported by the adapters or hoist plates. The correct lift points can be found in the vehicle's service manual. Always follow the manufacturer's instructions. Before operating any lift or hoist, carefully read the operating manual and follow the operating instructions.

Figure 6. Hazardous materials must be stored in proper containers and in proper locations.

	Class of Fire	Typical Fuel Involved	Type of Extinguisher
Class A Fires (green)	**For Ordinary Combustibles** Put out a class A fire by lowering its temperature or by coating the burning combustibles.	Wood Paper Cloth Rubber Plastics Rubbish Upholstery	Water*[1] Foam* Multipurpose dry chemical[4]
Class B Fires (red)	**For Flammable Liquids** Put out a class B fire by smothering it. Use an extinguisher that gives a blanketing, flame-interrupting effect; cover whole flaming liquid surface.	Gasoline Oil Grease Paint Lighter fluid	Foam* Carbon dioxide[5] Halogenated agent[6] Standard dry chemical[2] Purple K dry chemical[3] Multipurpose dry chemical[4]
Class C Fires (blue)	**For Electrical Equipment** Put out a class C fire by shutting off power as quickly as possible and by always using a nonconducting extinguishing agent to prevent electric shock.	Motors Appliances Wiring Fuse boxes Switchboards	Carbon dioxide[5] Halogenated agent[6] Standard dry chemical[2] Purple K dry chemical[3] Multipurpose dry chemical[4]
Class D Fires (yellow)	**For Combustible Metals** Put out a class D fire of metal chips, turnings, or shavings by smothering or coating with a specially designed extinguishing agent.	Aluminum Magnesium Potassium Sodium Titanium Zirconium	Dry powder extinguishers and agents only

*Cartridge-operated water, foam, and soda-acid types of extinguishers are no longer manufactured. These extinguishers should be removed from service when they become due for their next hydrostatic pressure test.

Notes:

(1) Freezes in low temperatures unless treated with antifreeze solution, usually weighs over 20 pounds, and is heavier than any other extinguisher mentioned.

(2) Also called ordinary or regular dry chemical. (sodium bicarbonate)

(3) Has the greatest initial fire-stopping power of the extinguishers mentioned for class B fires. Be sure to clean residue immediately after using the extinguisher so sprayed surfaces will not be damaged. (potassium bicarbonate)

(4) The only extinguishers that fight A, B, and C classes of fires. However, they should not be used on fires in liquefied fat or oil of appreciable depth. Be sure to clean residue immediately after using the extinguisher so sprayed surfaces will not be damaged. (ammonium phosphates)

(5) Use with caution in unventilated, confined spaces.

(6) May cause injury to the operator if the extinguishing agent (a gas) or the gases produced when the agent is applied to a fire is inhaled.

Figure 7. A guide to fire extinguisher selection.

Once you feel the lift supports are properly positioned under the vehicle, raise the lift until the supports contact the vehicle. Then, check the supports to make sure they are in full contact with the vehicle. Once you are satisfied that they are properly positioned raise the vehicle 6 to 10 inches from the floor and shake the vehicle to make sure it is securely balanced on the lift, and then raise the lift to the desired working height. Once you have achieved a comfortable working height, lower the lift until the mechanical locks are engaged.

JACK AND JACK STAND SAFETY

A vehicle also can be raised off the ground with a hydraulic jack. The lifting pad of the jack must be positioned under an area of the vehicle's frame or at one of the manufacturer's recommended lift points. Never place the pad under the floor pan or under steering and suspension components, as these are easily damaged by the weight of the vehicle. Always position the jack so the wheels of the vehicle can roll as the vehicle is being raised.

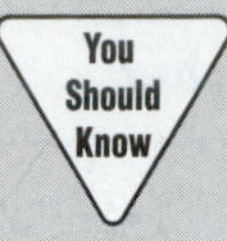
You Should Know *Safety stands should be placed under the vehicle immediately after the vehicle has been raised to the desired height.*

Safety (jack) stands are supports of different heights that sit on the floor. They are placed under a sturdy chassis member, such as the frame or axle housing, to support the vehicle. Once the safety stands are in position, the hydraulic pressure in the jack should be slowly released until the weight of the vehicle is on the stands. Like jacks, jack stands also have a capacity rating. Always use the correct rating of jack stand.

Never move under a vehicle when it is only supported by a hydraulic jack; rest the vehicle on the safety stands before moving under the vehicle. The jack should be removed after the jack stands are set in place. This eliminates a hazard, such as a jack handle sticking out into a walkway. A jack handle that is bumped or kicked can cause a tripping accident or cause the vehicle to fall.

BATTERIES

When possible, you should disconnect the battery of a car before you disconnect any electrical wire or component. This prevents the possibility of a fire or electrical shock. To properly disconnect the battery, disconnect the negative or ground cable first, then disconnect the positive cable. When reconnecting the battery, connect the positive cable first, then the negative.

You Should Know *Hydrogen gas is produced anytime that a battery is charging. Any spark created in the area of a charging battery could cause the battery to explode exposing you and your coworkers to personal injury from debris and electrolyte* ***(Figure 8)****.*

AIR BAG SAFETY AND SERVICE WARNINGS

The dash and steering wheel contain the circuits that operate the air bag system. Whenever working on or around air bag systems, it is important to be familiar with specific manufacturer system operation and service procedures. Failure to comply with these service precautions may result in an unintentional deployment of the air bag(s).

Figure 8. Sparks may cause a charging battery to explode.

You Should Know *The active chemical in a battery, the electrolyte, is basically sulfuric acid. Sulfuric acid can cause severe skin burns and permanent eye damage, including blindness, if it gets in your eye. If some battery acid gets on your skin, wash it off immediately and flush your skin with water for at least 5 minutes. If the electrolyte gets into your eyes, immediately flush them out with water, then immediately see a doctor. NEVER rub your eyes; just flush them well and go to a doctor. When working with and around batteries is a commonsense time to wear safety glasses or goggles.*

You Should Know *Disconnecting the battery does not protect you from accidental air bag deployment. Air bag systems retain the ability to deploy for several minutes after the battery has been disconnected.*

Summary

- True safe work practices are based on common sense and knowledge of safety procedures. Most major injuries and mishaps are preventable. Be aware of what is happening around you so that you will be in tune with potential hazards.
- Eye protection should be worn at all times when working in the automotive shop. A face shield will provide protection for your entire face.
- Many common shop procedures can cause loss of hearing.
- Your level of personal protection includes the clothing that you wear and the way in which is worn.
- Your work area is your responsibility and should be kept clean and safe.
- It is important that you know how to handle the chemicals that you come into contact with on a daily basis; this includes storage, usage, and disposal. The MSDS will provide you with information about chemical composition and precautionary information. All chemicals should be properly labeled.
- It is imperative that you know the location of all fire extinguishers and the procedures required to use them.
- Lift supports should be properly positioned before lifting a vehicle. Shaking the vehicle will help verify that the vehicle is stable. When using a floor jack, safety stands should be placed under the vehicle immediately after the desired height has been reached.
- When servicing air bags, all manufacturer's precautions should be strictly adhered to.

Review Questions

1. Explain when eye protection should be worn.
2. Your clothing is a significant part of your personal protection. Describe the safety aspects that are related to how you dress.
3. What information can be found on an MSDS?
4. The general intent of __________-___-__________ laws is for employers to provide a safe working place, as it relates to an awareness of hazardous materials.
5. Explain the correct procedure for putting out a fire with an extinguisher.
6. All of the following statements about lift operation are correct *except:*
   A. Every vehicle has designated lift points.
   B. The mechanical locks should be engaged once the vehicle has reached the desired working height.
   C. After the vehicle has been lifted approximately 6 feet from the ground, shake it to ensure that is properly positioned on the lift.
   D. Shaking the vehicle will help to ensure that the vehicle is properly positioned.
7. Vehicle air bags are being discussed. Technician A says that all manufacturer service procedures and precautions for that vehicle must be strictly adhered to. Technician B says that the air bag system may be capable of deploying even after the battery has been disconnected. Who is correct?
   A. Technician A
   B. Technician B
   C. Both Technician A and Technician B
   D. Neither Technician A nor Technician B

Chapter 2 Working as a Heating and Air Conditioning Technician

Introduction

An auto technician is a person who uses a variety of tools and equipment to diagnose and repair various automobile problems. As an auto technician, you have entered an incredibly demanding, yet satisfying field.

You are entering the auto repair industry at a time of significant positive growth and change. You are part of a new breed of technician. More will be expected of you and your counterparts than of any generation of technicians before you!

As an auto technician, you must first understand and accept the fact that the modern automobile is not merely a mechanical machine, made of nuts and bolts, but rather a high-tech instrument with highly integrated electronic systems and computers. Technicians must not only have mechanical aptitude but also have very strong analytical skills. You must have the desire and the ability to adapt constantly to ever-changing technologies. Employers also are looking for employees with excellent communication skills and the ability to understand what it takes to make the company successful **(Figure 1)**.

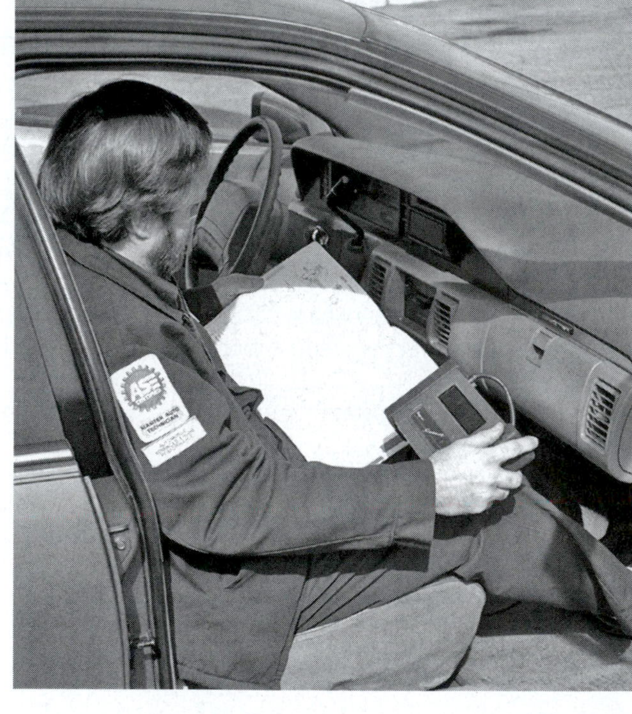

Figure 1. Technicians must have strong analytical, reading, writing, and communication skills in addition to strong mechanical aptitude.

CERTIFICATION

In many industries, service technicians are required by law to be licensed or certified. These include plumbers, electricians, and exterminators. However, certification for the auto technician is completely voluntary. The exception to this rule is for technicians who make vehicle repairs involving A/C refrigerant, who must be trained and certified according to **Environmental Protection Agency (EPA)** guidelines. Certification for any other area of automobile repair is completely voluntary. Today, however, many employers will prefer that you do have certification through the **National Institute for Automotive Service Excellence (ASE)**.

The ASE exams are broken down into several specialty areas, including heating and air conditioning and engine performance. Each area includes a comprehensive written

Figure 2. Certified technicians receive shoulder patches recognizing their achievement.

exam that is used to test your system knowledge as well as test your diagnostic reasoning skills. In addition to passing the exam, you must have completed two years of practical hands-on work experience in the transportation industry. Once a technician has passed at least one exam and met the experience requirement, he or she will be recognized as an ASE-certified technician in the specialty area in which he or she tested. A person that has successfully completed the specified series of exams in their vehicle type will earn the distinction of master technician, for example, master auto technician or master truck technician **(Figure 2)**.

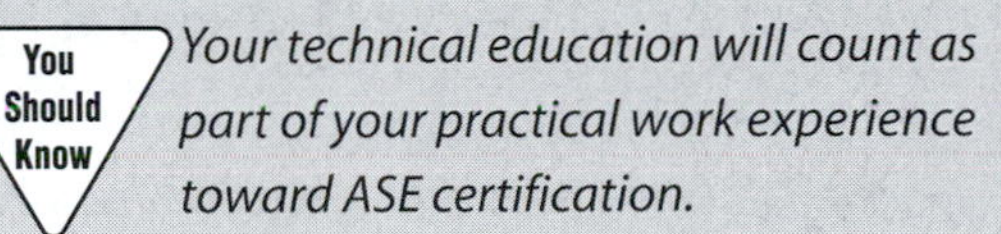

Your technical education will count as part of your practical work experience toward ASE certification.

EDUCATION

To be a successful well-rounded auto technician, you need to have a very strong academic background, especially in the areas of reading, physical science, English, and communication skills. For many years, service information was written on an eighth- to ninth-grade level, but within just a few years that has jumped to a college level. If your reading skills are inadequate, you need to work on improving those skills. Understanding the technical information provided is essential to your success.

The time you are spending on your education is only the beginning. Each year technology leaps forward; you must leap forward as well. It not unusual for technicians to take part in some form of training on a weekly basis, either at home or in the workplace. Educational opportunities are available from many outlets such as auto manufacturers, community colleges, parts manufacturers, and specialized providers. Many training programs are available in CD-ROM, video, and online formats.

JOB EXPECTATIONS

As an employee of a company, there are certain things that will be expected of you. You should always be on time or early to work. Tardiness is one of the largest problems faced by employers. If you are absent or late, work scheduled for you must be redistributed among other technicians. If you discover that you will be late for work or unable to attend for any reason, you will be expected to notify your employer before you are due to begin your shift.

Interesting Fact

Remember that the automotive service facility is a production-based profit center. Each day your service supervisor will schedule work in anticipation of the number of technicians expected to be there. The absence of just one technician will disrupt the profitability of the entire service facility.

When you are at work you will be expected to maintain a clean image. This includes your appearance as well as your language **(Figure 3)**. Almost all employers discourage the use of foul language. Because you will be required to talk with customers on occasion you must understand

Figure 3. Your personal image is a reflection on your employer.

that the impression that you leave with them will have an effect on their attitude toward you and your company. Remember that you are an employee of a company and, like it or not, you represent that company at all times, even when you are not at work.

CUSTOMER CONTACT

In the automotive industry, you will come into contact with customers on a regular basis. Each shop will have its own policies established for customer contact with technicians. You must always treat customers with the utmost respect. You should speak to them in a courteous manner and listen carefully to their questions and concerns.

SPECIALIZATION

As in many other professions, auto technicians are now often specialists. In years gone by it was typical for a technician to be comfortable working on any part of the vehicle. Because of the complexity of the modern automobile, it is commonplace for a technician to specialize in one or two system areas. That is not to say that a technician cannot competently repair a system that he or she is not specialized in; it just means they may not be as familiar with the intricacies of the other systems. This text will focus on the skills required by the heating and air conditioning specialist. The air conditioning specialist will repair concerns directly relating the performance of the heating, ventilation, and air conditioning systems. Other specialty areas include engine mechanical, transmission repair, brakes, engine performance, electrical, and chassis systems.

COMEBACKS

Comebacks are probably the biggest source for customer dissatisfaction in the auto repair industry. A comeback is a situation in which a customer has to make a return trip to the service facility either because the concern was not corrected or because another problem may have been created during correction of the original concern. Comebacks are something that every technician will experience from time to time. However, every good technician should be conscientious enough to make sure that comebacks are limited. In many facilities, comebacks are monitored and, in extreme cases, a technician with an excessive amount of comebacks may be terminated.

REPAIR ORDER

The repair order (RO) is your communication method with the customer. In most cases, the service manager or service advisor will be responsible for retrieving information from the customer. This will include:

- Name, address, and telephone information
- Vehicle make, model, type, and VIN
- A thorough description of the customer concern

The repair order is your tool in which to communicate with the customer. There are three "C's" that must exist on a repair order: concern, cause, and correction. Because this will be your direct line of communication with the customer, these must be written in as much detail as possible. The three "C's" are:

- *Concern:* the reason that the vehicle was brought in for service. The concern should be written with as much relevant information as possible, including when the concern may occur, how often the problem may occur, and any other specific information that might relevant to the customer concern.
- *Cause:* the second part of the work order that directly concerns the technician. This is where the technician fills in the **root cause** for the customer concern. The root cause is the primary factor that is causing the customer's concern. When a problem is diagnosed, you will be required to fill in the cause of the customer's concern. The cause should be written in as descriptive a manner as possible.
- *Correction:* the area in which you write in as much detail as possible about what you did to repair the vehicle. This should include any diagnostic steps that were performed and a brief description of the steps required to make the repair. A good description here allows the customer to get an idea of the amount of work that was required to repair their vehicle **(Figure 4)**.

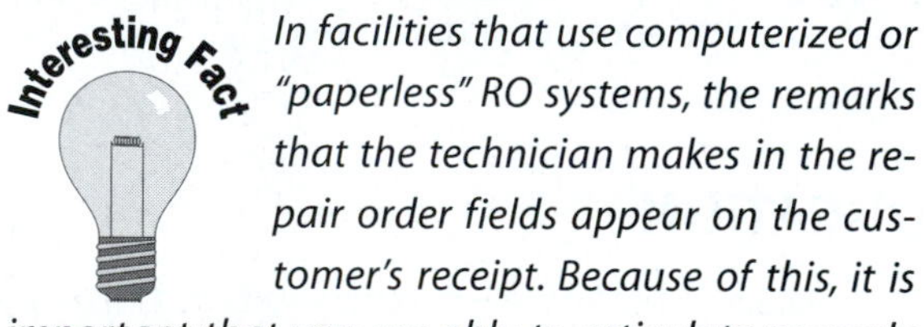

In facilities that use computerized or "paperless" RO systems, the remarks that the technician makes in the repair order fields appear on the customer's receipt. Because of this, it is important that you are able to articulate properly what was done to the vehicle.

ETHICS

Ethics is a set of moral rules or values that you apply when making decisions. Ethics in the automotive industry is extremely important because the customer trusts you with the safety and well-being of themselves and their passengers. Whether you realize it or not, ethics apply to every decision that you make. This includes the quality and attention that you exercise when repairing a customer's vehicle and the recommendations that you personally make to that customer. Furthermore, your ethics are extended not only to your customers but to your employer as well. You should treat everyone in the same way that you would want to be treated.

RHO 1020304 GRW MOTORS 100 N. MAIN ST SPOON RIVER, IL	Name: Tom Tompson Address: 1234 West Elm St. City: Spoon River State: Il Phone: Home: 603-5327 Business: 603-4327	Year: 1996 Model: Olds/Riv VIN: 1G4ED22K7T1000001 Color: Silver Lic. #: TT-500

Parts		Tech #	Customer comments	Lb. code
		#8	1. Check engine light "ON" Check for codes with scan tool. Call for authorization 2.Rattle in right rear of vehicle Check for loose shock mounting or tailpipe clearance	
		#12	Change oil and filter	

Figure 4. The repair order should contain specific information about the vehicle and the customer as well as the customer's concern, the cause of the concern, and the correction.

PRIDE IN YOUR WORK

From this point on, if you have not begun to think of yourself as a professional, it is time that you accept that mentality. Just like a doctor, you are obtaining the education that is necessary to diagnose and repair complex systems successfully. Contrary to what you may have been led to believe, at this point you are entering a highly technical and professional field. The amount of pride that you take in your career will directly affect the amount of success that you obtain.

You should strive to take an exceptional amount of pride in every job that you undertake. Whether it is a simple oil change or a major overhaul, you should treat every job with an equal amount of care and do the best job that is possible. There are no meaningless repairs made to an automobile. A repair to any vehicle system has potential safety implications and should not be taken lightly.

Summary

- Technicians must adapt to ever-changing technologies. Employers are looking for employees with strong communication skills who can make a significant contribution to making the company successful.
- Certification is not required to be an automotive technician, but increasingly it is expected. Certifications can be obtained through ASE by passing a comprehensive written exam and obtaining two years' verifiable work experience.
- An auto technician needs a very strong academic background to be successful. Your training will continue throughout your career.

- You will be expected to be punctual for work, to notify your supervisor if you will be late, and to maintain an acceptable appearance.
- Because of the increasing complexity of the automobile, technicians may often specialize in one or two areas.
- Comebacks are a large source for customer dissatisfaction. All technicians will have some amount of comebacks, but an excessive amount may lead to termination of employment.
- The RO will contain pertinent information about the customer, the vehicle, and the concerns that the customer has. Three elements that you will use to communicate with the customer through the use of the repair order are the concern, the cause, and the correction.
- Your ethics will be applied to every diagnosis and repair that you make. Your ethics will directly affect your customers, your employer, and yourself.
- You should take pride in your work no matter how large or how small the job is. Every repair made to an automobile is important.

Review Questions

1. Explain in your own words what is required of the modern automobile technician.
2. All of the following statements about ASE certification are true *except:*
   A. Federal law requires that all technicians be certified.
   B. Two years of hands-on experience are required to be fully certified in your specialty area.
   C. Your education will count toward your experience requirement.
   D. Exams are given in several specialty areas.
3. Explain why technicians must be better educated now than in the past.
4. Explain how you as an individual reflect on the image of your employer.
5. Technician A says that specialization has become commonplace because of the lack of formal education among technicians. Technician B says that specialized technicians are only competent in their specialty area. Who is correct?
   A. Technician A
   B. Technician B
   C. Both Technician A and Technician B
   D. Neither Technician A nor Technician B
6. Which of the following statements about comebacks is true?
   A. Comebacks are an opportunity to increase shop revenue.
   B. Are always preventable.
   C. Increase customer confidence.
   D. Are a primary source for customer dissatisfaction.
7. Explain what information each of the three "C's" represents.
8. True or False: As a technician you are responsible for making decisions that affect the safety and or well-being of your customers and employer.
   A. True
   B. False
9. Which of the following statements about taking pride in your work is true?
   A. All repairs deserve an equal amount attention to detail.
   B. The amount of pride that you put into your work is relative to your success.
   C. Automotive technicians are professionals.
   D. All of the above.

Chapter 3 Diagnostic Resources

Introduction

As technicians, we have only one fundamental task on our hands: to satisfy our customers. Your customers depend on you to diagnose and repair their vehicles accurately and in a timely manner. In order to do this, you must be competent in your skills; which means frequent training, a positive attitude, and a desire to fix cars. If any one of these is missing, the customer will not be satisfied. Many technicians are of the opinion that if a problem is not bothersome to them, then it is really not a problem. Remember that the customer knows his or her car best, and knows when something has changed. No matter how insignificant a problem may appear to you, the customer expects you to fix it!

INFORMATION RESOURCES

Repairing a vehicle today is all about the amount of information that you can gather. Information is the most important element in accurate diagnosis of the modern automobile. Information may come in the form of a description from a customer, a description of operation from a service manual, data that is downloaded from a car, or a suggestion from a help line. The technician who knows what his or her resources are and how to use them effectively will be more efficient, more accurate, and have greater earning potential. As a technician, you have many resources available.

Service Manuals

The service manual is the most important resource that a technician has. Service manuals are generally written for a specific vehicle model. The manual will generally contain a section devoted to every system used in that vehicle, with a description of system operation, information on the diagnosis of the system, and specific repair information. Service manuals also will contain wiring diagrams and schematics. Almost any piece of information that you will seek as a technician will be available in the service manual.

> **You Should Know** *As little as fifteen years ago, every vehicle model had one service manual volume. Today, it is not uncommon for one vehicle to have four or more dedicated volumes.*

There are generally two types of service manuals: factory and aftermarket. Factory service manuals are produced by the vehicle manufacturer and are the most specific; they will contain information about every system in the vehicle **(Figure 1)**. Aftermarket service manuals are usually based on the information found in a factory service manual; however, they usually are not as specific. Aftermarket manuals may have several makes and models

Figure 1. A manufacturer's service manual.

in one large volume and may only contain pertinent information about the most commonly serviced systems **(Figure 2)**. Aftermarket manuals also may be available for specific vehicles in small single-volume books.

Figure 2. An aftermarket service manual.

Service Bulletins

Service bulletins are an extremely helpful resource. Bulletins generally will come from three sources: auto manufacturers, parts manufacturers, and repair associations such as the Mobile Air Conditioning Society (MACS) or the Automatic Transmission Rebuilders Association (ATRA) **(Figure 3)**.

Service bulletins are written to assist the technician in the field to make cost-effective repairs. Bulletins usually are written to address a common problem with a specific vehicle or specific component. A bulletin also may be issued when a change in a particular service procedure is

ARTICLE BEGINNING

TECHNICAL SERVICE INFORMATION

FALSE DTC P0121 (REPROGRAM PCM)

Model(s): 1997 Buick Century, Skylark
1997 Chevrolet Lumina Monte Carlo Malibu, Venture
1997 Oldsmobile Achieva, Cutlass, Cutlass Supreme, Silhouette
1997 Pontiac Grand Am, Grand Prix, Trans Sport with 3100/3400 V6 Engine (VINs M, E - RPOs L82, LA1)

Section: 6E - Engine Fuel & Emission
Bulletin No.: 77-65-14A
Date: May, 1998

NOTE: This bulletin is being revised to add additional models and calibration numbers. Please discard Corporate Bulletin 77-65-14 (Section 6E - Engine Fuel & Emission).

CONDITION

Some owners may experience a MIL (Malfunction Indicator Lamp) light illuminated on the vehicle's instrument panel. Additionally, the engine's normal controlled idle speed may be slightly elevated when the MIL is illuminated.

CAUSE

The current DTC (Diagnostic Trouble Code) P0121 is too sensitive. The rational check that the diagnostic calibration performs has been changed. Part of those changes involve eliminating the defaulted higher idle.

CORRECTION

Check the calibration identification number utilizing a scan tool device. Re-flash with the updated calibration if the current calibration is not one listed in this bulletin. If the vehicle already has the most recent calibration, then refer to the appropriate service repair manual to diagnose and repair for DTC P0121. Test drive the vehicle after repair to ensure that the condition has been corrected. The new calibrations are available from the GM Service Technology Group starting with CD number 6 for 1998.

IMPORTANT: Do not attempt to order the calibrations from GMSPO. The calibrations are programmed into the vehicles PCM via a Techline Tool device.

Figure 3. A typical service bulletin.

suggested. Before a bulletin is released, it is heavily researched and tested to verify that the information is correct. It is a good idea to search for bulletins fairly early on in the diagnostic procedure, particularly when a problem appears to be out of the ordinary or extremely intermittent.

> **You Should Know** *Service bulletins will always address a specific problem for a specific vehicle. However, bulletins are not a substitute for proper diagnosis but an aid in speeding up your diagnosis.*

Labor Time Guide

The labor time guide is a book that is used to assist repair facilities in determining how much to charge a customer for a specific repair. The guide will list the most common service procedures performed and list suggested completion time for that repair **(Figure 4)**. This gives the repair facility a fair estimate of how long the repairs should take so they can advise the customer as needed on specific repair costs. Each vehicle make will have its own section and specific year models and vehicle models will be listed accordingly.

Many repair facilities base technician pay on the labor time guide. This means that the amount of money that you will be paid for making a specific repair is based on what is recommended in the time guide. This is referred to as the "Flat Rate"–based pay system. For example, if a repair time is listed as one hour, the customer is charged for one hour, and you get paid for one hour. If the repair takes three hours, the customer still only pays for one hour, and you still get paid for one hour. Should the repair take less time than quoted in the book the customer is still charged for one hour and you will be paid for one hour.

7-80

GENERAL MOTORS CORPORATION
AURORA : RIVIERA (1995 - 1998)

	Labor time	Service time
IGNITION		
Diagnose Driveability (A)		
1995-02 (.5)	.7	.7
Camshaft Interrupter, Replace (B)		
3.8L		
VIN 1 (3.2)	4.2	4.9
VIN K (2.9)	3.9	4.4
Replace camshaft sensor add	.1	.1
Camshaft Position Sensor, Replace		
1995		
3.8L VIN 1 (1.7)	2.3	2.6
3.8L VIN K (.8)	1.1	1.3
4.0L (.7)	1.0	1.1
1996-02		
3.5L (1.1)	1.5	1.7
3.8L (.8)	1.1	1.3
4.0L (.7)	1.0	1.1
Ignition Coil, Replace		
1995-97		
one (.2)	.3	.3
each additional, add	.2	.2
1998-02		
3.5L		
front (.3)	.5	.5
rear (.4)	.6	.6
both (.5)	.7	.7
3.8L, 4.0L		
one (.3)	.5	.5
each additional, add	.2	.2
Ignition Switch, Replace (B)		
1995-02 (1.0)	1.4	1.4
Spark Plug Cables, Replace (B)		
R3.8L (.8)	1.1	1.1
4.0L (1.0)	1.4	1.4

Figure 4. A labor time guide indicates how long a specific repair should take.

Computer-Based Information

Up to this point, all of the material referred to has been in the form of books or paper materials. Paper service manuals are rapidly becoming a thing of the past. Today, all of the major auto manufacturers support computer-based information systems as the primary means of information delivery. Computer-based systems provide access to all of the information previously found in a service manual as well as service bulletins, labor time guides, and other service materials. Computer-based materials are cheaper to produce and easier to update. Updates can occur through satellite or Internet links, and these mean that the technician has access to the most current material available.

Aftermarket manufacturers such as All Data and Mitchell On-Demand provide high quality information services to aftermarket repair centers. Their services are very similar to those provided by the manufacturers. However, aftermarket providers provide service information for all makes rather than for one specific manufacturer **(Figure 5)**. These systems can be delivered by CD-ROM, DVD, or through a high-speed Internet connection. Many of these systems are also coupled with shop and parts management software.

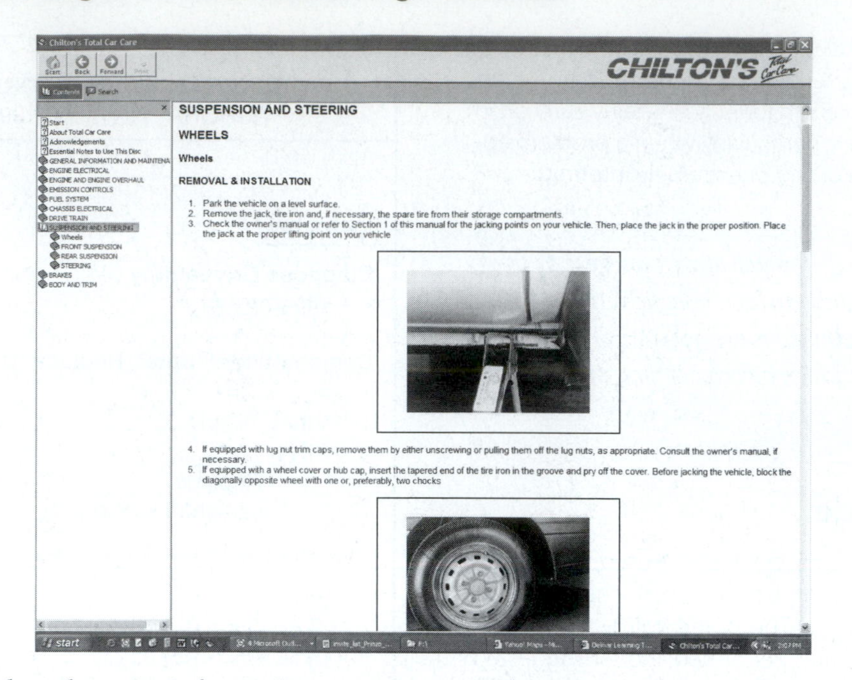

Figure 5. Computer-based service information systems give technicians access to the latest service information.

Technical Assistance

Technical assistance centers are set up to provide technical advice to technicians to help them solve difficult problems. The assistance center personnel receive their information from large databases comprised of repair information and engineering information. These services are available to dealership technicians through the manufacturers. Technicians not associated with a manufacturer have access to these services through parts manufacturers and automotive service associations. The latest trend in obtaining service information is the availability of Internet clearing houses such as the International Automotive Technicians Network (IATN). Technicians can search databases or ask for help by submitting a request to the network. The network will send out this inquiry to thousands of technicians around the world who are skilled in the area of the question. These technicians will send back ideas or suggestions. This service is free to technicians. This is a very useful source for information because it provides the ability to compare notes with thousands of other technicians. There also are hundreds of thousands of Internet Web sites and pages devoted to automotive service and information. This is an increasingly valuable resource.

Trade Associations

There are many different auto service associations. There is at least one major association for each major auto specialty area. Trade associations can prove to be extremely valuable to the auto technician, especially those not associated with a manufacturer. These associations provide valuable support to their members, many times including training, service bulletins, service information, publications, conventions, technical networks, and hotlines. For those specializing in automotive air conditioning, the Mobile Air Conditioning Society (MACS) is the most visible trade association. Founded in 1981, MACS is the leading nonprofit trade association for the mobile air conditioning, heating, and engine cooling system segment of the automotive aftermarket. MACS represents 1,600 members in North America and 47 countries around the world, and provides information and services to more than 60,000 industry shops, suppliers, and technicians. MACS serves the industry through information tools, self-paced educational materials, instructor-led training clinics, advocacy, and other member services. Since 1991, MACS has assisted more than 600,000 technicians to comply with 1990 Clean Air Act requirements for certification in refrigerant recovery and recycling to protect the environment. In addition, MACS also sponsors an annual convention and trade show. Information about MACS programs can be found on the Internet at http://www.macsw.org/.

You as a Resource

So, with all of the information, tools, and equipment available to the modern automobile technician, what is your most valuable resource? Yourself. You are still the most valuable resource; nothing can replace a qualified automobile technician. You must equip yourself to move through and analyze the amount of information presented to you. Many people think that connecting a car to a computer will tell you what is wrong. The truth of the matter is that there are no magic wands. Your tools do two things: they provide mechanical assistance and they provide information. It is still up to you to decide which one is best to use and to decipher what your tools are telling you. Because you are the most important resource, you owe it to yourself to take advantage of any training available. The more knowledge you can obtain, the more cars you will be able to repair, and, ultimately, the more money you will make and the more satisfied you will be with your career of choice!

Interesting Fact

Make it a point early in your career to explore all of your diagnostic resources and determine what is available. This may include reading your service manuals, exploring your computer-based information systems, or surfing the Internet. By learning what is available and how to access it when it is needed you will greatly increase your productivity.

DIAGNOSTIC PROCESS

A diagnosis is the process that is used by a technician to locate the root cause of a customer's concern. To diagnose a vehicle, the technician must gather information from the customer about the vehicle and from various service information sources in order to begin the process of finding the root cause of a customer concern. The skilled technician uses a diagnostic process that begins with the most simple ideas and solutions and incrementally increases in complexity. This is to help ensure that no simple problems are left unnoticed. A basic framework for a diagnostic process is provided here. The process can be customized to fit your own personal needs and diagnostic style.

A. Verify the customer's concern. All diagnostic procedures should begin by verifying what the customer has noticed. This should be accomplished using the information obtained from the customer and should include a thorough description of the problem, the conditions in which the problem occurs, and how often it occurs. Many times, it may be necessary to test drive the vehicle in order to experience what the customer has described; in some cases, it may be necessary to perform the test drive with the customer. When you are able to experience the problem for yourself, you will be able to identify two or three areas of the air conditioning or heating system that may be causing the concern.

Interesting Fact

Knowing the history of the customer's concern is always helpful in its diagnosis. This includes how often the problem occurs, how long the customer has been noticing the concern, if the problem is recurring, and if a repair has previously been made or attempted for this concern. Being informed often can make your diagnostic process more efficient by alerting you to the possibility of a previous misdiagnosis or the emergence of pattern failures; knowing this can help to prevent you from replacing good components.

B. Having verified the customer's concern, you need to perform a visual inspection of the air conditioning system. In this step, you are looking specifically for obvious problems such as loose electrical connectors, obvious oil and refrigerant leaks, or component damage **(Figure 6)**.

C. Once you have made sure that there are no obvious problems, you can begin to dig a little deeper into the diagnosis with the specific intent of isolating the concern. In this step, you may want to refer to service bulletins and diagnostic symptoms guides found in the

Figure 6. The intent of the visual inspection is to identify obvious problems such as loose connections or external component damage.

service manual. These will give you specific complaints and list the most common systems associated with them.

D. After you have isolated the concern, you will use specific diagnostic equipment and service information to locate the failure within the system. This process is often accomplished through the process of elimination by testing many components within the system. When this becomes necessary, start with the most likely component in the system that might cause the concern.

E. Once you have made a diagnosis, replace or repair the failed component.

F. When repairs are completed, the last step is to verify that the vehicle is fixed. This is accomplished by operating the vehicle in the same conditions in which the problem occurred originally.

SYMPTOMS

A symptom is the abnormal condition that a customer is experiencing with his or her vehicle. When a customer has a heating or air conditioning concern, he or she only knows that the vehicle interior does not get as hot or cold as it should, and often this may be the only description available to you. However, it is important to ask the customer a few questions so that you can actually duplicate the concern. Does the problem occur any time the "system" is on or is it intermittent in nature? If is intermittent, ask the customer to describe as specifically as possible when the problem occurs. Does the air flow from the proper selected vents and in the proper volume? Does the air seem to be tempered above or below ambient temperature? Knowing the answers to these questions will not only help you duplicate the concern but also will allow you to isolate the problem to a specific part of the heating and air conditioning system, which will help you to make an efficient and accurate diagnosis.

INTERMITTENT DIAGNOSIS

An **intermittent problem** is one that occurs at random intervals for a random period of time and then stops. From a diagnostic standpoint, there is no tougher problem to solve than an intermittent one; even the most experienced technicians have a difficult time finding the root cause of an intermittent problem. Intermittent problems do not follow distinct patterns, which makes them very difficult to verify. Intermittent concerns associated with the heat and air conditioning systems are most likely going to be directly related to the electrical or electronic controls systems. Mechanical failures are rarely intermittent.

> You Should Know
>
> *When an intermittent symptom is not occurring, it is likely that the systems associated with the symptom are operating as designed. Therefore, any tests that you perform to isolate an intermittent concern will likely indicate that the systems are operating properly.*

Summary

- Information is the most important resource in the repair of the modern automobile. The successful technician will be fully aware of all of his or her resources and will know how to use them.
- Service manuals are written with one vehicle in mind and will contain information on system operation, system diagnosis, and repair instructions.
- Service bulletins are written to address specific problems within an automobile system or component. Vehicle manufacturers, parts manufacturers, or repair associations may issue service bulletins.
- The labor time guide is used to help determine the cost for a specific repair. Common service materials are available in a computer-based format and may be supported by the vehicle manufacturer or an aftermarket provider.
- A technical assistance hotline may be able to provide technical advice to assist in the repair and diagnosis of difficult problems. Service associations provide a wealth of resources to their members.
- A well-thought-out diagnostic process starts with relatively simple tasks and will progress in complexity until the problem is solved. Familiarizing yourself with the symptoms that the customer's vehicle is experiencing will enhance your ability to properly diagnose the vehicle.
- Intermittent problems occur for brief periods of time and then disappear; this usually happens without the benefit of a distinct pattern.

Review Questions

1. Describe the type of information found in each of the following resources: service manual, service bulletins, and labor time guide.
2. Explain why the computer is the preferred method of delivery for service information.
3. Describe each of the six steps in the diagnostic process.
4. Describe why it is important to understand the vehicle symptoms before beginning diagnosis of the vehicle.
5. Which of the following statements about intermittent problems is true?
   A. Intermittent problems occur at regular intervals.
   B. The most difficult part of repairing an intermittent concern is being able to verify the concern.
   C. Most experienced technicians are able to locate intermittent problems very quickly.
   D. Very few symptoms can appear as intermittent concerns.

Chapter 4 Diagnostic and Service Tools

Introduction

Technicians involved with the service and repair of heating and A/C systems must use a variety of specialized tools and equipment. Aside from using common hand tools to make repairs, the technician will also use specialized equipment for system diagnosis, compressor service, and refrigerant service. The most commonly used tools in the air conditioning service industry are described here. It is not intended that you memorize each tool and its function at this point; however, you should familiarize yourself with the contents of this chapter and use it as a reference later when these tools are specifically referred to in a service section.

THERMOMETERS

Thermometers are used by the air conditioning service technician to measure a variety of temperatures, such as outlet temperature or condenser temperature. A/C service technicians will often use multiple thermometers simultaneously. Several styles of thermometers are practical for automotive use (see **Figure 1**).

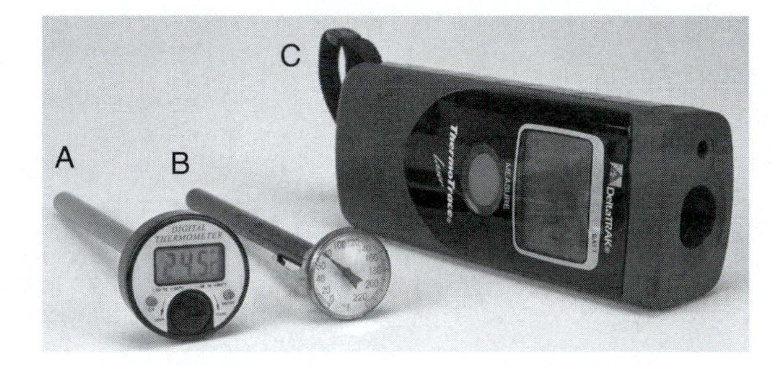

Figure 1. Typical thermometers: (A) digital, (B) dial, and (C) infrared.

The least expensive and the most straightforward thermometer is a dial thermometer. This thermometer has a long stem that is placed in contact with the surface of the object that is being measured; common locations include A/C and heating ducts or the surface of the condenser. The temperature reading is then read from the face of a dial. Dial thermometers used in A/C service work usually have a range of 0 degrees F to 220 degrees F (−18 to 104 degrees C) and require periodic recalibration.

A digital thermometer is almost identical to the dial thermometer. The difference is that a small liquid crystal screen replaces the dial. Because the temperature is displayed directly on the display, readings are easier to obtain. Additionally, digital thermometers maintain a greater degree of accuracy than dial thermometers.

An increasingly popular type of thermometer is the infrared thermometer, which is a noncontact thermometer that measures the infrared light emitted by an object. Because the infrared thermometer is a noncontact instrument, readings can be obtained from objects that are inaccessible because of high surface temperatures or physical obstructions. Some infrared thermometers are equipped with a laser that is used to target the object being measured.

LEAK DETECTION EQUIPMENT

One of the most common tasks performed by the A/C technician is to determine the source of refrigerant leaks. This can be accomplished by a number of different methods, including electronic detection, the use of fluorescent dye, liquid bubble testing, and ultrasonic tests, all of which are viable in the detection of specific leaks.

Electronic Leak Detectors

The electronic leak detector, sometimes called a "sniffer," uses a portable electronic module with a specially tipped wand to detect refrigerant leaks (see **Figure 2**). The wand is introduced into a suspected leak area. As the wand moves through the suspect area, the air is sampled and the amount of refrigerant in the sample is determined by the detector. When an excessive amount of refrigerant is detected, the operator may be notified by an audible signal, a visual signal, or both. Most modern electronic leak detectors can detect both R-12 and R-134A refrigerant leaks as slight as .5oz (14g) per year. However, some instruments may be able to detect leaks as small as .1oz (3g) per year. The electronic leak detector provides quick and easy test results for those refrigerant leaks that are consistently present. The disadvantages to this type of equipment is that the tip of wand must be in near contact with the test area. The farther away the tip, the less accurate the results. Second, because of the relative sensitivity of the electronic leak detector, small concentrations of moisture or traces from other substances may cause the detector to falsely indicate a leak. A great deal of experience is required to operate the electronic leak detector accurately. Always refer to the manufacturers' operators manual for proper procedures in the calibration and operation of your specific piece of equipment.

Because electronic leak detectors "sniff" leaking refrigerant, they are often referred to as "sniffers."

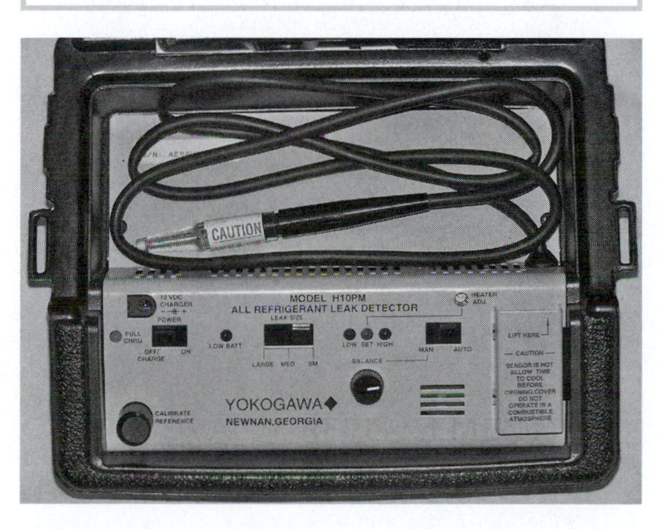

Figure 2. A typical halogen leak detector.

You Should Know

Because some designs of electronic leak detectors can produce high voltage sparks, the detector should only be used in well-ventilated areas free from any highly flammable materials or vapors.

Fluorescent Dye Systems

The fluorescent dye system is a visual method of identifying refrigerant leaks, and in recent years has gained wide acceptance as a highly dependable method of detecting hard-to-find refrigerant leaks (see **Figure 3**). The dye method is often used to locate leaks in areas that can be visually recognized but cannot be accessed with an electronic detector.

A specific amount of refrigerant dye is introduced into the refrigeration system. This is accomplished by mixing the dye with refrigerant oil or using special equipment to force the dye into a charged system. Once introduced, the system is then operated under normal conditions. As the refrigerant oil circulates throughout the system, the dye and oil become mixed and the mixture travels throughout the system. Because oil also mixes with the refrigerant, any time that refrigerant leaks from the system it will take a small amount of oil with it. This often leaves an oil film at the location of the leak. Leaks can then be spotted using a portable ultraviolet light. Under ultraviolet light, the dye will appear as a bright distinguishable fluorescent yellow or orange color. The main disadvantage of this method is that the system may have to be operated for several hours before a leak may appear, which often results in a return trip for the customer.

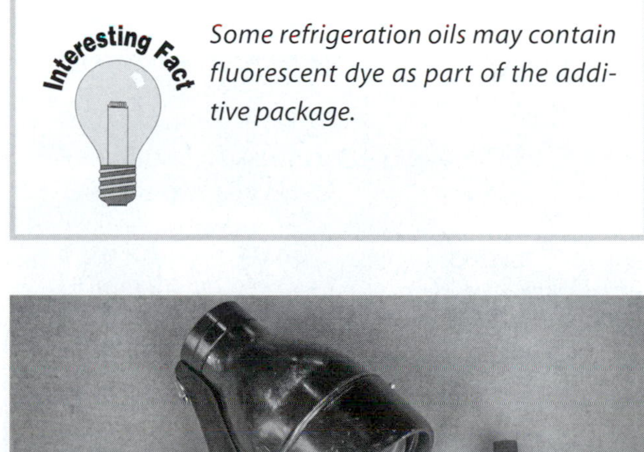

Some refrigeration oils may contain fluorescent dye as part of the additive package.

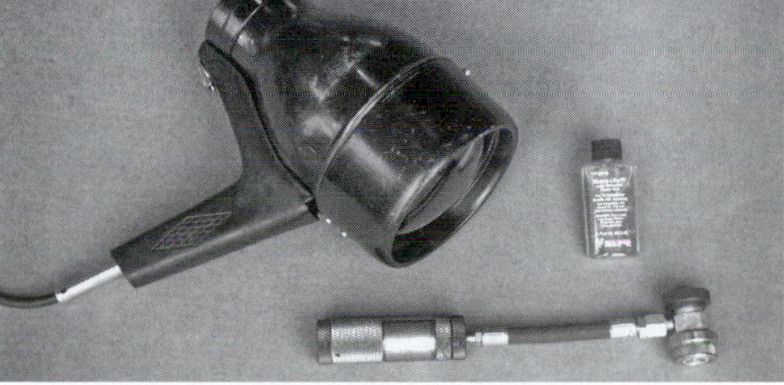

Figure 3. A typical fluorescent leak detector.

Bubble Testing

Bubble testing, another method of locating refrigerant leaks, often is used to verify the results of a suspected leak or is used in areas that may otherwise be inaccessible to a halogen leak detector. The test involves soaking the suspected leak area in a soap solution and watching for the appearance of bubbles. A commercially available leak detection solution is preferable, but a homemade solution can be mixed using water and dishwashing detergent. In order for this method to be effective, the system must have at least 50 psig (348 kPa) present in the system. If the system is lower than this, refrigerant must be added. With the system at or above the recommended pressure, soak the suspected components or fittings with the soap solution. If a leak is present, bubbles will form (see **Figure 4**). It should be noted that this method can only be used to detect relatively large leaks, but it is still used by some technicians as an alternative leak test method.

Ultrasonic

Some leaks can be found using an ultrasonic listening device. This device consists of an ultrasensitive microphone, an amplifier, and a set of headphones (see **Figure 5**). This type of equipment is able to hear noises that might not otherwise be heard by the human ear. In order for this method to be effective, pressure or a vacuum must be present in the system and the suspected leak area must be accessible to the wand-like microphone. The microphone is moved near the A/C components or lines while the operator listens for leaks through the headphones. With some practice, this can be used as a very effective method of finding otherwise difficult refrigeration leaks. This same piece of equipment may also be used to locate vacuum and other small leaks.

Figure 4. Leaks are exposed when bubbles form.

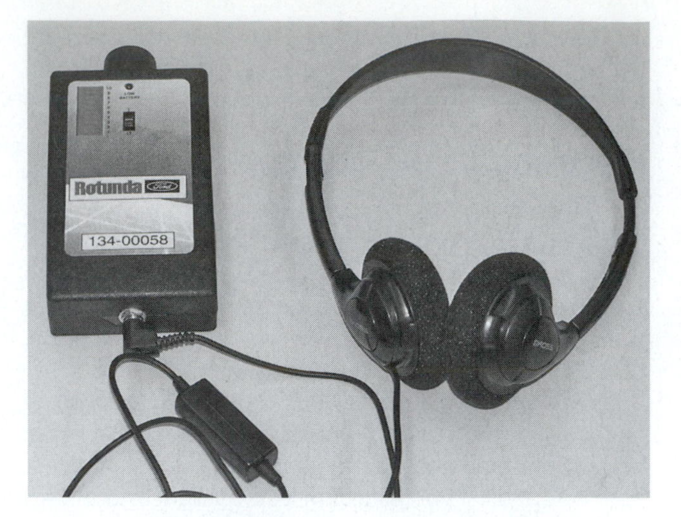

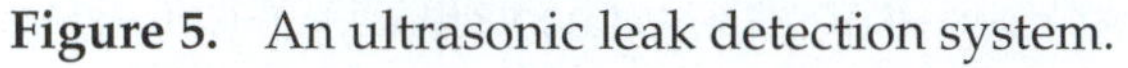

Figure 5. An ultrasonic leak detection system.

You Should Know

In order to reduce the possibility of personal injury, all leak detection services should be performed with the engine turned off.

SCAN TOOL

A scan tool is a hand-held bidirectional computer that is used to access on-board vehicle computer networks (see **Figure 6**). Once almost exclusively identified with the diagnosis of the power train systems, the scan tool is now an essential piece of equipment for technicians in all vehicle repair specialties. For the heating and A/C technician, the scan tool is used to access computer input and output information for various electronic controls that affect system operation. In addition to allowing the technicians to view system data, many scan tools allow the technician to use

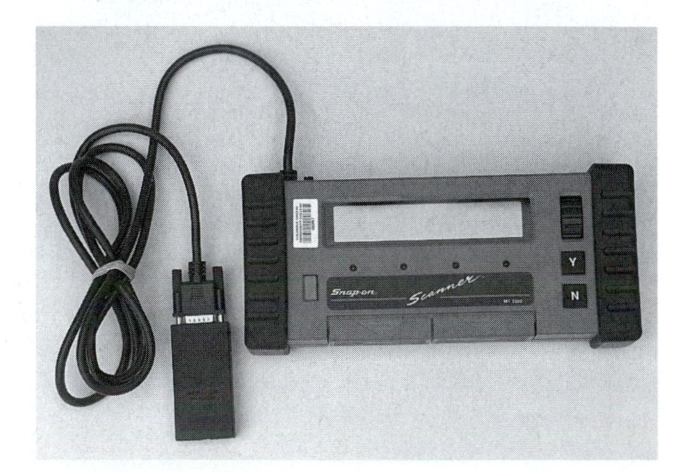

Figure 6. A typical scan tool.

bidirectional controls to activate or deactivate system **actuators** to aid in system diagnosis.

REFRIGERANT IDENTIFIERS

The refrigerant identifier is an essential piece of equipment used to determine the type of refrigerant or refrigerants that are present in the A/C system (see **Figure 7**). In today's service environment, there are many different refrigerants available. If refrigerant types are accidentally combined within a recycling unit, any refrigerant that is stored in the unit will be contaminated. When refrigerant becomes contaminated, it becomes unusable and must be disposed of.

The refrigerant identifier is connected to the A/C system through the service ports, and a small refrigerant sample is processed within the machine. The identifier can detect the amounts of R-12, R-134, R-22, air, and hydrocarbons. The amounts of each chemical will be displayed as a percentage. If hydrocarbons are present, an audible alarm will also sound, alerting the operator to their presence. When contaminated refrigerant is detected, it must be removed using special procedures and dedicated equipment. For more information, refer to Chapter 16.

Figure 7. A refrigerant identifier.

Sealant Detection Kits

Air conditioning system sealers are widely available in the market today as a quick fix to refrigeration system leaks. These products can be harmful to refrigeration service and test equipment and in some cases can void the warranty if damage should occur. There are several companies that manufacture sealant detection equipment. It is recommended that each system be tested for the presence of sealants before any test equipment is connected. Refrigerants that are found to be contaminated should not be recovered and should be handled in the same manner as other contaminated refrigerants.

MANIFOLD GAUGE SET

The manifold gauge set is a piece of equipment that allows the technician to tap into a fully charged and operational A/C system. The technician can monitor system pressures and perform routine service such as refrigerant recovery, recharging, and evacuation. To prevent contamination of the refrigeration system, different gauge sets are used for R-12 and R-134 systems. Although the function and construction of manifold gauge sets for R-12 and R-134 are virtually the same, the hoses and fittings are completely different, thus preventing the possibility of accidentally connecting the wrong gauge set to a refrigeration system.

The manifold gauge set consists of a **manifold** with two hand valves, two gauges, and three hoses (see **Figure 8**).

Figure 8. A typical manifold gauge set.

The manifold is a tube or block that provides multiple common passages that direct gases or liquids and provide a means of connecting the gauges, hoses, and valves (see **Figure 9**).

Two types of gauges are used in the typical manifold gauge set. A **compound gauge** is used to measure low-side system pressure. The compound gauge has the ability to measure both vacuum and pressure. The compound gauge will measure a vacuum in a range from 0 to 30 inches of mercury (in-Hg) and pressures from 0 pounds per square inch gauge (psig) or 0 kilopascal (kPa) up to 150 psig (1034 kPa), whereas some gauges may read above 150 psig (1034 kPa). This will prevent gauge damage in the event that the system pressure should exceed the gauge maximum. The accuracy of readings above this threshold are greatly reduced.

The second gauge is used to measure high-side system pressure. The pressure range capability of this gauge is typically 0 psig (0 kPa) to 500 psig (3447 kPa). Because the high side of the system should never pull into a vacuum, only pressures are displayed on the high-side gauge.

The gauges are threaded directly into the manifold. The low-side gauge is located on the left side and the high-side gauge is on the right. The manifold has passages that connect each gauge to the three hose connections. One hose is located below each gauge. A blue hose signifies the low-side connection and a red hose indicates a high-side connection. A yellow hose is located between the blue and red, this hose is used for refrigerant recovery, evacuation, or recharging. A hand valve is located on each side of the manifold. The hand valves are used to isolate the high- or low-side pressures from the yellow service hose. When the hand valve is open, the respective gas present in the hose has an open pathway to the yellow service hose. When the valve is closed, the pathway is blocked; however, the gauge is still able to read pressure. When both valves are open, the both the low and high sides of the system have a pathway to the yellow service hose. In this condition, the high and low side are effectively connected (see **Figure 9**).

Gauge Adjustment

To provide consistently accurate readings, the gauge may have to be recalibrated periodically. When properly calibrated, the needle should rest on zero with no pressure applied to the service hose. If the needle does not rest on zero, the gauge should be calibrated. This is done with the following steps:

1. Make sure that no pressure is applied to the pressure hose of the gauge that needs calibration.
2. Remove the lens retaining ring and lens.
3. Turn the adjusting screw in the proper direction to bring the needle back to zero.
4. Replace the lens and retaining ring.

VACUUM PUMP

Any time an A/C system is opened, air and moisture will become trapped in the lines and components. The longer the system is opened, the more moisture will become trapped. It is essential that all of the moisture and air is removed. This is accomplished using a vacuum pump (see **Figure 10**). Most vacuum pumps may be driven by an electric motor; however, there are less expensive models that are operated by compressed air.

The vacuum pump is lubricated with a specially blended vacuum pump oil. Vacuum pump oil is designed to resist high temperatures and to contain a minimal

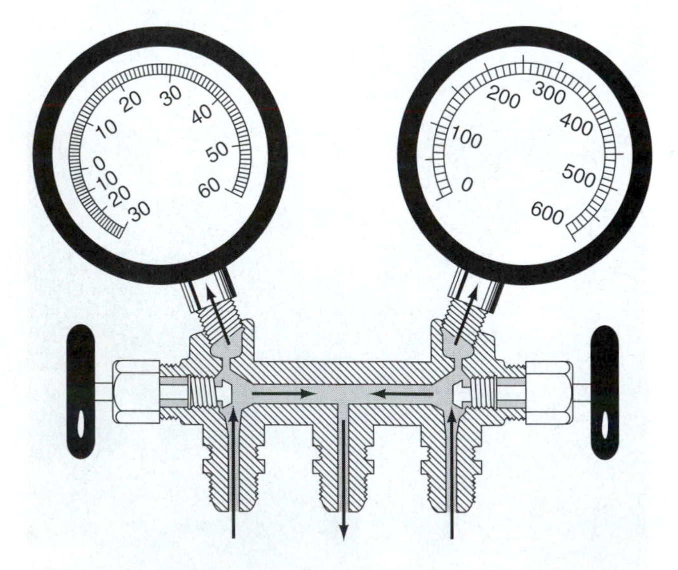

Figure 9. Refrigerant flow through a manifold gauge set.

Figure 10. A typical vacuum pump.

amount of moisture. The efficiency of the vacuum pump is reliant on the condition of the vacuum pump oil. For this reason, the vacuum pump must be serviced at regular intervals to maintain proper operation. Because some of the moisture and contaminants that are removed from the A/C system will end up in the vacuum pump, oil service intervals can vary. Under normal conditions, the pump should be serviced after 10 hours of service.

SCALE

An electronic scale is used to determine the amount of refrigerant that is being introduced into the A/C system (see **Figure 11**). Most of the scales that are in use today are electronic; however, there are some older dial or beam type scales still in use. Each scale has slightly different operating modes and procedures, but in summary the weight of the refrigerant cylinder is measured, and as the system is charged, the weight of the cylinder is monitored. When the weight of the cylinder is lightened by the same amount as the programmed refrigerant capacity, the charging sequence has been completed. Some electronic scales may be equipped with an electronically controlled valve that is connected between the refrigerant tank and the system being charged. When the programmed amount of refrigerant has been pulled from the tank, the valve will turn off and stop refrigerant flow.

RECOVERY MACHINE

Before federal laws went into effect, it was common practice to vent the capacity of the A/C system refrigerant into the atmosphere. It has since been determined that this practice is very harmful to the environment. Because of these regulations, any refrigerant removed from an A/C system must be captured or recovered. **Recovery** is a process in which A/C system refrigerant is captured and stored in a container. This is performed with a machine that pulls the refrigerant from the A/C system and pumps it into a storage vessel. Most automotive recovery units also possess the ability to **recycle** refrigerant; this is a process in which stored refrigerant is filtered to remove air, moisture, oil, and other impurities from the refrigerant. Once the refrigerant has been recycled, it can be reused in a refrigeration system.

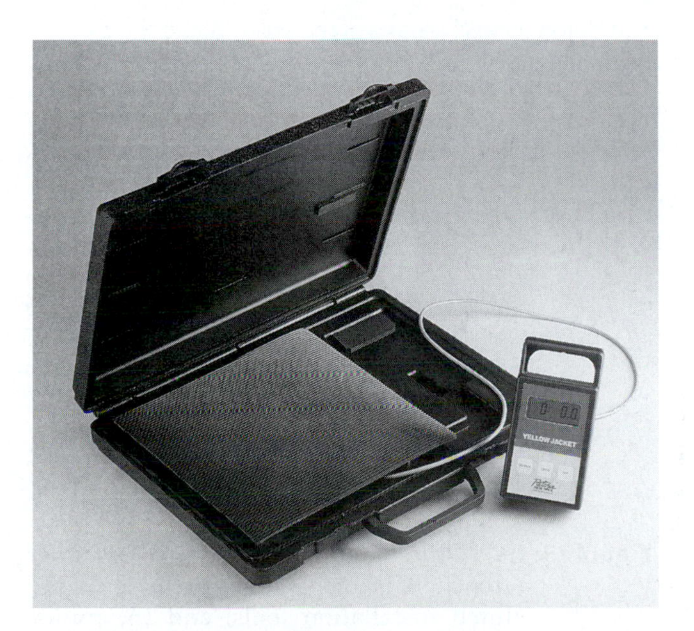

Figure 11. An electronic scale.

RECOVERY, RECYCLING, AND RECHARGING STATION

Many service facilities employ the use of recovery, recycling, and recharging systems. These systems include all of the equipment that is required to service the systems: a manifold gauge set, a recovery system, a vacuum pump, a recycling unit, and a scale (see **Figure 12**). Additionally, these systems are equipped with electronically controlled valves that control the flow of refrigerant based on the procedure being performed. The recovery, recycling, and recharging station reduces the amount of time and steps required in the repair of A/C systems and the required refrigerant service. Some higher-end equipment may also include other features such as a built-in refrigerant identifier, thermometer, and printer.

SPECIAL SERVICE TOOLS

For many A/C service procedures, special tools are required to complete the service without component damage.

Spring Lock Tools

Spring release tools are used to disassemble fittings that use a spring lock to retain them rather than a typical threaded fitting (see **Figure 13**). The spring release tool

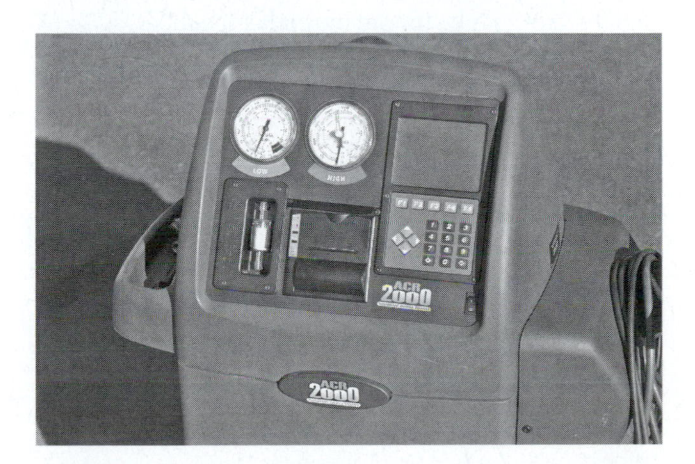

Figure 12. A typical refrigerant recovery/recycling station.

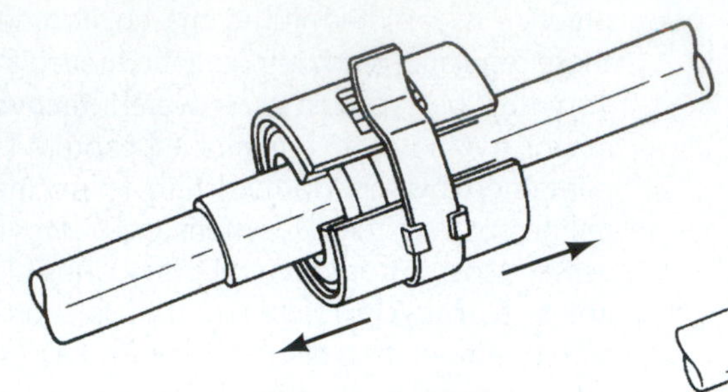

Figure 13. The release of a snap lock fitting.

slides under the garter spring to spread the spring. When the tool is still inserted beneath the spring, the fitting can then be pulled apart.

Orifice Removal Tool

In systems that use orifice tubes, the tube is often recessed within a rigid line; this makes removal of the tube relatively difficult. After the tube has been in service for any amount of time, debris tends to collect around the seals of the tube; this makes removal of the tube even more difficult. An orifice tube removal tool reaches into the tube to engage a set of plastic lugs that are made as part of the tube (see **Figure 14**). The shaft of the remover has external threads; a hub is threaded around this shaft. As the hub is screwed down, it contacts the tube opening and forces the shaft and orifice tube to pull toward the end of the A/C line.

In some cases, a difficult tube can pull apart, making removal seemingly impossible. A special removal tool is made for these situations. The tool is similar in design to the regular removal tool. The difference is that instead of the shaft being equipped with a device to engage the lugs of the tube, the shaft is equipped with tapered threads that screw into the broken tube. The outer hub is then threaded down the shaft and against the rigid line, which forces the shaft and the broken tube toward the end of the tube.

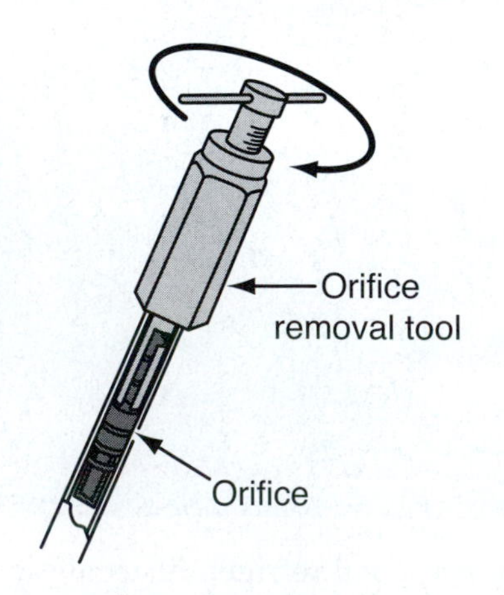

Figure 14. An orifice tube remover.

Compressor Service Equipment

Many service procedures will require the removal of various compressor components. Although in today's service industry a damaged compressor is replaced rather than repaired, there are still a few service procedures that are performed without replacement of the compressor. Most of these procedures deal with the removal of the clutch components. Without the required special service tools, proper clutch service is impossible. The following are some of the most common compressor service tools (see **Figure 15**):

- Spanner wrench: this tool is used to stop the clutch plate from turning when service is performed.
- Clutch plate remover: this tool is splined into the front of a pressed-on clutch plate. The center bolt is turned in to push against the shaft and pull the plate off.
- Clutch plate installer: this tool is used to press the clutch plate back onto the compressor shaft. The center bolt of the tool threads onto the compressor shaft and the outer hub turns to press the plate back on.
- Pulley puller: this tool is used to pull off the compressor clutch pulley. The jaws of the puller fit into the grooves of the pulley; as the center bolt is tightened, the body of the puller is pushed away from the compressor and the pulley comes off. Another style of puller works in the same fashion, except the puller engages into the

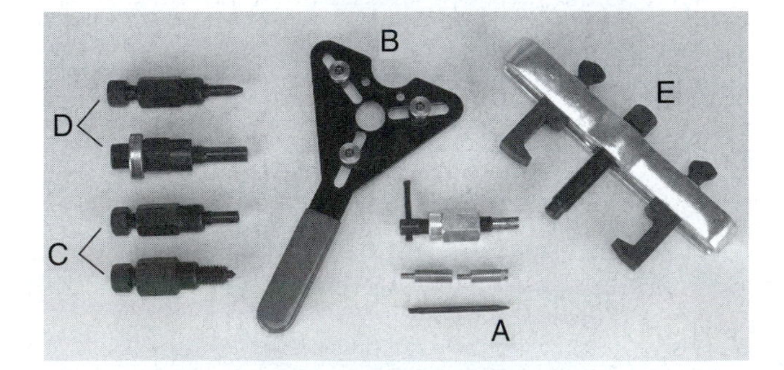

Figure 15. (A) Orifice removal tools, (B) spanner wrench, (C) clutch plate removal tools, (D) clutch installation tools, and (E) pulley puller.

grooves located on the pulley face. This tool is more compact and easier to use.

- Seal tools: these tool sets are made for a specific seal design and consist of seal removers that grip the body of the seal. Specially designed seal installers press against the body of the seal to press it into the compressor. Additionally, the set can include a tapered seal protector that slides over the threads of the compressor shaft to protect the seal against abrasion; the protector is usually slightly tapered to assist the seal in sliding over the shaft and prevent abrasion damage.

Summary

- Technicians may use multiple thermometers when diagnosing an A/C system problem. There are three common types of thermometers: dial, digital, and infrared.
- Leak detection is among the most common tasks performed by A/C repair technicians. Leak detection methods include halogen, fluorescent dyes, ultrasonic tests, and the use of a bubble solution.
- A scan tool provides data about the operation of electronic A/C system components.
- The refrigerant identifier determines the amount of specific chemicals present in a closed A/C system.
- The manifold gauge set allows the technician to monitor pressures within an operating A/C system. Additionally, the gauge set allows the technician to add refrigerant, remove refrigerant, or pull a vacuum on the system.
- The vacuum pump is used to remove air and moisture from the refrigeration system. The efficiency of the vacuum pump is directly related to the condition of the vacuum pump oil; therefore, oil must be serviced regularly.
- The scale is used to accurately measure the amount of refrigerant being introduced into the A/C system. Some scales have the ability to stop refrigerant flow when the proper amount has been introduced.
- A recovery machine is used to capture A/C system refrigerant and store it in a separate storage tank. Most recovery stations have the ability to recycle refrigerant.
- Recovery, recycling, and recharging stations have the ability to perform all related refrigerant storage tasks: recovery, recycling, evacuation, and recharging.
- Many special service tools are required to perform specific A/C system repair tasks. The most common tools include spring lock tools, orifice tube removal tools, and compressor service equipment.

Review Questions

1. List three different styles of thermometers and the advantages of each.
2. Technician A says that excessive moisture in a test area may cause an electronic leak detector to indicate false results. Technician B says that bubble solution will mix with the refrigerant oil to indicate a leak. Who is correct?
   A. Technician A
   B. Technician B
   C. Both Technician A and Technician B
   D. Neither Technician A nor Technician B
3. Which of the following substances can a refrigerant identifier detect?
   A. R-12
   B. R-22
   C. Hydrocarbons
   D. All of the above
4. All of the following statements about a manifold gauge set are true *except:*
   A. A compound gauge can read both pressure and vacuum.
   B. The hand valves isolate each respective pressure from the service hose.
   C. The yellow hose is used for service access.
   D. When the hand valves are closed, the gauges will not read pressure.
5. Which of the following statements concerning a vacuum pump are true?
   A. A vacuum pump uses specially developed oil.
   B. Moisture from the A/C system collects in the vacuum pump oil.
   C. Vacuum pump oil must be serviced only when the oil level is low.
   D. Both A and B
6. Which of the following is not the function of a recovery, recycling, and recharging station?
   A. To detect refrigerant leaks.
   B. To measure the amount of refrigerant recovered from the system.
   C. To evacuate the A/C system.
   D. To recharge the A/C system.
7. Which of the following tools should be used to install an A/C compressor clutch?
   A. An orifice tube remover
   B. A hammer
   C. Spring lock remover
   D. A clutch plate installer

Section 2

Fundamentals of Heating and Refrigeration

SECTION OBJECTIVES

At the conclusion of this section you should be able to:

- Explain the basics of heat and heat transfer.
- Identify various forms of heat transfer.
- Identify the types of heat transfer that are used in the refrigeration system.
- Identify how the human body uses heat transfer to cool itself.
- Define a pressure-to-temperature relationship.
- Understand how the manipulation of heat and pressure affect the temperature of a chemical.
- Describe what a refrigerant is.
- Identify and select various refrigerants.
- Identify and select various lubricants.
- Explain a basic refrigeration cycle.

The operation of the refrigeration system is fully dependent on the principles of heat transfer and the pressure-to-temperature relationships of a refrigerant.

Chapter 5 Matter, Heat, and Comfort

Introduction

As a heating and air conditioning technician it is important that you understand the physical principles of matter, and heat, and their effects on personal comfort. These principles are the backbone of how the heating, air conditioning, and engine cooling systems operate. Although you may not consciously think about these principles when diagnosing or repairing one of these systems, understanding these principles will help you to develop a deeper understanding of what occurs within each of these systems. What must be considered during all automotive technical studies is that your main objective is not necessarily learning how to repair a system, but instead how to diagnose a system effectively. This cannot be accomplished without a thorough understanding of a system's operation.

MATTER

The first thing that we must identify is **matter**. Matter is defined as anything that occupies space and has mass. Matter is composed of single elements or a combination of elements that occur in nature. Some of the more common elements that we can easily identify with are also widely used in the construction and operation of the automobile heating, A/C, and cooling systems:

- Carbon: found in some quantity in all organic compounds. Carbon is found in the fuels, lubricants, and materials used in the construction of automobiles. Carbon is also used in the manufacture of air conditioning system refrigerants.
- Chlorine: a highly reactive chemical that easily combines with other elements to form compounds. Because it easily combines with other elements, it is not found in an uncombined state in nature. In order to obtain pure chlorine, chlorine-containing compounds are broken down to extract the chlorine. One of these common compounds is sodium chloride, commonly known as salt. Chlorine is used in the manufacture of some automotive refrigerants.
- Aluminum: the most common metal found within the crust of the earth. Aluminum is always found combined with another material, such as clay or mica, and it must be separated from its host compound to be used. Once refined, pure aluminum is a soft, lightweight, and ductile material that is also an excellent conductor of heat and electricity and provides excellent resistance to corrosion. The strength and rigidity of aluminum can be improved significantly by combining it with other materials such as copper, manganese, magnesium, or zinc to form an **alloy**. An alloy is a material that is manufactured from a combination of two or more materials. This is done to provide specific desirable characteristics from metals. Aluminum alloys are common in the manufacture of many automotive components.

Atoms

The atom is the smallest particle of an element. An atom still retains all of the physical characteristics of the element from which it was extracted. Atoms from different elements have different characteristics; for example, iron atoms are different from aluminum atoms. Atoms are formed of yet smaller particles called protons, electrons, and neutrons. Although we will not go into the specific construction of the atom, you should know that it is

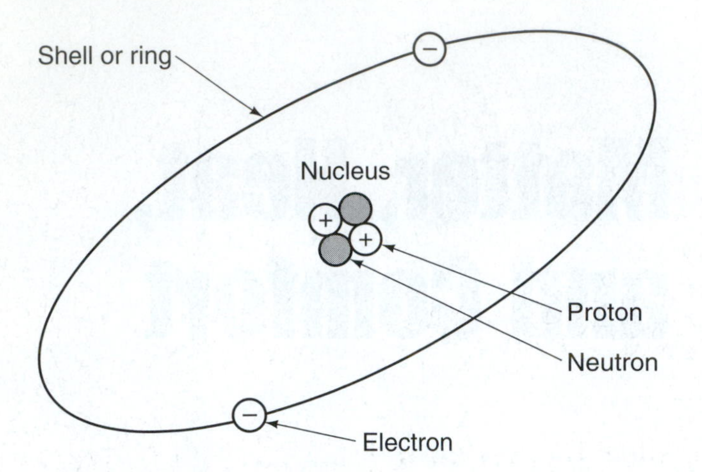

Figure 1. Different combinations of electrons, neutrons, and protons give the atom its specific characteristics.

the different combinations of protons, electrons, and neutrons that give an element its specific characteristics (see **Figure 1**).

Compounds

Compounds are combinations of two or more elements that, when combined, form a totally different material. The resulting compound often will not even resemble the elements from which it was formed. For example, water, or H_2O, is comprised of two hydrogen atoms and one oxygen atom. In its natural state, hydrogen is a highly flammable gas, and oxygen is an oxidizer that supports combustion. But when combined, these two elements form water, a liquid that resembles neither of its base elements (see **Figure 2**). Most if not all materials used in the construction and operation of the automobile are compounds. Compounds can be manufactured to provide predictable, desirable characteristics.

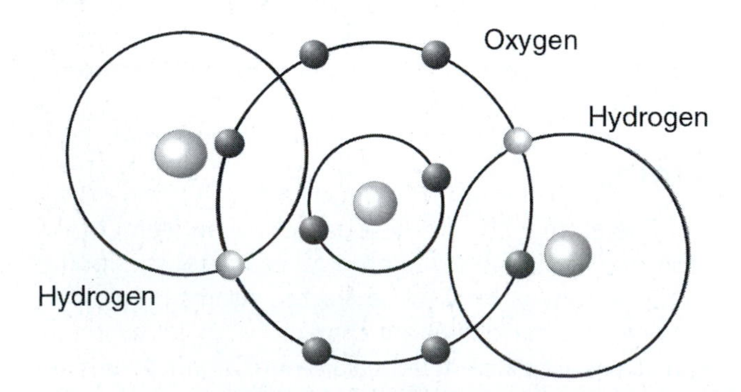

Figure 2. Two atoms of hydrogen and one atom of oxygen bond together to form the chemical compound water.

Molecules

A molecule is the smallest particle of a compound that retains all of the characteristics of the compound. Molecules assume very distinctive structures and appear in many different sizes, shapes, and weights, and are in a state of perpetual motion.

We now know that molecules are constantly in motion, even in those objects that appear to be solid. When a substance is solid, the molecules are held together by their mutual attraction to one another. This is called **cohesion**. Even though appearances may suggest otherwise, there is some space in between the molecules, which allows for the molecules to move about, albeit with some restriction. As the molecules become warmer, they begin to move faster. If a material becomes hot enough, the rapid movement of the molecules can overcome their cohesive bond to one another and the substance can change state. In a liquid, the molecules are loosely bound and are allowed to move about more freely.

HEAT

The basic principles of automotive heating, air conditioning, and engine cooling are based on the state of matter, whether it is the addition of heat, the removal of heat, or the transformation of a refrigerant. This all revolves around the reaction of specific molecules.

As the molecules of a material become warmer, the molecules begin to move faster. But how do we make molecules warmer? This is accomplished by adding heat. Heat is a nonmechanical energy. Like other forms of energy, heat cannot be made or destroyed but can only be converted or transferred.

Contrary to what we typically think, heat and temperature are not the same thing. Temperature is a measurement that is relative to heat; in other words, temperature measures the effects of heat being applied to a material (see **Figure 3**). All molecules have some amount of heat within them. The exceptions to this rule are those molecules that are at a temperature of −459.67 degrees F (−273.15 degrees C); this is called **absolute zero**. Even materials that we refer to as cold contain heat, and there are some forms of heat that exist in materials that do not change the relative temperature of the material itself.

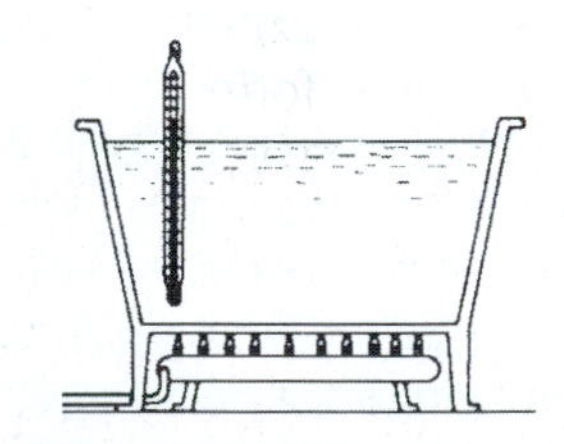

Figure 3. Temperature measures the effects that heat has on a material.

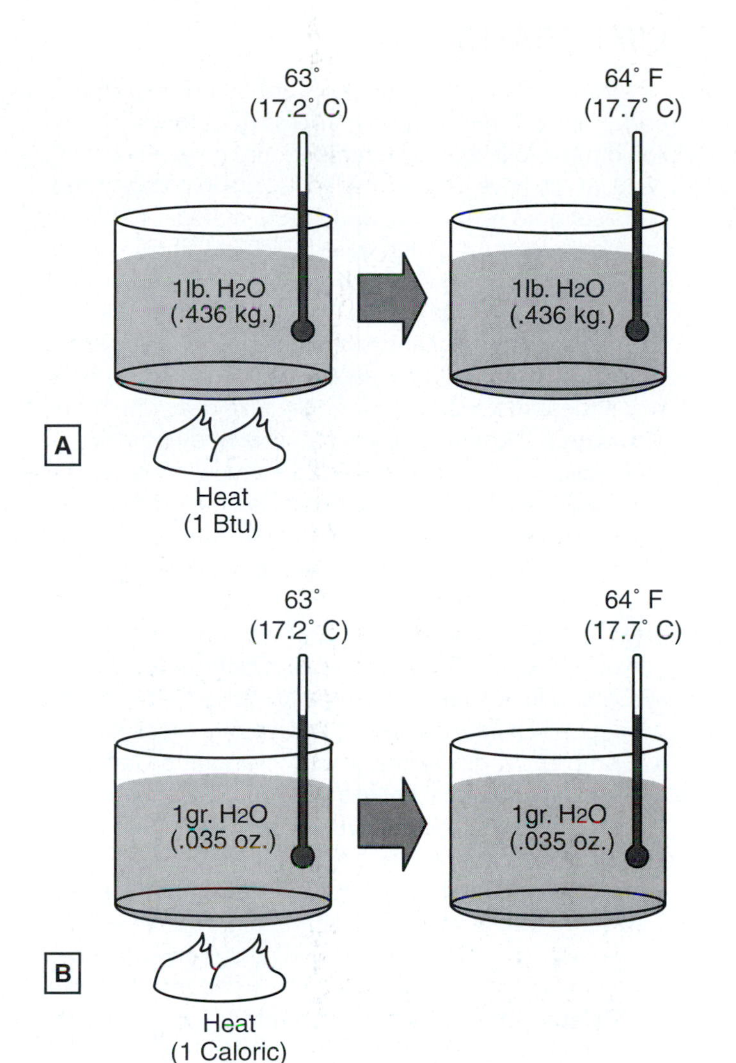

Figure 4. (A) It takes one Btu to raise the temperature of water 1 degree F. (B) It takes 1 calorie to raise the temperature of 1 gram of water 1 degree C.

Heat is measured in either British Thermal Units (Btu) or the metric unit of calories. The calorie is a unit of heat that is most often associated with the amount of energy stored in food items. It takes one Btu of energy to raise the temperature of 1 pound (.45 kg) of water 1 degree F (.56 degree C). It takes 1 calorie of energy to raise the temperature of 1 gram (.035 oz) of water 1 degree C (see **Figure 4**). There are three forms of heat that are important to understand: sensible heat, latent heat, and specific heat.

Sensible Heat

Sensible heat is any amount of heat that can be felt and can be measured with a thermometer. A quick way to remember this is that sensible heat can be "sensed."

Latent Heat

Latent heat, which is hidden heat, is the amount of heat that is applied to a material that causes a change of state. It cannot be measured with a thermometer. For instance, if we bring a pot of water to a boil on a medium heat setting, the water will boil at 212 degrees F at sea level (100 degrees C). When the water boils, some of the molecules will naturally be transformed into a vapor. After the water has begun to boil, we switch the burner to high. This adds even more heat; however, the temperature of the water does not change. Evidence of the additional heat is apparent in the behavior of the molecules. The amount of heat that is present above what is necessary to cause the water to boil is latent heat (this is demonstrated in **Figure 5**).

Specific Heat

Specific heat is the amount of heat that is required to cause a rise in temperature of a specific material (see **Figure 6**). This is important because this tells us a material's

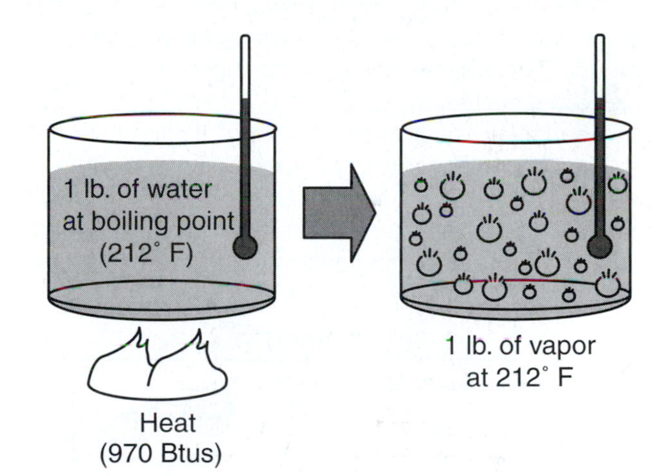

Figure 5. Latent heat is the amount energy that is applied to a substance that causes a change of state, but does not cause the temperature to rise.

Material	Value	Material	Value
Air	0.240	Nitrogen	0.240
Alcohol	0.600	Oxygen	0.220
Aluminum	0.230	Rubber	0.481
Brass	0.086	Silver	0.055
Carbon dioxide	0.200	Steel	0.118
Carbon tetrachloride	0.200	Tin	0.045
Gasoline	0.700	Water, fresh	1.000
Lead	0.031	Water, sea	0.940

Figure 6. Specific heat values of selected materials.

relative ability to be a conductor or an insulator. Specific heat can be explained as the amount of heat required to raise the temperature of 1 pound of a specific material 1 degree F. For example, it takes 1 Btu to raise 1 pound of water 1 degree F, so the specific heat is one. By contrast, the specific heat for air is .240, which means that it takes approximately .25 of a Btu to heat 1 pound of air 1 degree F.

CHANGE OF STATE

A change of state occurs when a material changes from a solid to a liquid, a liquid to a solid, a liquid to a vapor, or a vapor to a liquid. A change of state occurs when the amount of heat within the material is manipulated. As heat is added to a solid material, the molecules overcome their attraction to one another and begin to move more freely. When enough heat has been added to a material to increase its temperature to the melting point it will begin to change from a solid to a liquid. If we continue to add additional heat, the liquid material can turn to a vapor.

When heat is removed from a vapor material, the motion of the molecules slows considerably. When a sufficient amount of heat is removed from the substance, the molecules cannot overcome their mutual attraction to one another and the material becomes a liquid. If enough heat is removed from a material, it can turn to a solid. This is illustrated in **Figure 7**. The chemical composition and the environment to which a material are exposed have a significant effect on when a material changes state.

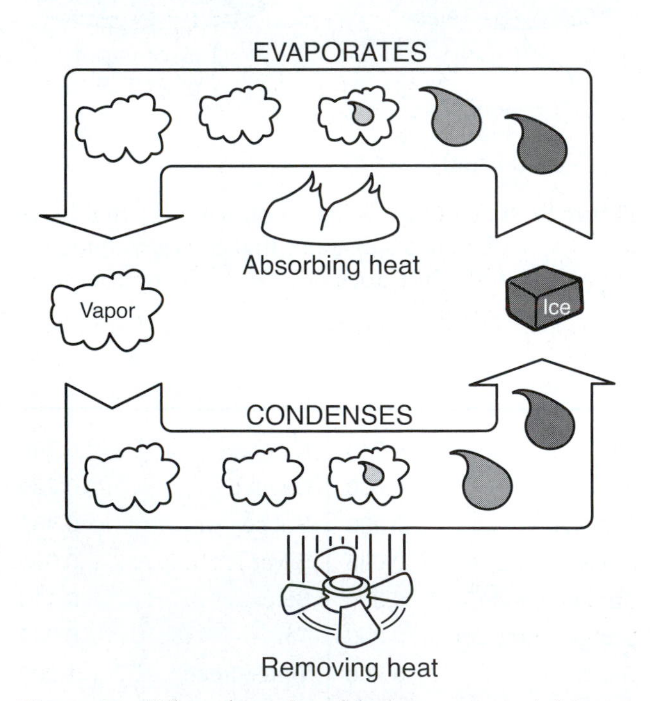

Figure 7. When heat is added to a material it can change from a solid to a liquid and then to vapor. When heat is removed the vapor can change to a liquid and then to a solid.

EVAPORATION

Many refrigeration principles are based on evaporation. Evaporation is a natural process that occurs when a liquid turns to a vapor at a temperature below its boiling point. As we have learned, molecules are in constant motion; the more heat that is added to a material, the faster the molecules move. When water, which has a boiling point of 212 degrees F (100 degrees C), is stored in a container at room temperature, the molecules near the top have less resistance to movement and can generate enough speed to overcome cohesion and escape into the atmosphere. When these molecules escape, they take heat with them, providing a slight cooling effect to the body from which they escaped. The ease at which a material evaporates is typically inversely proportional to its boiling point. For example, if we place an equal amount of water, which has a boiling point of 212 degrees F (100 degrees C) and rubbing alcohol, which has a boiling point of 180 degrees F (82 degrees C) in identical containers exposed to identical conditions, the alcohol will evaporate faster. External factors that affect evaporation rates are temperature, humidity, altitude, wind speed, and the amount of surface area exposed.

- Ambient temperature affects evaporation rates by adding additional heat to the liquid. When temperatures are high, the molecular speed is increased relative to speeds at lower temperatures.
- Humidity is the amount of moisture that is present within the air. When humidity is high, the air will not readily absorb additional moisture; however, when humidity is low, it easily accepts additional moisture. **Relative humidity** is the term most often used to describe how moist the air is. Relative humidity is expressed as a percentage. The percentage represents how much moisture is in the air relative to how much moisture the air can hold. For example, if the relative humidity is 70 percent, this means that the air is holding 70 percent of the moisture that it could hold at a given temperature.
- As surface area is increased, there are more molecules that are exposed to the top layer of the liquid, which gives more molecules the opportunity to escape.
- As altitude is decreased, atmospheric pressure is lowered, and vice versa. At low altitudes, the molecules have more resistance applied to them and therefore have more difficulty escaping, which decreases the evaporation rate. When altitude is increased, the pressure applied to the body is decreased and the molecules have less resistance to movement.
- Wind speed creates an environment that helps molecules overcome cohesion and rise from the surface of the host liquid into the atmosphere. The higher the wind speed, the easier it is for those molecules near the surface to be pulled from the liquid and create a space for other energy-laden molecules to occupy.

BOILING AND VAPORIZATION

As more and more heat is introduced into a substance, the molecules begin to move more rapidly. When the molecules have absorbed enough energy, they will gain enough momentum to break away from their host liquid and turn to vapor (see **Figure 8**). This is referred to as boiling. The temperature at which a material boils is called the **boiling point**. When liquid reaches its boiling point, the temperature of the liquid no longer increases. As the liquid boils, any additional energy beyond what is required to maintain the boiling temperature is absorbed by the molecules and rapidly turns them to vapor; as the vapor enters the atmosphere, so does the heat that was stored within them. This extra energy is called **latent heat of vaporization**. The latent heat stored in these molecules is enough to allow them to break their bonds easily and escape into the atmosphere as vapor or gas. In the case of water, the vapor is called steam. Vaporization is similar to evaporation with the exception that it takes place after a liquid reaches its boiling point and it occurs at a much more rapid rate.

Boiling Point and Pressure

The boiling point of a material is also relative to the pressure applied to it. At sea level, water will boil at 212 degrees F (100 degrees C). If water is placed in a sealed container and pressure is applied, the boiling point will rise approximately 3 degrees F (16.11 degrees C) for every 1 pound per square inch (psi) (6.89 kPa) of pressure applied to the water. If 10 psi (68.95 kPa) of pressure were applied to the container the boiling point would increase to approximately 242 degrees F (117 degrees C). If pressure is reduced, such as water that is exposed to a vacuum, the boiling point of the water would drop in relation to the amount of vacuum applied (see **Figure 9**).

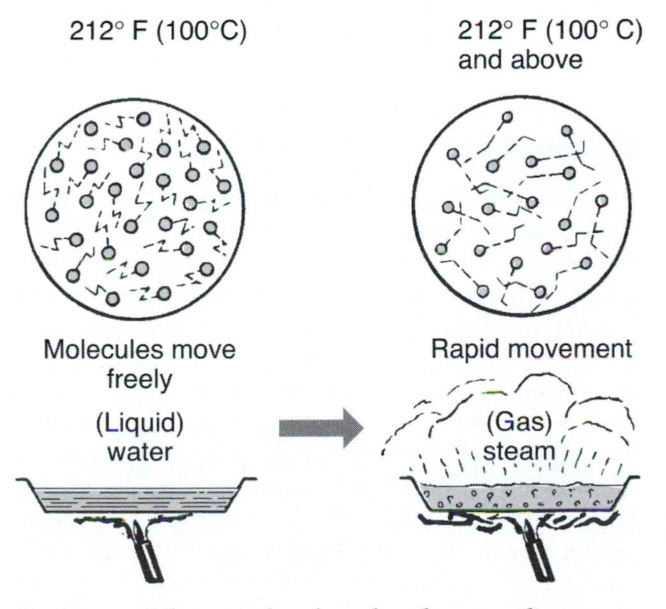

Figure 8. When molecules absorb enough energy they will break away from their host liquid.

Water Boiling Points at Various Vacuums and Gauge Pressures

Inches Vacuum (Hg)	Temp.	Gauge Pressure	Temp.
29"	71°F (22°C)	5 Lbs.	227°F (108°C)
25"	133°F (56°C)	10 Lbs.	240°F (116°C)
20"	161°F (72°C)	15 Lbs.	250°F (121°C)
15"	179°F (82°C)	25 Lbs.	267°F (131°C)
10"	192°F (89°C)	35 Lbs.	281°F (138°C)
5"	203°F (95°C)	50 Lbs.	298°F (148°C)
0"	212°F (100°C)	75 Lbs.	320°F (160°C)

Figure 9. This chart shows the various boiling points of water under specific pressure and vacuum levels.

When heat is added to a liquid, the molecules begin to move about rapidly. When the vessel is uncovered, those molecules with enough energy are allowed to escape. But when a liquid is heated within a sealed container, the molecules cannot escape. However, they still expand away from one another, increasing the volume of the liquid and vapor; as this occurs, the pressure inside the container increases. The amount of pressure and the rate in which pressure is built is relative to the amount of heat that is applied to the container. As pressure within the container increases, so does the boiling point of the liquid within the container. If the pressure was suddenly released, the boiling point of the liquid would immediately return to its original temperature and the liquid would boil instantly. This often results in a violent explosion (see **Figure 10**).

Condensation

When latent heat is added to a liquid, the result is either evaporation or vaporization. But what happens when that heat is lost? We will again use water as an example. When moisture-laden air molecules begin to lose their latent heat, the movement of the water molecules slows down and the molecular bonds once again draw the molecules together. As the molecules come together, they bond to one another and form droplets of water. This process is

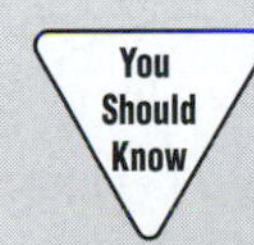

Because of the sheer amount of energy stored in steam, it may become hotter than the boiling point of the liquid in which it originated. When steam condenses on the surface of an object, the energy stored in the molecules is transferred to that object.

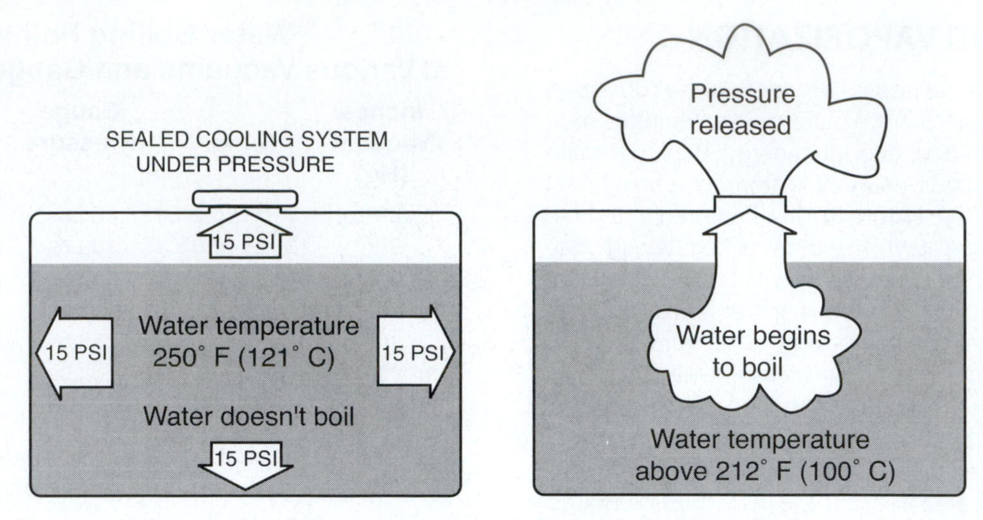

Figure 10. This illustrates the relationship of pressure and the boiling point of water and what can occur if pressure is suddenly lost.

called condensation. This is a basic operating principle of the automotive air conditioning system (see **Figure 11**).

HEAT TRANSFER

To completely understand heat and understand how it affects the operation of automotive systems, we must understand how heat moves and is transferred. Heat transfer is the process by which thermal energy moves from one material to another. Heat transfer will only occur between two materials at different temperatures; specifically, heat will travel from a warm object or material to a cooler object or material. When the two objects reach the same temperature, heat transfer stops. Heat can be transferred in any combination among solids, liquids, and gases. The operating principles of all automotive heating, cooling, and air conditioning systems are based on the process of heat transfer.

There are three ways in which heat is transferred: radiation, conduction, and convection.

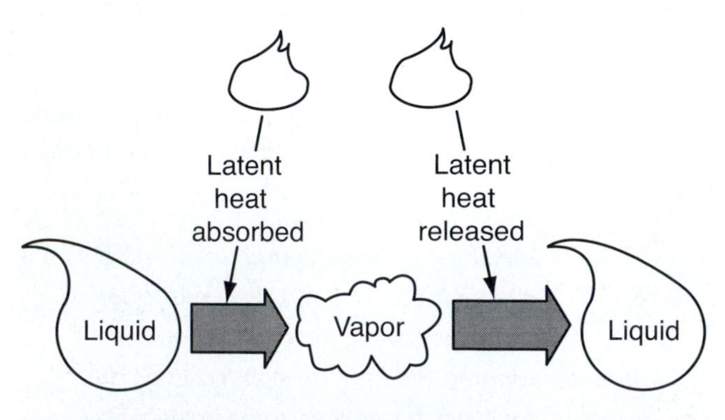

Figure 11. This is a graphic representation of vaporization and condensation.

Radiation

Radiation is a method of heat transfer that occurs by an object emitting heat rays. As the heat rays pass from one object to another, they do so without heating the medium in which they passed. Radiation is typically associated with light energy.

Conduction

Conduction is a form of heat transfer in which the heat energy moves from molecule to molecule. In this form of heat transfer, the molecules must be touching. For example, inside an internal combustion engine heat is generated in the cylinders. The heat that is created within the cylinder is transferred by conduction throughout the engine block and cylinder heads.

Convection

Convection is a process that relies on the movement of a liquid or gas to remove heat from an object. There are two types of convection: natural convection and forced convection. When gas or liquid comes in contact with a hot object, heat will be transferred into the medium. As the medium heats up, the molecules become lighter and rise, leaving a void. Cooler, more dense, molecules rush in to fill the void left by the warm molecules. A natural circulation of air occurs as this process is repeated, with cool molecules replacing those warmer lighter molecules (see **Figure 12**). This is called natural convection. Forced convection occurs when the liquid or gas is circulated by a mechanical means such as a fan or water pump (see **Figure 13**).

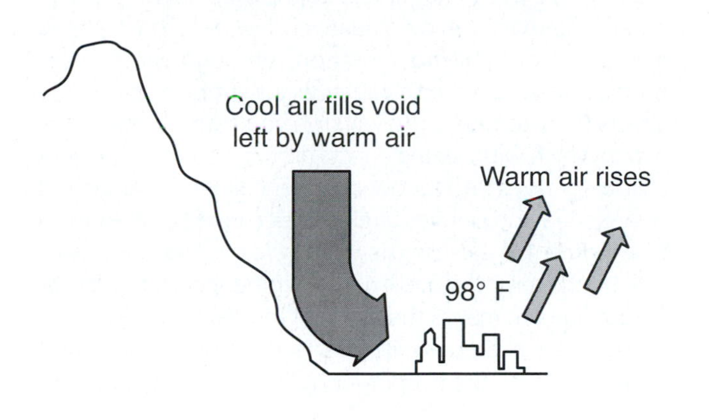

Figure 12. Convection occurs as light, warm molecules rise and cool dense molecules move in to fill the void creating natural circulation.

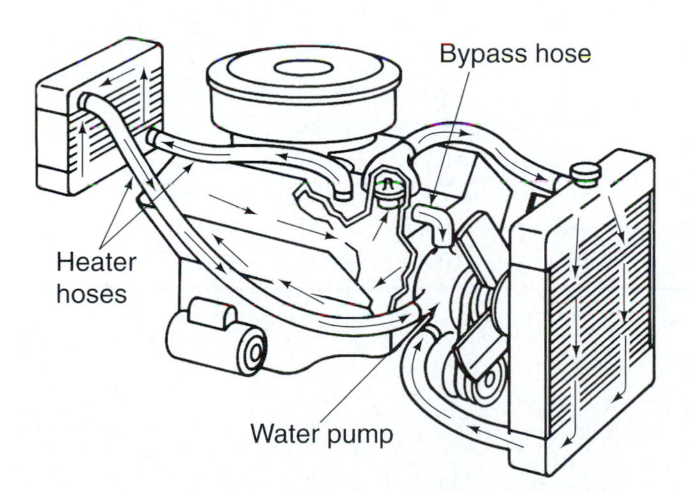

Figure 13. Conduction and convection are used in the automotive cooling system to keep the engine at the desired operating temperature.

Conduction and Convection in the Automobile

Conduction and convection are the major forms of heat transfer used in automobile systems. In an automotive system such as the engine, transmission, or air conditioning system, a cooling medium is circulated through the area from which we want to remove heat. The heat is transferred to the cooling medium through conduction as it is circulated through the system. The cooling medium is then transferred to a heat exchanger, where the heat is released into the atmosphere through convection (see Figure 13).

HUMAN TEMPERATURE CONTROL

Normal human body temperature is 98.6 degrees F (37 degrees C), and we each have built-in mechanisms to maintain that temperature. The human body is continually producing heat. When our bodies change food into energy, heat is created; when our muscles move, heat is created; even breathing creates heat. When the heat from all of these processes is combined, the body produces more heat than is required. Because there is a surplus of heat, it must be reduced. The human body rids itself of heat in three different ways: radiation, convection, and evaporation (see **Figure 14**).

Evaporation is probably the most important method for automotive technicians to understand. The human body continually produces perspiration on the surface of the skin; under normal circumstances, the perspiration will evaporate, taking heat away with it. When the body produces more heat than can be removed by evaporation, convection, and radiation, perspiration will appear on the surface of the skin.

Comfort

All of the principles that we studied in this chapter are a precursor to the study of human comfort. For all practical purposes, this is what the heating and air conditioning systems are designed to do: keep the occupants of the automobile comfortable. As a technician, you may find that no matter how well the system is mechanically functioning, if the occupants aren't comfortable then the perception is that the system is broken.

There are several things that affect human comfort: temperature, humidity, and air movement (see **Figure 15**). Temperature is obvious to most of us: the hotter the ambient temperature, the more uncomfortable we become.

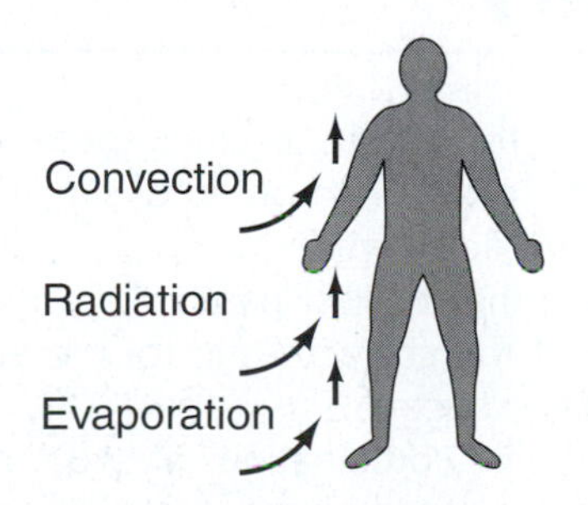

Figure 14. The human body uses convection, radiation and evaporation to maintain a desirable temperature.

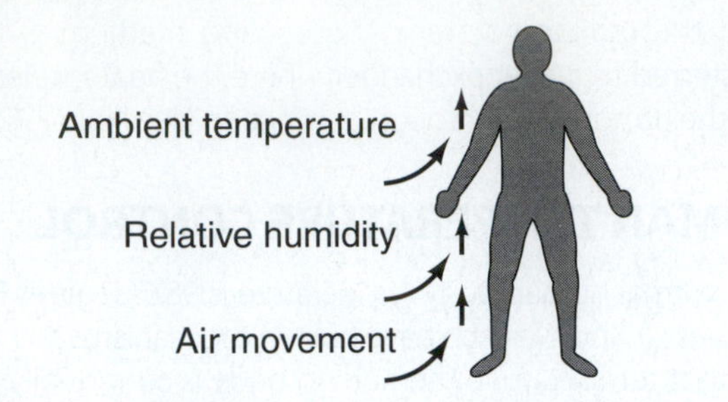

Figure 15. Human comfort is directly related to ambient temperature, relative humidity and air movement.

Humidity affects body temperature by regulating the amount of perspiration that can be absorbed in the atmosphere. If the humidity is high, less moisture can be absorbed into the atmosphere, which forces the perspiration and heat to stay on the skin. If humidity is low, the atmosphere will readily absorb the moisture that is being produced by the body.

Air movement aids the evaporation process; even a slight breeze will greatly speed up the evaporation process, enough so that even on a relatively hot day a breeze may give us a slight chilling sensation. Although wind chill is normally associated with winter weather, observing how it affects body temperature will give you a good idea of the way in which wind speed affects human comfort. The wind chill factor takes into account ambient air temperature and wind speed and derives a relative perceived temperature (see **Figure 16**). We use the word *perceived* because, even though the temperature is not lower, the sensation that we feel on our skin makes the ambient temperature feel much cooler than it actually is. This is because higher wind speeds increase the evaporation rate of perspiration from our skin.

AIR TEMP (°F)	WIND SPEED (MPH)							
	5	10	15	20	25	30	35	40*
40	37	28	23	19	16	13	12	11
30	27	16	9	4	1	–2	–4	–5
20	16	3	–5	–10	–15	–18	–20	–21
10	6	–9	–18	–24	–29	–33	–35	–37
0	–5	–22	–31	–39	–44	–49	–52	–53
–10	–15	–34	–45	–53	–59	–64	–67	–69
–20	–26	–46	–58	–67	–74	–79	–82	–84
–30	–38	–58	–72	–81	–88	–93	–97	–100
–40	–47	–71	–85	–95	–103	–109	–113	–115

*Winds above 40 mph (64 km/h) have little additional effect on windchill factor.

Figure 16. The effects of the wind chill factor.

Summary

- Matter is anything that occupies space and contains mass. Matter can be composed of single elements or combination of elements.
- The atom is the smallest particle of an element that retains all of the characteristics of the element from which it was extracted.
- Compounds are combinations of two more elements that when combined form a totally different material. Compounds can be manufactured to obtain specific results.
- A molecule is the smallest part of a compound that retains all of the characteristics of the compound from which it was extracted. Molecules can be many different shapes and sizes. Molecules stay in perpetual motion.
- Heat is a nonmechanical energy. As heat is added to molecules, they become more active. Heat is measured either in Btus or calories. Temperature is a measurement of the effects of heat.
- Sensible heat is any amount of heat that can be measured with a thermometer. Latent heat is hidden heat and cannot be measured with a thermometer. Specific heat is the rating given to a material to signify how much heat energy is required to raise the temperature of 1 pound of material 1 degree F.

- A change of state occurs when a material changes from a liquid to a solid, a solid to a liquid, or a liquid to a vapor or a vapor to a liquid. A material change of state is manipulated by varying the amount of heat within the material.
- Evaporation occurs when a liquid material changes from a liquid to a vapor at a temperature below its boiling point. Evaporation rates are affected by ambient temperature, humidity, exposed surface area, altitude, and wind speed.
- Boiling and vaporization occur when the molecules of a liquid are heated to the material's boiling point and the molecules overcome their molecular bond to escape the host body of liquid. The boiling point of a liquid is relative to the amount of pressure applied to the liquid.
- Condensation is a process that occurs when latent heat is removed from vapor molecules and the vapor returns to a liquid state.
- Heat moves from warm molecules to those containing less heat. Heat can be transferred through conduction, convection, and radiation.
- Radiation is a method of transfer that occurs through the emission of heat rays and occurs without heating the medium in which the rays pass. Conduction occurs when heat moves from one molecule to another. Convection occurs through a natural movement of cool dense air replacing warmer air molecules.
- Normal body temperature is 98.6 degrees F (37 degrees C). All body functions create heat. The body regulates temperature through convection, radiation, and evaporation.
- Temperature, humidity, and air movement affect human body comfort.

Review Questions

1. Which of the following statements about matter is true?
   A. Matter is anything that occupies space and possesses mass.
   B. Matter can be composed of single elements.
   C. Matter can be a combination of elements.
   D. All of the above.
2. Technician A says that an atom is the smallest particle of a compound. Technician B says that a molecule is the smallest part of a compound. Who is correct?
   A. Technician A
   B. Technician B
   C. Both Technician A and Technician B
   D. Neither Technician A nor Technician B
3. Describe the relationship between heat and temperature.
4. List and describe three types of heat.
5. Technician A says that when a material gains enough heat it will turn to a solid. Technician B says that when material loses a significant amount of heat it will turn to a vapor. Who is correct?
   A. Technician A
   B. Technician B
   C. Both Technician A and Technician B
   D. Neither Technician A nor Technician B
6. Technician A says that the evaporation rate is affected by the surface area of material that is exposed to the atmosphere. Technician B says that evaporation occurs below the designated boiling point of a material. Who is correct?
   A. Technician A
   B. Technician B
   C. Both Technician A and Technician B
   D. Neither Technician A nor Technician B
7. All of the following statements about boiling and vaporization are correct *except:*
   A. The temperature of a material will no longer increase once the boiling point is reached.
   B. As heat is increased, the molecules move more rapidly.
   C. As the external pressure applied to a material is increased, the boiling point is decreased.
   D. When enough energy is removed from a vapor, it will return to a liquid state.
8. List and describe the three methods of heat transfer.
9. Explain how evaporation is used to cool the human body.
10. What are three things that affect human comfort?

Chapter 6 Pressure and Temperature Fundamentals

Introduction

A refrigerant is any liquid or vapor material that is readily able to absorb heat from an area or object, transfer that heat, and release it in another location (see **Figure 1**). Nature has provided us with many natural refrigerants, such as air and water; both are used in the engine cooling system. Although air and water meet the needs required to keep the engine at the proper operating temperature, different refrigerants must be selected based on the requirements of the refrigeration system. For each specific refrigeration application, the refrigerant used must meet the stringent needs of the system. In order for you as an automotive technician to properly analyze and diagnose an automotive air conditioning system, it is important for you to understand the pressure-to-temperature relationships of a refrigerant, what a refrigerant is, and how a refrigerant is used.

PRESSURE AND TEMPERATURE

In order to understand how a refrigerant can remove heat from an area or object, you need to understand a few basic principles about pressure and temperature relationships. The boiling point is a temperature representation of the amount of heat that is required to make a liquid turn in to a vapor. Although a liquid will turn to a vapor at a temperature lower than the boiling point, once the boiling point is reached the change of state from liquid to vapor becomes very rapid. It is also at this temperature that the material will condense (see **Figure 2**). The specific boiling point is based on the premise that the material has plenty of room for expansion and that the material is in its pure state. The addition of other chemicals or impurities will alter the boiling point by either increasing or decreasing the boiling temperature. When heat is added to a material in an

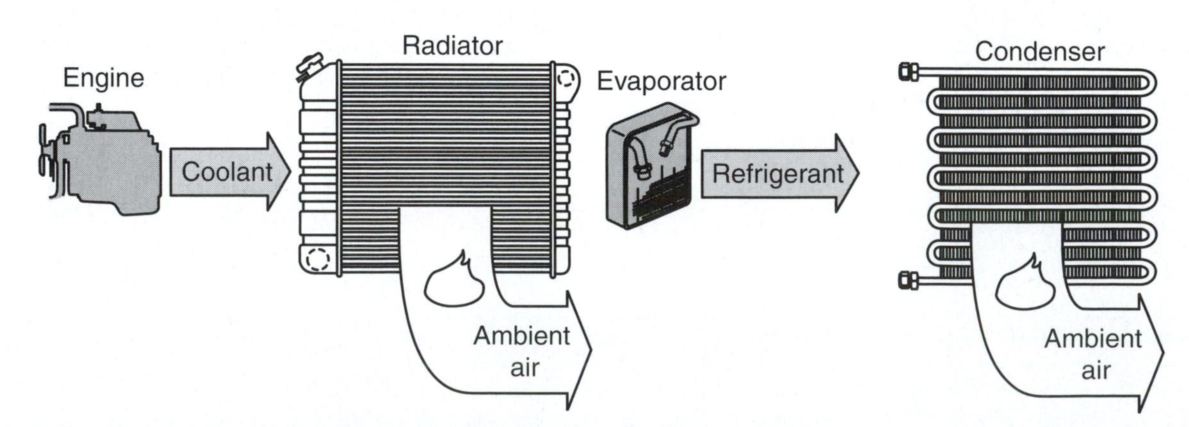

Figure 1. A refrigerant has the ability to readily absorb and release heat.

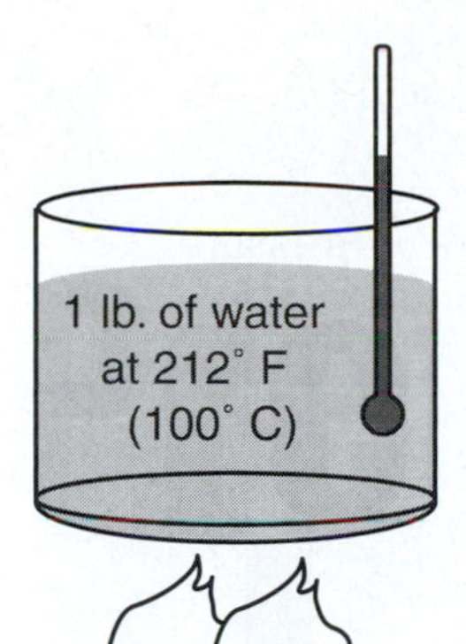

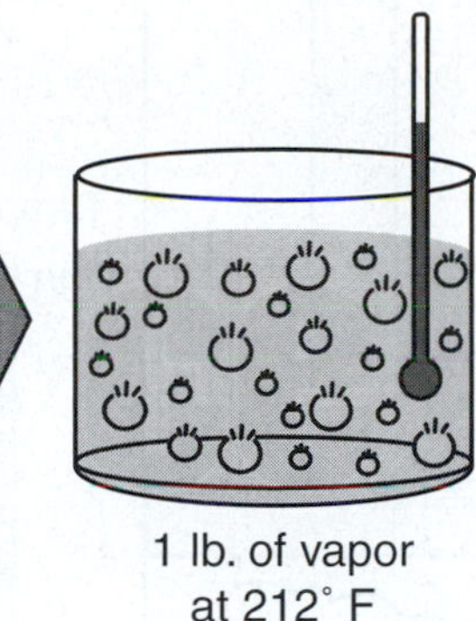

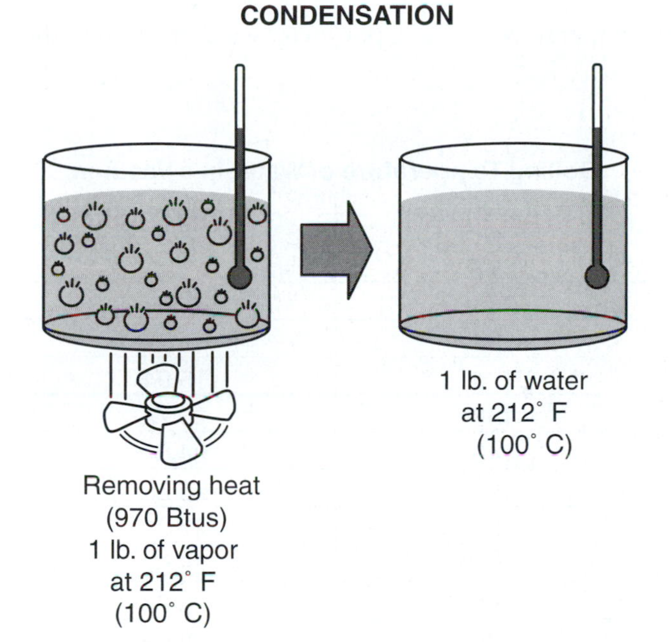

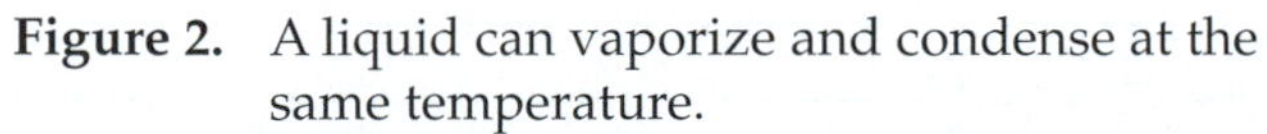
Figure 2. A liquid can vaporize and condense at the same temperature.

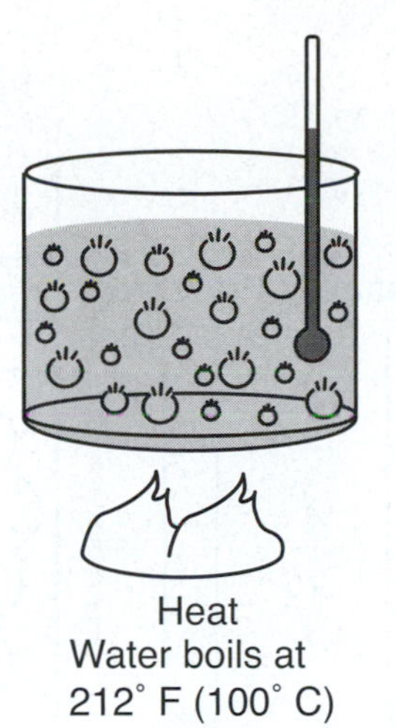

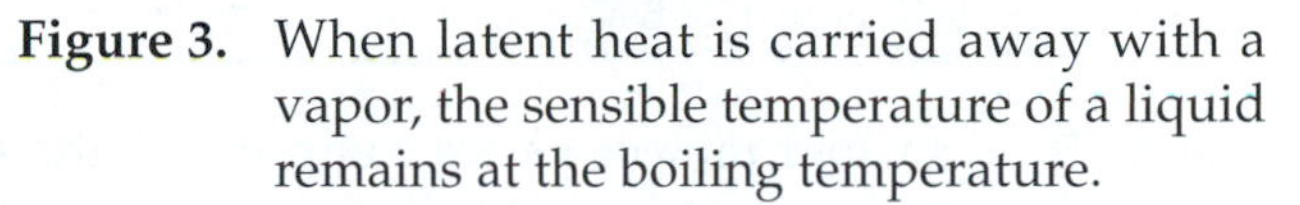
Figure 3. When latent heat is carried away with a vapor, the sensible temperature of a liquid remains at the boiling temperature.

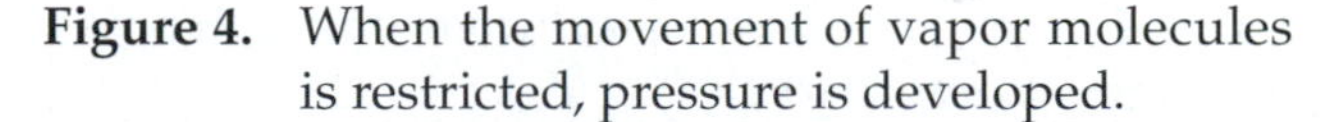
Figure 4. When the movement of vapor molecules is restricted, pressure is developed.

open container, the movement of the molecules intensifies to the point that some of the molecules will have acquired enough energy to escape into the atmosphere as vapor. When molecules of a liquid change to vapor, they carry a large amount of latent heat away. Because much of the latent heat is carried away by the escaping molecules, the sensible temperature of the liquid does not rise, and the temperature remains at the boiling point (see **Figure 3**).

Temperature and Pressure Increase

When a liquid is heated within a sealed container, the movement of the molecules is restricted. As heat is added to the container, the molecules still absorb heat, but once some of the molecules vaporize, any empty space within the container quickly becomes saturated with vapor molecules and vaporization essentially stops. At this point, the pressure within the vessel begins to increase in direct proportion to the amount of heat applied to the container. As the pressure within the container is increased, the boiling point of the liquid continues to increase as well (see **Figure 4**).

The same temperature-to-pressure relationship exists when the pressure applied to a liquid is changed rather than the heat. For example, if a liquid is stored in a sealed container or system and the pressure is increased, the temperature of the liquid will be increased in proportion to the amount of pressure applied (see **Figure 5**). Because of the increase in applied pressure, the boiling point of the liquid is also increased.

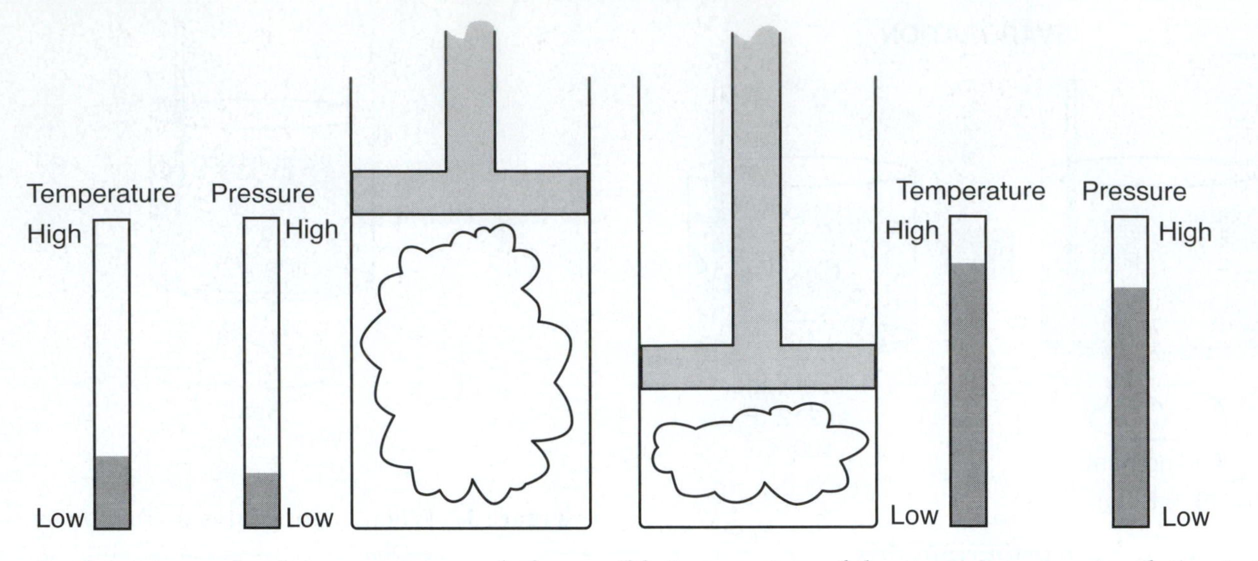

Figure 5. As vapor molecules are compressed, the sensible temperature of the vapor increases in relation to the amount of pressure applied.

Temperature and Pressure Decrease

Just as the addition of heat increases the pressure of a liquid in a sealed container, the opposite occurs when heat is removed; the pressure will decrease proportionately to the amount of heat expelled. In addition, as the pressure is decreased, the boiling point will decrease in relation to the remaining pressure applied to the vessel.

Although the pressure of a liquid in an open container cannot be reduced, when placed in a sealed container or sealed system not only can the pressure be lowered but, in some situations, it is possible and desirable to pull a low vacuum on the system or container. This considerably lowers the boiling point and temperature of a chemical. When a liquid is cool, it has much more capacity to absorb heat. As we look at the behavior of refrigerants, these principles become very important.

Interesting Fact

In normal conditions, water will boil at approximately 212 degrees F (100 degrees C) at sea level. However, when placed in a vacuum, water can actually boil at room temperature. For example, when placed in an environment in which a vacuum can be maintained, at 29 in-Hg water will boil at 76 degrees F (24.4 degrees C). (See ***Figure 6****.)*

Boiling Temperature of Water in a Vacuum

Temperature °F	Vacuum in-Hg
212	0
205	4.92
194	9.23
176	15.94
158	20.72
140	24.04
122	26.28
104	27.75
86	28.67
80	28.92
76	29.02
72	29.12
69	29.22
64	29.32
59	29.42
53	29.52
45	29.62
32	29.74
21	29.82
6	29.87
−24	29.91
−35	29.915
−60	29.919
−70	29.9195

Figure 6. The boiling temperature of water in a vacuum.

REFRIGERANT

When a liquid turns to vapor, the molecules escaping the main body of the liquid carry heat away with them. In the case of boiling water, they carry enough heat away to prevent the temperature of the liquid from increasing. But let's suppose that we substitute a chemical that will boil at a temperature much lower than that of the ambient temperature. In this scenario, enough heat is provided by the ambient temperature to cause the chemical to boil. As the molecules escape, they carry heat away with them and because no additional heat is added, enough heat will escape to actually reduce the temperature of the remaining body of liquid and the container in which it is stored.

These cooling effects can be observed using a commercially available aerosol dust remover, for example, one typically used to clean electronic equipment. Many of these products are actually filled with the same R-134a refrigerant that is used in most automobiles. R-134a refrigerant has a very low boiling point of –15 degrees F (–26 degrees C). Because the refrigerant is stored in a sealed container, it won't boil and it will remain in a liquid state. However, when the trigger on the can is depressed, the pressure within the can suddenly decreases and the refrigerant begins to boil. When this occurs, the molecules expand and create a burst of pressurized vapor that is used to blow away dust particles. As the refrigerant escapes from the nozzle, it removes heat with it. This action can be viewed by observing the tip of the nozzle. As refrigerant exits the can, heat is removed and the nozzle becomes very cold and frost will often form on the nozzle. If the trigger is depressed long enough, the escaping refrigerant will remove enough heat from the can to actually make it feel cold to the touch. This is basically what occurs within a refrigeration system (see **Figure 7**).

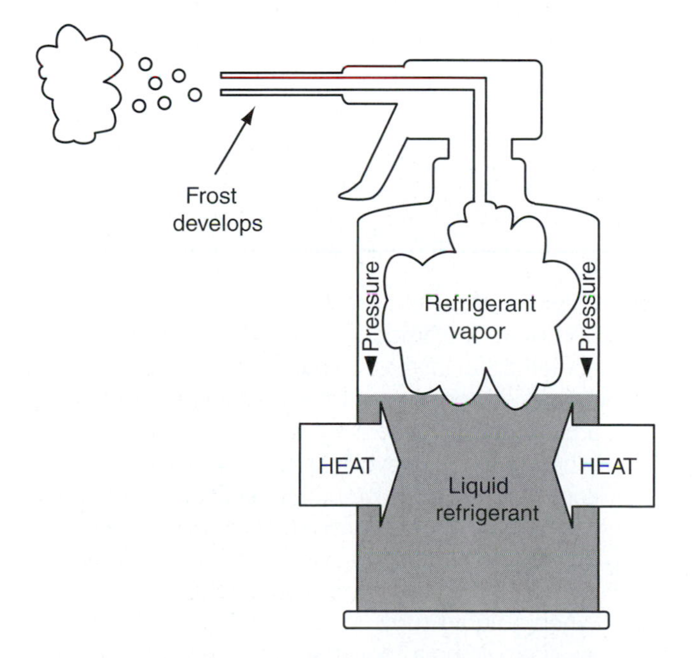

Figure 7. Refrigerant absorbs heat from its container when it is changed from a liquid to a vapor.

> **You Should Know** *Because a refrigerant removes heat from any surface with which it comes in contact, it is extremely important that gloves and eye protection are worn at all times. Failure to do so can result in the possibility of frostbite to the skin as well as to sensitive eye tissues.*

The opposite effect occurs when a chemical condenses. When vapor molecules are cooled enough that they condense, latent heat is transferred from the vapor to an object or liquid. This occurs when steam condenses on human skin; all of the latent heat remaining in the steam is deposited on the surface of the skin and can cause a burn.

REFRIGERATION CYCLE

To harness the potential of a refrigerant, a controlled operating environment must be provided. This environment is provided by the refrigeration system. The refrigeration system provides a sealed system in which refrigerant is separated into a high-pressure, high-temperature state in one section of the system and a low-pressure, low-temperature state within another section of the system (see **Figure 8**).

Understanding the temperature-to-pressure relationships of refrigerant within a refrigeration system is of the utmost importance in understanding the operation of the system. When the pressure of a refrigerant within a sealed system is known, the temperature of the refrigerant can be calculated. This is useful in the analysis and diagnosis of the refrigeration system.

As the refrigerant is circulated throughout each section of the system, it changes pressure, temperature, and state to

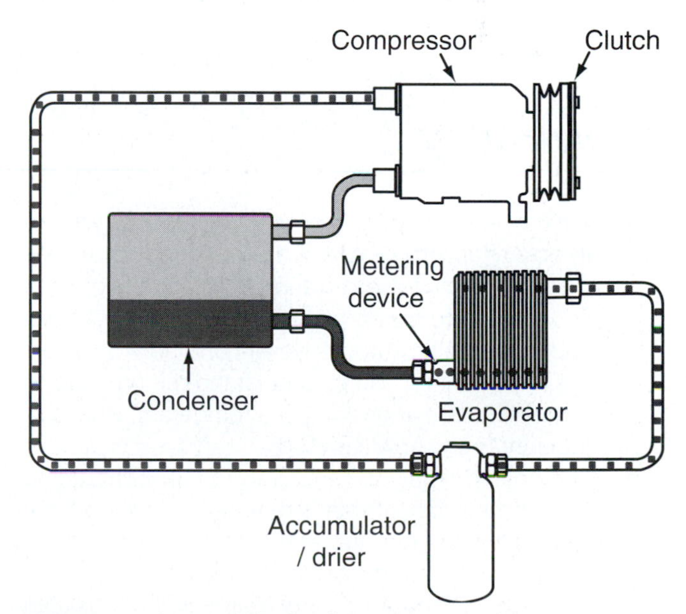

Figure 8. A simplified refrigeration cycle.

facilitate both the absorption and release of heat energy. This is called the refrigeration cycle, and it is made possible by the use of a **heat exchanger**. A heat exchanger is a component that is similar in construction to a radiator that facilitates the transfer of heat between ambient air and the chemical contained within the exchanger. When a chemical is contained within a sealed system or container, the container will assume the temperature of the chemical. When contained in a heat exchanger, the principles of temperature to pressure relationship, latent heat absorption, and release are used to remove heat from an area or add heat to an area.

The basic refrigeration cycle uses the principles mentioned earlier to transfer heat energy. The refrigerants used in the automotive air conditioning system have extremely low boiling points and operate over a wide range of pressures and temperatures. The specific temperature and pressure will vary depending on which section of the refrigeration system they currently occupy. A simplified refrigeration cycle is described here. A more precise description will be given in a later chapter.

The refrigeration cycle begins at the outlet of the compressor where high-pressure, high-temperature refrigerant in a vaporized state leaves the compressor outlet and is directed to a heat exchanger called the **condenser**. The condenser is a heat exchanger located at the front of the vehicle in front of the radiator that is used to transfer heat from the refrigerant to the atmosphere. Inside the condenser, much of the heat within the refrigerant is transferred to the surface of the condenser; as air passes over the condenser, the heat is released into the atmosphere. Enough heat is removed from the refrigerant that the pressure of the refrigerant is lowered slightly and it changes state from a vapor to a liquid.

Once it exits the condenser, the high-pressure liquid refrigerant is routed to a **metering device**. The metering device is located between the condenser and another heat exchanger called the **evaporator**. The evaporator is installed in a position in which it is exposed to the air within the vehicle's interior. The metering device controls the temperature of the evaporator by limiting the amount of refrigerant that is allowed to enter the evaporator. Because the amount of refrigerant entering the evaporator is reduced, once the molecules enter the evaporator inlet they have room for expansion. This causes the refrigerant pressure on the evaporator side of the metering device to decrease. With the decrease in pressure comes a decrease in temperature as well. This is important because when the temperature of the refrigerant is reduced, its capacity to absorb heat is increased.

Because the evaporator is exposed to the passenger compartment air, its surface temperature is roughly the same as the ambient air. As the cool liquid refrigerant passes through the heat exchanger, the heat absorbed from the surface of the evaporator is absorbed by the refrigerant causing the refrigerant to vaporize, thus cooling the surface of the evaporator and the air surrounding it. Consequently, as the surface of the evaporator is cooled, the moisture present within the air surrounding the evaporator is condensed on the surface of evaporator. This lowers the humidity of the air within the passenger compartment.

As the low-pressure vapor exits the evaporator, it is drawn into the compressor inlet. Inside the compressor, the vapor is compressed, which raises the pressure and temperature of the refrigerant. The increase in pressure is necessary to facilitate circulation of the refrigerant. In addition to the required circulation, the increase in temperature causes the refrigerant to condense more rapidly when it enters the condenser. The refrigerant leaves the compressor as a high-pressure, high-temperature vapor, and the cycle is repeated.

Summary

- A refrigerant is any liquid or vapor material that is readily able to absorb heat from an area or object, transfer that heat, and release it to another location. In order for you as an automotive technician to properly analyze and diagnose an automotive air conditioning system, it is important for you to understand the pressure-to-temperature relationships of a refrigerant, what a refrigerant is, and how it is used.
- The boiling point is a temperature representation of the amount of heat that is required to make a liquid turn to a vapor. It is also at this temperature that the material will condense. When molecules of a liquid change to a vapor, they carry a large amount of latent heat away.
- Pressure within a sealed container is increased in proportion to the amount of heat that is applied to the container. An increase in pressure to a liquid or vapor in a sealed container will cause an increase in temperature proportional to the amount of pressure applied.
- When heat is removed, the pressure within a sealed container decreases proportionately to the amount of heat expelled. When pressure is reduced, the boiling point is also reduced. Lowering the boiling point increases a liquid's ability to absorb heat.
- When a liquid turns to vapor, latent heat is carried away, providing a cooling effect to the body or object from which it left. When a vapor condenses, heat is transferred to the object or liquid to which it condensed.

- The refrigeration system provides an operating environment in which the potential of a refrigerant can be harnessed. Refrigerant within a refrigeration system undergoes changes in pressure, temperature, and state to facilitate heat absorption and release. Heat exchangers are used both to absorb and expel heat.

Review Questions

1. A liquid will both boil and condense at the same temperature.
   A. True
   B. False
2. Technician A says that an unpressurized liquid will never rise above its boiling temperature. Technician B says that when a liquid vaporizes, a large amount of heat is carried away with the vapor. Who is correct?
   A. Technician A
   B. Technician B
   C. Both Technician A and Technician B
   D. Neither Technician A nor Technician B
3. Explain the relationship of temperature and pressure in a liquid by describing what occurs when a change takes place in either pressure or temperature.
4. Describe how an object or a body of liquid becomes cooler when the liquid begins to vaporize.
5. Technician A says that a refrigerant used in a refrigeration system is subject to high pressure. Technician B says that a refrigerant used in a refrigeration system is subject to low pressure. Who is correct?
   A. Technician A
   B. Technician B
   C. Both Technician A and Technician B
   D. Neither Technician A nor Technician B
6. All of the following statements are true *except:*
   A. A heat exchanger provides a means for refrigerant to absorb heat.
   B. A heat exchanger provides a means for refrigerant to expel heat.
   C. The evaporator is exposed to passenger compartment air.
   D. The refrigerant condenses within the evaporator.
7. Technician A says that refrigerant is in a high-pressure state as it passes through the evaporator. Technician B says that moisture will condense on the surface of the evaporator. Who is correct?
   A. Technician A
   B. Technician B
   C. Both Technician A and Technician B
   D. Neither Technician A nor Technician B

Chapter 7 Refrigerants and Lubricants

Introduction

As we have learned, a refrigerant is any liquid or vapor that is able to absorb heat, transfer heat, and disperse that heat. Additionally, a refrigerant must meet the requirements of the operating environment provided by the specific refrigeration system. At this time, to be accepted as an automotive refrigerant, a material must have relatively low operating pressures, it must be stable over a wide range of temperatures and pressures, it must not be reactive with system components, and it should have a low environmental and human impact. In the history of mobile air conditioning systems, two refrigerants have been used to fill the needs of most all automotive air conditioning systems: dichlorodifluoromethane, commonly known as R-12, and tetrafluoroethane, known as R-134A. Although other refrigerants have made appearances and others are on the horizon, for all practical purposes all vehicles sold throughout the world have been equipped with one of these two refrigerants. In this chapter, you will learn what each of these materials is composed of, what their operating pressures and characteristics are, how these materials should be stored and handled, which lubricants are recommended for use with each refrigerant, how these materials affect the environment, and the regulations that govern the usage and service of systems equipped with these refrigerants.

> **Interesting Fact**
>
> *Refrigerant is often referred to as Freon; however Freon is a registered trade name of refrigerants produced by the DuPont Corporation and should only be used in reference to those products.*

R-12

The first widely used automotive refrigerant was a chemical compound known as dichlorodifluoromethane ($CCl2F2$), otherwise known as R-12. R-12 is a member of a family of organic chemicals known as chlorofluorocarbons, which are most often referred to by the abbreviation CFC. Chemicals within the CFC family are primarily composed of chlorine, fluorine, and carbon; other elements may be added to obtain specific characteristics. Chemicals of the CFC family have been successfully used as refrigerants in both mobile and stationary applications, as plastic foam materials, aerosol-propellants, sanitizers, and solvents. CFCs were widely used because their chemical compositions were extremely stable and nonflammable, and it was believed that they provided very little if any human or environmental impact. However, discoveries were made in the early 1970s that proved otherwise. It was determined that chlorine, one of the primary chemicals in CFC composition, was harmful to the earth's ozone layer. Because of these harmful effects, production of R-12 was halted for good on December 31, 1995. Vehicles with alternative refrigerants began to appear in the marketplace in 1992 and by 1994 all new vehicles were equipped with an alternative refrigerant, mainly R-134A. Although R-12 is no longer the refrigerant of choice, many R-12–equipped vehicles still remain on the road today, and for the near future these vehicles will still need to be properly maintained and serviced. However, these numbers diminish significantly each year as older vehicle refrigeration systems are converted to alternate refrigerants or as the vehicles themselves are removed from service.

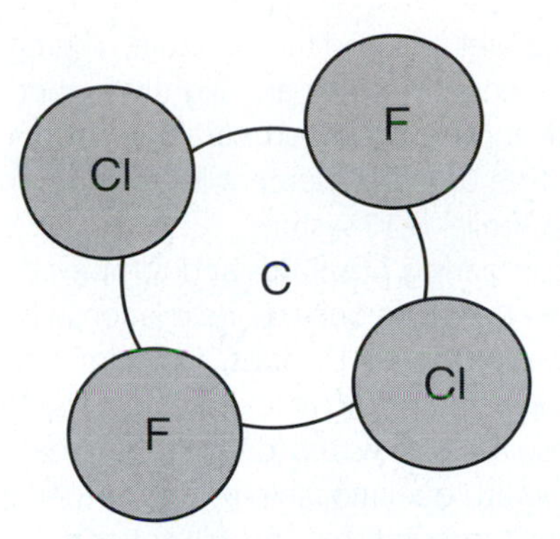

Figure 1. The chemical structure of an R-12 molecule.

Characteristics

Each molecule of CCl2F2 or R-12 consists of one atom of carbon, two atoms of chlorine, and two atoms of fluorine (see **Figure 1**). R-12 has a boiling point of –21.62 degrees F (–29.79 degrees C) at sea level. Because of this very low temperature boiling point, R-12 must be kept in a sealed container. When exposed to normal atmospheric conditions, R-12 will evaporate immediately. However, a low boiling point coupled with the ease at which R-12 can be vaporized and condensed make R-12 an excellent heat transfer agent. When introduced as a low-pressure liquid into the evaporator, R-12 has the ability to absorb heat, even at very low ambient temperatures. R-12 had many desirable characteristics that made it an attractive choice as an automotive refrigerant. Up until the ban of many CFC products, R-12 was considered to be the ideal refrigerant for almost all mobile air conditioning units, including cars, trucks, and construction and agricultural machinery. The perceived benefits of R-12 were as follows:

- Relatively low operating pressures, as compared to other available refrigerants.
- Excellent stability at both high and low pressure.
- Excellent stability at high and low temperatures.
- Nonreactive with air conditioning system components.
- Low perceived human and environmental impacts.
- Readily available and inexpensive.

Pressure-to-Temperature Relationship

When servicing an air conditioning system, it is desirable to know the **pressure-to-temperature relationship**. This can be defined as the specific temperature of the refrigerant at a given pressure. In an operating refrigeration system, the pressure of the refrigerant is a direct indicator of how well the refrigeration system is operating. All refrigerants have a direct pressure-to-temperature correlation. R-12 has a unique pressure-to-temperature relationship in that at pressures of between 20 and 70 psi the temperature on Fahrenheit scale is almost identical to the pressure. This is extremely useful in determining evaporator temperature.

R-12 REFRIGERANT

R-12 PRESSURE-TEMPERATURE CHART

Pressure kPa (psi)		Temperature °F (°C)	Pressure kPa (psi)		Temperature °F (°C)
127 (18)		16 (−9)	808 (117)		100 (38)
136 (20)		18 (−8)	833 (121)		102 (39)
145 (21)		20 (−7)	859 (125)		104 (40)
155 (22)	—	22 (−6)	893 (129)		106 (41)
165 (24)	EVAPORATOR RANGE	24 (−4)	917 (133)	—	108 (42)
175 (25)		26 (−3)	940 (136)	CONDENSER RANGE	110 (43)
185 (27)		28 (−2)	969 (140)		112 (44)
196 (28)		30 (−1)	997 (145)		114 (46)
207 (30)		32 (0)	1027 (149)		116 (47)
219 (32)		34 (1)	1057 (153)		118 (48)
230 (33)		36 (2)	1087 (158)		120 (49)
249 (36)		38 (3)	1118 (162)		122 (50)
255 (37)		40 (4)	1150 (167)		124 (51)
287 (42)		45 (7)	1182 (171)		126 (52)
322 (47)	—	50 (10)	1215 (176)		128 (53)
359 (52)		55 (13)	1248 (181)		130 (54)
398 (58)		60 (16)	1334 (194)		135 (57)
440 (64)		65 (18)	1425 (207)		140 (60)
485 (70)		70 (21)	1519 (220)		145 (63)
531 (77)		75 (24)	1618 (235)		150 (66)
580 (84)		80 (27)	1721 (250)		155 (68)
633 (92)		85 (30)	1828 (265)		160 (71)
688 (100)		90 (32)	1940 (281)	—	165 (74)
746 (108)		95 (35)	2057 (298)		170 (77)

Figure 2. The pressure-to-temperature relationship of R-12.

Unfortunately, this relationship does not extend to the Celsius scale. Additionally, as pressures rise above 70 psi, the pressure-to-temperature relationship moves farther apart (see **Figure 2**).

R-134A

When it was determined that R-12 was harmful to the environment, a replacement had to be identified. That replacement was tetrafluoroethane (CH2FCF3), known as R-134a. R-134a has been accepted by all major automobile manufacturers as the replacement refrigerant in both production systems and as retrofit applications. R-134a is from a family of chemicals called hydrofluorocarbons, or HFCs. HFCs are organic compounds that contain hydrogen, carbon, and fluorine, but no chlorine. HFCs have many of the same characteristics as CFCs and are used in many of the same applications, but because there is no chlorine in HFCs, they are not harmful to the ozone layer. However, HFCs do contribute to **global warming**; for this reason, manufacturers are continually exploring new and safer chemicals that can be used as refrigerants. However, for the near future, R-134a will remain the refrigerant of choice.

Characteristics

One molecule of R-134a contains two atoms of carbon, two atoms of hydrogen, and four atoms of fluorine (see **Figure 3**). R-134a has a boiling point of −14.9 degrees F (−26.1 degrees C), and it will boil immediately if exposed to the atmosphere, necessitating the need for storage in a closed container. Like R-12, R-134a has the ability to easily change states from a liquid to a vapor and back to a liquid again, making it an ideal heat transfer agent. Because of its low temperature boiling point, R-134a has the ability to easily absorb heat, even at low ambient temperatures that are often encountered at the evaporator. Other characteristics that make R-134a a desirable automotive refrigerant include:

- It is nonchlorinated, making it an ozone-safe alternative.
- It has excellent compatibility to A/C system materials.
- It has relatively low operating pressures and temperatures.
- It has excellent chemical stability at both high and low pressures.
- It is nonflammable under normal conditions.
- It is readily available and inexpensive.
- It has a single-source chemical composition.

Pressure-to-Temperature Relationship

Although the characteristics of R-12 and R-134a are very similar, there are some significant operational differences. When directly compared to R-12, R-134a has a lower low-side operating pressure and temperature, whereas high-side pressures and temperatures are slightly higher than R-12. This is significant in that in comparable systems an R-134a system will have up to 15 percent less refrigerant capacity than that of a similar R-12 system.

When comparing pressures and temperatures, R-134a does not share the same one-to-one relationship of pressure and temperature that R-12 does. Although the pressure-to-temperature relationship at low pressures is close, it is not nearly as close as R-12. As temperatures increase, the pressure-to-temperature relationship moves farther apart. When comparing the temperatures, you will notice a significant difference between the temperatures and pressures of each refrigerant (see **Figures 2 and 4**).

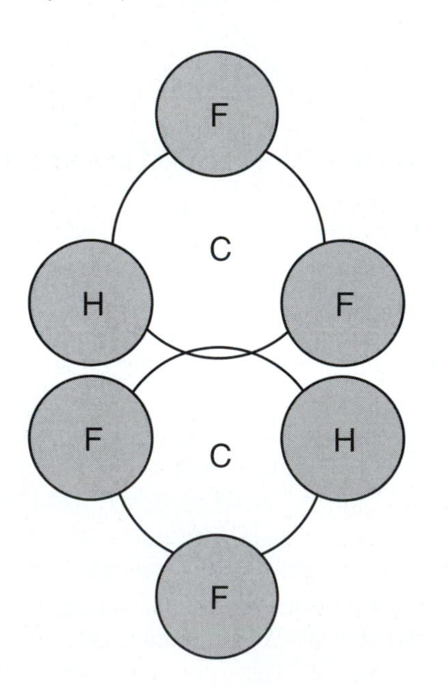

Figure 3. The chemical structure of an R-134a molecule.

OTHER ALTERNATIVES

We have discussed the most popular automotive refrigerants on the market today, including the two that you are most likely to encounter. It should be recognized that all major automotive manufacturers have selected R-134a

R-134A REFRIGERANT

R-134A PRESSURE-TEMPERATURE CHART

Pressure kPa (psi)		Temperature °F (°C)	Pressure kPa (psi)		Temperature °F (°C)
106 (15)		16 (−9)	857 (100)		100 (38)
115 (17)		18 (−8)	887 (129)		102 (39)
124 (18)		20 (−7)	917 (133)		104 (40)
134 (19)	——	22 (−6)	948 (137)		106 (41)
144 (21)		24 (−4)	980 (142)	——	108 (42)
155 (22)	EVAPORATOR RANGE	26 (−3)	1012 (147)		110 (43)
166 (24)		28 (−2)	1045 (152)		112 (44)
177 (26)		30 (−1)	1079 (157)		114 (46)
188 (27)		32 (0)	1114 (162)		116 (47)
200 (29)		34 (1)	1149 (167)		118 (48)
212 (31)		36 (2)	1185 (172)	CONDENSER RANGE	120 (49)
225 (33)		38 (3)	1222 (177)		122 (50)
238 (35)		40 (4)	1260 (183)		124 (51)
272 (40)		45 (7)	1298 (188)		126 (52)
310 (45)	——	50 (10)	1337 (194)		128 (53)
350 (51)		55 (13)	1377 (200)		130 (54)
392 (57)		60 (16)	1481 (215)		135 (57)
438 (64)		65 (18)	1590 (231)		140 (60)
487 (71)		70 (21)	1704 (247)		145 (63)
540 (78)		75 (24)	1823 (264)		150 (66)
609 (88)		80 (27)	1948 (283)		155 (68)
655 (95)		85 (30)	2079 (301)		160 (71)
718 (104)		90 (32)	2215 (321)	——	165 (74)
786 (114)		95 (35)	2358 (342)		170 (77)

Figure 4. The pressure-to-temperature relationship of R-134a.

as the refrigerant of choice for both retrofit applications and as the successor to R-12. However, the Environmental Protection Agency (EPA) has certified several other refrigerants as acceptable R-12 substitutes. R-134a is the only single-source composition refrigerant available as an R-12 replacement.

Blends

Refrigerants fall into two categories: single-source refrigerants and blends. Single-source refrigerants are composed of only one type of refrigerant material, for example, R-12 and R-134a. Blends are manufactured by combining multiple single-source refrigerants in various quantities to achieve a desired set of characteristics (see **Figure 5**). For example, Freeze 12 is an EPA-approved refrigerant that is comprised of 80 percent R-134a and 20 percent R-12. However, the new refrigerant will take on characteristics of both base refrigerants and can even develop some unique characteristics. For this reason, blended refrigerants can be somewhat less predictable than single-source refrigerants in how they respond at different temperatures and pressures.

Although some refrigerants may be advertised as "drop-in" replacements for R-12, indicating that they will directly replace R-12, it should be noted that there is no direct replacement for either R-12 or R134a. Any time one refrigerant

Motor Vehicle Air Conditioning Substitutes for R-12 Reviewed under EPA's SNAP Program as of May 1, 2001

Acceptable Subject to Use Conditions (2)

Name (1)	Date	Manufacturer	Components						
			HCFC-22	HCFC-124	HCFC-142b	HFC-134a	Butane (R-600) (3)	Isobutane (R-600a) (3)	HFC-227ea
HFC-134a	3/18/94	Several	—	—	—	100	—	—	—
FRIGC FR-12	6/13/95	Intercool Distribution 800-555-1442	—	39	—	59	2	—	—
Free Zone/ RB-276 (4)	5/22/96	Refrigerant Management Services of Georgia 800-347-5872	—	—	19	79	—	—	—
R-406A/ GHG (5)	10/16/99	People's Welding 800-382-9006	55	—	41	—	—	4	—
GHG-HP (5)	10/16/99	People's Welding 800-382-9006	65	—	31	—	—	4	—
GHG-X4/ Autfrost/ Chill-It (5)	10/16/96	People's Welding 800-382-9006 McMullen Oil Products 800-669-5730	51	28.5	16.5	—	—	4	—
Hot Shot/ Kar Kool (5)	10/16/96	ICOR 800-357-4062	50	39	9.5	—	—	1.5	—
Freeze 12	10/16/99	Technical Chemical 800-527-0885	—	—	20	80	—	—	—
GHG-X5 (5)	6/3/97	People's Welding 800-382-9006	41	—	15	—	—	4	40

1. Many refrigerants, including R-401A (made by DuPont), R-401B (DuPont), R-409A (Elf Atochem), Care 30 (Calor Gas), Adak-29/Adak-12 (TACIP Int'l), MT-31 (Millenia Tech), and ES-12R (Intevest), have not been submitted for review in motor vehicle air conditioning, and it is therefore illegal to use these refrigerants in such systems as an alterntive to CFC-12.
2. See text for details on legality of use according to status
   - Acceptable Subject to Use Condtions rgarding fittings, labeling, non drop-in, and compressor shutoff switches.
   - Unacceptable; illegal for use as a CFC-12 substitute in motor vehicle air conditioners.
3. Although some blends contain flammable components, all blends that are Acceptable Subject to Use Conditions are nonflammable as blended.
4. Freezone contains 2% of a lubricant
5. HCFC-22 content results in an additional use condition: must be used with barrier hoses

Figure 5. Common R-12 substitutes and their compositions.

is removed, there is a level of modification that must be completed before a different refrigerant is introduced.

As a technician, it is imperative to understand the chemical makeup of the refrigerants that you are using. Some unapproved blends contain materials that can be dangerous to the service technician, the environment, and the consumer. Common blend materials that should be considered dangerous are hydrocarbon gases such as butane or isobutene. If blended in sufficient quantities, these materials present a possible fire or explosion hazard to service technicians or vehicle occupants. Under no circumstances should an unapproved refrigerant be used in any refrigeration system. Additionally, you should adhere strictly to all replacement guidelines.

MIXING REFRIGERANTS

Under no circumstances should different refrigerants be mixed. Although R-12 and R-134a are chemically compatible, this merely means that they will not react negatively with one another. What actually occurs is that the two chemicals form a type of compound known as an **azeotrope**. What this means to you is that the new compound assumes chemical properties and characteristics that are different from either R-12 or R-134a. If these two refrigerants are mixed, it can lead to undesirably high compressor head pressures. This can cause serious compressor and refrigeration system component damage. Once these two refrigerants are combined, it is impossible for them to be separated using common on-site refrigeration and recycling equipment, which means that any quantity of mixed refrigerant must be disposed of properly.

PACKAGING AND HANDLING

Handling and storage techniques for R-12 and R-134a are virtually the same. Automotive refrigerants are typically found in two quantities: 1 lb cans and 30 lb drums (13.61 kg). Although other quantities can be purchased, these are typical of what is found in service and parts facilities. It should be noted that whereas the small 1 lb cans are often found at retail outlets, they only contain either 12 oz (.340 g) or 14 oz (.397 kg) of refrigerant. This is important to note when refilling a system using small cans (see **Figure 6**).

Automotive refrigerants are packaged and sold in disposable containers; each container has a blow-off protection device that prevents the entire cylinder from rupturing in the event that internal pressure should become excessive. Excessive cylinder pressure is commonly caused by extreme heat being applied to the storage vessel. Refrigerants should never be stored or used in areas that can exceed 125 degrees F (52 degrees C), and direct heat should never be applied to a refrigerant container. Additionally, these containers should never be reused for any purpose. When containers are empty, they should be

Figure 6. Typical disposable refrigerant containers.

completely evacuated of any remaining refrigerant using the proper equipment for the refrigerant type and disposed of in accordance with local regulations.

In order to help prevent system cross-contamination, each type of refrigerant container is equipped with a specific style of dispensing valve and a detailed label with a specific background color. Thirty-pound R-12 containers use a fitting that measures $^{7}/_{16}$ in. in diameter and has 20 threads per inch. R-12 labels have a white background. The 30 lb R-134a containers use a fitting that measures $^{1}/_{2}$ in. in diameter and uses a #16 Acme fitting. The R-134a label is sky blue in color (see **Figure 7** for various

Refrigerant	Background
CFC-12	White
HFC-134a	Sky Blue
Freeze 12	Yellow
Free Zone/RB-276	Light Green
Hot Shot	Medium Blue
GHG-X4	Red
R-406A	Black
GHG-X5	Orange
GHG-HP	not yet developed*
FRIGC FR-12	Grey
SP34E	Tan

Figure 7. Designated refrigerant label background colors.

> **You Should Know** *Liquid refrigerant may cause frostbite to sensitive human tissues. Of particular concern are the eyes and skin. Protective eyewear and gloves should be worn at all times when working with any type of automotive refrigerants. Never apply direct heat to a refrigerant container or expose to an open flame. Refrigerant should never be exposed to temperatures higher than 125 degrees F (52 degrees C). Disposable containers should never be refilled with substances.*

label colors). Other measures to prevent cross-contamination include the mandatory installation of specific vehicle fittings; this is to ensure that only dedicated service equipment may be used with a particular refrigerant (see **Figures 8 and 9**). Refer to Chapter 19 for complete refrigerant retrofitting guidelines and procedures.

LUBRICANTS

Although there are very few moving parts within an automotive A/C system, the most critical of all components, the compressor, has to have constant lubrication. The automotive compressor has no internal pump to pressurize and circulate lubricant and must depend on the circulation of refrigerant for lubrication. Compressor lubrication is very similar to that of a two-cycle engine, in which the lubricant is mixed in with the fuel. The internal components of the compressor are lubricated as the refrigerant passes across the friction surfaces. In order for this to occur, the lubricant must be mixed with the refrigerant; this is called **miscibility**. Miscibility is defined as the ability of a liquid to completely mix with another liquid (see **Figure 10**). Although the

Refrigerant	Contact	High-Side Service Port			Low-Side Service Port			30-lb. Cylinders			Small Cans		
		Diameter (inches)	Pitch (threads/ inch)	Thread Direction	Diameter (inches)	Pitch (threads/ inch)	Thread Direction	Diameter (inches)	Pitch (threads/ inch)	Thread Direction	Diameter (inches)	Pitch (threads/ inch)	Thread Direction
CFC-12 (post-1987)	multiple	6/16	24	Right	7/16	20	Right	7/16	20	Right	7/16	20	Right
CFC-12 (pre-1987)		7/16	20	Right	7/16	20	Right	7/16	20	Right	7/16	20	Right
HFC-134a	multiple	quick connect			quick connect			8/18	16 Acme	Right	8/16	16 Acme	Right
Freeze 12	Technical Chemical 800-527-0885	7/16	14	Left	8/16	18	Right	8/16	18	Right	6/16	24	Right
Free Zone /RB-276	Refrigerant Gases 888-373-3066	8/16	13	Right	9/16	18	Right	9/16	18	Right	6/16	24	Left
Hot Shot	ICOR 800-357-4062	10/16	18	Left	10/16	18	Right	10/16	18	Right	5/16	24	Right
GHG-X4 Autofrost	People's Welding 800-382-9006	.305	32	Right	.368	26	Right	.368	26	Right	14mm	1.25mm spacing	Left
GHG-X5		8/16	20	Left	9/16	18	Left	9/16	18	Left	not sold in small cans		
R-406A		.305	32	Left	.368	26	Left	.368	26	Left	8/16	20	Left
GHG-HP		not yet developed											
FRIGC FR-12	Intercool 800-555-1442	quick-connect, different from HFC-134a			quick-connect, different from HFC-134a			8/16	20	Left	7/16	20	Left
SP34E	Solpower 888-289-8866	7/16	14	Right	8/16	18	Left	8/16	18	Left	5/16	24	Left

Figure 8. Designated refrigerant service fitting specifications.

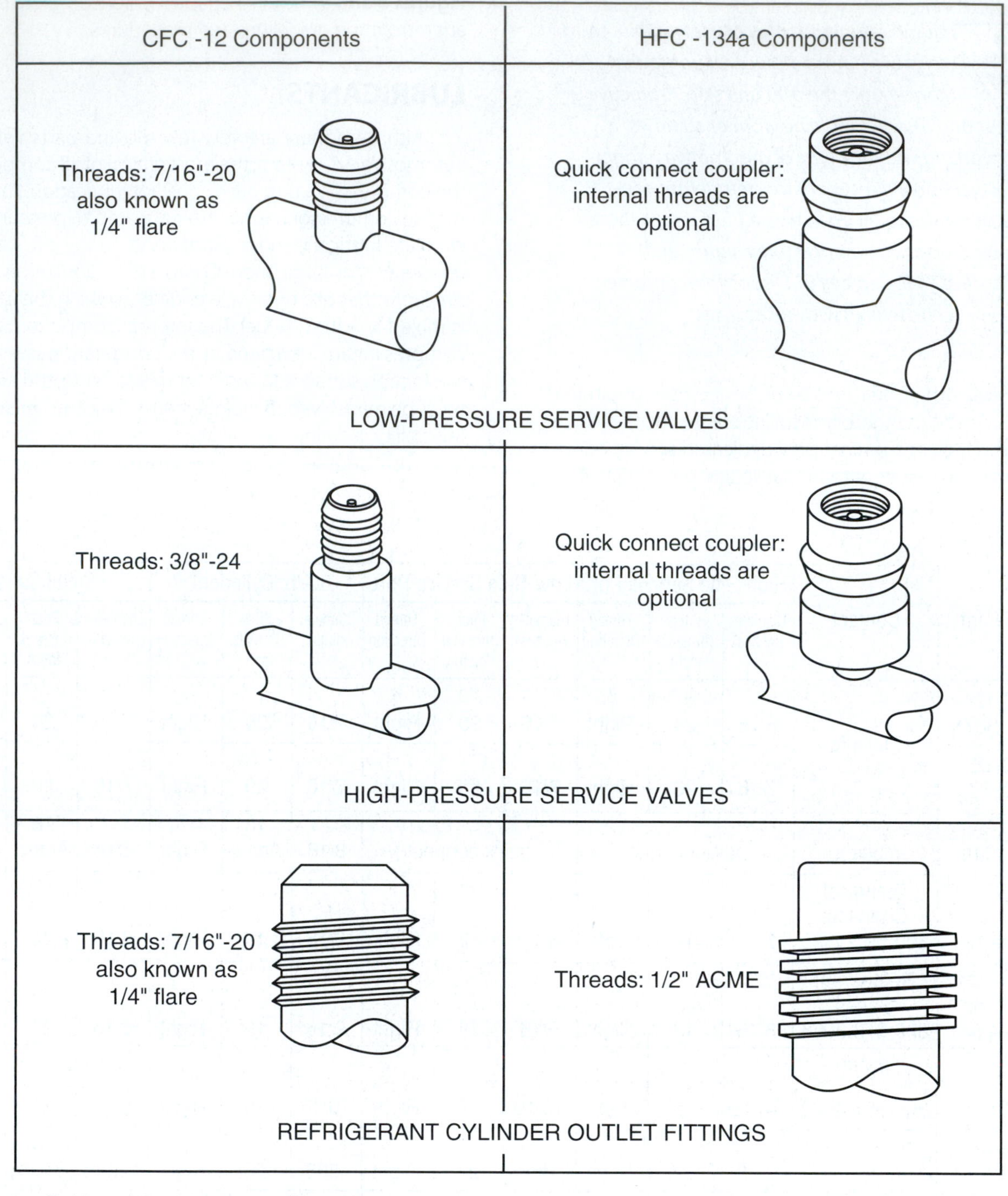

Figure 9. R-12 and R-134a service fittings.

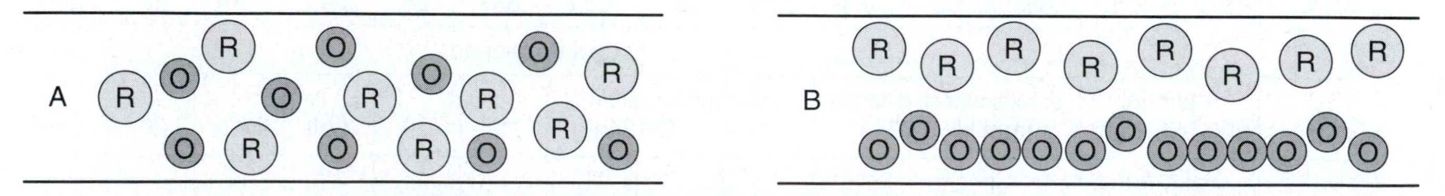

Figure 10. (A) When oils are miscible, oil and refrigerant mix easily. (B) When oils are not miscible, the oil will not mix and will eventually fall out of suspension.

compressor manufacturers determine what lubrication qualities are desirable for their compressor, they must also consider the miscibility of the lubricant with the intended refrigerant. Although there are many lubricants available, the majority of manufacturers have settled on a select few lubricant types for a majority of the compressor applications. Although R-12 and R-134a have many of the same characteristics, their chemical compositions make it impossible to use the same type of lubricants.

Lubricants used in refrigeration systems are known as refrigeration oils and are classified using four factors:

- Viscosity: this is the resistance of fluid to flow; typically, the higher the viscosity number, the thicker the lubricant is. Refrigeration oils are typically in the 500 to 525 viscosity range for mineral oil, whereas synthetic products can be found in a range between 46 and 150.
- Pour point: this is the temperature at which refrigeration oil will begin to flow. This is particularly important given the drastic range of temperature fluctuation experienced within a refrigeration system. The pour point for refrigeration oils is usually in the −40 degree F (−40 degree C) to −10 degree F (−23.3 degree C) range.
- Compatibility: the lubricant used must be compatible with the system components and the refrigerant used. R-12 refrigeration systems use mineral-based refrigerant oil, whereas R-134a systems use synthetic-based refrigeration oil.
- Additives: as with engine oils, refrigeration oils may have specific additives that help prevent wear and add to the overall benefit of the product. These additives will vary greatly between manufacturers.

Mineral Oils

Mineral oils are used in systems containing R-12 refrigerants. Mineral-based refrigeration oils are clear to light yellow in color and have very little, if any, odor. Any discoloration or pungent odor is an indication of contamination either from external tampering or compressor failure. Mineral oils are not miscible in R-134a refrigerant; when used with R-134a refrigerant, mineral oils fall out of suspension and will eventually accumulate in various air conditioning system components. This will starve the compressor for oil and will lead to damage and subsequent compressor failure. Additionally, when large amounts of unsuspended oil accumulate within the condenser or evaporator, the ability of the refrigerant to transfer heat from the refrigerant to the heat exchanger is greatly reduced.

Synthetic Oils

There are two types of refrigeration oil that are recommended for use with R-134a refrigerant: polyalkylene glycols, also known as PAG oil, and polyol ester oil, or POE. Each of these oils is synthetic in composition and the color of oils may vary depending on the manufacturer. Although mineral oils are practically odorless, synthetic oils may have pungent odors, making it impossible to judge by smell if a lubricant is contaminated. Although both types are synthetic, their properties are quite different, so the two should never be mixed. Remember that the lubricant type is largely determined by the manufacturer of the compressor, so always refer to the service manual to ensure that the proper lubricant is being used.

There are downsides to synthetic oils, however; both PAG and POE oils are extremely **hygroscopic**, which means that they have a tendency to attract and absorb water. This in itself is a critical fault. Moisture invasion is inevitable when the system is opened for service. If moisture is not properly removed, **hydrolysis** can occur. Hydrolysis is a chemical reaction that takes place when two chemicals combine to form one or more other substances. When moisture is trapped within air conditioning systems, it can mix with refrigerants and some POEs to form hydrochloric acid, causing severe damage to air conditioning system components.

Although PAG oil has been selected as the refrigerant lubricant of choice by many automobile manufacturers, a distinct disadvantage that is associated with its use is that when it is combined with chlorine, such as that found in R-12, a chemical reaction can occur that will cause lubricant degradation and the formation of sludge. Lubricant degradation will eventually lead to compressor damage and subsequent failure, whereas the formation of sludge can lead to the clogging of metering devices. For this reason, some manufacturers may recommend a POE-based oil in R-12 to R-134a conversions.

Packaging

Refrigeration oil is packaged either in nonpressurized bulk containers or in a pressurized oil charge. Bulk containers range in size from 8 oz (.237 L) to 5 gallons (18.93 L). An oil charge contains a 2 oz (.060 L) quantity of oil packaged in a sealed container pressurized with a 2 oz (.057 kg) charge of refrigerant. The refrigerant is used as a propellant to force the oil into the refrigeration system. The use of an oil charge provides a means for the technician to add small amounts of oil using a normal can tap and charging hose, without having to evacuate and open the system for service. These units are available for both R-12 and R-134a using the proper lubricant for each. However, the most common way of restoring the oil charge is when the system is open for service. Various quantities of oil products are shown in **Figure 11**.

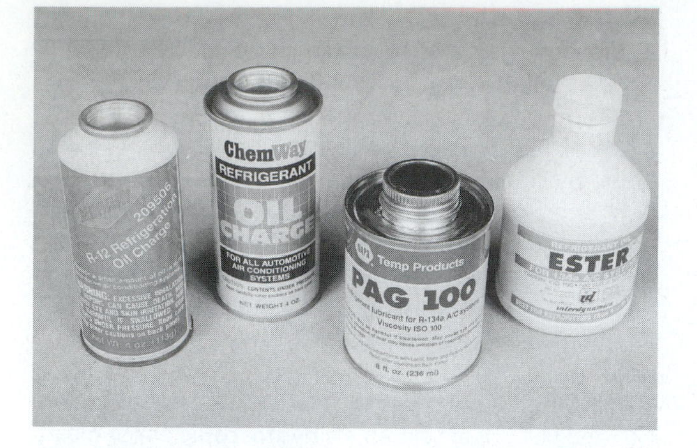

Figure 11. Several types and grades of refrigeration oil.

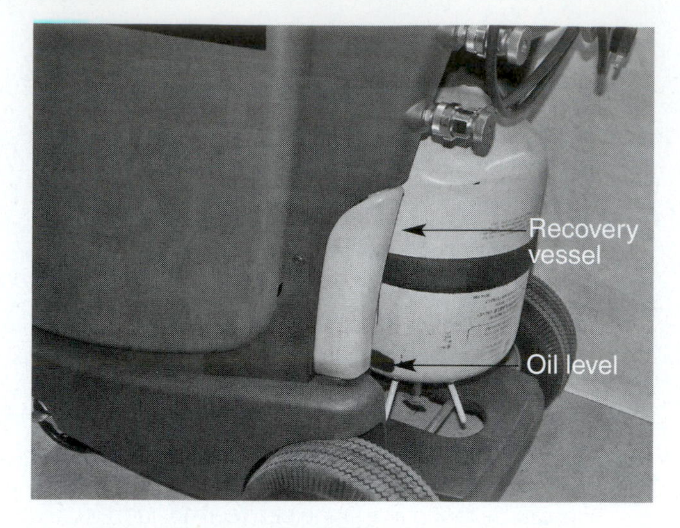

Figure 12. The level of oil in the oil recovery vessel should be checked before and after refrigerant recovery.

SERVICE

In a properly operating air conditioning system, lubricant loss should not be of significant concern. However, any time that a system has a leak or is serviced, some lubricant will be lost. It is of critical importance to make sure that the system is filled with the proper amount of oil; too little oil will eventually lead to compressor failure, whereas an overabundance of lubricant will lead to inefficient transfer of heat energy within the heat exchangers. Because the air conditioning system is not equipped with a method in which to periodically check system oil level, oil balance must be carefully maintained through proper service procedures. Maintaining the oil balance in the system can be accomplished by carefully estimating and replacing only the amount of refrigeration oil that has been lost.

If a refrigeration system is low or completely out of refrigerant, it is impossible to determine how much lubricant has been lost. Often, the amount is insignificant and you should not try to compensate for the lost amount. However, each component within the system will hold a predictable amount of lubricant. When these components are replaced, a specific amount of oil should be restored to the system to compensate for the amount stored within the component. The vehicle service manual will typically give specifications on how to maintain oil balance when system components are replaced.

When refrigerant is recovered from a system, some oil will be drawn out with the refrigerant. Most recovery machines will separate the oil from the refrigerant and dispense the oil into a container (see **Figure 12**). The level of this container should be checked before and after the residual oil is purged. When the purge is complete, the same amount of fresh clean oil should be restored into the system in addition to the amount required with component replacement. Refrigeration oil should never be reused, no matter how clean it appears to be. When servicing the oil level in an air conditioning system, the replacement amount of oil should be measured precisely.

> **You Should Know** *Because PAGs and POEs are very hygroscopic, they must be handled with extra care. The most important precaution is to make sure that the lid is securely replaced once the desired amount of oil is retrieved from the bottle, which will help keep out excess moisture. Old quantities or bottles that have been left open should be discarded, as they are likely to have absorbed a critical amount of moisture.*

There are three ways in which oil can be added to the system. The first is using the provisions of the recovery and recharging station. This method is best used when performing routine services in which only small amounts of oil need to be replaced, usually 2 oz (.060 L) or less (see **Figure 13**). It is sometimes difficult to add large amounts of oil into the system using this method. Keep in mind, however, that larger amounts of oil are usually only required when a component has been replaced.

The second method is accomplished by pouring the oil directly into a component of the system (see **Figure 14**). This method is used when a component is being replaced or has been drained or for some reason a large amount of oil needs to be added. The oil should be added directly to the component being replaced when possible.

The third method of adding refrigerant oil is accomplished using the pressurized oil charge (see **Figure 15**). This method typically is used when the system is suspected

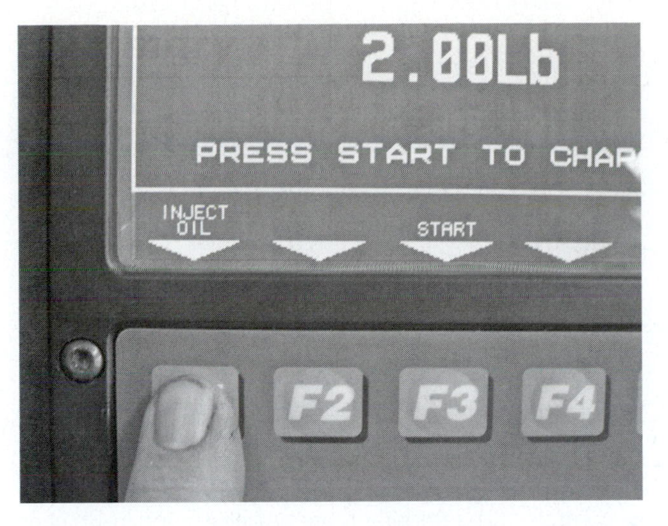

Figure 13. Most recovery/refilling stations have provisions to add refrigerant oil to the system.

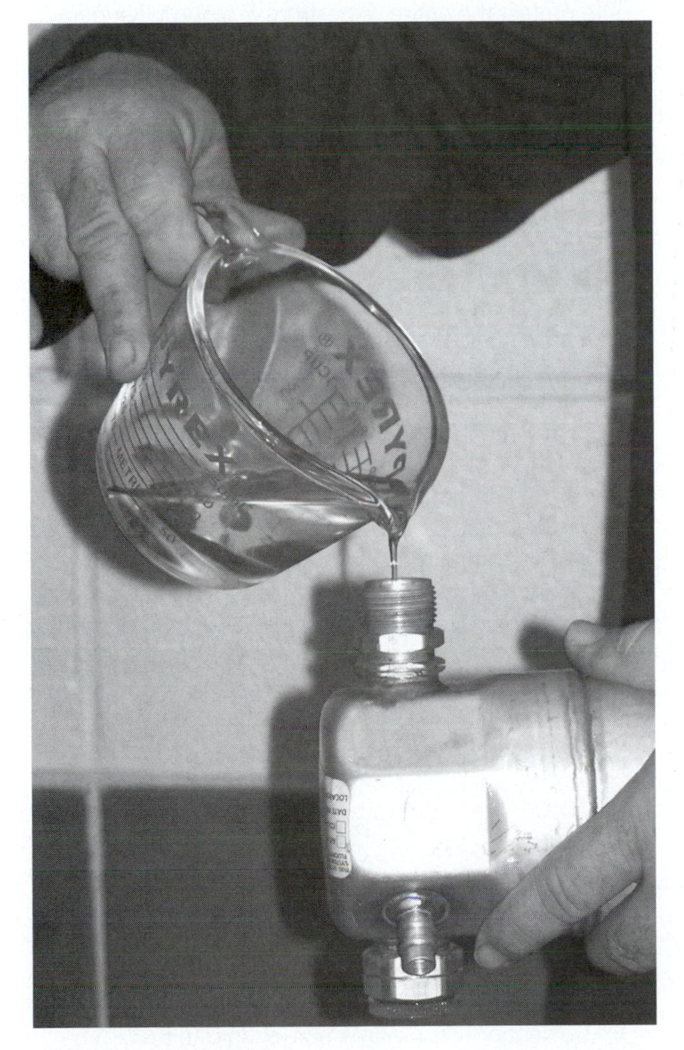

Figure 14. Refrigeration oil can be poured directly into the component.

Figure 15. An oil charge can be used when the system is fully charged.

to be low on oil or when the recovery/recycling unit is not being used to service the system. This method is often used by do-it-yourselfers. The oil charge connects directly to the same can tap that is used for installing small cans of refrigerant. When charging with this method, the can tap should always be connected to the low-pressure side of the system.

You Should Know *If can tap equipment is inadvertently connected to the high-pressure side of a refrigeration system with the system engaged, the high-pressure within the system can be enough to cause the can to burst. This can cause serious injury to anyone near the vehicle.*

> **Interesting Fact**
>
> *It was once commonplace to add "a couple of ounces" of oil for "good measure" each time a refrigeration system was opened for service. This would seem to be a harmless practice, but consider the results if the system has been serviced on multiple occasions. Within a few installments of "good measure," the system could potentially have double the amount of required refrigerant oil. This could make diagnosis of a poor cooling condition difficult at best. For this reason, it is absolutely critical that you try to maintain the oil balance in the system by only replacing what you know has been lost.*

Summary

- A refrigerant is any material that is able to absorb, transfer, and dispense heat energy. Refrigerants must meet the needs required of the operating environment. R-12 and R-134a are the most common automotive refrigerants.
- It has been determined that the chlorine found in R-12 is harmful to the earth's ozone layer. Production of R-12 was halted on December 31, 1995. R-12 has relatively low operating pressures and temperatures, excellent chemical stability, is nonreactive, was perceived to be safe for humans and the environment, and was readily available.
- R-134a has been the predominant replacement for R-12. R-134a has many of the same operating characteristics as R-12. R-134a contains no chlorine, which makes it safer for the environment. R-134a has relative low operating pressures and temperatures, it is nonflammable under normal conditions, and it has a single-source composition.
- Other refrigerants are available as R-12 and R-134a alternatives, but all require some amount of system modification. Hydrocarbons are common in the manufacture of some alternative refrigerants.
- Refrigerants of different compositions should never be mixed. Mixed refrigerants are inseparable using on-site equipment. Mixed refrigerants can cause undesirable operating temperatures and pressures.
- Automotive refrigerants are typically found in 1 lb cans or 30 lb drums. Refrigerants are packaged in disposable containers and should never be reused for any purpose. Refrigerant containers have blow-off protection devices to protect the entire vessel from explosion. Refrigerants should never be stored where they will be exposed to temperatures greater than 125 degrees F (52 degrees C). Each different refrigerant type is assigned its own specific dispensing valve and specific label.
- Compressor lubrication is dependant on the miscibility of the refrigerant and the lubricating oil. Different refrigerants require different refrigeration oils. Refrigeration oils cannot be interchanged or mixed. Refrigeration oils are classified using four factors: viscosity, pour point, compatibility, and additives.
- Refrigerant oils for automotive use are grouped into two categories: mineral oils and synthetic oils. Synthetic oils are categorized as PAGs or POEs. All synthetics are very hydroscopic. Refrigerant oils are packaged in various sized bottles or are available as oil charges.
- Any time refrigerant is leaked or removed, oil will also be lost. Oil balance must be maintained for proper system operation. Each component within the system will hold a specific amount of oil. Oil can be added to the system using three methods: using a recovery/recharging station, pouring oil directly into components, and using a pressurized oil charge.

Review Questions

1. Describe what makes R-12 a good automotive refrigerant.
2. Technician A says that when R-12 is at a pressure of 50 psi, its temperature is also at or near 50 psi. Technician B says that R-12 refrigerant pressures and temperatures measured on the Fahrenheit scale are nearly identical in the 71–200 psi range. Who is correct?
   A. Technician A
   B. Technician B
   C. Both Technician A and Technician B
   D. Neither Technician A nor Technician B
3. R-12 and R-134a have many similar characteristics. Describe the major differences between R-12 and R-134a.

4. Technician A says that when two different refrigerants are combined in an air conditioning system, the two chemicals can be separated using on-site refrigerant recycling equipment. Technician B says that using mixed refrigerants can cause higher system operating pressures, which might result in system damage. Who is correct?
   A. Technician A
   B. Technician B
   C. Both Technician A and Technician B
   D. Neither Technician A nor Technician B
5. Technician A says that empty 30 lb disposable refrigerant tanks can safely be converted to air tanks. Technician B says that 30 lb R-12 tanks use a #16 Acme dispenser fitting. Who is correct?
   A. Technician A
   B. Technician B
   C. Both Technician A and Technician B
   D. Neither Technician A nor Technician B
6. Technician A says that an air conditioning compressor uses a pump to pressurize and circulate system lubricant. Technician B says that if the refrigeration lubricant fails to mix with the refrigerant, compressor damage can occur. Who is correct?
   A. Technician A
   B. Technician B
   C. Both Technician A and Technician B
   D. Neither Technician A nor Technician B
7. Which of the following statements about refrigerant oils is correct?
   A. Any lightweight petroleum-based lubricant can be used in an automobile refrigeration system.
   B. Adding additional oil above what is required is a good way to protect the air conditioning compressor.
   C. Mineral oils are miscible with R-134a refrigerant.
   D. If oils are not miscible, they will accumulate in various system components.
8. Technician A says that PAGs and POEs are synthetic blends of oils. Technician B says that PAGs and POEs have the same characteristics and can be freely mixed and interchanged. Who is correct?
   A. Technician A
   B. Technician B
   C. Both Technician A and Technician B
   D. Neither Technician A nor Technician B
9. Technician A says that the refrigeration oil within an air conditioning system is distributed among various components within the system. Technician B says that the refrigeration oil is stored in a sump and circulated throughout the system. Who is correct?
   A. Technician A
   B. Technician B
   C. Both Technician A and Technician B
   D. Neither Technician A nor Technician B
10. Technician A says that when replacing a large amount of refrigeration oil, the oil should be poured directly into the component that has been drained or is being replaced. Technician B says that small amounts of refrigeration oil, usually 2 oz (.060 L) or less, can be distributed using the recovery/recycling station. Who is correct?
    A. Technician A
    B. Technician B
    C. Both Technician A and Technician B
    D. Neither Technician A nor Technician B

Chapter 8 Refrigerants and the Environment

Introduction

For decades, R-12 was the refrigerant of choice for almost all mobile air conditioning applications. R-12 appeared to be a very safe but effective refrigerant; it had relatively low operating pressures, it was nonreactive with system components, it had excellent stability at both high and low pressures and temperatures, it was readily available and inexpensive, and it was believed to cause little if any damage to humans or the environment. However, studies that began in the early 1970s indicated that CFC compounds, including R-12 and other refrigerants, were in fact harmful to the earth's ozone layer. These studies eventually led to the complete halt in production of R-12 refrigerant and resulted in the implementation of tight regulations surrounding automobile air conditioning service.

You are entering the automotive industry at a time of much-heightened awareness to the environmental impact that results directly from automotive service. The last decade has increased the expectations of the industry and the technicians that serve the industry. As an automobile technician, it is important for you to understand how the usage and handling of the chemicals and materials that you use on a daily basis affect you and the environment. Furthermore, it is critical that you understand the regulations that govern the industry and the procedures that you use in the performance of your job duties. Ignorance or avoidance of these policies could result in serious fines or, in the case of serious infractions, jail time. The air conditioning sector currently is the most regulated and monitored of the automotive service industry.

THE ATMOSPHERE

The earth's atmosphere is a blanket of gases that surrounds the globe and extends outward for hundreds of miles. The lowest part of the atmosphere, which extends about 7 miles (11 km) from the earth's surface, is called the **troposphere**. Above the troposphere, out to a distance of about 30 miles (48 km), is the stratosphere (see **Figure 1**). The atmosphere consists of 12 different gases. The most plentiful of those gases is nitrogen, comprising 78 percent of the

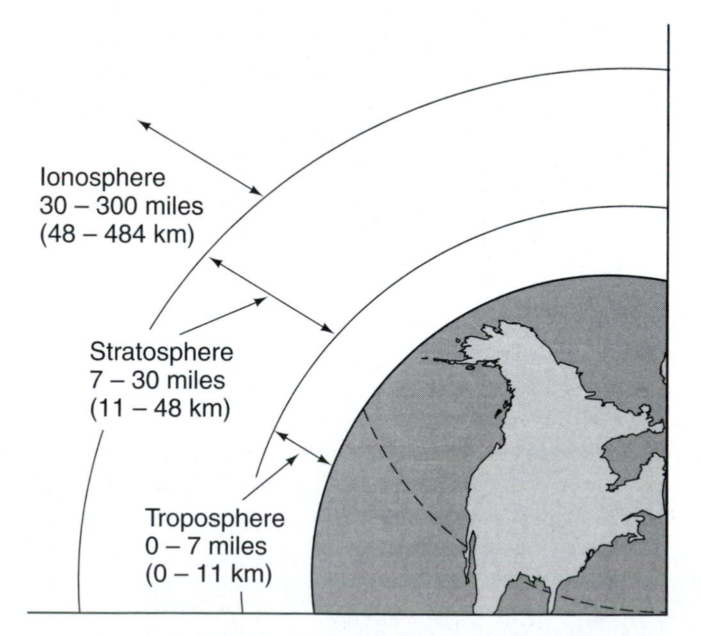

Figure 1. The earth's atmosphere has three different layers.

GAS	PPM by VOLUME	PERCENTAGE
Nitrogen (N)	780,840	78
Oxygen (O)	209,460	21
Argon (Ar)	9,340	0.0934
Carbon dioxide (CO_2)	350	0.0035
Neon (Ne)	18.18	0.0002
Helium (He)	5.24	0.00005
Methane (CH_4)	2.00	0.00002
Krypton (Kr)	1.14	0.00001
Hydrogen (H)	0.50	0.000005
Nitrous oxide (N_2O)	0.50	0.000005
Ozone (O_3)	0.40	0.000004
Xenon (Xe)	0.09	0.0000009

Figure 2. Composition of the atmosphere.

earth's atmosphere, and oxygen, comprising up to 21 percent of the atmosphere. The remaining 1 percent is comprised of the other 10 gases, called trace gases (see **Figure 2**).

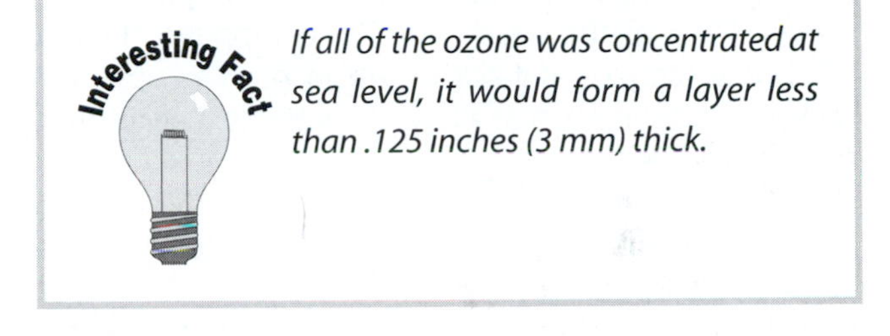

If all of the ozone was concentrated at sea level, it would form a layer less than .125 inches (3 mm) thick.

The Ozone Layer

Ozone is a trace gas that makes up approximately .000004 percent of the volume of atmospheric gases. Although most trace gases reside in the troposphere layer of the atmosphere, 90 percent of ozone is concentrated in the stratosphere at altitudes of 9 to 22 miles (15 to 35 km). The purpose of ozone is to block potentially damaging ultraviolet beta rays (UVB). Ozone is a molecular form of oxygen. A normal oxygen molecule contains two oxygen atoms, whereas an ozone molecule contains three oxygen atoms. Ozone can be formed by the action of electrical discharges and its pungent irritating odor can sometimes be detected in an area where an electrical discharge has occurred, such as near an electric motor or after a lightning strike. More often, however, ozone is formed by the action of the sun's ultraviolet (UV) rays on oxygen. When concentrated in large enough volume, ozone can appear as a light blue gas. Because sunlight is necessary for ozone formation, most of the ozone in the atmosphere is formed over the equatorial region of the earth, where solar radiation is highest. From there, wind currents distribute the ozone throughout the stratosphere. Ozone concentrations are highest near the equator and lowest toward the North and South Poles. Since the discovery of the hole in the ozone layer over Antarctica, there has been a great deal of concern about the effects of ozone depletion and the greenhouse effect. The laws and regulations now imposed on the service of automotive air conditioning are a direct result of the earth's ozone depletion and global warming.

Although it only makes up a very thin layer of gases within the atmosphere, the ozone layer acts as a shield and blocks most of the sun's harmful UV rays from reaching the earth's surface while allowing enough UV rays to reach the surface of the earth to sustain human, animal, and plant life (see **Figure 3**). A reduction in ozone means an increase in

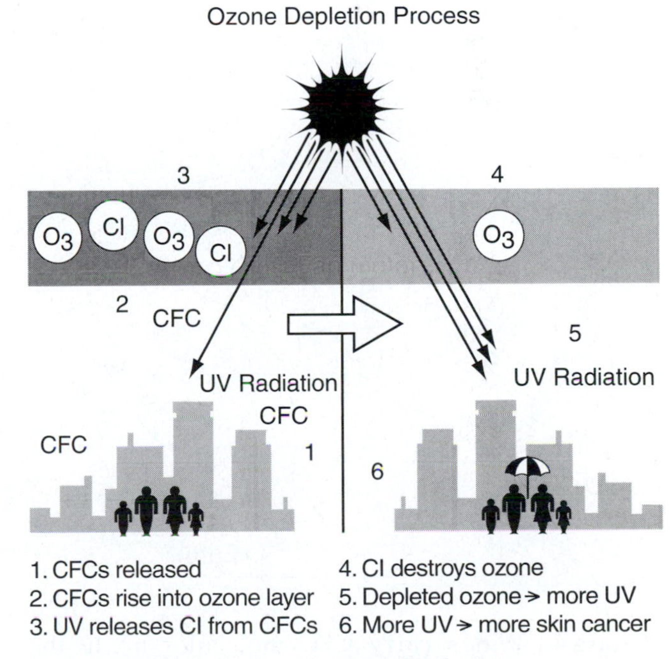

Figure 3. The effects of ozone depletion.

ultraviolet rays reaching the earth's surface, which can in turn lead to increased risks of skin cancer, cataracts, and immune system damage.

Through extensive research, it has been discovered that chlorine is chiefly responsible for depletion of ozone molecules. Many natural sources of chlorine emission are present on the surface of the earth, such as that found in sea water, as well as the hydrochloric acid that is thrust high into the air when a volcano erupts. However studies concluded first, that chlorine from the sea dissolves when it rains, bringing the chlorine back down to the earths' surface, and second, that any hydrochloric acid from volcanoes could never reach into the stratosphere. After eliminating these natural sources of chlorine emission, the question that had to be answered was: How is that much chlorine reaching that far into the earth's atmosphere? The answer to this important question focused primarily on CFC refrigerants. CFCs were at that time used extensively in both stationary and mobile air conditioning units as well as in many other industrial processes. One of the desirable characteristics of CFCs was the stability of CFC compounds. In fact, CFCs are so stable that they can remain intact for up to 120 years. It is this stability that allows CFCs ample time to reach the stratosphere.

Although CFC molecules are heavier than air, they are swept up by winds and eventually carried up into the atmosphere. This process can take as long as ten years (see **Figure 4**). Once the CFC molecules reach the ozone, UV rays from the sun react with the CFC compounds, ultimately releasing the chlorine from the compound. The free chlorine molecules then combine with ozone molecules to form chlorine monoxide, which effectively depletes the ozone molecule (see **Figure 5**). The process, however, does not stop there; the chlorine is then freed again and continues to bond with other ozone molecules. This process can occur thousands of times over. It is believed that one chlorine molecule can destroy up to 100,000 ozone molecules.

Since determining that CFCs were a major cause of ozone depletion, it became critical to determine where the bulk of the CFC emissions originated. The source was directly tied to the service of refrigeration units, both stationary and mobile. Before policies were enacted forbidding the release

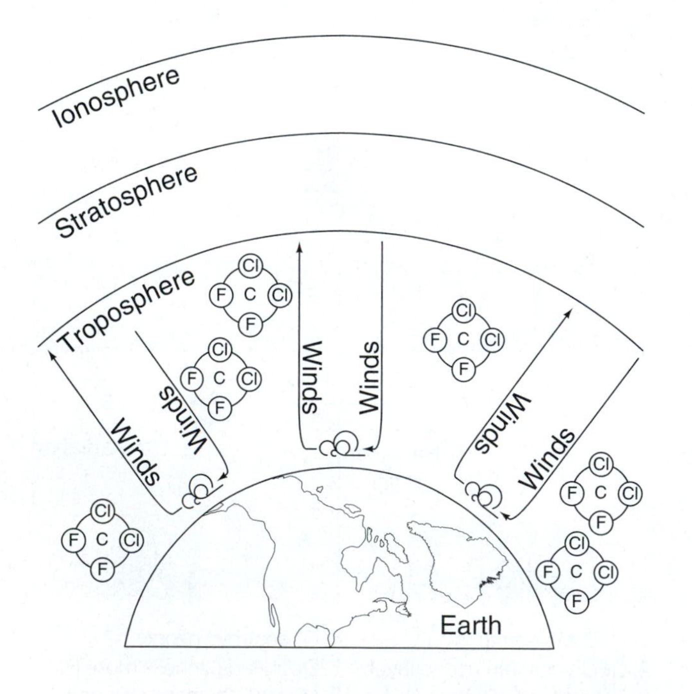

Figure 4. Winds carry CFC molecules up to the stratosphere.

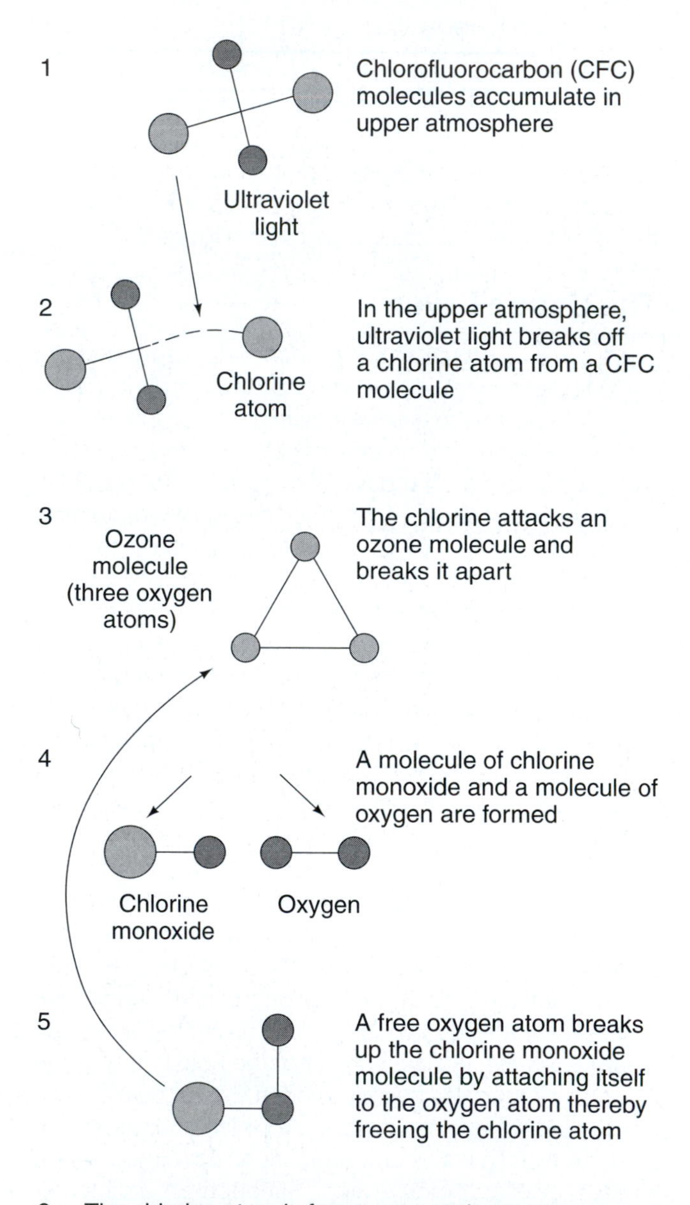

Figure 5. The process in which CFCs destroy ozone molecules.

of refrigerant into the atmosphere, it was common practice to freely release the refrigerant into the atmosphere when a refrigeration system had to be opened up for service. This practice dumped literally millions of tons of CFC gases into the atmosphere. Using this as an example, you can see how the delicate balance of ozone could be easily upset if major steps were not enacted to prohibit such practices. Recognizing the seriousness of ozone depletion, twenty-five nations ratified the **Montreal Protocol**, formally known as the **Protocol on Substances That Deplete the Ozone Layer**, on September 16, 1987. Currently, 168 nations have agreed to the protocol. The protocol officially laid out guidelines for the worldwide phase-out of CFC materials as well as of other ozone-depleting materials.

THE GREENHOUSE EFFECT

A greenhouse is a building in which the walls and ceiling are constructed of glass or translucent plastic material. The glass allows the sun's radiant energy to pass through without allowing the air to escape. The end result is that the air is both heated and retained within the building. A similar phenomenon occurs in the earth's atmosphere, known as the greenhouse effect. This is a desirable effect necessary to the maintenance of life as we know it here on earth. The greenhouse effect takes place through a blanket of gases in the atmosphere, and allows radiant energy to pass through and heat the earth's surface. A certain amount of heat energy is also retained in the lower atmosphere. Without this effect, the earth would become some 60 degrees F (33 degrees C) cooler.

Although the greenhouse effect is a completely separate phenomenon from ozone depletion, the release of CFCs are a common link between the two. Gases such as CFCs and HFCs rise into the atmosphere and combine with existing greenhouse gases, forming additional layers of "insulation," which results in more heat being trapped in the lower atmosphere than is desired (see **Figure 6**). This is coupled with the fact that ozone depletion allows additional UV rays to reach earth, generating additional heat that results in accelerated **global warming**. Global warming is the gradual warming of the earth's atmosphere. This can result in the gradual melting of ice caps and the alteration of weather patterns. Although both the greenhouse effect and some ozone depletion are normal occurrences, the release of pollutants into the atmosphere increases the rate at which these phenomena occur.

> **You Should Know** *Greenhouse gases are those chemicals that contribute to global warming. Examples include carbon dioxide, methane, nitrous oxide, HFCs, and CFCs. The largest contributor to the greenhouse effect is carbon dioxide, which is released into the atmosphere through the burning of fossil fuels. This led to regulations to control industrial and automobile emissions.*

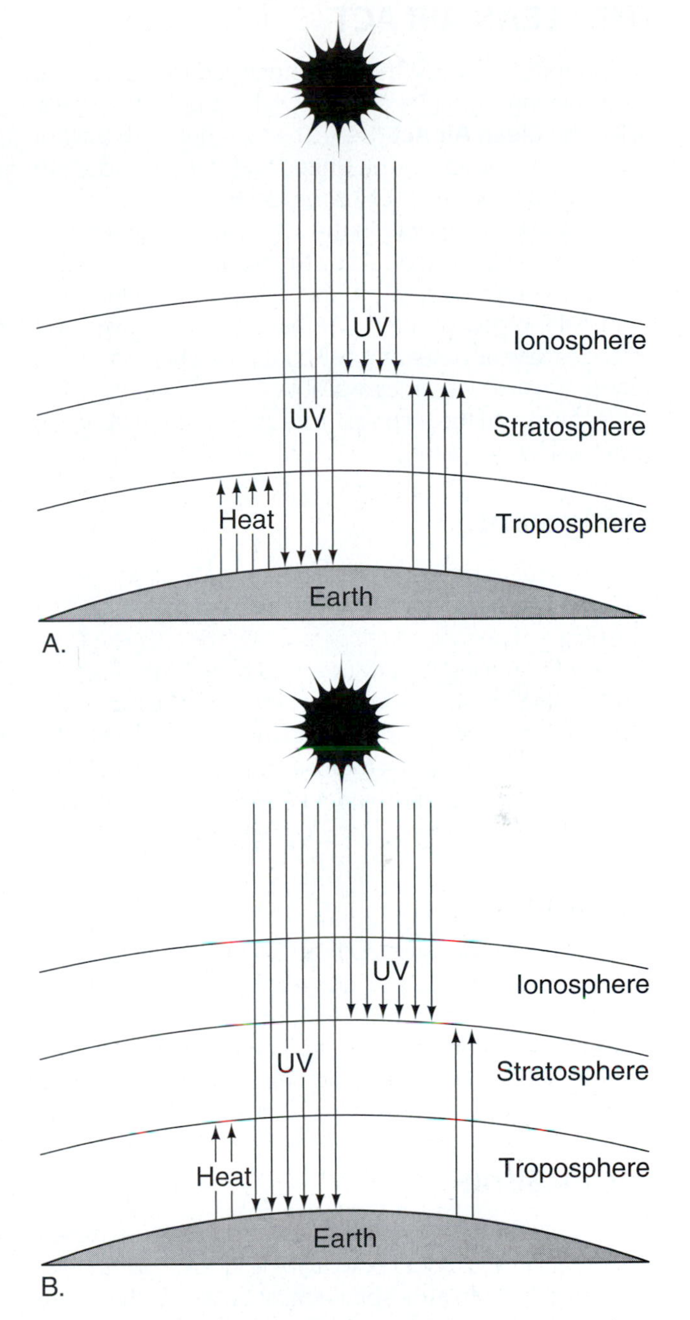

Figure 6. (A) A normal atmosphere allows radiant energy in and allows some heat to escape. (B) The addition of greenhouse gases coupled with ozone depletion allow in excessive amounts of UV rays and trap in additional heat energy.

THE CLEAN AIR ACT

Adoption of the Montreal Protocol led to the passage of a sweeping legislative package within the United States, called the **Clean Air Act** (CAA). The CAA put forth a set of polices and procedures the United States would adopt to deal with the problem of ozone depletion. Most of the rules and regulations within the act were a direct result of the recommendations made by the Montreal Protocol. The CAA was signed into law on November 15, 1990. Both the Montreal Protocol and the CAA have been reviewed and amended several times and will continually be updated as new information becomes available.

As far as you the technician are concerned the CAA set forth these major points:

Refrigerant

It is illegal to intentionally vent any type of refrigerant from a mobile air conditioning unit into the atmosphere. Any time service is performed, refrigerant must be recovered and recycled. This means that any time R-12 or R-134a is removed from a system, it must be recycled before it is reused, even if it is reinstalled in the vehicle from which it was drawn. The rule governing blended refrigerant is slightly different because at this time blends cannot be recycled on-site and should only be recovered using dedicated equipment.

Equipment

Facilities must have dedicated equipment for each type of refrigerant serviced. This means dedicated recovery/recycling equipment for R-12 and R-134a and dedicated recovery equipment for each type of approved blend. All recovery and/or recycling equipment must be certified to meet SAE standard specifications.

Certification

Any technician who handles refrigerant or makes vehicle repairs that could potentially lead to refrigerant being vented into the atmosphere must be certified through an EPA-certified agency. This includes the operation of recovery and recycling equipment and once-common procedures such as adding or topping off a refrigeration system. There are many agencies that offer EPA certification programs. Among the most visible agencies are the Mobile Air Conditioning Society Worldwide (MACS) and the National Institute for Automotive Service Excellence (ASE). It should be noted that normal ASE certification in ASE test area A-7 Heating and Air Conditioning does not meet EPA certification requirements for the refrigerant recovery program. EPA certification requires the successful completion of an EPA-approved course, which also can be obtained through ASE.

Service

Some common misconceptions often surround the service procedures performed on leaking systems. It is legal under federal guidelines to add refrigerant to a system with a preexisting leak. Furthermore, federal laws do not require that refrigerant be removed from a leaking system nor do they require that the system be repaired.

Important Dates

The CAA set forth many important dates that affected the automobile industry:

- January 1, 1992: All CFC refrigerants must be contained and recycled.
- November 15, 1992: All persons purchasing R-12 containers less than 20 lbs must have EPA section 609 certification.
- November 14, 1994: Any technician purchasing ozone-depleting chemicals of any size must be certified by an EPA-approved agency.
- July 1995: Air conditioning systems that have been converted to an alternate R-12 refrigerant must have unique service fittings.
- November 15, 1995: Recovery and recycling of any substitute R-12 refrigerant is required.
- December 31, 1995: The production of R-12 refrigerant was permanently halted in the United States.
- January 1, 1996: After this date, it was illegal to import new CFC refrigerant in the United States. Recycled refrigerant can, however, be imported.

You Should Know *Because the EPA had the foresight to enlist the assistance of the Society of Automotive Engineers (SAE) when the agency was developing standards for equipment specifications and service procedures, many of the standards that you commonly see associated with equipment and retrofit procedures have an SAE designation. The SAE designation is usually four numbers preceded by the letter "J," for instance J1989.*

You Should Know *Some state and local laws are more stringent than those set forth by the CAA. It is important for you to be familiar with the laws that are enforced in your local area.*

Summary

- The earth's atmosphere consists of a blanket of gases that surround the globe and extend outward for hundreds of miles. The most plentiful gases in the atmosphere are nitrogen and oxygen. Approximately 1 percent of the atmosphere is made up from several gases, called trace gases.
- Ozone is a trace gas that makes up about .000004 percent of the earth's atmosphere. Ninety percent of the ozone is located in the stratosphere. Ozone blocks the earth from harmful UV rays produced by the sun.
- Ozone is a molecular form of oxygen and is formed by UV radiation acting on oxygen molecules. Most ozone is formed over the equator and distributed throughout the stratosphere by winds.
- Some UV radiation is required to sustain life on earth. Too much radiation can increase human health risks and be damaging to plant life.
- Chlorine is the leading cause of ozone depletion. Many natural sources of chlorine exist but do not have the potential to reach the stratosphere. CFCs are swept up by winds and carried to the stratosphere where they react with UV radiation and eventually attack ozone molecules.
- Before laws were enacted, millions of tons of CFCs were released directly into the atmosphere. The Montreal Protocol officially laid out guidelines for the phase-out of CFC materials.
- The greenhouse effect is a phenomenon in which atmospheric gases allow sufficient radiation to reach the earth to sustain life, while also retaining the heat in the lower atmosphere. Pollutants accelerate and intensify the greenhouse effect. Acceleration of the greenhouse effect is the direct cause of global warming.
- The Clean Air Act details the policies and procedures the United States would use to deal with problem of ozone depletion. The Society of Automotive Engineers assisted the EPA with the development of many standards and procedures related to the CAA.
- It is illegal to vent any refrigerant. R-12 and R-134a must be recovered and recycled. Alternate blend refrigerants must be recovered.
- Facilities must have dedicated equipment for each type of refrigerant serviced. All recovery and/or recycling equipment must be certified.
- Any technician making vehicle repairs that could result in the release of refrigerant must have EPA certification. ASE certification in area A-7 does not qualify as EPA certification.
- It is legal to add refrigerant to a leaking system. Refrigerant does not have to be removed from a leaking system. Some state and local laws may be more stringent than federal laws.

Review Questions

1. The atmosphere is comprised of which three elements?
2. How much of the atmosphere does ozone represent?
3. Technician A says that the ozone layer blocks harmful UVB rays produced the sun. Technician B says that the ozone layer blocks all of the UV rays produced by the sun. Who is correct?
   A. Technician A
   B. Technician B
   C. Both Technician A and Technician B
   D. Neither Technician A nor Technician B
4. Technician A says that ozone can be formed through the action of electrical discharges such as those that occur within an electric motor. Technician B says that ozone is formed through a reaction between ultraviolet radiation and oxygen molecules. Who is correct?
   A. Technician A
   B. Technician B
   C. Both Technician A and Technician B
   D. Neither Technician A nor Technician B
5. Describe how CFCs deplete ozone molecules.
6. Which of the following chemicals is not a source of chlorine?
   A. Hydrochloric acid
   B. R-134a
   C. Sea water
   D. R-12
7. Explain how the release of CFCs and HFCs into the atmosphere contributes to global warming.
8. Technician A says that the alternate blend refrigerants must be recycled. Technician B says that the Clean Air Act states that all automotive technicians must be EPA certified. Who is correct?
   A. Technician A
   B. Technician B
   C. Both Technician A and Technician B
   D. Neither Technician A nor Technician B

9. Technician A says that by law refrigerant must be removed from a leaking system. Technician B says that it is legal to add refrigerant to an air conditioning system with a preexisting leak. Who is correct?
   A. Technician A
   B. Technician B
   C. Both Technician A and Technician B
   D. Neither Technician A nor Technician B

10. Explain how implementation of the Clean Air Act affects the automotive service industry.

Section 3

Automotive Heating Systems

SECTION OBJECTIVES

At the conclusion of this section you should be able to:

- Understand how a typical cooling system operates.
- Understand different antifreeze compositions.
- Explain the flow of coolant through the cooling system.
- Identify various cooling system components.
- Explain the construction and operation of cooling system components.
- Understand the construction and operation of the heater system.
- Select the proper replacement coolant for an application.
- Determine coolant condition using various test methods.
- Drain and refill an engine cooling system.
- Diagnose accessory belt wear and replace as needed.
- Diagnose hoses and replace as needed.
- Diagnose cooling fan concerns.
- Locate cooling system leaks and repair as needed.
- Diagnose engine temperature concerns.
- Replace engine cooling system components.
- Diagnose poor heating concerns.
- Replace various heater system components.

The engine cooling system is among the most neglected of automotive systems.

Chapter 9 Engine Cooling Systems

Introduction

The engine cooling system is responsible for removing heat from the engine. This can be accomplished in two ways: either by circulating air around the engine or by using a liquid cooling medium. Although air cooling is a popular method used in lawn equipment, motorcycles, and all terrain vehicles (ATVs), it is used in only a few automobiles, most notably the rear engine Volkswagen Beetle. Although the liquid cooling system is still rather simple in nature compared to the other essential systems such as the fuel and ignition systems, cooling system maintenance and repair are among the most critical and common services performed in the automotive shop.

THE COOLING SYSTEM

To understand the role of the engine cooling system in the modern automobile, we should first adjust our way of thinking about the system. To gain a more definite understanding, let's think about it as a temperature control system, because we not only desire to keep the engine from overheating but we also need to keep the engine within a specific operating range. The cooling system must allow the engine to reach operating temperature as quickly as possible and maintain the engine temperature within its operating range without significant fluctuations. The desired operating temperature is a compromise that is based largely on the efficiency combustion process. For example, most vehicles will have a normal cooling system temperature of about 200 degrees F (93 degrees C). This is based on the fact that that engine calibrations in this temperature range provide the best compromise of thermal efficiency, drivability, and exhaust emissions. If temperatures vary greatly either above or below the "calibrated" range, engine drivability and emissions will suffer.

Combustion temperatures of a modern automobile engine can exceed 4500 degrees F (2482 degrees C). Sixty-five percent of the heat generated by the engine is removed with the exhaust gases, absorbed by the metal components of the engine, and dissipated through the engine oil. The remaining 35 percent is removed by the engine cooling system. The liquid filled cooling system consists of the following components as shown in **Figure 1**.

- Coolant
- Cooling system passages
- Coolant pump
- Radiator
- Pressure cap
- Thermostat
- Overflow/coolant recovery system
- Cooling fan

COOLANT

The engine coolant is the lifeblood of the cooling system. Coolant is circulated through the engine where through conduction it absorbs the heat that is transferred to the metal engine components during the combustion process. The coolant is then transferred to the radiator where the heat energy is given up to the atmosphere through convection. Coolant is comprised of a mixture of 50 percent water and 50 percent antifreeze. A good engine coolant should provide the engine with freeze protection, boilover protection, corrosion prevention, and good heat transfer.

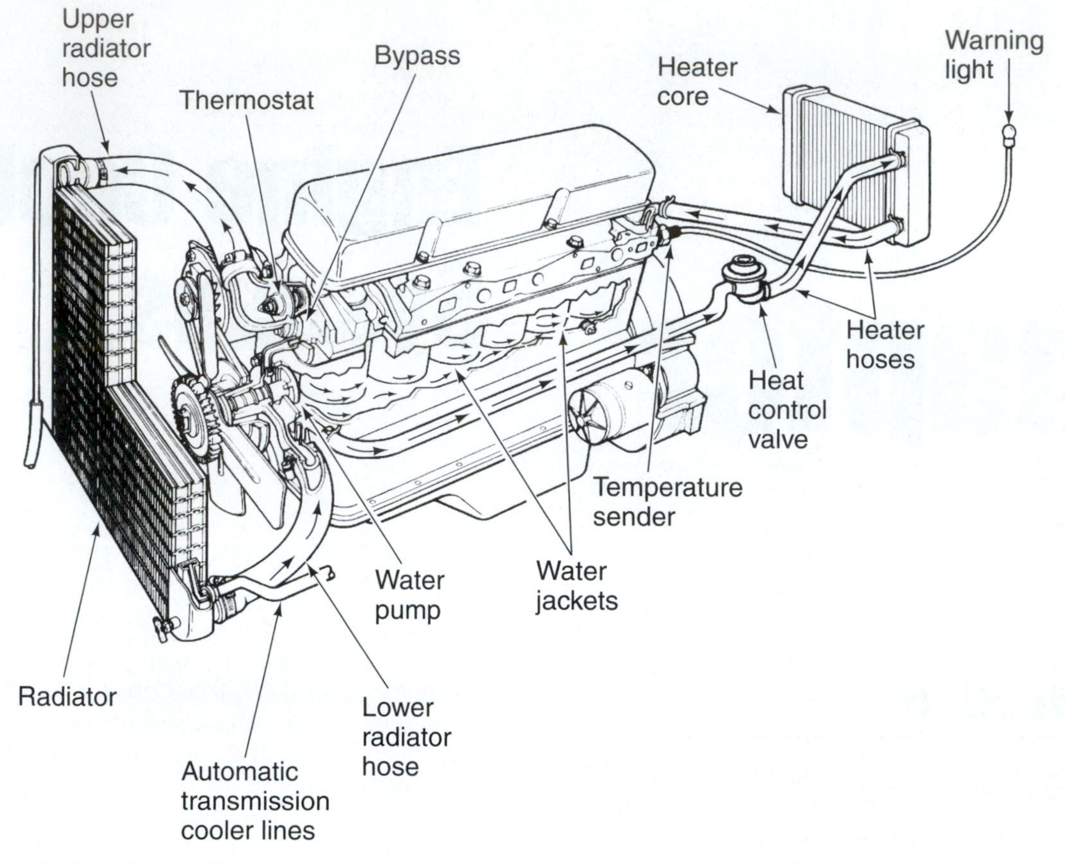

Figure 1. A typical engine cooling system.

Water by itself is a very good coolant that possesses excellent heat transfer qualities. However, water has several undesirable characteristics that make it impossible to use as a stand-alone engine coolant. The most obvious fault is its relatively high (32 degrees F [0 degrees C]) freezing point; this alone makes water an unsuitable coolant to use in cold weather. In addition, pure water provides the cooling system with no corrosion prevention and, when used alone, not only promotes corrosion of metal engine components but also becomes acidic rather quickly, leading to electrolysis.

Electrolysis

Electrolysis can be caused by two things. The first is the movement of stray electrons traveling through the cooling system searching for a ground. This is most often caused by an electrical component that has a poor ground circuit. The second cause is when the engine coolant becomes acidic. The process of electrolysis works like this: when two dissimilar metals are submerged in an acid, electrical currents build up on the surface of the metal components, basically forming a battery. Electrolysis can attack virtually all cooling system components; however, radiators, heater cores, and hoses are most susceptible because of their relatively light construction and their ability to conduct electricity. Electrolysis can cause corrosion and pitting of the cooling system and engine components (see **Figure 2**). Damage caused by electrolysis is not visible from the exterior of the engine and only becomes visible once a leak has developed or when service

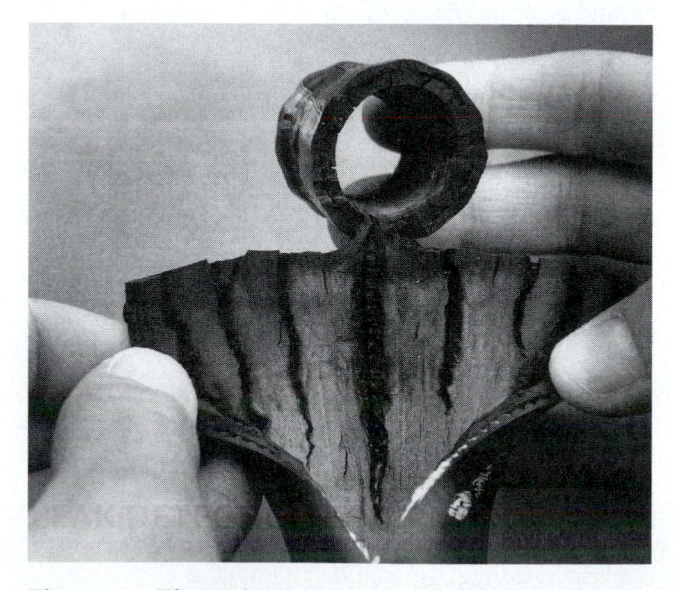

Figure 2. Electrolysis can cause damage to cooling system components.

is performed. When left unchecked, electrolysis can actually eat through components such as the engine block, causing irreparable damage. Electrolysis is a potential problem in all engine cooling systems; however, it is more problematic in engines that use aluminum cylinder heads or intake manifolds and cast-iron blocks. Electrolysis is especially problematic in diesel engines that use steel cylinder liners and cast-iron blocks.

Antifreeze

To compensate for all of the shortcomings of water, antifreeze is mixed with water to form engine coolant. Antifreeze lowers the freezing point of water, raises the boiling point of water, provides additives that prevent corrosion, and adds lubricants for coolant pump seals. There are many different types of antifreeze on the market today and each vehicle manufacturer has specific recommendations as to which antifreeze should be used in their vehicles.

Antifreeze is the generic term used for the additive that is mixed with water to prevent freezing and boilover. In the early years, a methyl alcohol solution was mixed with water to form engine coolant. Although the alcohol did provide the necessary freeze protection, it offered no corrosion protection and actually lowered the overall boiling point of the coolant. Since the early 1960s, ethylene glycol has been the antifreeze of choice.

Ethylene glycol is the base stock for most automotive antifreeze products on the market today. Although ethylene glycol does not transfer heat as well as water, it provides other necessary elements required of engine coolant. In its pure form, ethylene glycol is a colorless, viscous liquid that has a sweet taste that can be attractive to small children and animals. In addition to ethylene glycol, antifreeze products also have additive packages that prevent corrosion and provide water pump lubrication. Most antifreeze products consist of 90 to 93 percent ethylene glycol. The remaining 7 to 10 percent consists of other additives.

It is recommended that all antifreeze products be mixed as a 50-50 antifreeze to water solution. The freezing point of the resulting coolant mixture is –34 degrees F (–36.7 degrees C). Although, antifreeze concentrations of 40 percent to 70 percent are considered to be acceptable. Maximum freeze protection for ethylene glycol occurs at 67 percent solution. This concentration will freeze at –84 degrees F (–64.4 degrees C). However, as mixtures exceed this concentration, the ability of the coolant to transfer heat greatly suffers. Below 40 percent concentration, the freezing point increases and additives become diluted, effectively shortening the service life of the antifreeze (see **Figure 3**).

Figure 3. A coolant percentage/protection chart.

Almost all automobile manufacturers use some form of ethylene glycol antifreeze as a factory fill. However, there are three different types of ethylene glycol antifreeze: **IAT, OAT,** and **HOAT**. These types are separated by the use of their additive packages. IAT antifreeze has a short service life, whereas OAT and HOAT are considered long-life antifreeze products. Because all of these materials have the same base chemical, they can be mixed; however, this practice is discouraged because any long-life benefits provided by OAT or HOAT will be lost. Antifreeze colors are also different but do not necessarily signify the type of antifreeze that is being used. Before dyes are added, all antifreeze is basically clear in nature. Antifreeze is colored to provide the technician with an easy way to identify if the proper antifreeze is in the vehicle and to easily identify the fill level within the overflow tank. Colors are also selected to contrast with the color of other fluids to help prevent accidental cross-contamination.

IAT

IAT stands for Inorganic Additive Technology. This is the traditional antifreeze that was used exclusively as **factory fill** from the early 1960s until 1995. IAT antifreeze is an ethylene-glycol–based antifreeze that uses inorganic additives to protect against corrosion. IAT additives are fast acting and have excellent protection properties but have short service life. Therefore, coolant should be changed every two years or 30,000 miles (48,280 km). IAT antifreeze is manufactured in several different colors, usually green, yellow, or blue. IAT antifreeze was used by some major manufacturers until 2001. At this time, most major manufacturers no longer use IAT antifreeze as factory fill.

OAT

OAT stands for Organic Additive Technology. OAT antifreezes use additive packages that are typically comprised of organic acids. These acids provide excellent protection and have excellent service life but are slow acting when exposed to cooling system conditions. These additives also fail to adequately protect the solder joints found in some radiators and heater cores. The first and most recognizable of the long-life antifreeze products is Dex-Cool. Beginning in 1995, the General Motors Corporation began using Dex-Cool as factory fill in some vehicles. Dex-Cool is orange in color and has a service life of five years or 150,000 miles (241,401 km). It is worth noting that although Dex-Cool is an OAT antifreeze, its composition is different from other OAT antifreeze products. Toyota and Honda both use an OAT antifreeze in their vehicles, but it is not the same as Dex-Cool.

Interesting Fact

Dex-Cool is a registered trademark of the General Motors Corporation. Although Dex-Cool was first produced by Texaco/Havoline, there are several brands of antifreeze that carry the Dex-Cool label. This signifies that the antifreeze has met General Motors specifications.

You Should Know

Coolant selection in the last few years has become a complicated process with many different types, colors, and recommendations. It is important that you consult the manufacturers' recommendations for coolants. Even within families of coolants such HOAT or OAT, different compositions exist, and some manufacturers are very explicit about what kind of coolant should or should not be used in their vehicles.

HOAT

HOAT stands for Hybrid Organic Additive Technology. HOAT uses additive packages that combine both organic and inorganic additives. This allows for a broader range of protection and an extended service interval. HOATs can be orange, yellow, or red. HOAT antifreeze has been used as factory fill by Ford, Chrysler, and Mercedes. HOAT antifreeze also may carry the designation of G-05.

Propylene Glycol

Propylene glycol is another type of antifreeze. Although it is not used as a factory fill, it can be used as a replacement. Propylene glycol and ethylene glycols are very similar in nature and have many of the same characteristics. The benefit of propylene glycol is that it is not as harmful to the environment as ethylene glycol.

Interesting Fact

All used coolant mixtures should be considered hazardous material because as coolant circulates through a cooling system it collects particles of the metals that are used in the cooling system and may contain aluminum, iron, copper, brass, and, in some cases, lead.

WATER JACKETS

The engine's cylinder block provides the structure that houses the engine's rotating components, the cylinders that house the pistons, and mounting location for the surface mount components such as the cylinder heads, water pump, and other accessories. Additionally, the cylinder block provides provisions to lubricate and cool the internal engine components. Although the engine oil removes some heat that is created within the engine, it is the engine coolant that removes most of the heat that remains after the exhaust gases have left the cylinders. In order for coolant to absorb heat from the engine, it must somehow be circulated through the engine.

Engines that use a liquid-filled cooling system provide passages within the cylinder block and cylinder head to circulate coolant; these are called **water jackets.** The water jackets surround the engine cylinders and extend from the bottom of the cylinder to the top of the block (see **Figure 4**). Water jackets within the cylinder head surround the combustion chamber; these connect to the water jackets in the cylinder block so that water can pass back and forth between the block and cylinder head.

The water jackets in the block and cylinder head are formed as part of the casting process. Melted cast iron or aluminum is poured into and around molds that are made of sand and other materials. When the metal has cooled, the sand must be removed; when removed, what remains is a passage in which coolant can flow. To remove the sand, holes are cast in the sides of the block. Once the sand has been removed and the engine block cleaned, steel or brass plugs are pressed into the holes to seal the block. These are often called freeze plugs, expansion plugs, or core plugs. These can be a common source of coolant leaks.

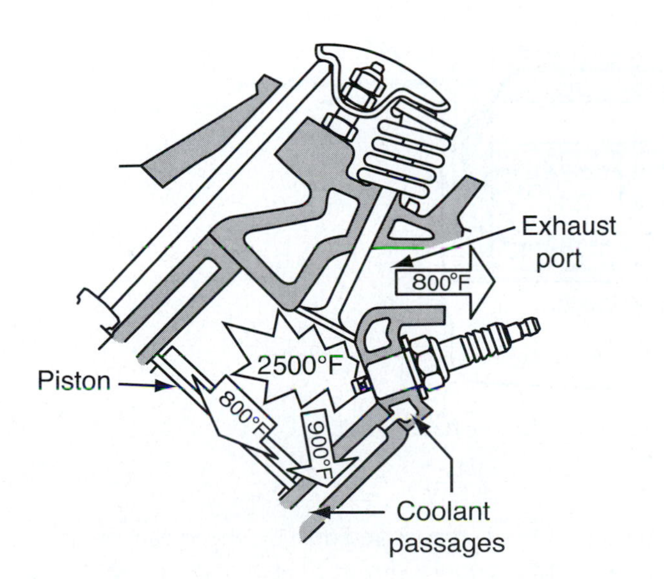

Figure 4. Engine water jackets provide cooling to the engine.

Circulation

Most of the heat that is created in an internal combustion engine is generated in the combustion chamber of the cylinder head and in the top one-third of the engine's cylinders. Because of the large amounts of heat generated in these areas, these are the most critical to cool. The internal combustion engine uses two methods to circulate coolant through the engine block and cylinder heads. Depending on the design, the engine may use conventional cooling or reverse flow cooling.

Conventional Flow

Most engines use a conventional flow cooling system. In this type of cooling system, the coolant from the radiator is circulated around the bottom of the block. The coolant is then pumped to the top of the block and into the cylinder head. From the cylinder head, coolant is transferred back to the radiator (see **Figure 5**). The downside to this method is that the coolant absorbs heat as it passes through the lower end of the cylinder block; this limits the amount of heat that the coolant can effectively absorb from the cylinder head area.

Reverse Flow

A reverse flow cooling system starts by moving coolant through the cylinder head first and then pumps it to the lower portion of the engine block and back to the radiator (see **Figure 6**). This method allows the coolant to absorb considerably more heat from the cylinder head, allowing the cylinder head to run at a considerably cooler temperature. The advantages to lower cylinder head temperatures are increased combustion efficiency, lower emissions, and reduced engine detonation or "ping." Cooler cylinder head temperatures also provide manufacturers with the ability to raise compression ratios in some instances.

COOLANT PUMP

The coolant pump used in automotive engines is of the centrifugal impeller design. The pump can be located at the front of the engine or on the side of the engine. The pump consists of a sealed, two-piece housing with an inlet and an outlet. Inside the housing resides the impeller. The impeller consists of a flat plate with a series of blades that somewhat resemble a fan. The impeller is press-fit to a shaft that extends through the front of the housing and is supported by a sealed bearing unit in the front of the housing. The water pump uses power from the engine to turn the impeller shaft.

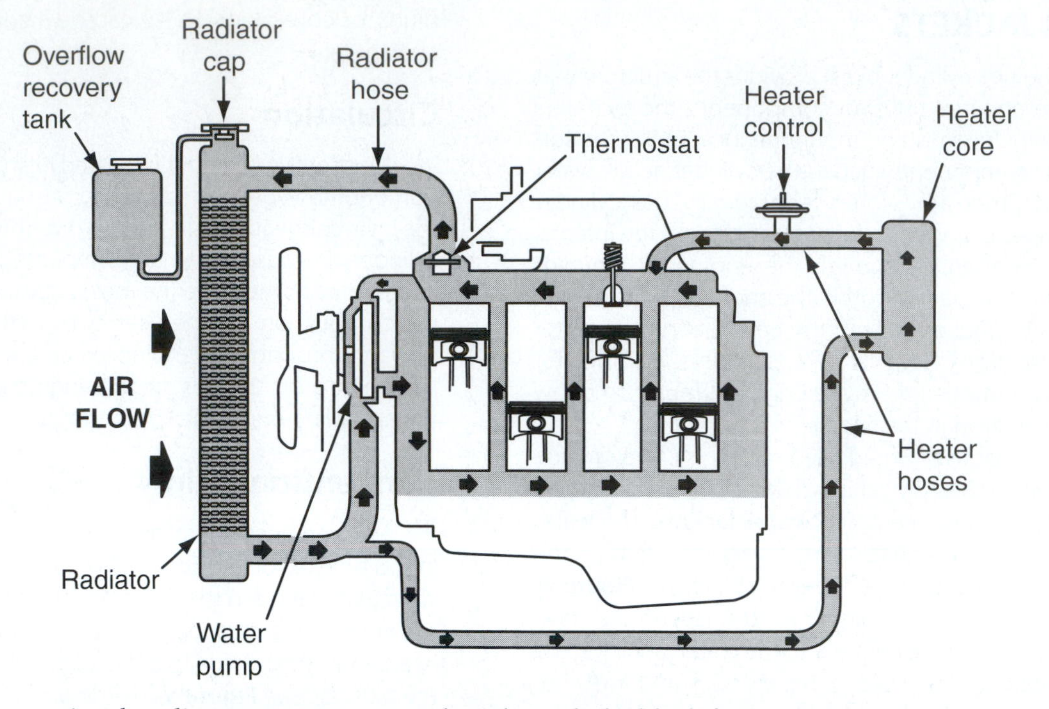

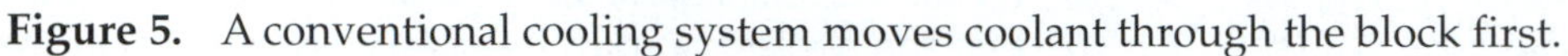
Figure 5. A conventional cooling system moves coolant through the block first.

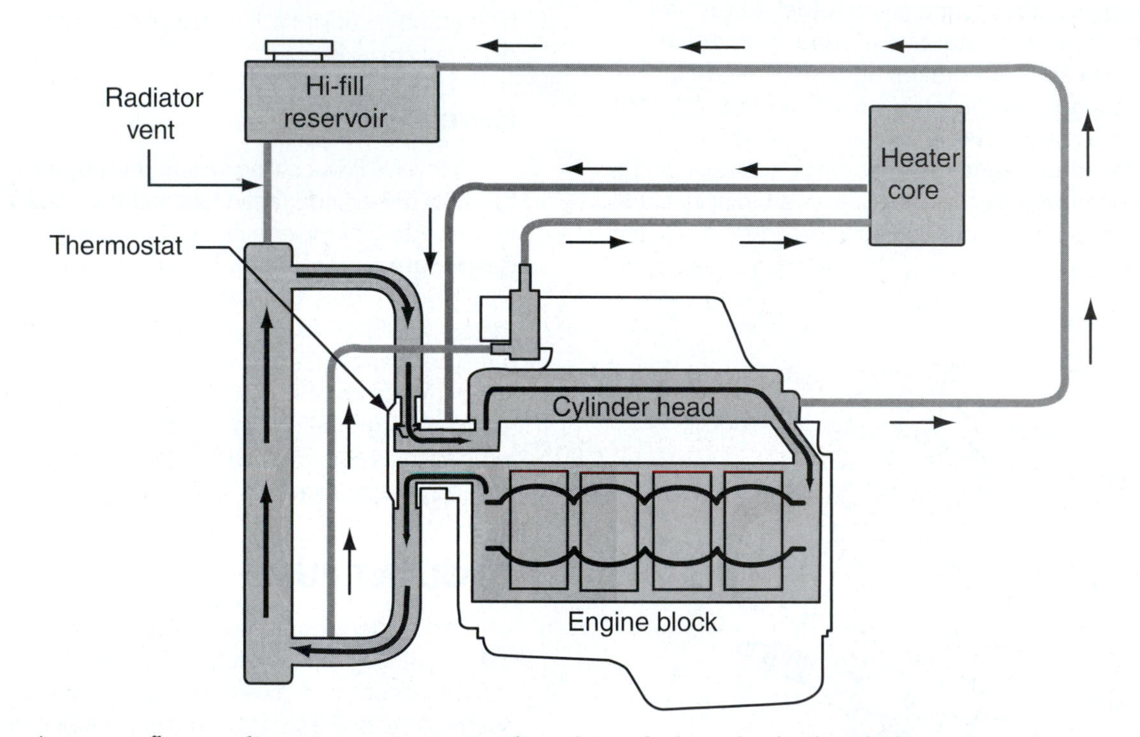

Figure 6. A reverse flow cooling system moves coolant through the cylinder heads first.

A pulley or gear is attached to the impeller shaft opposite the impeller and can be driven by an accessory belt, timing belt, or timing chain and, in some cases, driven by the camshaft (see **Figure 7**). The accessory belt drive method is the oldest method, but as engines became more compact and dual overhead cam engines became more popular, driving the pump using the timing belt or chain became the method of choice for turning the water pump on most four-cylinder engines and many six-cylinder engines (see **Figure 8**). There are a few

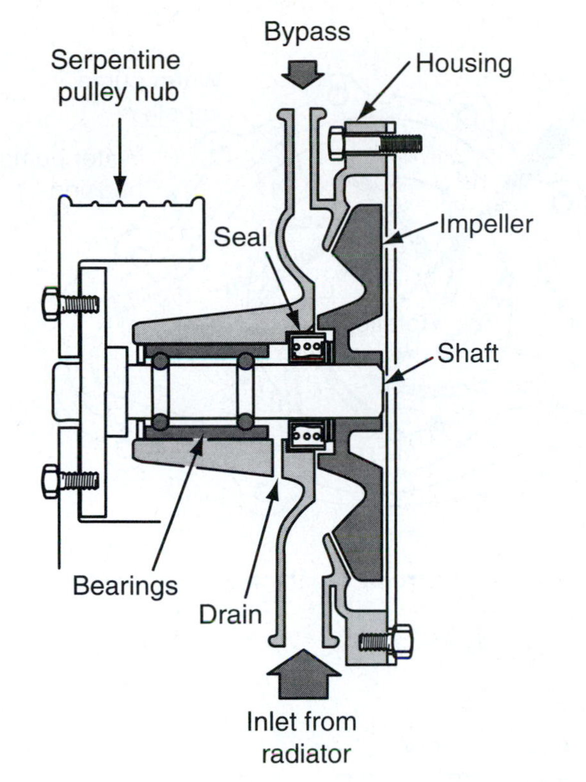

Figure 7. Typical coolant pump construction.

engines in which the coolant pump is directly driven through the use of a short stub shaft connected directly to the engine's camshaft, most notably the General Motors–produced LT1.

The inlet of the cooling pump is connected to the radiator outlet using a formed radiator hose. As the impeller turns, a low-pressure area is created near the center of the impeller, coolant from the radiator is forced through the radiator hose and fills the low-pressure area formed at the impeller; as the impeller continues to spin, the coolant is picked up by the fins and forced outward, where it is forced through the outlet of the coolant pump (see **Figure 9**). Coolant is fed into the engine block through passages in the pump and engine. The passages are joined using a mechanical connection between the pump and the engine block rather than through a hose. Coolant is circulated through the block and cylinder head and goes back to the radiator. As the speed of the impeller changes so does the volume of coolant that is circulated. Because of its design, the restriction of coolant flow caused by the thermostat does not cause damage to the coolant pump. The pump is also equipped with a bypass port. This port is used to allow coolant to circulate through the engine block when the thermostat is closed. Doing this helps to equalize coolant temperature in the engine block during engine warmup.

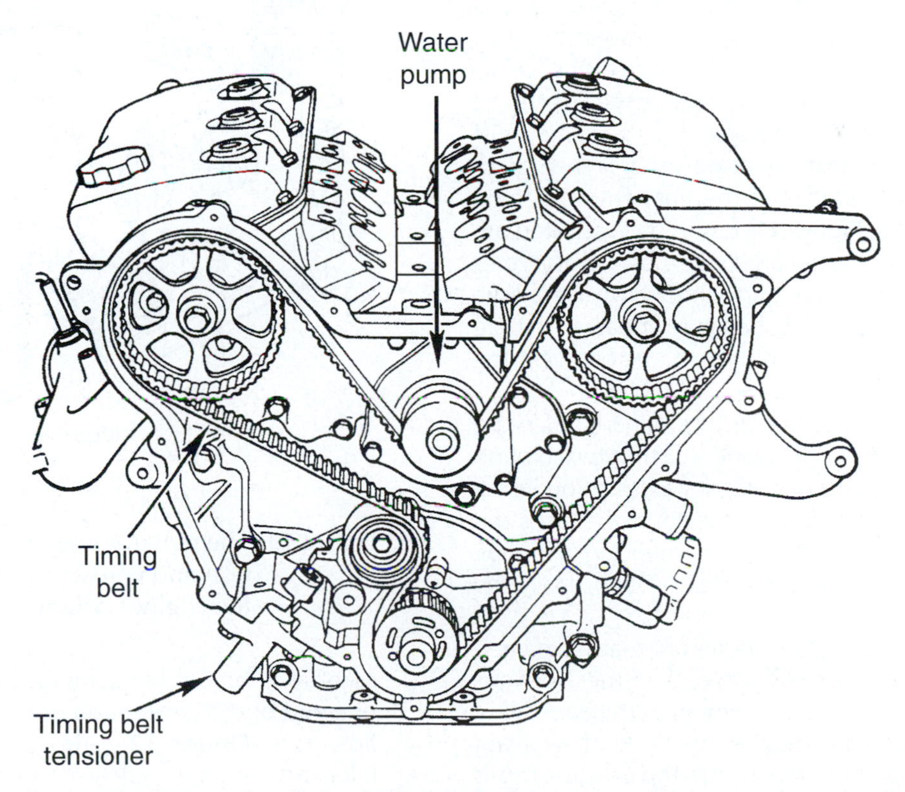

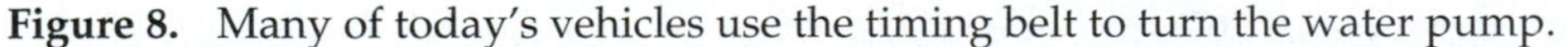

Figure 8. Many of today's vehicles use the timing belt to turn the water pump.

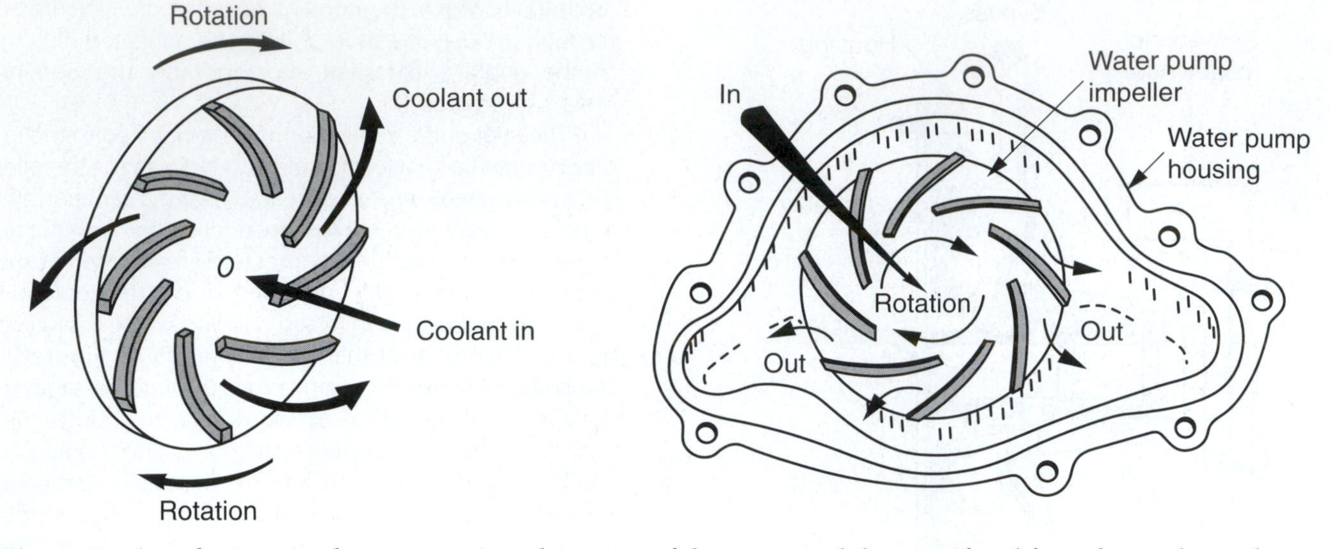

Figure 9. A coolant pump draws water into the center of the vanes and the centrifugal force forces the coolant out toward the outlet.

RADIATOR

The radiator is a heat exchanger in which heat that is absorbed by the coolant is carried away from the engine and is given up to the atmosphere through the processes of conduction and convection (see **Figure 10**). Automotive radiators consist of two tanks and a core. The core is really the heart of the radiator. The core is built from several rows of flat or oblong tubing that connect between the two tanks. Between each row of tubes, very thin fins are attached perpendicular to each of the tubes; the fins are the same width as the tube. The fins provide a large amount of surface area in which heat can disperse. Air is then circulated between the surfaces of the fins removing large amounts of heat (see **Figure 11**). Automotive radiators are made from a variety of materials. Today, most radiator cores are constructed from aluminum cores and use plastic tanks. These style radiators first began to appear in the late 1980s. Before this, copper and brass radiators were the norm.

As coolant is pumped out of the engine, it enters the inlet tank. Some of the hot coolant is distributed through each of the core's tubes; as coolant passes through the tubes, heat is transferred from the coolant to the tubes and from the tubes to the fins. As air is circulated across the fins, heat is removed from the fins, the tubes, and, finally, the coolant.

There are two designs of radiators that have been widely used throughout the years; each of these designs depends primarily on the arrangement of the core tubes. When the core tubes are arranged vertically, the coolant flows from the top of the radiator to the bottom. This is called a vertical flow radiator. When the tubes of the core are arranged horizontally, the coolant flows from one end of the core to the other. This is called a cross-flow (see **Figure 12**). The location of the radiator inlet and outlet is solely determined by the engine requirements.

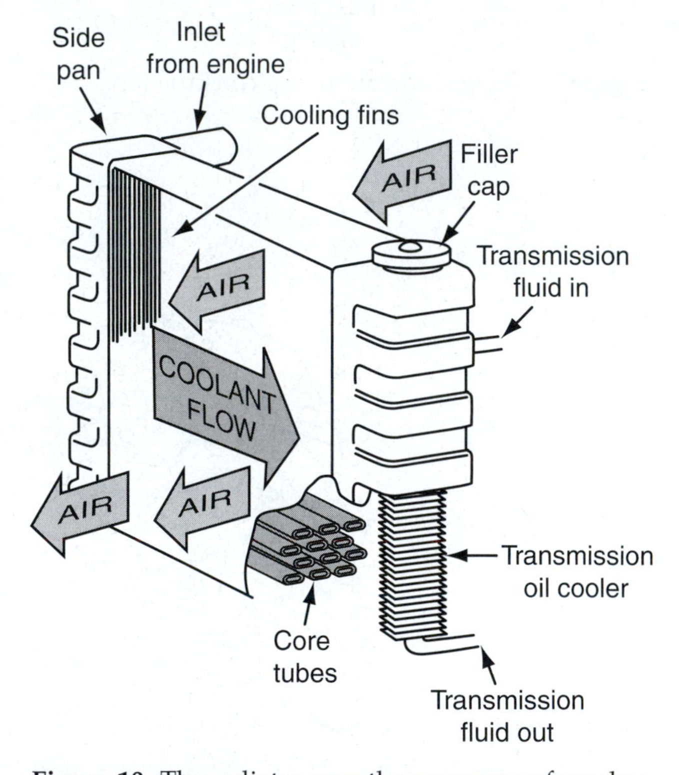

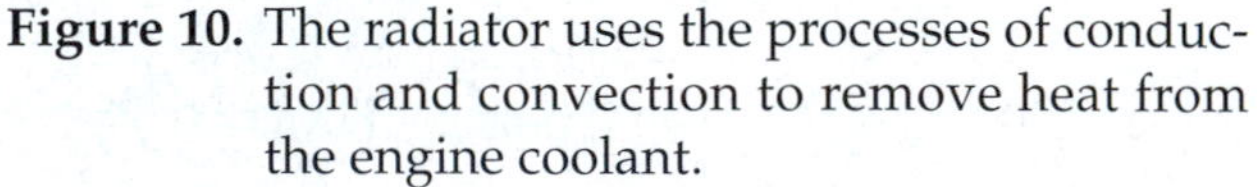
Figure 10. The radiator uses the processes of conduction and convection to remove heat from the engine coolant.

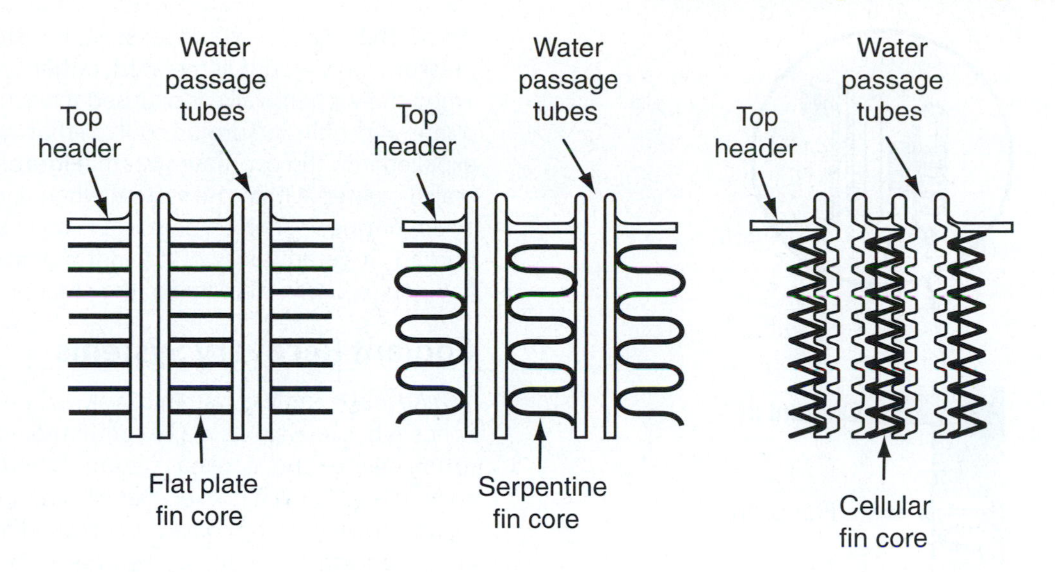

Figure 11. Typical types of radiator construction.

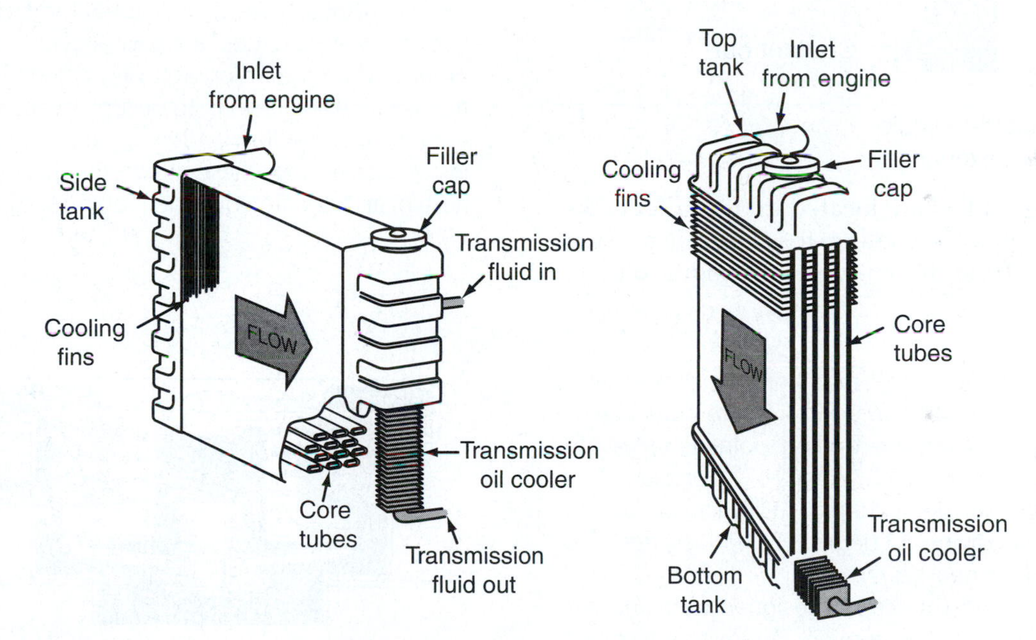

Figure 12. Cross-flow and vertical flow radiators.

Most radiators also make use of oil-to-water coolers. These are small tanks that are located within the tanks of the radiator. Hot oil is circulated through the cooler, causing the fluid to give up some of its heat to the engine coolant. In vehicles equipped with automatic transmissions, an oil cooler is placed in the outlet tank radiator (see **Figure 13**). Some vehicles also make use of an engine oil cooler placed in one of the tanks, usually the inlet tank. In vehicles that use both transmission and engine oil coolers, each radiator tank has an oil cooler inside.

RADIATOR PRESSURE CAP

One aspect of engine cooling is that in a pressurized system the boiling point of the coolant can be increased, which provides an additional measure of boilover protection that is necessary with the relatively high operating temperatures found in today's engines. As the temperature of coolant—or any other liquid for that matter—increases, the liquid expands. Under normal operating temperatures, engine coolant can expand as much as 10 percent; this figure is increased when engine overheating occurs.

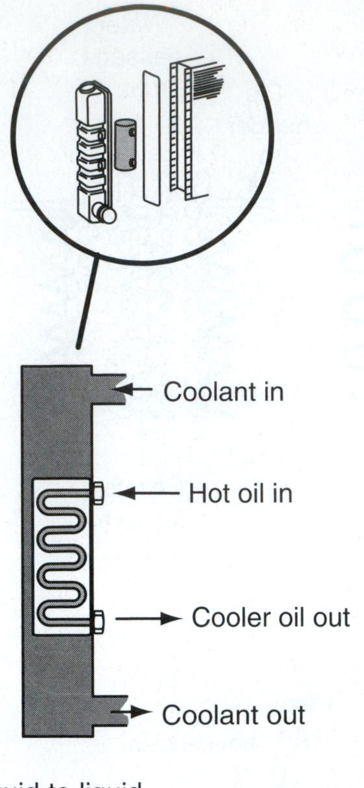

Figure 13. Oil coolers are located in the outlet tanks to provide cooling for the transmission fluid and in some cases the engine oil.

When a liquid expands within a sealed system, pressure is created. For each pound (6.9 kPa) of pressure that is applied to the coolant, the boiling point is increased approximately 3 degrees F (1.7 degrees C). For example, if pure water is pressurized to 15 psi (103.4 kPa), the boiling point will increase from 212 degrees F (100 degrees C) to 257 degrees F (125 degrees C).

The radiator pressure cap is designed to maintain a consistent pressure on the cooling system, to maintain a consistent level of coolant in the radiator, and to help to purge pockets of air from the cooling system. To accomplish this, the pressure cap is equipped with two calibrated valves: a pressure valve and a vacuum valve. When pressure increases above a predetermined level, usually 15 psi, the pressure valve is lifted off its seat and allows some coolant to escape the radiator through a small tube connected to the side of the filler neck, thus relieving some of the cooling system pressure. Once excess pressure is relieved and system pressure drops below the opening point of the valve, it closes, and the pressure is again allowed to build (see **Figure 14**).

The vacuum valve works in the opposite way. As the coolant temperature drops, the hot coolant will contract. When this occurs within a sealed system, negative pressure or vacuum is created rather than pressure. When the vacuum valve is exposed to vacuum, the valve opens and pulls in surplus coolant that was previously expelled into the overflow system (**Figure 14B**). Vacuum is also created within the system when large pockets of air are trapped. When these pockets of air are eventually purged, large amounts of coolant are displaced and a low-pressure area is created in the radiator.

Coolant Recovery Systems

Air in the cooling system has always been a source of concern because air is a major contributor to rust and corrosion within the cooling system. When the radiator coolant level is low, a pocket of air fills any space not occupied by coolant. As the coolant is circulated, the coolant becomes **aerated** and air is taken deep into the cooling system. When air is present in the cooling system, it contributes to the formation of corrosion. The coolant overflow system and expansion systems help to keep the radiator full of coolant and air pockets out. Before overflow systems became standard, coolant was allowed to spill on the ground when expansion occurred. When the coolant contracted, air was drawn in to occupy the space vacated by the coolant. Overflow and expansion systems have been installed to remedy these shortcomings. Overflow systems have been standard equipment on most cars since the late 1960s.

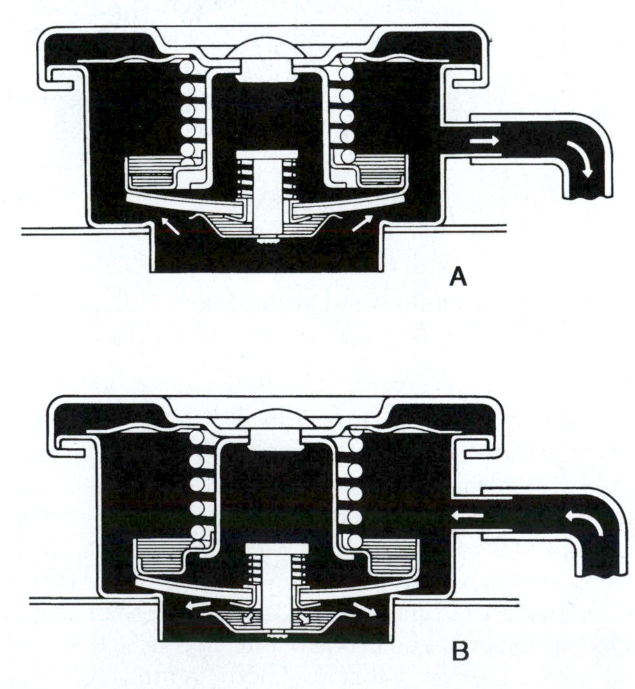

Figure 14. Operation of the radiator cap pressure and vacuum valves.

Overflow System

The overflow system consists of a small plastic tank that is connected to the barb on the radiator filler neck using a rubber tube. The tank is placed in close proximity to the radiator. The tube is installed in such a manner that the tip is constantly submerged in coolant unless the tank is empty. When coolant inside the radiator expands, pressure is created and the pressure valve inside the radiator cap opens, releasing coolant into the overflow tank. When the system cools down and the coolant contracts, vacuum is created in the radiator causing the vacuum valve to open and coolant in the expansion tank is pulled back into the radiator (see **Figure 15**). This keeps the radiator full of coolant and helps to expel any trapped air from the system.

Expansion System

The expansion tank is similar in appearance but is completely different in construction and operation than an overflow system. The expansion tank is part of the pressurized cooling system. It is installed in a position so that it becomes the highest point in the cooling system. A hose is used to connect the expansion tank to the radiator. In this type of system, the pressure cap is installed on the expansion tank rather than the radiator and the cooling system filled at the tank rather than the radiator (see **Figure 16**). The expansion tank is constructed of reinforced plastic and holds enough coolant so that the cooling system fluid level can expand and contract without releasing any coolant from pressurized portion of the system. Because air in the cooling system attempts to rise to the highest point of the system, any air bubbles that may be trapped will purge into the expansion tank. When trapped air is purged from the cooling system, it rises to enter the expansion tank and the displaced air is replaced by coolant. When system pressure exceeds the pressure rating of the cap, pressure is released to the atmosphere. When a vacuum is present within the system, the vacuum valve opens and allows atmospheric pressure to enter the tank. This does not cause a problem, because the tank is located above the radiator. In this position it becomes nearly impossible for the coolant inside the tank to become agitated and therefore coolant will not aerate.

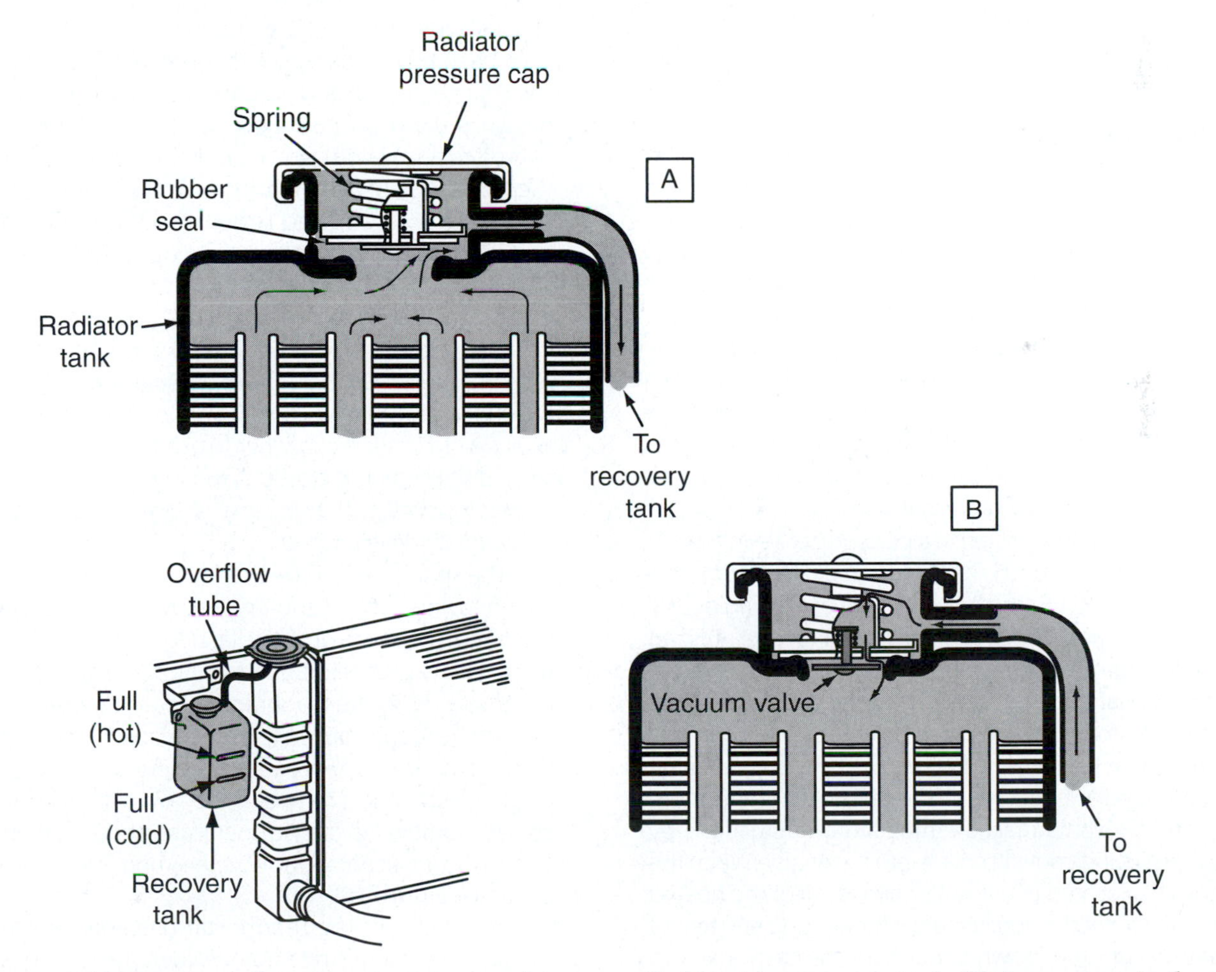

Figure 15. (A) When the radiator is pressurized coolant is expelled into the recovery tank. (B) As pressure in the radiator decreases, a vacuum is created which pulls coolant out of the recovery tank and back into the radiator.

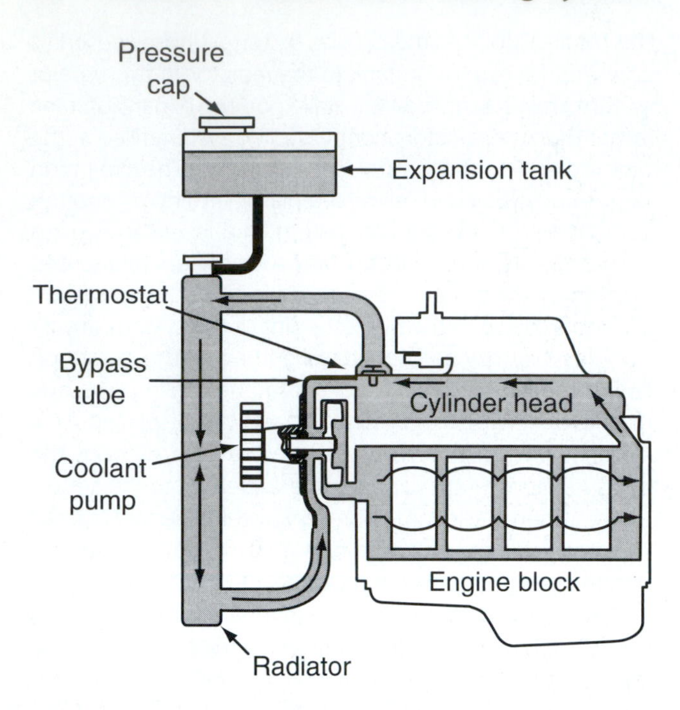

Figure 16. An expansion tank allows the cooling system to purge air and provides a place for expansion.

THERMOSTAT

The thermostat is a temperature-sensitive control valve and has several important functions in the cooling system. The thermostat allows the engine to reach operating temperature in the shortest possible time, maintains consistent temperature within the cooling system, and helps to prevent coolant pump **cavitation**. Cavitation is a condition in which the coolant pump is starved for coolant, which results in poor coolant pump efficiency and the introduction of air into the system. If allowed to persist for extended periods of time, cavitation can cause damage that is similar in appearance to electrolysis and corrosion.

The thermostat is located in a position between the engine and the radiator and is held in place by the thermostat housing. When located on the outlet side of the engine, the thermostat can usually be found in the cylinder head on an inline style engine or in the intake manifold on a V-style engine. When located on the inlet side of the engine, the thermostat can be found between the coolant pump outlet and the engine inlet (see **Figure 17**). Locating the thermostat at the engine coolant inlet helps the engine to warm more evenly and prevents drastic temperature fluctuations by slowly introducing cooled coolant into the engine. When the thermostat is located in the outlet, almost all of the heated coolant has to exit the engine before the sensing element of the thermostat cools down enough for the thermostat to close. When located in the inlet, the coolant is constantly moving across the sensing bulb causing the opening point

Figure 17. The thermostat is typically located in the coolant outlet.

of the thermostat to be more gradual and consistent than a thermostat that is located in the outlet.

Thermostat construction has remained virtually unchanged through the years. The one aspect that has changed is the method used to open the thermostat. There have been two primary designs used in the last forty years: a wax pellet design and a bimetallic spring. The wax pellet design is the most common. The thermostat consists of two discs; a large disc with a hole in the center serves as the main body of the thermostat. The second disc is used to cover the hole in the larger disc. The thermostat has additional support structures on both sides of the disc to support the operating piston and return spring. The smaller disc is equipped with a sealed copper control capsule that is filled with wax and sealed using a rubber diaphragm. A spring surrounds the control capsule and is trapped between the disc and the lower support structure. The spring is tensioned in such a manner to exert pressure against the small disc, keeping it closed. A rod is installed between the upper support structure and is seated against the diaphragm of the chamber.

The spring side of the thermostat is installed with the spring facing toward the engine. When the coolant surrounding the control capsule begins to heat, the wax within the capsule expands, forcing the diaphragm to expand upward. As the diaphragm expands, it exerts its force against the rod. Because the rod is unable to move, it forces the disc to move down against the pressure of the spring, creating separation between the two discs and allowing coolant to flow. As the temperature of the coolant is reduced, the wax contracts and the spring forces the disc to close (see **Figure 18**).

When closed, the thermostat prevents coolant from circulating through the engine. When the temperature at the base of the thermostat reaches the thermostat opening temperature, the thermostat opens and coolant can

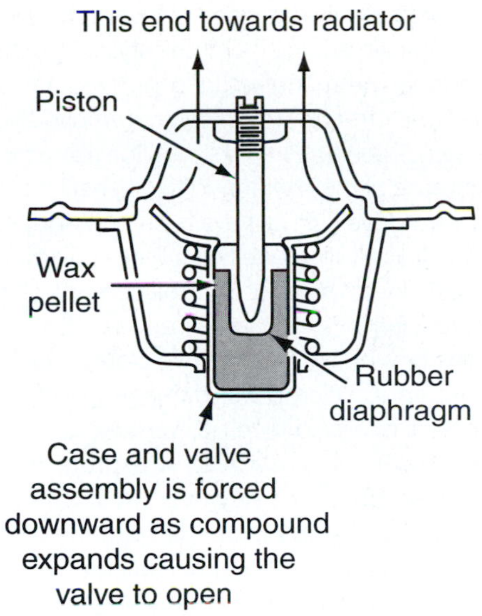

This end towards radiator
Piston
Wax pellet
Rubber diaphragm
Case and valve assembly is forced downward as compound expands causing the valve to open

Figure 18. Operation of a solid expansion thermostat.

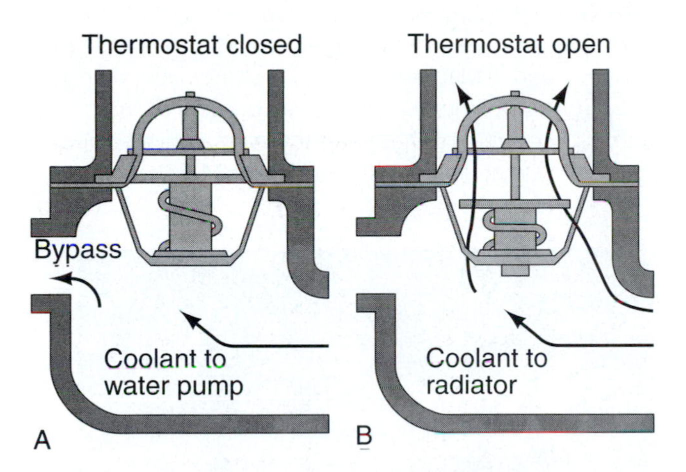

Figure 19. Typical thermostat operation.

then flow through the gap between the two discs and into the radiator where it is cooled (see **Figure 19**). The thermostat is rated according to the temperature at which it begins to open. This temperature can usually be found stamped into the thermostat body. Most thermostats in use today have an opening temperature of 195 degrees F (91 degrees C) and will be fully opened at 220 degrees F (105 degrees C). The thermostat is designed to open and close gradually to prevent drastic temperature fluctuations within the engine. Some thermostats can be equipped with a small notch or "jiggle pin" that relieves air pockets that may be trapped beneath the thermostat during service (see **Figure 20**).

The cooling system can be divided into two operating sections: the area between the thermostat outlet and the water pump inlet, including the radiator. The second section exists between the water pump outlet and the

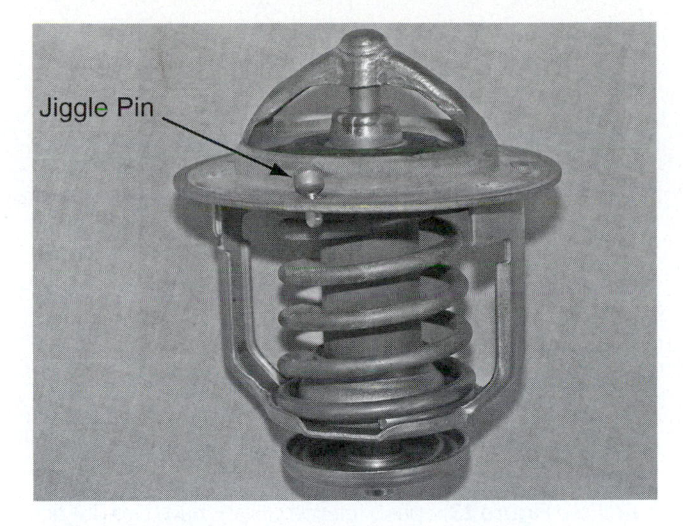

Figure 20. The jiggle pin is used to prevent air bubbles developing under the thermostat.

inlet side of the thermostat. When the thermostat is closed there is very little coolant flow, only that provided by the water pump bypass. When the thermostat opens and coolant begins to leave the engine, the pressure within the engine drops slightly, decreasing the pressure at the water pump outlets. However, during this time the pressure on the water pump inlet remains constant. The result is that the pressure on the water pump inlet is greater than at the water pump outlet. The difference in pressure encourages coolant to enter the coolant pump inlet helping to decrease cavitation and sustain coolant flow. If the thermostat was left out, coolant pressure could equalize and hinder the coolant flow through the pump.

HOSES AND CLAMPS

The cooling system uses various hoses to connect the radiator and heater core to the engine. These hoses are constructed of reinforced synthetic rubber. Modern hose compositions are resistant to heat, oil, and electrolysis. Most of the cooling system hoses that are used in the automotive applications are preformed hoses (see **Figure 21**).

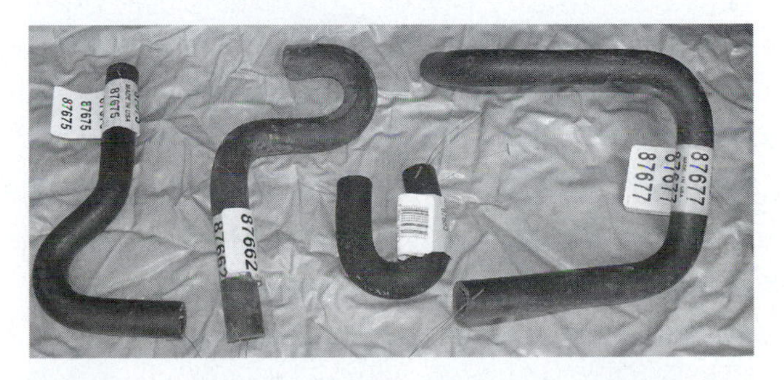

Figure 21. Most automotive applications require the use of preformed hoses.

Preforming of the hoses allows the hose to maintain very tight radius bends that are often required to connect the various cooling system components found in today's more compact engine compartments.

Automotive cooling system components are connected together using various hoses, pipes, and clamps. Each component in the cooling system has a short connecting pipe where the coolant hoses slide over. Each connecting pipe is equipped with a raised barb near the end of the pipe that helps to secure the hose to the pipe (see **Figure 22**). When the hose clamp is placed behind this barb, a very strong connection is formed.

Clamps used in automotive cooling systems fall into two categories: constant tension spring clamps and worm drive clamps. Each of these styles can be found as original equipment (see **Figure 23**). Spring tension clamps have two distinct advantages and are the most commonly used clamp today. The spring clamp applies a constant amount of tension to the hose and connecting pipe. This means that clamps do not have to be periodically tightened as the hose relaxes. This is an especially important feature when used with plastic radiator tanks that can be easily damaged by overtightening a hose clamp. The other advantage is that this style of clamp is easier to install during the manufacturing process. Although this style of clamp is often more difficult to remove and replace during service procedures, they should be reused whenever possible because of their self-tensioning benefits.

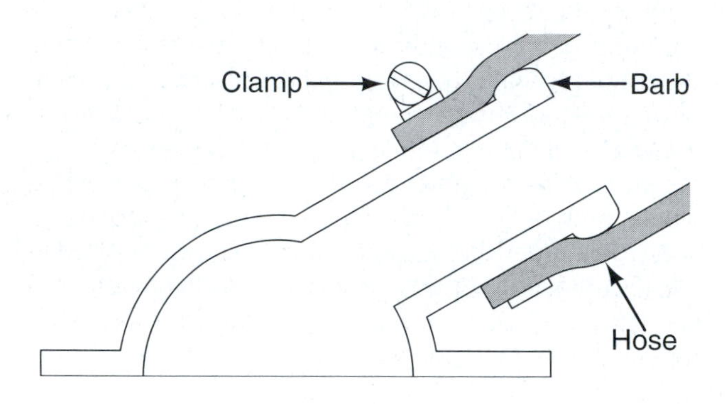

Figure 22. A hose barb located on most cooling system hose fittings works in conjunction with the clamp to form a tight connection.

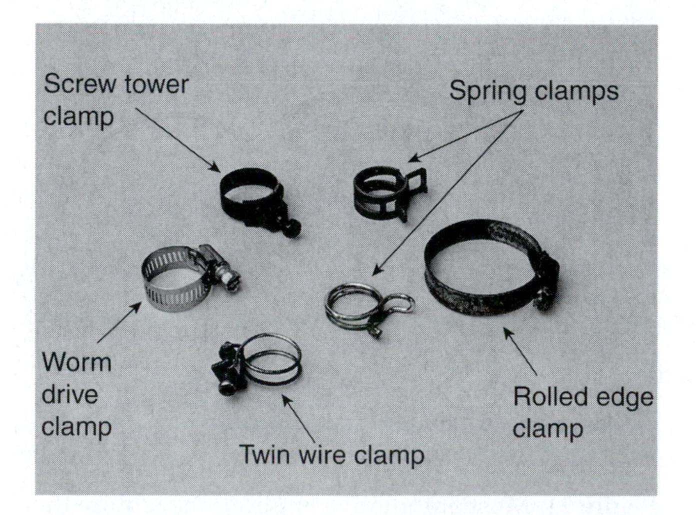

Figure 23. Various types of automotive hose clamps.

The worm drive clamp is available in various sizes to fit hoses as small as 1/2 in. to well over 6 in. in diameter. These are the most common type of replacement clamps. The worm drive clamp is a viable replacement for those applications in which a worm drive clamp was originally installed or a spring clamp was damaged. In many service applications, it is very difficult to remove and install the spring clamps. You may wish to substitute worm drive clamps in these situations; however, great care should be taken so that the clamp is not overtightened, which can cause damage to the radiator or heater core.

Some manufacturers have made widespread use of a third type of hose connection method, the snap connect fitting. Typically found on smaller hoses such as the heater hoses, the snap connect fitting uses a specially formed pipe that slides down into a pipe fitting and is sealed with an O ring and held in place with a plastic retainer. Depending on the application, the hose itself may utilize a permanently crimped connection between the hose and the pipe or may use a clamp to retain it (see **Figure 24**).

Figure 24. Many pipe connections use snap connect fittings.

You Should Know *Most lower radiator hoses have a coiled wire insert in the hose. This insert keeps the lower radiator from collapsing under the lower pressure conditions that are often experienced in the coolant pump inlet. If the original hose is equipped with a wire insert, the replacement should be equipped as well.*

BELT-DRIVEN COOLING FANS

The engine cooling fan is used to create artificial air flow through the radiator particularly at low engine speeds. Cooling fans used today are either belt-driven directly by the accessory belt drive system or can be driven by an electric motor.

The belt-driven fan has been used since the advent of the automobile and is still used in many applications today. The fan is mounted to the coolant pump shaft and shares a belt drive pulley with the coolant pump. The fan blade is precisely balanced to provide vibration-free operation. Since the fan is mounted to the coolant pump shaft, any imbalance that exists in the fan blade will be transferred to the coolant pump and will eventually cause bearing damage. Fan blades can be found in various sizes, blade pitch, and blade count, and are constructed from various materials such as steel, aluminum, and plastic.

Interesting Fact

Some fleet vehicle applications such as police cars are equipped from the factory with high-temperature cooling system hoses. High-temperature hoses may also be available for many aftermarket applications. These hoses can often be identified by their green or blue color.

Early belt-driven fans were driven directly and spun at a speed relative to that of the engine. Although this provided a great deal of airflow, it created a lot of noise especially at higher engine speeds. Turning the fan at 100 percent of coolant pump speeds creates a great deal of parasitic drag on the engine and is largely inefficient at higher engine speeds. This is because when the vehicle is moving even at very moderate speeds enough airflow is created to remove sufficient heat from the radiator and keep the engine cool. To solve this problem, a fan clutch was installed on most vehicles beginning in the early 1970s. The fan clutch is a mechanism that bolts to both the cooling fan and the coolant pump but its operation allows each component to spin at a different speed. There are three primary styles of fan clutches used: thermostatic, centrifugal, and electronically controlled. Each of these three styles of fan clutch uses silicon fluid to control the fan speed. The silicon fluid is transferred to different parts of the clutch to change the amount of slippage that the clutch allows. The most common type of clutch is the thermostatic type.

You Should Know

Extreme caution should be used working around the engine cooling fan and belt-driven accessories. Clothes or jewelry may become entangled, causing serious personal injury.

Thermostatic Clutch

The thermostatic fan clutch is comprised of four major components: a fan hub, an input shaft, an operating chamber, and a clutch plate (see **Figure 25**). The fan hub is part of the clutch to which the fan attaches. A chamber is formed inside the hub where the silicon fluid resides. The input shaft attaches to the coolant pump and pulley and passes through the back of the clutch hub where it is supported by a sealed bearing. The last major component is the clutch plate. The clutch plate is free-floating within the fan hub and is attached to the input shaft. The clutch plate separates the chamber within the fan hub into two chambers, a storage chamber and a working chamber. The silicon fluid is stored within the storage chamber. When fluid is directed from the storage chamber to the working chamber, the clutch and the hub speeds begin to equalize. It should be understood that there is always some drag between the fan hub and clutch plate and under normal circumstances should never spin completely independent of one another. A thermostatic spring that is located at the front of the clutch is attached to a control valve that allows the silicon fluid to move between the two chambers. When the clutch is disengaged, fan speed is limited to approximately 800 to 1400 rpm. When the temperature is above 160 degrees F (71 degrees C) the spring uncoils and allows additional fluid to flow into the working chamber of the clutch, causing the clutch to engage, allowing fan speeds to increase up to a maximum of about 2000 rpm. When the temperature cools down below 160 degrees F (71 degrees C), the fluid is allowed to return to the storage chamber and the fan slows.

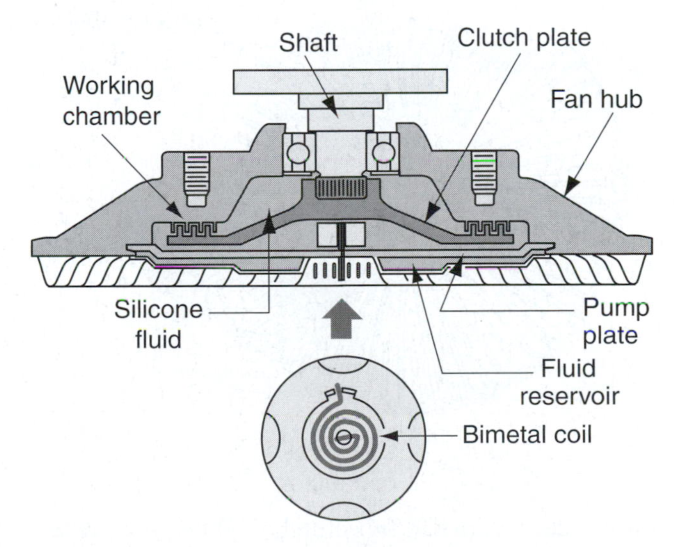

Figure 25. A cutaway of a thermostatic fan clutch.

Centrifugal Clutch

The centrifugal clutch is similar in construction to the thermostatic clutch but operates based primarily on the speed of the cooling pump shaft; engine compartment air temperature can affect the clutch operation slightly. When the engine is operating at low speeds, the cooling fan is allowed to spin at practically the same speed as the coolant pump up to a maximum of about 800 rpm. As the shaft speed increases, the ability of the clutch to transfer energy is decreased and the fan is allowed to spin up to a maximum of 1100 to 1350 rpm when the engine is cold and 1500 to 1750 rpm when the engine is warm. The maximum speed of the fan is limited to about 2000 rpm regardless of engine speed.

Electronic Clutch

Many components of the modern vehicle are controlled by an on-board computer called the Powertrain Control Module (PCM). Computer control of some belt driven cooling fans is made possible by the use of an electronic solenoid that controls the flow of fluid within the clutch and a **Hall Effect Switch** to monitor fan speed (see **Figure 26**). Internal operation is similar to a thermostatic clutch. The PCM uses the following input signals to determine the proper fan speed:

- Engine coolant temperature
- Transmission fluid temperature
- Ambient air temperature
- Air conditioning system pressure
- Vehicle speed
- Cooling fan speed

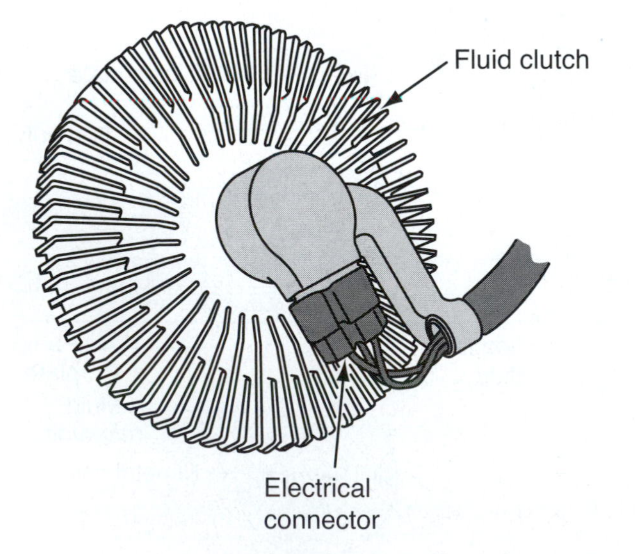

Figure 26. Electronic fan clutches allow for precise control of belt-driven fan applications.

ELECTRIC COOLING FANS

Belt-driven fans were the only type of fans used until the early 1980s. When transverse-mounted engines began to appear, electric fans became necessary to cool the radiator. Electric cooling fans use a small electric motor to spin the fan blade (see **Figure 27**). The fan can be located in the engine compartment and pulls air across the radiator. In this arrangement the fan is a puller fan. The fan can also be placed on the grill side of the radiator and pushes air across the radiator. This is called a pusher fan. Most applications use puller fans; however, many heavy-duty cooling systems often make use of two puller fans or, in some cases, use a pusher fan to assist the puller fan (see **Figure 28**). The fan

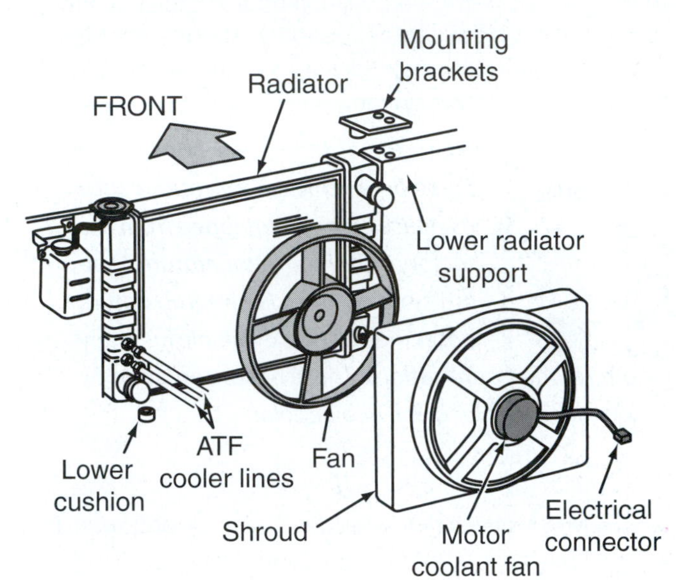

Figure 27. An expanded view of a typical electric cooling fan.

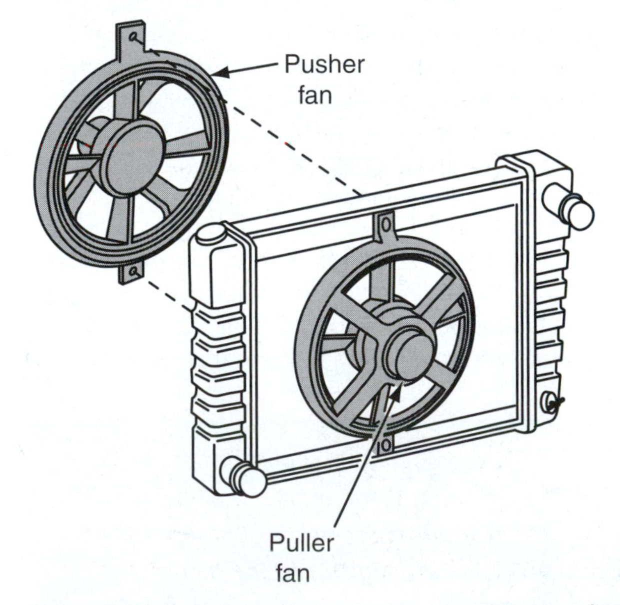

Figure 28. Some applications require the use of multiple fans.

used during most normal cooling modes is called the primary fan. The other fan, usually called the secondary fan, is typically used when additional cooling is desired.

The electric fan is controlled by the PCM through the use of a relay. The fan motor is supplied with a permanent ground, whereas power is supplied by the relay. Power is supplied to the coil side of the relay through the ignition switch and the PCM controls the operation of the relay by providing a ground to the coil side of the relay. Electric fans may be single speed or dual speed. Fans that use single-speed motors make use of additional wiring that allows a **ballast resistor** to be installed between the relay and the fan allowing for two-speed operation as shown in **Figure 29**. The PCM uses the following input signals to determine proper cooling fan operation:

- Engine coolant temperature
- Transmission fluid temperature
- Ambient air temperature
- Air conditioning system pressure or system request
- Vehicle speed

The operation of the cooling fan is relatively consistent across manufacturers' vehicle lines. In most cases, the cooling fan is turned on any time that the air conditioning compressor clutch is engaged; this usually includes the defrost mode. Additionally, the cooling fan is usually disengaged when the vehicle reaches a threshold speed that is set by the manufacturer. In heavy-duty applications, the primary fan is used most of the time and the secondary fan is usually requested only when temperatures surpass a preset threshold temperature or air conditioning pressure exceeds a preset limit.

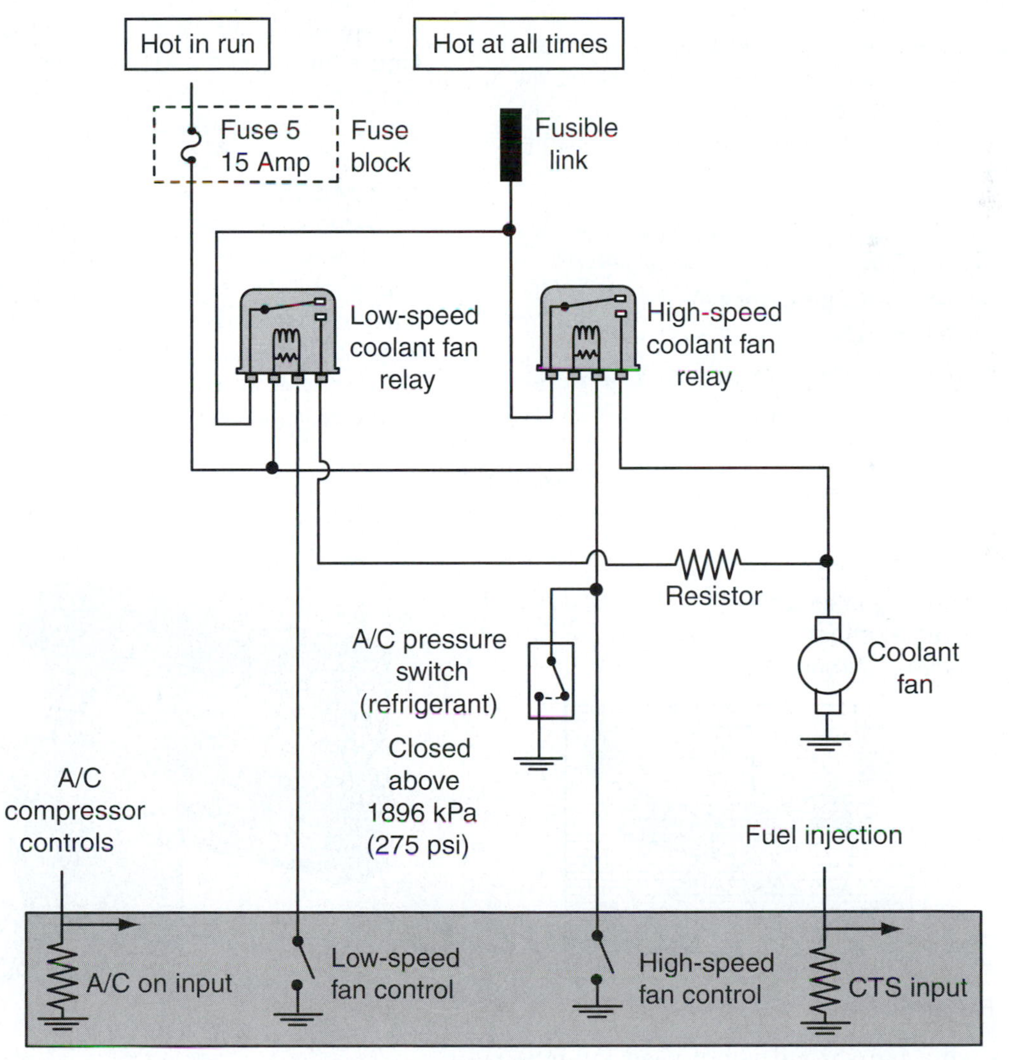

Figure 29. A typical electric cooling fan control circuit.

Some belt-driven fans use flexible blades. The purpose is to allow the blade to pitch in relation to the engine speed.

FAN SHROUDS AND DEFLECTORS

All vehicles use a combination of fan shrouds and air deflectors. Fan shrouds dramatically increase the efficiency of the fan and are essential to proper radiator cooling. Essentially, the fan shroud is used to maximize the amount of airflow that is directed across the radiator. Although the fan shrouds used with belt-driven fans and those found on electric fans are different in appearance, they perform the same essential function. If a fan blade has no shroud, it has the ability to easily pull air from the engine compartment without creating any significant airflow across the radiator (see **Figure 30**). Because most fans used in automotive applications are usually puller fans, it is essential that the fan pull air from a concentrated area of the radiator. This ensures that all of the air that is being moved by the fan is pulled across the radiator.

You Should Know *Some electric fans can start at any time in some cases, even when the ignition is in the off position. Great care should be exercised when working on or around electric fans.*

A fan shroud is basically a dome or box that is placed across the surface of the radiator. A hole is cut in the top of the dome that is slightly larger than the fan. The fan is forced to pull air across the radiator in the area that is isolated by the shroud. Because belt-driven fans are relatively large and have the ability to move large amounts of air, the shrouds used with these fans cover the entire surface of the radiator (see **Figure 31**). Because electric fans are smaller and move less air, they use much smaller fan shrouds that often only cover approximately one-half of the radiator's surface (see **Figure 32**). Some electric fans do not use a shroud but instead have a specially designed blade that has a skirt around the bottom of the blade. The blade is placed close to the surface of the radiator where it pulls air across a very specific area of the radiator (see **Figure 33**).

Air deflectors are used on many automobiles today. Air deflectors are little more than wide strips of plastic that are placed under the radiator. These pieces of plastic catch the air as the vehicle is moving and force it upward from the bottom of the vehicle to the front of the radiator (see **Figure 34**). As automobiles have become more aerodynamic, the available surface area for air to enter the grill area and ultimately the radiator has drastically decreased, making the use of air deflectors an absolute necessity in some cases.

Although often neglected, the seals around the hood and those found between the radiator and fenders are very important. These seals help to insure that the maximum amount of air is moved through the radiator. If these seals were not installed, air turbulence in front of the radiator and inside the engine compartment would be increased and could actually result in decreased air flow.

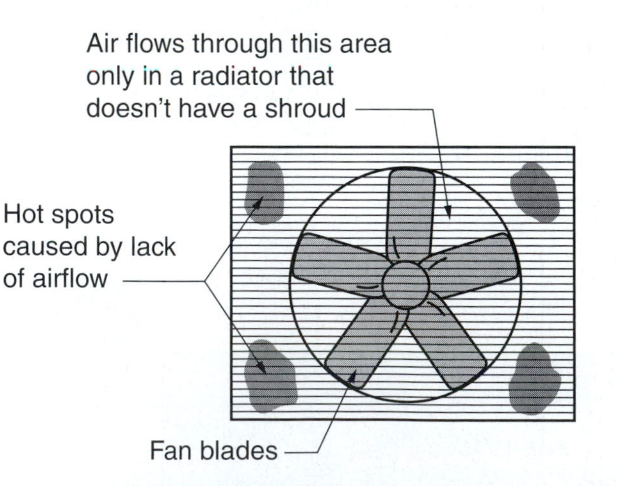

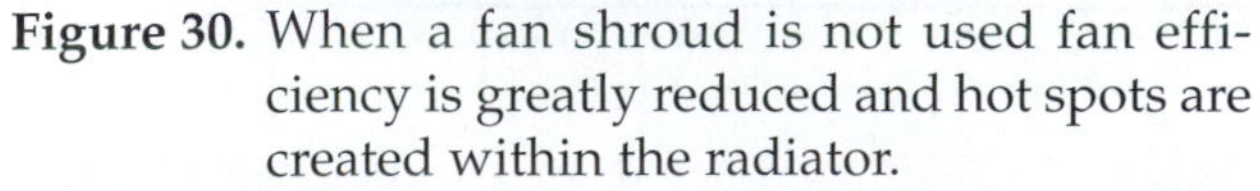
Figure 30. When a fan shroud is not used fan efficiency is greatly reduced and hot spots are created within the radiator.

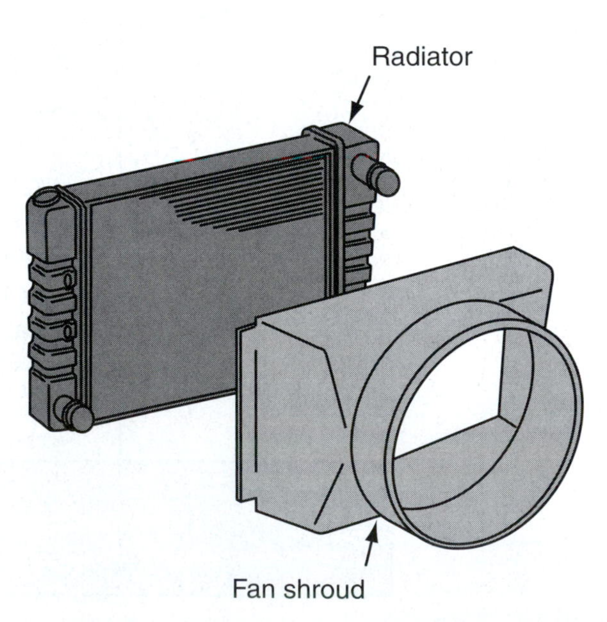

Figure 31. This style of fan shroud is used with a belt driven fan.

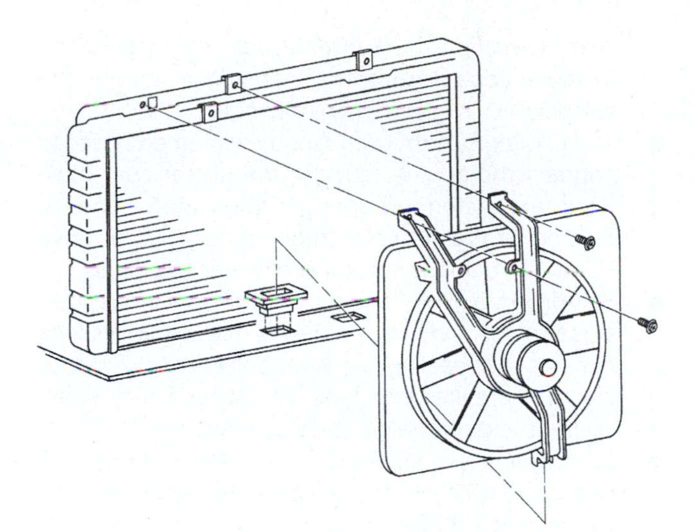

Figure 32. A typical electric cooling fan.

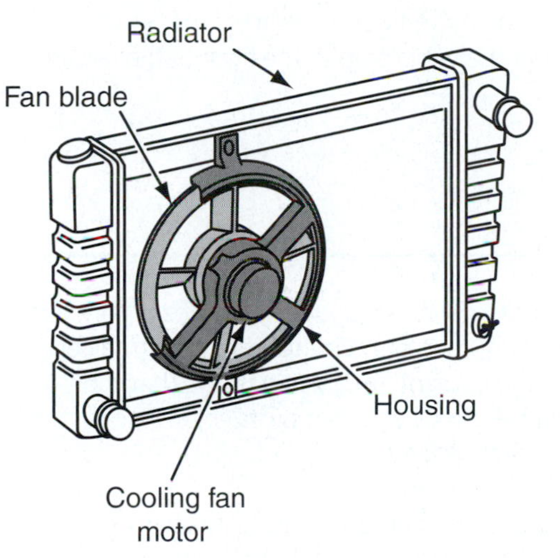

Figure 33. This style of fan blade does not require the use of a fan shroud.

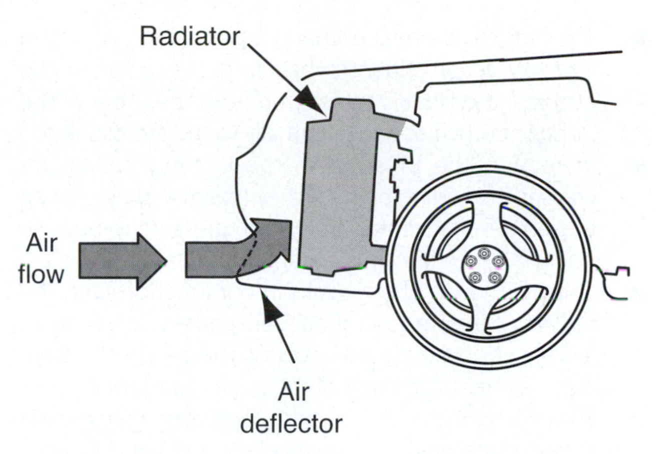

Figure 34. Air deflectors force air into the radiator.

Interesting Fact

Air deflectors also may be called air dams because they "dam" up the air and redirect it.

Summary

- The cooling system really acts as a temperature control system. The cooling system should allow the engine to heat up as quickly as possible and maintain the desired operating temperature without major fluctuations. Fluctuations above or below the calibrated temperature will cause poor drivability and increased tail-pipe emissions.
- A good engine coolant should provide freeze protection, boilover protection, corrosion protection, and adequate heat transfer abilities. Automotive coolant is comprised of a mixture of 50 percent water and 50 percent coolant.
- The base material for most automotive antifreeze is ethylene glycol. IAT, OAT, and HOAT are the three major types of automotive antifreeze. Different types of antifreeze differ greatly in their color, additives, service intervals, and applications.
- Water jackets are the passages that exist in the cylinder block and cylinder head for the circulation of engine coolant. There are two methods in which coolant is circulated: conventional flow and reverse flow.
- The coolant pump can be located on the front or side of the engine. The coolant pump can be driven one of several ways: by an accessory drive belt, timing belt or chain, and, in some engines, driven directly by the camshaft. The spinning impeller creates a low-pressure area at the coolant pump inlet. The bypass port is used to circulate coolant when the thermostat is closed.

- The radiator is a heat exchanger that uses conduction and convection to transfer heat from the coolant to the atmosphere. Heat is transferred from the tubes of the radiator to the fins and eventually to the atmosphere.
- Increasing the pressure in the cooling system increases the boiling point of the coolant. A pressure cap is used to regulate cooling system pressure, remove air pockets, and maintain a consistent coolant level.
- The overflow system works in conjunction with the radiator pressure cap to maintain coolant level and alleviate air pockets in the radiator. The overflow system is not a pressurized part of the cooling system.
- The expansion system is a pressurized part of the cooling system that helps reduce air pockets and maintain radiator coolant level. In those systems in which an expansion is used, the tank is located at the highest point in the system. The pressure cap is located on the expansion tank.
- The thermostat allows the engine to heat up in the shortest possible time, maintains a constant cooling system temperature, and helps to reduce coolant pump cavitation. The thermostat acts as a temperature-sensitive control valve. The thermostat rating is the temperature in which it begins to open.
- The cooling system uses various reinforced hoses to connect the engine, radiator, and heater core. Most hoses are preformed. Spring clamps apply constant tension to the hose. Overtightening a worm drive clamp can damage radiator and heater core pipes.
- Belt-driven cooling fans can be built in various sizes, blade counts, and from a variety of materials. The speed of belt-driven cooling fans is controlled using a viscous clutch. There are three types of viscous fan clutches: thermostatic, centrifugal, and electronic.
- Electric fans can be placed in a position to either pull or push air. Most electric fans are computer controlled by the PCM. Vehicles that use electric fans are typically equipped with either one or two fan assemblies.
- Fan shrouds are used to direct the air flow through the radiator. Air deflectors redirect the air that is moving under the vehicle up to the front of the radiator.

Review Questions

1. All of the following statements about cooling system operation are correct *except:*
   A. The cooling system should keep the engine from overheating.
   B. The cooling system should allow the engine to run as cool as possible.
   C. The cooling system should allow the engine to warm up as quickly as possible.
   D. The cooling system should be able to maintain the engine temperature within a specific operating range.
2. An engine's desirable operating temperature is a compromise of ________________, ________________, and ________________.
3. List four desirable characteristics that an engine coolant should possess.
4. Technician A says that water has excellent heat transfer properties. Technician B says that water has better heat transfer properties than ethylene glycol. Who is correct?
   A. Technician A
   B. Technician B
   C. Both Technician A and Technician B
   D. Neither Technician A nor Technician B
5. Technician A says that antifreeze concentrations below 40 percent diminish the heat transfer ability of engine coolant. Technician B says that the proper antifreeze can be selected by its color. Who is correct?
   A. Technician A
   B. Technician B
   C. Both Technician A and Technician B
   D. Neither Technician A nor Technician B
6. Technician A says that IAT, OAT, and HOAT antifreeze are made from polypropylene. Technician B says all used coolants should be handled as hazardous materials. Who is correct?
   A. Technician A
   B. Technician B
   C. Both Technician A and Technician B
   D. Neither Technician A nor Technician B
7. Explain the difference between conventional coolant flow and reverse coolant flow and explain why reverse flow is more efficient.
8. Explain how a coolant pump circulates coolant.

9. Technician A says that the radiator transfers heat from the radiator fins to the atmosphere. Technician B says that radiation is used to transfer heat from the radiator tubes to the atmosphere. Who is correct?
   A. Technician A
   B. Technician B
   C. Both Technician A and Technician B
   D. Neither Technician A nor Technician B
10. Technician A says that a faulty vent valve in the radiator cap can cause cooling system pressure to become excessive. Technician B says that a faulty vent valve in the radiator cap can cause the engine coolant to boil prematurely. Who is correct?
    A. Technician A
    B. Technician B
    C. Both Technician A and Technician B
    D. Neither Technician A nor Technician B
11. Describe the differences between a coolant overflow system and an expansion system.
12. List three things that the thermostat accomplishes within the cooling system.
13. Technician A says that the thermostat should be fully opened by the time the coolant reaches the rated temperature of the thermostat. Technician B says that a vehicle equipped with a thermostat that is rated at 195 degrees F (91 degrees C) can allow the engine to operate at a temperature as great as 220 degrees F (105 degrees C). Who is correct?
    A. Technician A
    B. Technician B
    C. Both Technician A and Technician B
    D. Neither Technician A nor Technician B
14. Technician A says that a ruptured diaphragm in the thermostat control capsule will cause the thermostat to stick in the open position. Technician B says that a broken thermostat spring will cause keep the thermostat from opening. Who is correct?
    A. Technician A
    B. Technician B
    C. Both Technician A and Technician B
    D. Neither Technician A nor Technician B
15. Explain how the thermostat helps prevent coolant pump cavitation.
16. Technician A says that belt-driven cooling fans are more efficient at higher engine and vehicle speeds. Technician B says that enough airflow is created at moderate vehicle speeds to keep the engine and radiator properly cooled with little or no help from the cooling fan. Who is correct?
    A. Technician A
    B. Technician B
    C. Both Technician A and Technician B
    D. Neither Technician A nor Technician B
17. Technician A says that the use of a fan clutch allows the cooling fan and coolant pump to spin at different speeds. Technician B says that at temperatures below 160 degrees F (71 degrees C) the fan can spin at speeds of up to 2000 rpm. Who is correct?
    A. Technician A
    B. Technician B
    C. Both Technician A and Technician B
    D. Neither Technician A nor Technician B
18. Which of the following is *not* an input used to control the electric cooling fan?
    A. Vehicle speed
    B. Air conditioning system request
    C. Engine coolant temperature
    D. Fan speed
19. Technician A says that the PCM controls the electric cooling fan by providing a ground directly to the motor. Technician B says that the PCM controls the fan relay by supplying power to the fan relay. Who is correct?
    A. Technician A
    B. Technician B
    C. Both Technician A and Technician B
    D. Neither Technician A nor Technician B
20. Explain how fan shrouds and air deflectors are used to increase the efficiency of the cooling system.

Chapter 10 Heater System Operation

Introduction

Although air conditioning is still an option on some vehicle models, the heater is an essential system that is required on every vehicle built. The heater system is a component of both the engine cooling system and passenger comfort systems. Although the heater can provide for a slight amount of engine cooling, its primary function is to heat the air within the passenger compartment.

HEATER CORE

The heater core is the main component of the heater system. The heater core is a water-to-air heat exchanger that operates exactly the same as a radiator, but whereas the radiator is charged with merely dispersing the heat energy that is absorbed from the coolant, the heat that is absorbed and then disbursed by the heater core is used to warm the passenger compartment (see **Figure 1**).

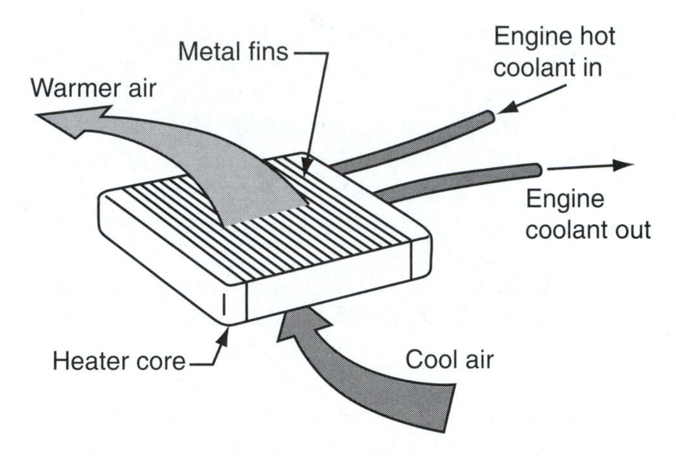

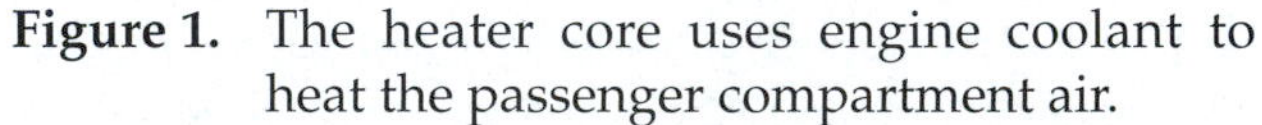
Figure 1. The heater core uses engine coolant to heat the passenger compartment air.

The heater core consists of two tanks: an inlet tank and an outlet tank. Fitted in between the two tanks is the actual core (see **Figure 2**). The core consists of rows of flat or oblong tubes placed next to one another. In between the rows of tubes, thin fins are placed perpendicular to the tubes. The fins provide a large amount of surface area in which heat can be transferred to the air surrounding the core (see **Figure 3**). Heater cores can be manufactured from brass and copper; however, most modern heater cores utilize composite plastic tanks with aluminum cores.

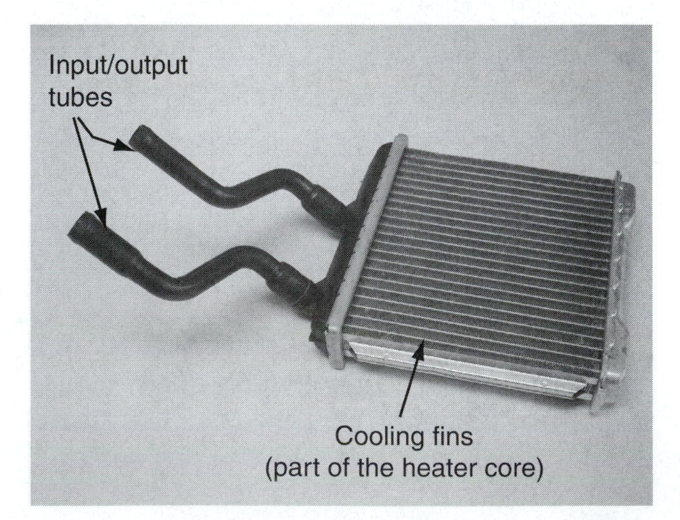

Figure 2. A typical modern heater core.

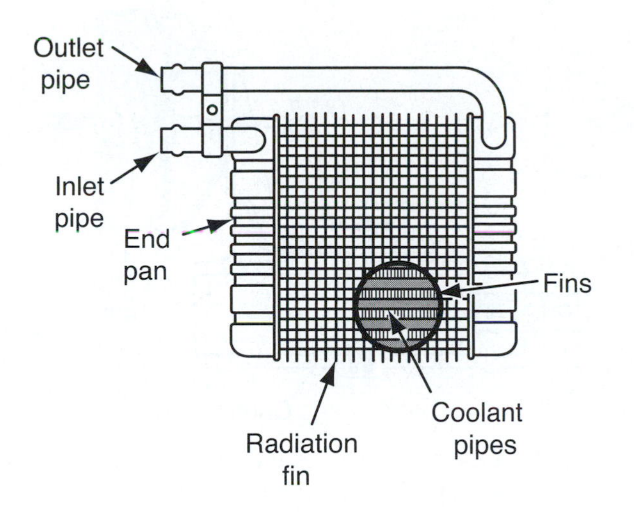

Figure 3. Typical heater core construction.

CIRCULATION

The heater core is connected to the cooling system using a parallel connection. This allows the flow of coolant through the heater core to be switched off and on as required without affecting the flow of coolant through the cooling system. The heater core is connected to the engine with two hoses, an inlet hose and an outlet hose. The heater inlet hose connects the heater core and the engine. Water is sourced from the engine at a location in which water at engine operating temperature is forced through the inlet hose and through the core. This can be considered a pressure hose. Typical heater inlet connections can be found in the top of the cylinder head on an inline engine or at the coolant crossover on the intake manifold of a V-type engine. The heater outlet hose, also called the return hose is connected either to the radiator or the water pump. Each of these locations is lower in relative pressure than the pressure hose readily allowing coolant to flow through the core (see **Figure 4**).

The hoses used to connect the heater core and the engine are typically either $\frac{5}{8}$ in. (16 mm) or $\frac{3}{4}$ in. (19 mm) diameter hoses. Heater hoses can greatly vary in length, depending solely on the engine and heater core installation locations. Hoses may be straight, but in many cases tight engine compartment locations require the use of preformed molded hoses to avoid crimping and abrasion.

Thermostat

Because the heater is an auxiliary unit of the cooling system, it is the operation of the thermostat that controls the temperature of the coolant flowing to the heater core. In a properly operating system, the coolant available to the heater core will be at or near the same temperature as the

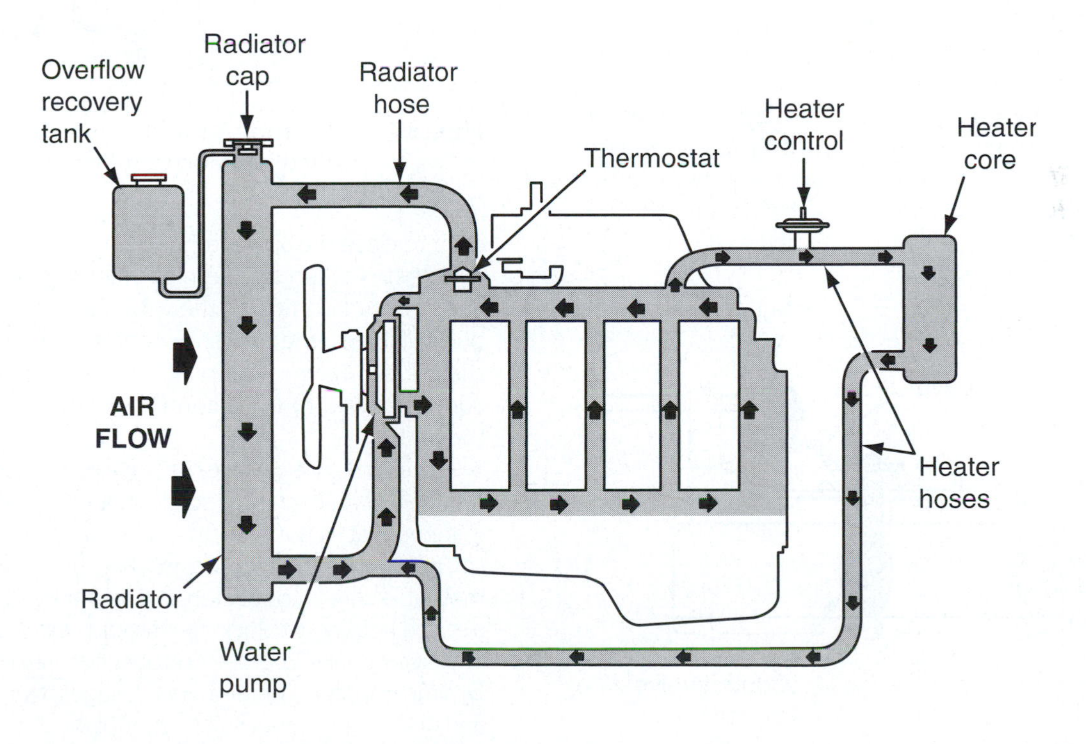

Figure 4. The heater core receives heated coolant directly from the cooling system. Some heater systems may use a valve to control coolant flow through the core.

engine coolant. Malfunctioning thermostats often may be recognized by poor heater operation. For complete thermostat operation, refer to Chapter 9.

HEATER CONTROL VALVE

The heater control valve is used to control the flow of coolant through the heater core. Any time that heated coolant is allowed to flow through the heater core, heat will be transferred to the passenger compartment. Heater control valves are installed for one of two reasons: to completely stop the flow of warm air into the passenger compartment when heater operation is not desired and to control the heater core surface temperature when heater operation is desired. Depending on the design of the system, the control valve can be installed in either the heater inlet hose or the heater return hose (see Figure 4). It should be noted that heater control valves may not be used on all vehicles.

There are three common methods for controlling the operation of the control valve: cable, vacuum actuated, and electrically actuated. All three methods can still be found in use today. The cable method is one of the most common and makes use of a cable that is connected directly to the temperature control on the dash. As the cable is moved, a lever that is connected to a flap within the valve is moved either allowing or blocking the flow of coolant (see **Figure 5**).

There are two styles of vacuum actuated heater control valves. The first style uses an internal flap to block coolant, much like the cable-operated valve. However, this style of valve makes use of an external vacuum chamber to move the lever rather than using a cable (see **Figure 6**). The second style uses a relatively large vacuum diaphragm that is connected to a pintle that is lifted up when vacuum is applied (see **Figure 7**). In either case, when vacuum is applied to the valve, it moves to the closed position. In either style valve, the vacuum signal is used to close the valve rather than open it. This is a measure that is taken by the manufacturers to ensure that if the control portion of the valve fails, then heat will still be available to the passenger compartment. Vacuum for the control valve is supplied by the engine and to the control valve through a control switch that is located on the dash.

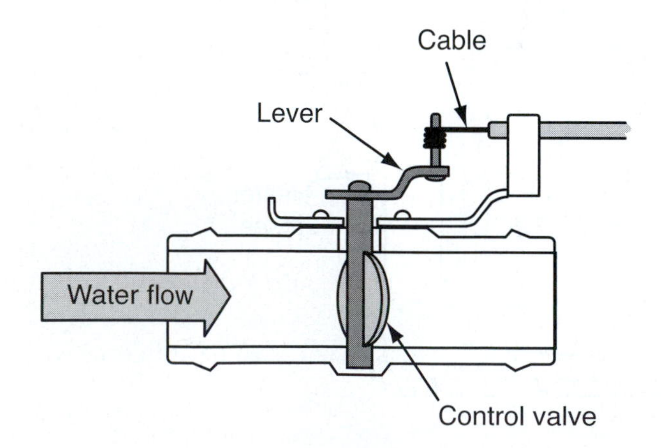

Figure 5. Operation of a cable-operated heater control valve.

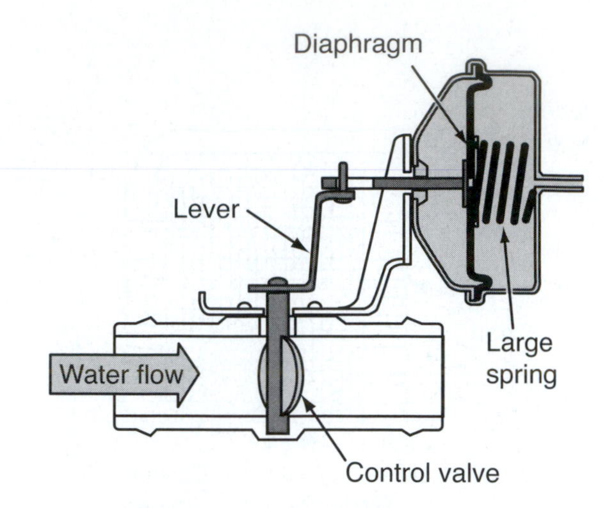

Figure 6. Operation of a vacuum controlled lever actuated heater control valve.

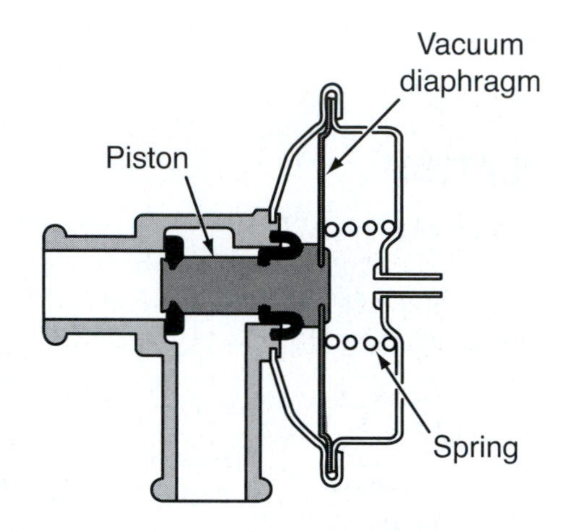

Figure 7. Operation of a vacuum diaphragm controlled heater control valve.

The third style of valve uses an electric **solenoid** to control the flow of coolant. A solenoid is essentially an electromagnet that is allowed to move within a coil. When a solenoid is used in the construction of a heater valve, the moveable core is connected to a diaphragm that fits within a sealed housing to block the flow of coolant (see **Figure 8**). The electric valve can be controlled in three different ways: by simply applying or removing voltage to turn the valve off and on, by using a rheostat to control the amount of current that is allowed to reach the solenoid, or by using **pulse width modulation (PWM)**. Pulse width modulation is a

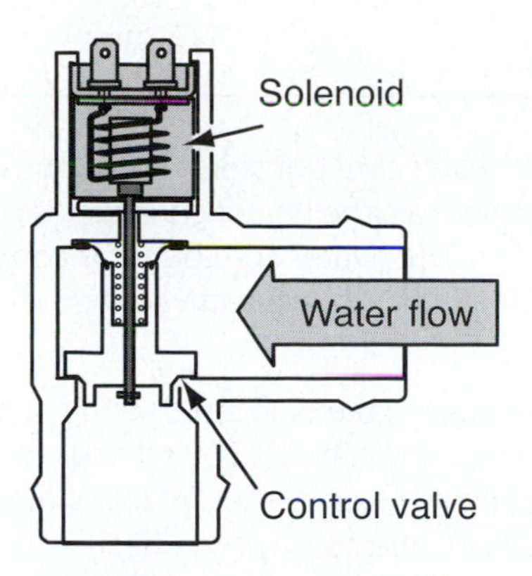

Figure 8. Operation of a electronically activated heater control valve.

method of controlling electronic devices by rapidly turning the voltage on and off. This can occur several times per second. The longer the average time that electricity is applied to the solenoid in each 1-second increment determines the relative position of the solenoid. Both rheostat and PWM control allow for an infinite number of valve positions.

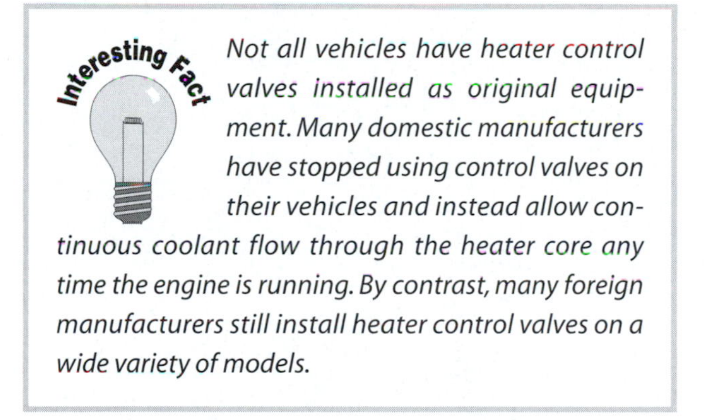

Not all vehicles have heater control valves installed as original equipment. Many domestic manufacturers have stopped using control valves on their vehicles and instead allow continuous coolant flow through the heater core any time the engine is running. By contrast, many foreign manufacturers still install heater control valves on a wide variety of models.

AIR CIRCULATION

The heater core is located within a system of ducts that directs passenger compartment air across the surface of the heater core to different locations throughout the vehicle. An electric blower motor is used to pull outside air into the vehicle and force it across the surface of the heater core and out through various ducts in the passenger compartment. In vehicles that are equipped with air conditioning, the heater core shares a common duct system with the air conditioning system (see **Figure 9**). Although air conditioning is installed on a majority of vehicles, there are a still a few vehicles that are equipped with heat only. Vehicles that are equipped with only a heater have three basic modes of operation: heat, defrost, and vent. More complex duct systems will be covered extensively in Section 6 of this book.

In either configuration, the heater core can be located in a variety of positions within the duct system. When larger vehicles were common, the heater core was installed in a sealed case within the engine compartment. As engine compartments have grown increasingly more crowded, the heater core and its associated ductwork have been moved into the vehicle interior. Because the space within the vehicle is very tight, the heater core can be found installed in either a vertical or horizontal position to meet the specific needs of the installation.

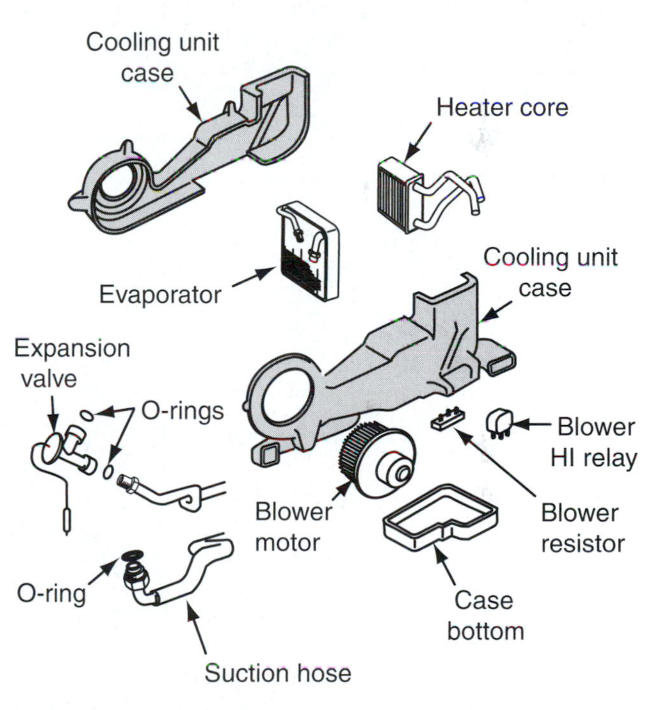

Figure 9. A simplified HVAC air distribution system.

Summary

- The heater core is comprised of two tanks and a core. Common heater core materials are brass, copper, aluminum, and composite plastic. The heater core is connected parallel to the cooling system.

- The heater control valve can be used to completely block coolant flow or to modulate coolant flow. The heater control valve can be cable, vacuum, or electrically controlled.

- The heater core transfers heat to the passenger compartment air. The heater core inlet is connected to the engine. The heater core outlet is connected either to the water pump or directly to the radiator.

- The heater core is located within a system of ducts where a blower motor is used to direct passenger compartment air across the surface of the core. The heater has three operation modes: heat, defrost, and vent.

Review Questions

1. Describe the construction of the heater core.
2. List three ways in which a heater control valve can be controlled.
3. Technician A says that some heater cores may be located in the engine compartment. Technician B says that the heater core can be located in the passenger compartment. Who is correct?
   A. Technician A
   B. Technician B
   C. Both Technician A and Technician B
   D. Neither Technician A nor Technician B
4. Which of the following is not a heater operation mode?
   A. Vent
   B. Heat
   C. Mix
   D. Defrost

Engine Cooling System Diagnosis

Introduction

The cooling system is one of the most important systems found on any automobile; it is also one of the most neglected systems. Often the cooling system receives no attention unless a problem becomes apparent. As coolant replacement intervals have increased, up to 150,000 miles (241,401 km) in some cases, the perception of the motoring public is that the cooling system is maintenance free. Although this might be feasible in controlled environments, this can be unrealistic in the field. As long as a cooling system is always properly serviced or never tampered with, these extra-long intervals may be realistic. However, each time the cooling system is serviced, the potential for cooling system contamination exists.

COOLANT

One of the most common service procedures in any service facility is topping off the cooling system. However, with the number of different antifreeze products on the market today, this can become treacherous territory. Ten years ago, coolant replacement was a straightforward issue because there was only one type of coolant, but each year since 1995 that issue has become slightly more complex. With three different types of coolant, which one should we use to top off the system?

The best solution is to use exactly what the manufacturer specifies, especially if the vehicle is still under factory warranty. If the incorrect antifreeze is mixed with the factory fill, it can be considered contamination. This in itself can affect the warranty coverage of subsequent cooling system repairs. Contamination can occur even if it is the right antifreeze but the wrong color. However, from an economic standpoint, it does not make good sense to stock every coolant from every manufacturer. So what should you add if the factory fill is not on hand? A sound policy for topoff applications is to have an IAT antifreeze for those vehicles that use traditional green antifreeze. You also should have a Dex-Cool compatible OAT for General Motors vehicles and a non-Dex-Cool OAT for the other manufacturers. Last but not least, you should have a HOAT antifreeze available for those G-05 applications (see **Figure 1**).

Mixture

Engine coolant should be comprised of 50 percent antifreeze and 50 percent water. This ratio provides the best balance of freeze protection, boilover protection, corrosion protection, and heat transfer (see **Figure 2**). However, all water is not equal. There are some important facts about

Figure 1. There are many different types of coolant in today's vehicles.

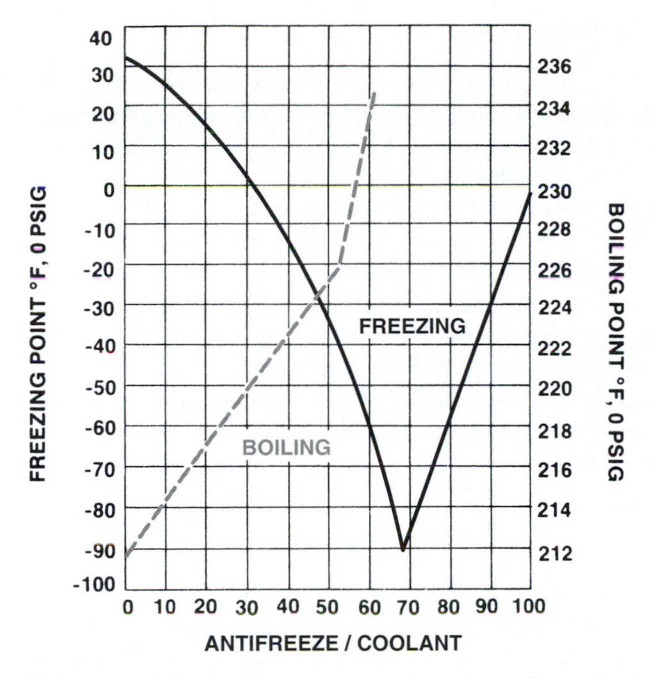

Figure 2. The antifreeze concentration affects both the boiling point and freezing point.

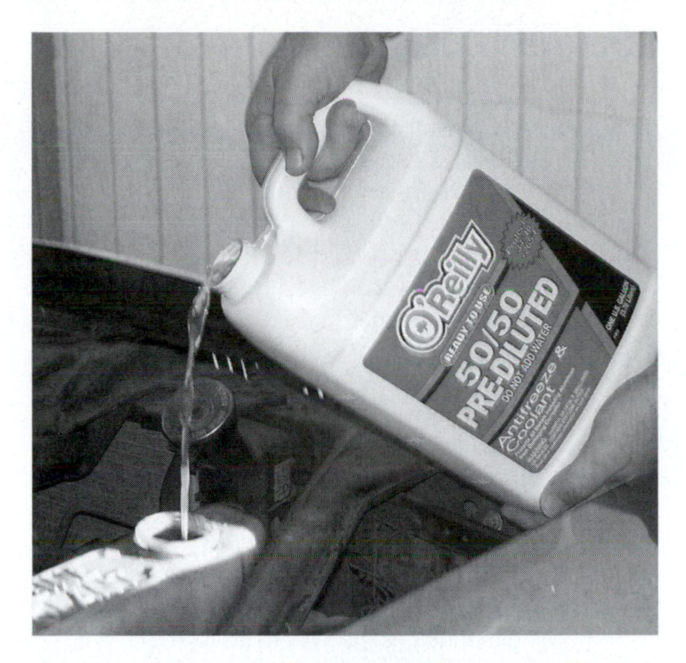

Figure 3. Many antifreeze manufacturers offer premixed coolant solutions.

water that should be considered. All tap water contains some amount of calcium and magnesium. Water that contains large amounts of these minerals is called **hard water**. Although the presence of minerals is not a health issue even in hard water, the presence of minerals can lead to the clogging of cooling system passages. When tap water is used in the cooling system, the minerals tend to separate from the water and collect on the surfaces of the cooling system, often on the end of the radiator and heater core tubes. These deposits can eventually lead to clogging of the tubes and result in very poor system performance. The mineral content found in tap water varies greatly between various geographic regions. For this reason, it is recommended that cooling systems be filled with **distilled water**. Distilled water goes through a process in which the normal tap water is heated and the water molecules are vaporized and then condensed. This process effectively separates the water from the minerals. Although this may not be practical in some instances, it is definitely recommended in areas in which the water is notably hard.

To ensure the proper mixture of antifreeze and water, the coolant should be premixed before filling or topping off a cooling system. It is good practice to obtain an empty antifreeze jug and premix antifreeze and water ahead of time so that it is available when needed. Just make sure that the jugs are labeled as premixed so that they are not mistakenly overdiluted. There are several companies that offer ready-to-use premixed coolants (see **Figure 3**). Just be aware of whether the coolant you are using is full strength or premixed, and handle it accordingly.

MAINTENANCE

Regular maintenance is required for all automotive systems, but maintenance does not necessarily mean service. In many cases, it may be nothing more than a thorough inspection. Cooling system inspection should take place any time the vehicle is serviced. The process takes just a few minutes and is a benefit to both you and the customer. The customer receives peace of mind that the cooling system has been inspected and that no obvious faults are present, and you have the opportunity to inform the customer of any pending problems that potentially can lead to additional service work. The basic cooling system inspection should also be the first step performed any time a cooling system concern is suspected. Basic cooling system inspection should consist of inspecting coolant level and condition, accessory belt condition, cooling system hose condition, and inspection of fan blades, shrouds, and air deflectors.

Any time a consumer takes a vehicle in for service, you as the technician have the opportunity to sell additional services. This is part of your job, but it is also your responsibility to inform the customer of the situation and present the facts as they exist without embellishment. All suggested vehicle repairs should be made with consumer benefit in mind, not the benefit of the technician or service facility. When this approach is taken, more often than not it will build a level of trust and confidence between the consumer and the service facility that can benefit the consumer, the service facility, and the technician for many years.

COOLANT CONDITION

The coolant should be checked for proper level, freeze protection, and overall appearance. In most vehicles, the condition and level of coolant can be observed by taking a sample from the expansion or overflow tank. We will refer to each of these as the reservoir. There are usually two marks on the reservoir: a cold level and hot level. The coolant should be at the proper level as determined by the temperature of the coolant. A good rule of thumb to follow is that if the hoses are not pressurized, the level should be at the cold mark; if pressure is present within the hoses, the coolant should be at the hot level. Relative pressure in the cooling system can be determined by squeezing the upper radiator hose. If the hose feels hot to the touch and is firm, pressure exists within the system. If the hose is cool to the touch and the hose can be easily collapsed so that your fingers nearly touch, there is very little or no pressure within the system.

Assuming that no cooling system leaks are present, the reservoir should remain at a relatively stable level, allowing for slight losses as a result of evaporation. If the reservoir is extremely low or empty, the coolant level should be checked in the radiator and the system pressure tested and inspected for a leak.

You Should Know *Removal of the radiator cap from a pressurized cooling system can result in the explosion of hot pressurized coolant that can cause serious burns to the face, eyes, and skin. Before removing any radiator cap, the technician should ensure that the cooling system has been fully relieved of its internal pressure.*

After verifying proper coolant level, the coolant can be tested for overall appearance and proper antifreeze concentration. The engine coolant should appear clean, have a bright color, and possess the tint of the intended coolant type. If coolant has a brown tint or appears muddy or murky, contamination is indicated. Contaminated coolant also may have a pungent or sour odor; this is an indication that the cooling system should be serviced.

Freeze protection and antifreeze concentration go hand in hand. If a coolant has the proper concentration of antifreeze, it should be able to protect the system from freezing down to –34 degrees F (–36.7 degrees C). This measurement represents a 50-50 solution of water and antifreeze, which provides a desirable balance of freeze protection, corrosion protection, and heat transfer abilities (see Figure 2). Antifreeze concentration can be measured with three different tools: a hydrometer, a refractometer, and an alkaline test strip.

Hydrometer

A hydrometer is a tool that uses a calibrated float to measure the specific gravity of a liquid. Because the specific gravity of coolant changes in relation to the concentration of antifreeze, a hydrometer can be an effective tool for determining coolant protection levels. As the concentration of antifreeze increases, so does the specific gravity of the coolant. The hydrometer that is used for testing antifreeze consists of a sealed tube that has either several floating discs, colored balls, or a calibrated float. A bulb is located at the top of the tube, when squeezed coolant is drawn into the sealed tube (see **Figure 4**). The protection level is gauged by the number of discs or balls that are floating within the tube. A scale is usually printed on the side of the tube to use as a convenient reference. In the case of the weighted tube, coolant protection level is determined by observing the coolant level in reference to markings on the floating tube (see **Figure 5**). Although the hydrometer is by far the most popular and easiest tool to use to determine protection level, it is not the most accurate. Furthermore, standard hydrometers intended for ethylene glycol cannot be used to test propylene glycol coolants.

Testing Procedure

Because the contents of the coolant in the reservoir are often unknown, the best sample can be retrieved directly from the radiator; however, great care must be taken when removing the radiator cap from a pressurized system.

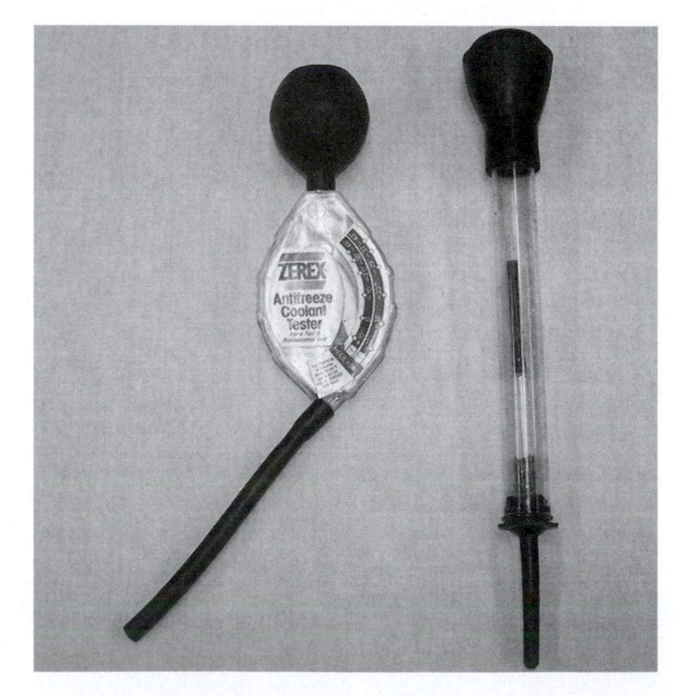

Figure 4. Two common types of coolant hydrometers.

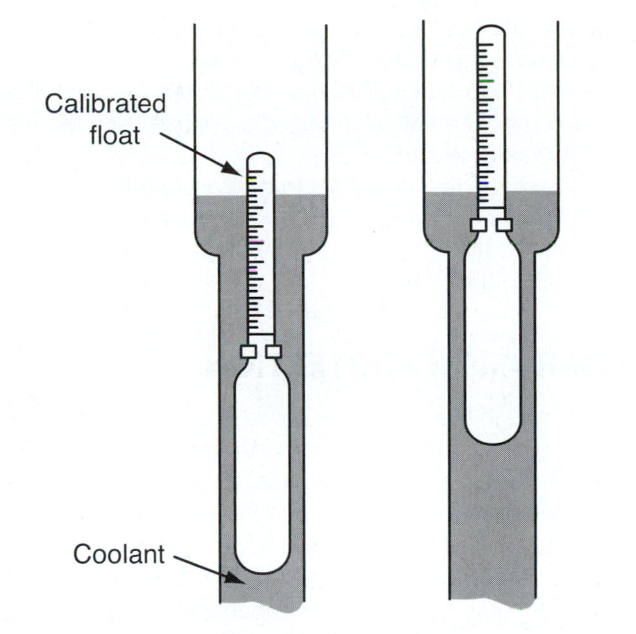

Figure 5. The specific gravity of the engine coolant can be read directly from the float.

1. Carefully remove the pressure cap.
2. Submerge the hydrometer tube in the coolant to be tested and fully squeeze and release the bulb to draw a full sample of coolant.
3. Remove the tube from the sample and gently pinch or curl the tube to prevent coolant from dripping from the tube.
4. Make sure that coolant level in the hydrometer completely fills the tube, allowing a small pocket of air at the top to displace any trapped air bubbles.
5. Gently tap the side of the hydrometer to release any trapped air and ensure that the float or the discs are not hanging up.

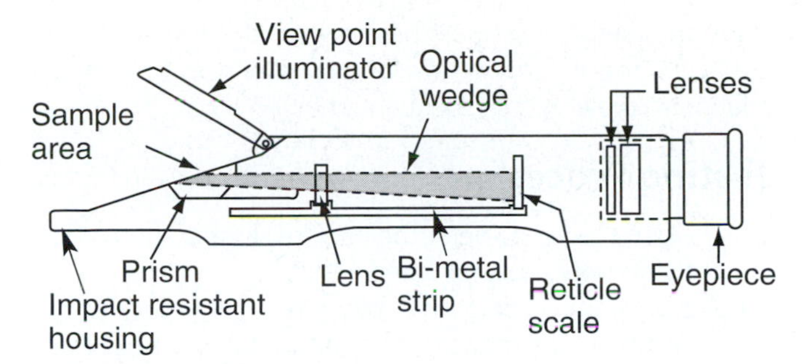

Figure 6. Refractometer operation.

6. Observe the number of discs that are floating and compare to the chart; this is typically located on the hydrometer.

Refractometer

The refractometer is a tool that is widely gaining acceptance in the automotive industry as an extremely accurate method of testing coolant protection. Although it is more expensive than a hydrometer, a refractometer is more accurate and can measure both ethylene glycol and propylene glycol solutions. A refractometer essentially measures how a light beam is refracted or bent as it passes through a coolant sample (see **Figure 6**). Light slows down and bends slightly when passed through a liquid. The refractor uses a prism and a system of lenses and a calibrated scale to generate a visual image of protection level. The refractometer determines coolant protection levels by focusing a light beam that is passed through the coolant sample and the optics of the refractometer onto the scale. When viewed through the eyepiece, a sharp contrasting line can be clearly seen on the scale indicating the protection level of the coolant (see **Figure 7**). As the concentration of antifreeze in a coolant

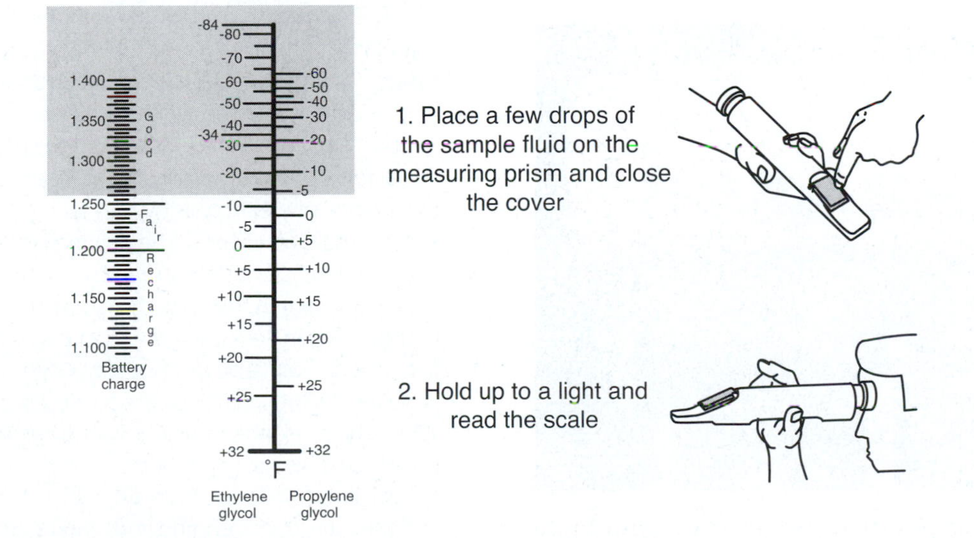

Figure 7. Measuring antifreeze concentration levels using a refractometer.

solution changes, so does the amount of refraction created; thus, the shadow line will move up or down in direct relation to the antifreeze concentration.

Testing Procedure

As with the hydrometer, the coolant sample should be removed from the radiator when possible.

1. Obtain a small coolant sample using the "dipstick" provided with the refractometer and place it under the illuminator.
2. Look through the eyepiece and gently press down on the illuminator to create a thin film of the sample.
3. The reading is read directly from the scale where the sharp contrasting line intersects the scale (shown in Figure 7).

Test Strips

Test strips are another method of testing antifreeze condition. In addition to freeze protection level, some strips can also test the acidity of the coolant sample, which is a relative indication of the corrosion protection of the coolant. Test strips are merely plastic strips of material that have chemically treated pads on the end. The number of pads will vary depending on how many different tests the strip can perform. When exposed to the coolant sample, the pad will change in color. The colors of the various pads are compared to a chart that indicates the results (see **Figure 8**).

Testing Procedure

For the best test results, the coolant should be removed from the radiator and the temperature should be between 40 and 110 degrees F (4.44–43.33 degrees C). Instructions for various products may vary.

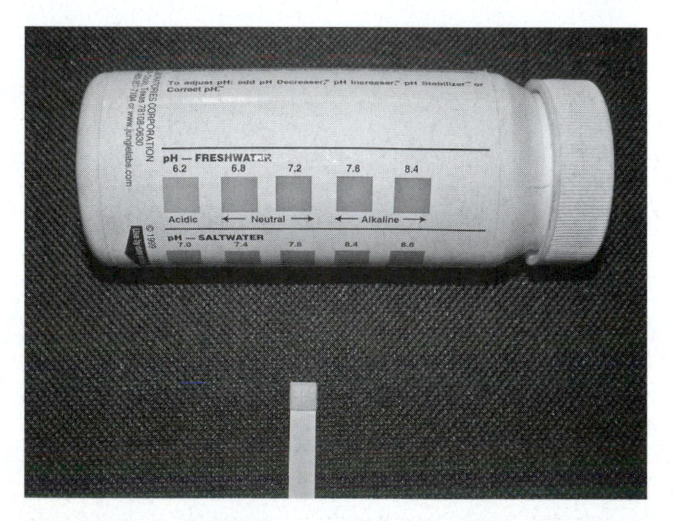

Figure 8. Test strips can be used to check antifreeze concentration levels as well as pH.

1. Immerse the test strip into the sample for 1 second.
2. Remove and shake off excess coolant.
3. Wait 1 to 5 minutes (depending on the specific product used) and observe the colors of the test pads and compare to chart.
4. Match the test strip pads to the corresponding colors of the nearest match on the chart that accompanied the test strips.
5. Discard the test strip.

CORROSION PROTECTION

Although the clean appearance of engine coolant and proper protection level are good indicators of coolant condition, these measures should not be used as an absolute clean bill of health. Coolant that has been in service for some amount of time may be very clean in appearance and even have the proper freeze protection; however, this does not mean that coolant is capable of protecting and lubricating the cooling system components as it should. As the coolant goes through cycles of heating and cooling, the additives in the antifreeze become depleted and the coolant becomes acidic. The level of acidity of coolant can be determined by measuring the **pH level** of the coolant. The pH level is a measurement of how acidic the coolant has become and is a good indicator of overall coolant condition. The pH level is measured on a scale of zero to 14. A pH reading of zero indicates pure acid, whereas a reading of 14 indicates an alkaline substance. The pH level can be measured using commercially available test strips (see Figure 8). The desired pH for engine coolant is a range from 8.5 to 11. When levels are either above or below these ranges, the cooling system should be completely flushed and refilled with fresh coolant.

Detecting Electrolysis

The formation of electrolysis within the cooling system is a normal occurrence that, if left unresolved, can lead to severe cooling system damage. Electrolysis primarily occurs when coolant becomes acidic or when an electrical component is poorly grounded. Measuring the pH level will indicate the acidity of the coolant, but we can test for the actual occurrence of electrolysis by measuring the amount of stray voltage that is present in the cooling system. The formation of electrolysis can be measured by observing the amount of voltage present in the cooling system. This test is also recommended when severely damaged parts are removed from the vehicle or repeated cooling system failures occur. Repeated failures may indicate an electrical system problem. When the presence of electrolysis is suspected, a thorough underhood visual inspection should be performed, paying special attention to the installation of aftermarket accessories and the integrity of engine and chassis grounds. The following test will check for stray voltage in the cooling system.

1. First set the Digital Volt Ohm Meter (DVOM) to the direct current (DC) millivolt scale.
2. Connect the negative probe to the negative post of the battery and submerge the positive lead into the coolant at the radiator filler neck, making sure the probe does not touch any metal.
3. Note the meter reading. The voltage reading should be below 0.10 volts.
4. Start the vehicle, turn on all of the accessories, and retest. If the voltage was above .010 in step 3 and remains the same with the engine and accessories on, cooling system service is required. If the voltage increases above 0.10 volts with the engine running and the accessories on, an electrical system problem is indicated.
5. Methodically shut off electrical components one at a time while observing the voltage reading. If the voltage drops when an accessory is turned off, a problem exists in that system.

COOLANT MAINTENANCE

Coolant maintenance is one the most frequent services performed by entry-level auto technicians. The engine coolant should be serviced:

- According to the manufacturer's recommended intervals
- When freeze protection falls outside of the desired range
- When the pH balance falls outside of the desired range
- When obvious contamination is present

Typical coolant services include drain and refill, manual flush, and machine flushing or exchange.

Drain and Refill

The drain and refill procedure consists of draining the coolant that is present within cooling system and refilling the system with the proper 50-50 mixture of coolant. This method can be used for normal maintenance or when component replacement requires that the engine coolant be removed.

The drain and refill procedure should be performed as follows:

1. Make sure the engine is cooled and pressure has been relieved; this can be determined by squeezing the upper radiator hose.
2. Place a drain pan underneath the radiator petcock and engine block drain plugs. These are screw-in plugs that are usually located in the cooling jacket just above the oil pan (see **Figure 9**). Do not remove at this time. They should not be confused with the "pressed in" core plugs.
3. Leave the radiator pressure cap in place and open the petcock (see **Figure 10**).
4. Leaving the pressure cap in place will drain the coolant from the reservoir.
5. When the coolant has been drawn from the reservoir, remove the radiator cap. The reservoir hose can be removed and any debris or deposits can be rinsed out at this time.
6. Remove block plugs to drain any remaining coolant from the block.
7. When the coolant flow has stopped, close the radiator petcock and replace the block plugs.
8. Open any air bleed valves that may be present on the engine (see **Figure 11**). Refer to the service manual for specific locations.
9. Fill the cooling system with a premixed 50-50 antifreeze and water solution. Observe the air bleed valves while filling. Once the coolant running out of the valve is free of air bubbles, close the valve. Continue until all of the valves have been closed.
10. Continue to fill the cooling system until water reaches the base of the radiator fill neck.
11. Fill the reservoir to the hot mark.
12. Replace the pressure cap and start the engine. Run the engine until the thermostat opens. This can be

Figure 9. Removal of the block plugs is the only way to drain coolant from the engine block.

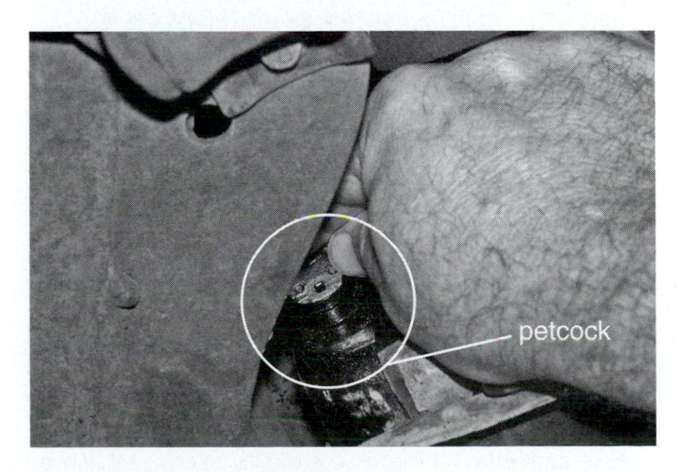

Figure 10. Opening the radiator petcock.

Figure 11. Opening the air bleed valves helps to avoid the formation of air pockets.

detected by feeling the upper radiator hose; when the thermostat has opened, the entire length of the hose will become hot. Be sure to monitor the coolant temperature to ensure that the engine does not overheat during this procedure.

> **You Should Know** *Because of the relative low mounting position of the radiator in many modern vehicles, the radiator filler may not be the highest point in the cooling system. This makes it very difficult to purge air pockets from the cooling system. It may be necessary in some instances to jack up the front of the vehicle to temporarily raise the filler neck to a point higher than the rest of the cooling system. Refer to the specific vehicle service manual for specific refill procedures.*

13. Once the thermostat has opened, turn the engine off and allow the pressure to decline. Remove the cap and check the coolant level. Adjust as needed. Replace the cap.
14. Test drive the vehicle for several miles to ensure that overheating does not occur and recheck the coolant level.

Manual Flush

If the coolant in the cooling system is in poor condition, or signs of scale and mineral buildup exist within the cooling system, it should be flushed before refilling with fresh coolant. Flushing can be performed with water alone; however, a flushing agent is often used to remove stubborn deposits. When using a flushing agent, follow the manufacturer's recommendations to avoid cooling system damage. The following is a list of steps required to flush the cooling system.

1. Make sure the engine is cooled and pressure has been relieved.
2. Place a drain pan underneath the radiator petcock and engine block drain plugs.
3. Leave the radiator pressure cap in place and open the petcock. Leaving the pressure cap in place will drain the coolant from the reservoir.
4. When the coolant has been drawn from the reservoir, remove the radiator cap.
5. Remove the block plugs.
6. When the coolant flow has stopped, close the radiator petcock and replace the block plugs.
7. Open any air bleed valves that may be present on the engine (refer to Figure 11). Refer to the service manual for specific locations.
8. Completely fill the engine with clean water. If a flushing agent is used, it should be added with the water (see **Figure 12**). Refer to the manufacturer's instructions.
9. Observe the air bleed valves while filling. Once the coolant running out of the valve is free of air bubbles, close the valve. Continue until all of the valves have been closed.
10. Set the heater to its highest setting and start the engine and monitor the engine temperature and thermostat opening. When the thermostat has opened, run the engine for 5 minutes. If using a flushing agent, refer to the manufacturer's instructions for specific run times.

Figure 12. There are many radiator flush products on the market.

11. Repeat steps 1–7.
12. If no flushing agent was used, the system is ready to be refilled with fresh coolant. Continue to step 12. If a cooling system flush was used, it may be necessary to repeat steps 1–11 to remove any residue that might reside in the system. This information can be found in the manufacturer's instructions.
13. Fill the cooling system with a premixed 50-50 antifreeze and water solution. Observe the air bleed valves while filling. Once the coolant running out of the valve is free of air bubbles, close the valve. Continue until all of the valves have been closed.
14. Continue to fill the cooling system until water reaches the base of the radiator fill neck.
15. Fill the reservoir to the hot mark.
16. Replace the pressure cap and start the engine. Run the engine until the thermostat opens. This can be detected by feeling the upper radiator hose; when the thermostat has opened, the entire length of the hose will become hot. Be sure to monitor the coolant temperature to ensure that the engine does not overheat.
17. Once the thermostat has opened, turn the engine off and allow the pressure to decline and remove the cap and check the coolant level. Adjust as needed. Replace the cap.
18. Test drive the vehicle for several miles to ensure that overheating does not occur and recheck the coolant level.

Automated Flush Systems

Many automated flushing systems are available on the market today and are widely used in automotive service facilities (see **Figure 13**). Each specific automated system has specific functions and operating procedures; for that reason, each unit cannot be covered in the scope of this text. The most common features found in the marketplace today are listed here. Some equipment may perform one or several of the following functions:

- Coolant filtering and recycling
- Back-flush the cooling system
- Automated coolant replacement

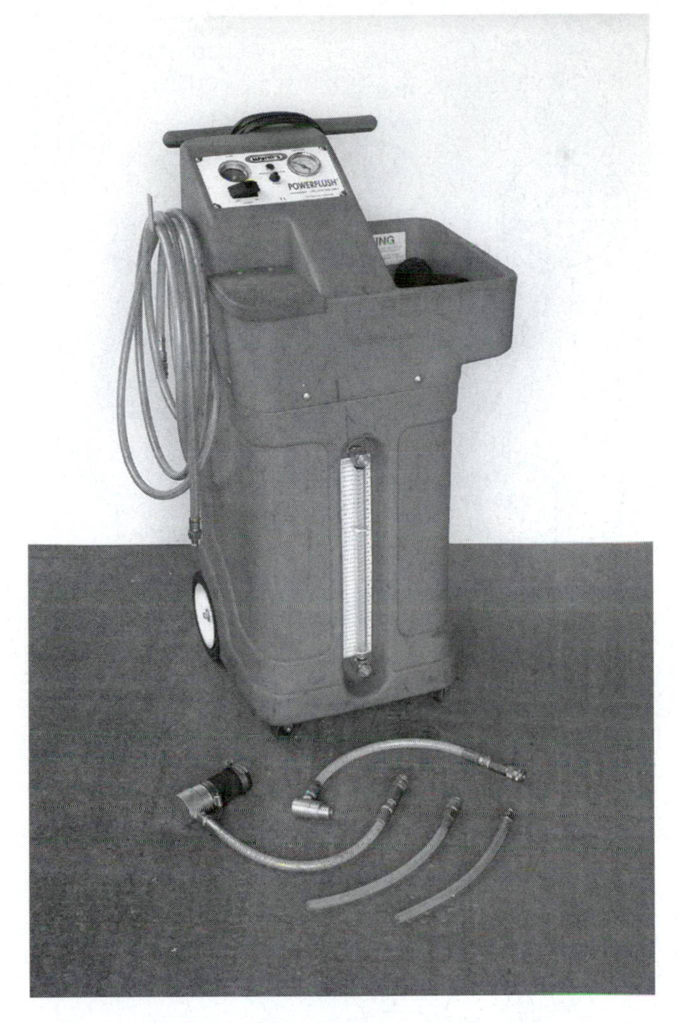

Figure 13. Commercial flushing equipment may be used in some high-volume facilities.

ENGINE ACCESSORY DRIVE SYSTEMS

Service of engine accessory systems is another job that is often performed by the entry-level technician. Service of the accessory systems can range from routine tasks such as the inspection and replacement of the accessory drive belt system to the diagnosis of unusual noises and vibrations. The accessory drive belts are used to drive the engine accessories such as the alternator, A/C compressor, and, additionally, the water pump and the cooling fan. There are two styles of accessory drive belt systems. The oldest style is a multiple v-belt system that uses one or more single ribbed v-shaped belts that are driven off a crankshaft pulley and drive the various accessories (see **Figure 14**). The belts in this configuration require periodic adjustment as the belts wear and become stretched. The serpentine belt system uses a single multiribbed belt that is driven by the crankshaft and extends to all of the accessories (see **Figure 15**). The serpentine belt system uses an automatic tensioning system that keeps constant tension applied to the belt. This system requires no periodic adjustment. The serpentine belt is used on most all automobiles built in the last 15 years. The first step in the diagnosis of the accessory drive system is the inspection of the belts. The condition of the belt can give some insight into the condition of the accessory drive system. The accessory drive belt(s) and their associated pulleys should be inspected any time that the hood is open. The belts should be inspected for proper alignment, cracking, glazing, and correct tension. Small cracks in the belts' surface will not impair performance and are not a cause for replacement. However, belts should be replaced if large cracks are present or chunks of rib are missing from the drive surface (see **Figure 16**).

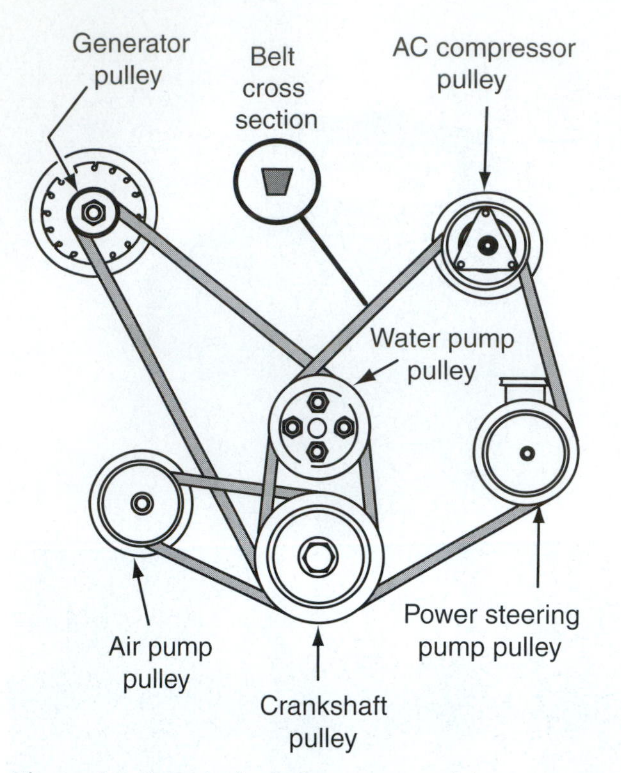

Figure 14. A typical v-belt accessory drive system.

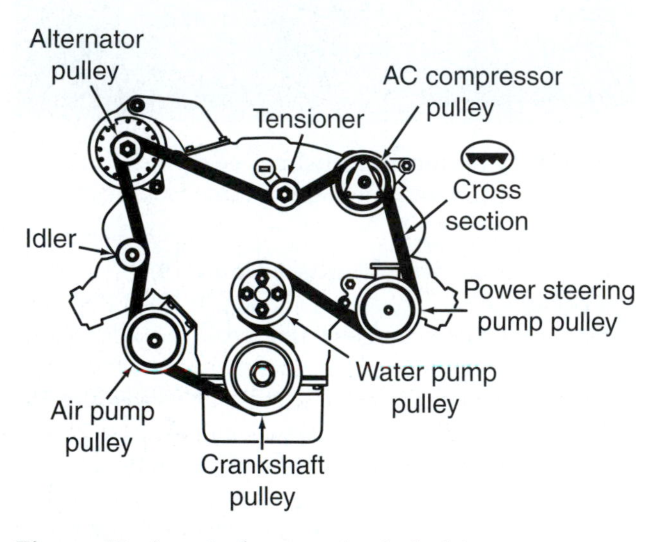

Figure 15. A typical serpentine belt drive system.

A serpentine belt system uses one or more idler and tensioner pulleys. Although most pulleys are attached to and drive an accessory component, idler and tensioner pulleys are only used to guide and position the belt. These pulleys are bolted directly to the accessory brackets and rotate on a sealed bearing. These pulleys are often made of plastic and are prone to severe wear and bearing failure. The pulleys should be inspected for severe wear, cracking, and proper alignment (see **Figure 17**). When any of these conditions occur, the pulley should be replaced.

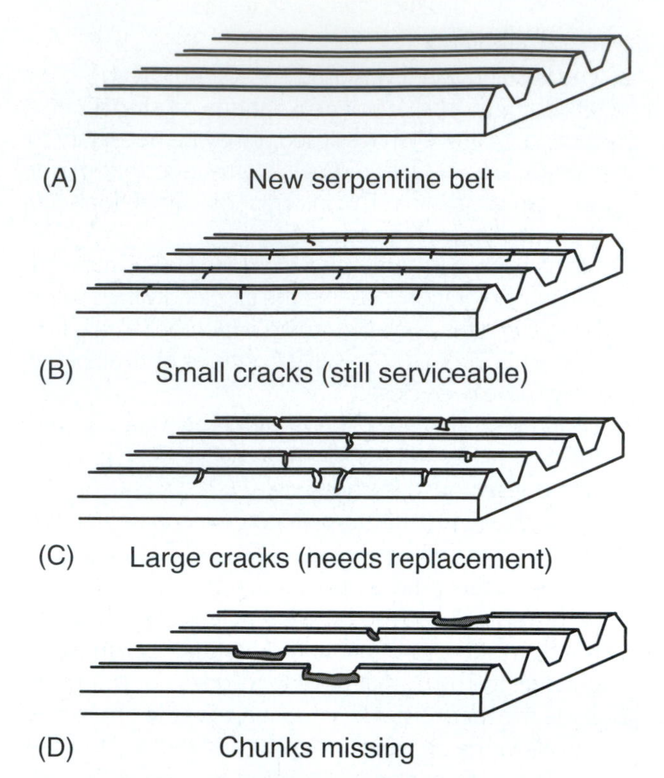

Figure 16. A new serpentine belt (A). Small cracks will not affect the operation of the belt (B). Belts should be replaced when large cracks begin to appear (C). The belt should be immediately replaced if chunks are missing from the belt (D).

Serpentine Belt Replacement

1. Identify the belt routing diagram (see **Figure 18**). This is usually a decal located under the hood. If a diagram is not present, draw your own before the belt is removed. This will help in reassembly.
2. Locate the tensioner. Release the tension following the manufacturer's specific instructions.
3. Remove the belt.
4. Inspect pulleys for nicks, cracks, corrosion, and other damage. Service as required.
5. Install the new belt in reverse order.

You Should Know *When installing the serpentine belt, the belt must be routed correctly. If not, the water pump may rotate in the wrong direction, causing the engine to overheat.*

Figure 17. Idler and tensioner pulleys should be closely inspected for excessive wear.

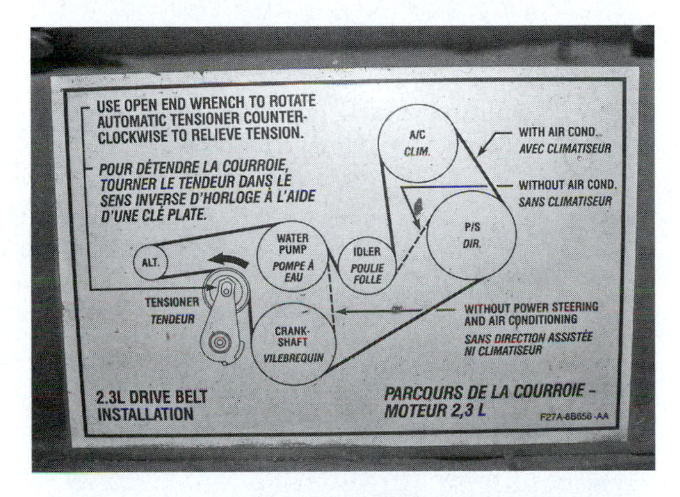

Figure 18. The belt routing diagram can often be found under the hood of the vehicle.

SERVICING COOLING SYSTEM HOSES

Engine cooling system hoses should be replaced every few years. This replacement should become part of a good preventive maintenance program. If done on a periodic schedule, more expensive repairs, such as those caused by an overheating engine, are not as likely to occur. Carefully check all cooling system hoses when a vehicle is being serviced. The following is a simple checklist for this service:

1. Check for leaks, usually noted by a white, green, or rust color at the point of the leak.
2. Check for swelling, usually obvious when the engine is at operating pressure.
3. Check for chafing, usually caused by a belt or other nearby component.
4. Check for a soft or spongy hose that would indicate chemical deterioration.
5. Check for a brittle hose indicating repeated overheating.
6. Squeeze the hose (see **Figure 19**). If its outer layer splits or flakes away, replace the hose.
7. Squeeze the lower radiator hose. If the reinforcing wire is missing (due to rust or corrosion), replace the hose.

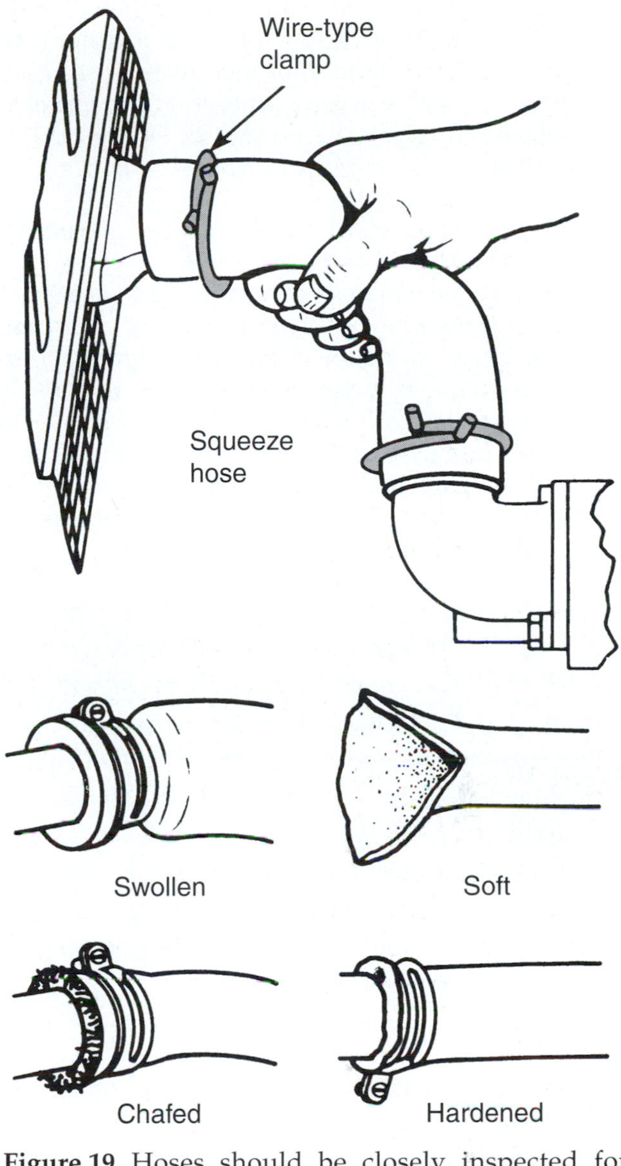

Figure 19. Hoses should be closely inspected for defects.

You Should Know *When performing cooling system service, be certain to check the condition of by-pass and other auxiliary hoses.*

Replacement

It is good practice to replace all the radiator and heater hoses if any of them are found to be defective. Generally speaking, if the hoses are the same age, they have all been exposed to the same conditions and are most likely in the same relative condition.

1. Drain the cooling system as required.
2. Loosen the hose clamp at both ends of the hose.

3. Firmly but carefully twist and turn the hose to break it loose from the coolant pump or radiator (see **Figure 20**). If the hose is not going to be reused, carefully split the hose where it connects to the radiator or engine and peel the hose away (see **Figure 21**). This will help alleviate damage to vehicle components.
4. Remove the hose.
5. Inspect the hose connectors for corrosion and pitting. Clean or replace as needed.
6. Compare the replacement hose to the original hose. The new hose should be similar in length and shape although it may not be identical (see **Figure 22**). On many occasions, it may be necessary to trim a new hose to fit the application.
7. Inspect clamps to verify their condition. Constant tension clamps should be inspected for proper tension and shape. If the clamp appears to be damaged, it should be replaced.

You Should Know *Although it is common practice to replace constant tension clamps with adjustable clamps, the constant tension clamps should be used whenever possible. Although they are significantly more difficult to both remove and install, the constant tension clamp alleviates the need to periodically tighten hose clamps and virtually eliminates damage to heater cores and radiators that routinely occurs from overtightening worm drive clamps.*

8. Loosely install the clamps over the ends of the hoses.
9. Reinstall hoses in the reverse order of removal. Inspect to make sure that the hose is free of kinks and twists, and also ensure that it does not come into contact with any components that could cause damage.

Figure 20. Carefully remove the hose with a gentle twisting motion.

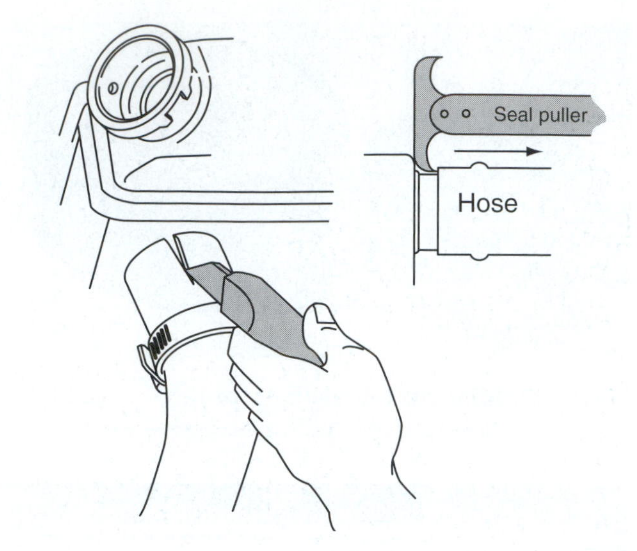

Figure 21. Splitting stuck hoses can ease removal and prevent damage to the radiator.

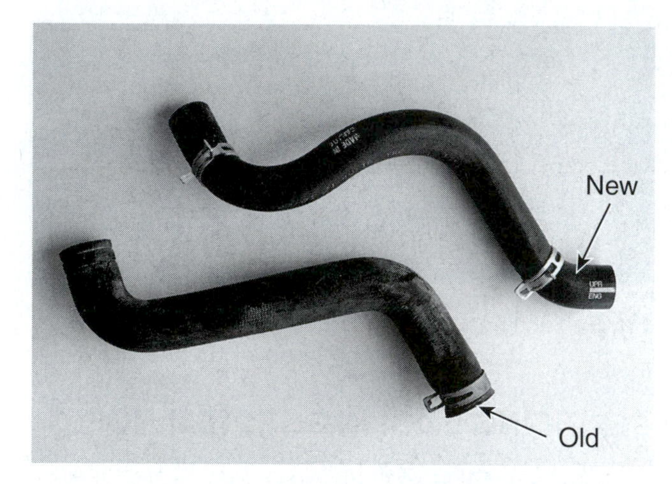

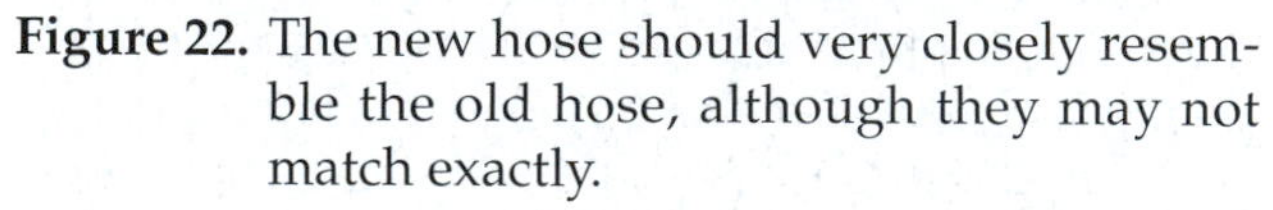

Figure 22. The new hose should very closely resemble the old hose, although they may not match exactly.

10. Slide the clamps just past the barbs found on the hose connectors and tighten as needed.
11. Refill the cooling system to the proper level.

You Should Know *Many hoses found on a modern vehicle are formed to a specific shape. This is done to avoid various obstacles that can be found on most all vehicles. It is imperative that formed hoses are replaced with formed hoses of the correct shape. Failure to do so will result in the hoses being crimped and will restrict the flow of coolant.*

FANS, SHROUDS, AND AIR DEFLECTORS

The placement and operation of the fans, shrouds, and air deflectors is critical to the operation of the cooling system. A vehicle may have several fan shrouds and deflectors in a variety of locations. These should be inspected for proper mounting and general overall condition. If any problems are noted with these components, they should be repaired as needed. The following sections will outline diagnosis of clutch type and electric fans.

Belt-Driven Fans

Coolant pump-mounted fans occasionally require service. They can be damaged as a result of metal fatigue, collision, road hazards, and abuse. It is important that fans be in good shape. Any condition that causes an out-of-balance pump-mounted fan will result in early coolant pump-bearing failure. A check for fan problems is a rather simple task.

1. Remove the belt(s).
2. Visually inspect the fan for cracks, breaks, loose blades, or other damage (see **Figure 23**). Is the fan sound? If yes, proceed with step 3. If no, replace the fan.
3. Hold a straightedge across the front of the fan. Are all blades in equal alignment? If yes, proceed with step 4. If no, replace the fan.
4. Slowly turn the fan while looking for any out-of-true condition or any other damage.
5. Turn the fan fast and look for out-of-true conditions.
6. If the fan fails either test (step 4 or 5), it must be replaced. Because of high operating speeds, it is not recommended that repairs be attempted on an engine cooling fan blade.

Figure 23. The fan blade should be inspected for obvious signs of damage.

Fan Clutch

Most vehicles with a coolant pump-mounted fan use a fan clutch to regulate speed. There are several basic tests for troubleshooting a fan clutch. First, make certain that the engine is cold and that it cannot be accidentally started while inspecting the fan clutch.

1. Visually inspect the clutch for signs of fluid loss. If fluid loss is apparent, the clutch should be replaced.
2. Check the condition of the fan blades.
3. Check for a slight resistance or roughness when turning the fan blades. Gently rock the blade back and forth (see **Figure 24**). The clutch should have only a slight amount of fore and aft movement. If the clutch has very little or no resistance, feels rough when turned, or has an excessive amount of movement, the clutch should be replaced.
4. Spin the fan. If it rotates more than twice on its own, replace the clutch.
5. Check for looseness in the shaft bearing.
6. Install a timing light and tachometer to the engine and place the thermometer between the radiator fins and fan clutch. Start the engine and turn on the air conditioner.
7. Bring the engine speed up to 2,000 rpm, and observe movement of the fan using the timing light. It should be rotating slowly. Place a piece of cardboard in front of the radiator to speed up the warm-up time. Note the temperature at which the fan begins to rotate faster, generally 150–195 degrees F (65–90 degrees C). Compare this temperature to the manufacturer's specifications for engagement temperature.
8. If the fan clutch fails any of these tests, it should be replaced. There are no repairs for a faulty fan clutch.

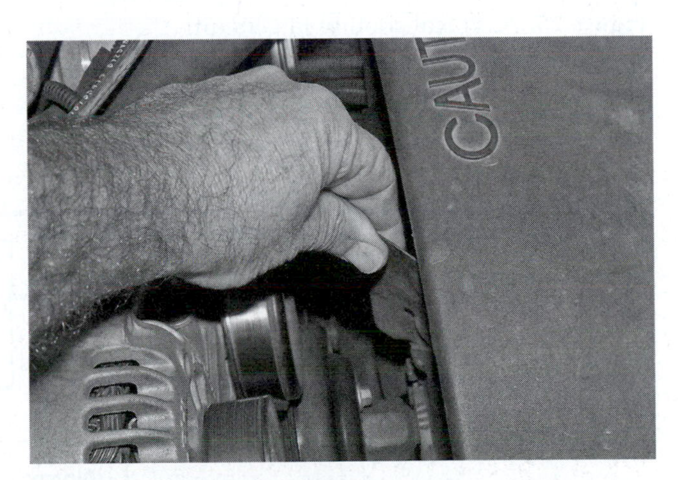

Figure 24. Rock the fan and clutch to test for excess movement. Spinning the fan will test the bearing operation.

Electric Fans

Electric cooling fans are used because they can be more precisely controlled. They may be turned on and off by temperature- and pressure-actuated switches, thereby regulating engine coolant and air conditioning refrigerant temperatures more precisely. Follow the schematic in **Figure 25** for testing and troubleshooting a typical engine cooling fan system.

You Should Know *Electric engine cooling fans may start and operate at any time and without warning. This may occur with the ignition switch off or on.*

1. Start the engine and bring the coolant up to operating temperature.
2. Turn on the air conditioner.
3. Disconnect the cooling fan motor electrical lead connector (see **Figure 26**).
4. Make sure that the ground wire is not disturbed. If the ground wire is a part of the electrical connector, establish a ground connection with a jumper wire.
5. Connect a test lamp from ground to the hot wire of the connector (see **Figure 27**). Make sure the lamp is good before performing the test.
6. Did the lamp light? If yes, proceed with step 7. If no, check for a defect in the fan relay control circuit. It is

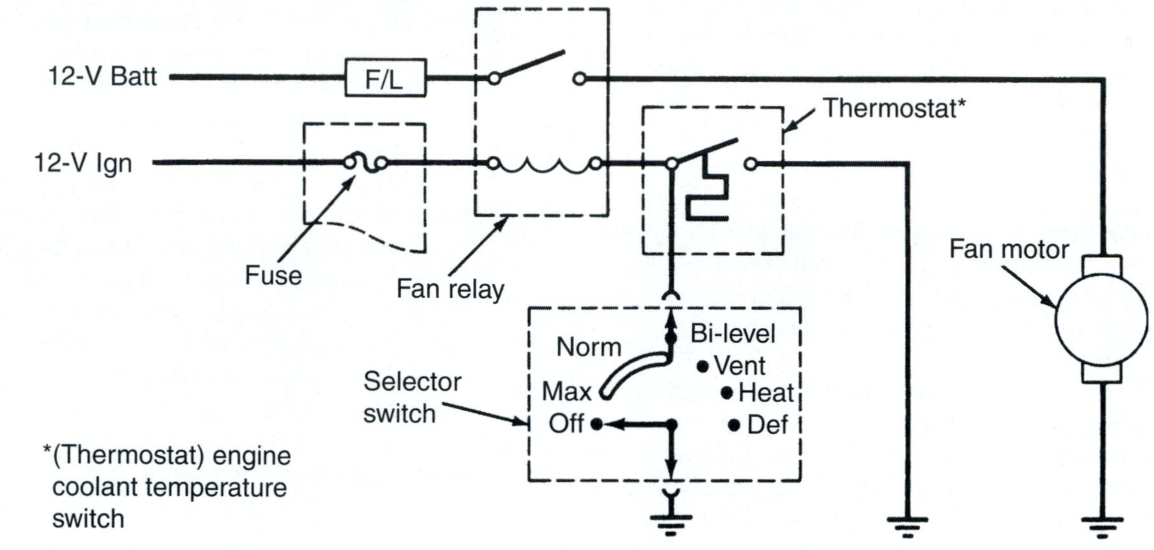

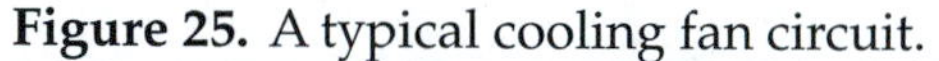

Figure 25. A typical cooling fan circuit.

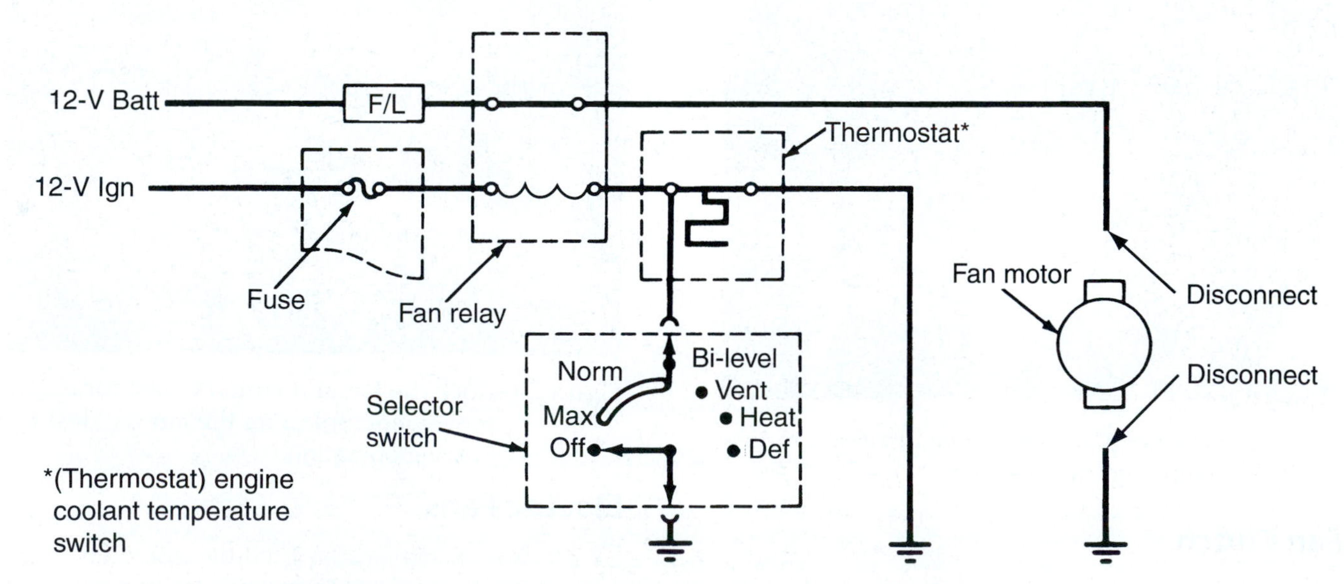

Figure 26. Disconnect the fan motor.

also possible that the engine is not up to sufficient temperature to initiate cooling fan action.

7. Connect a fused jumper wire from the battery positive (+) terminal to the cooling fan connector (see **Figure 28**). Make sure the fuse is good.
8. Did the fan start or run? If yes, the fan is all right. Check for poor connections at the fan motor and repair them if necessary. If no, proceed to step 9.
9. Replace the cooling fan motor.

LOCATING COOLING SYSTEM LEAKS

The most common cooling system problems are a result of a leaking system. Leaks are generally easy to find using a pressure tester. Several types are available. The following is a typical procedure for pressure testing a cooling system:

1. Allow the engine and coolant to cool to ambient temperature.

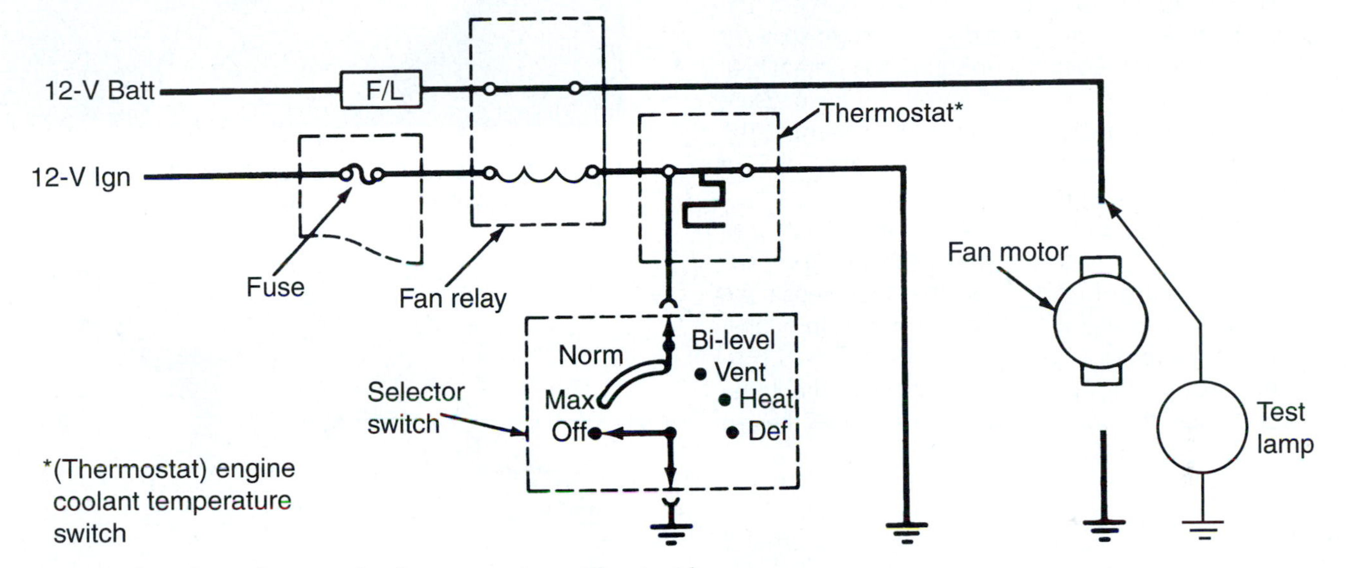

Figure 27. Test for voltage at the fan connector with a test lamp.

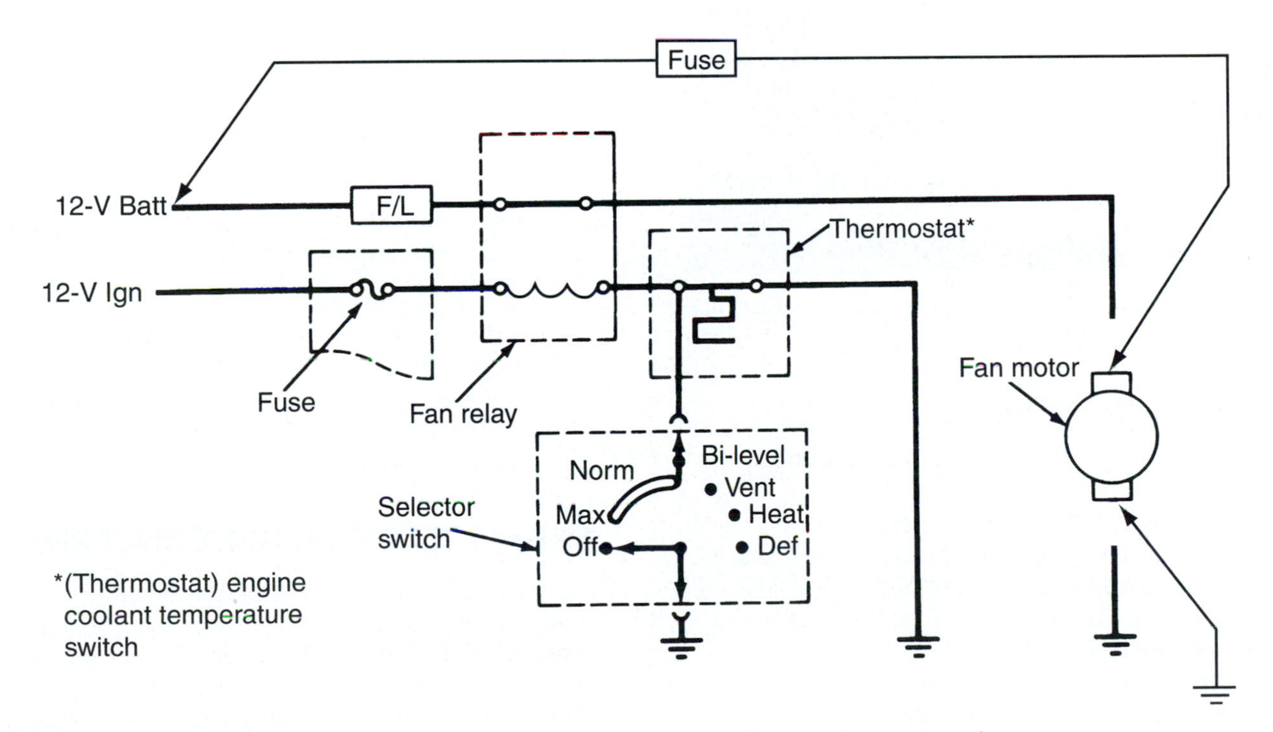

Figure 28. The fan motor can be tested by connecting a power and ground source to the motor.

2. Remove the pressure cap. Note the pressure range indicated on the cap (see **Figure 29**).
3. Adjust the coolant level to a point just below the bottom of the fill neck of the radiator.
4. Attach the pressure tester (see **Figure 30**).
5. While observing the gauge, pump the tester until a pressure equal to the cap rating is achieved. If the pressure can be achieved, proceed with step 6. If the pressure cannot be achieved, make a visual inspection for leaks.
6. Let the system stand for 5 minutes. Recheck the gauge. If pressure drops rapidly, proceed to step 7. If the pressure is the same as in step 5, the system is all right, and it is safe to proceed to step 8. If the pressure has dropped slightly, repressurize the system to the top end of the pressure rating and proceed to step 7. If no leak is detected, proceed to step 8.
7. Locate the source of the leak by visibly inspecting all connections. Also inspect the passenger compartment for signs of coolant on the floor. Repair the source of the leak, and then repeat steps 4 through 6 to verify the repair.
8. Check the radiator pressure cap. The cap should be able to hold the pressure noted on the cap. If the cap fails the test, replace the cap.

Figure 29. The pressure cap rating is usually imprinted on the cap.

Pressure Cap Testing

Pressure caps may be tested using a cooling system pressure tester and an adapter. The requirement is that a pressure cap must not release at a pressure below what it is rated and that it must open at a pressure above what it is rated. A pressure cap is usually designed to operate in the 14–17 psi (97–117 kPa) pressure range. To pressure test a radiator cap, proceed as follows:

1. Attach the adapter to the pressure tester.
2. Install the pressure cap to be tested.
3. Pump the pressure tester to the value marked on the pressure cap (see **Figure 31**).

Figure 30. Attach the pressure tester to the radiator neck.

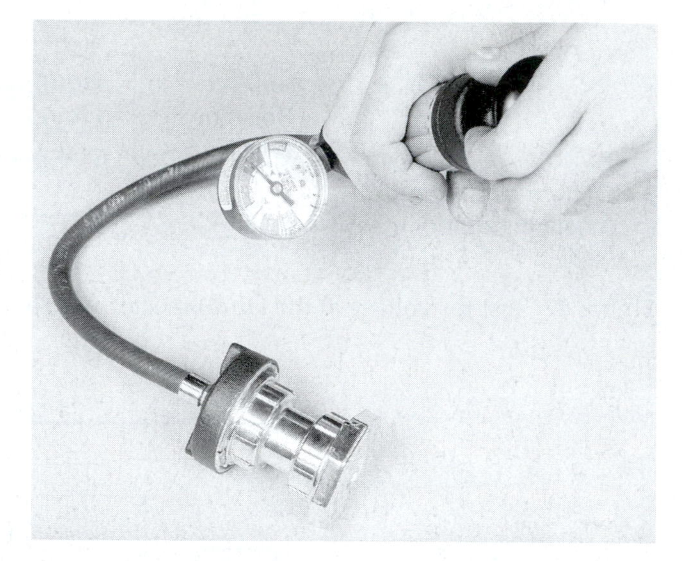

Figure 31. Pump the tester until the pressure rating of the cap is reached.

4. Did the cap hold pressure? If yes, proceed with step 5. If no, replace the cap.
5. Pump to exceed the pressure rating of the cap.
6. Did the cap release pressure? If yes, the cap is good. If no, replace the cap.

ENGINE TEMPERATURE CONCERNS

There are two basic engine temperature concerns faced by the automotive technician: overheating and overcooling. Overheating is a condition that can cause serious damaged if left undetected, whereas overcooling will affect heater system performance and can lead to drivability concerns, increased emissions, and decreased fuel economy.

Overheating

The first step in the diagnosis of an overheating concern is to determine when the problem occurs. There are three basic patterns of engine overheating:

- The engine overheats at all times: this condition is typically related to low coolant level, a thermostat that is stuck in the closed position, or a loss of pressure.
- The engine overheats only when the vehicle is sitting still: this concern is typically related to an airflow problem. This could result from an inoperative cooling fan or radiator restriction.
- The engine overheats after it has been driven for a while: this condition is typically related to a flow restriction within the radiator or a loose or missing air dam.

The following are some basic steps for the basic diagnosis of an engine overheating concern:

1. Check and adjust the coolant level. If the coolant is low, inspect the entire system for leakage.
2. Inspect the coolant pump drive belt for proper tension.
3. Inspect the radiator for external airflow restrictions.
4. Inspect the radiator for internal flow restrictions.
5. Test the radiator pressure cap to make sure that it is operating properly.
6. Check the thermostat for proper operation.
7. If these steps fail to isolate a probable cause, a possible internal engine problem may exist.

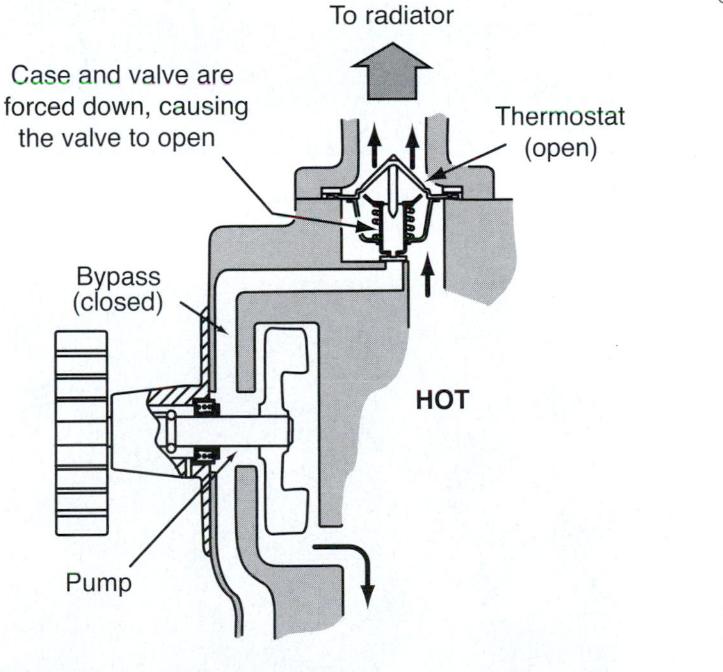

Figure 32. A thermostat stuck in the open position will cause an overcooling condition.

Overcooling

Engine overcooling typically only exists when the thermostat is stuck in the open position (see **Figure 32**), or the thermostat is of the wrong opening temperature. This condition allows coolant to constantly circulate through the system, removing any heat that may have been built up within the system. This will lead to engine temperatures that are significantly below normal.

RADIATOR DIAGNOSIS

Radiator problems are typically recognized as leaking, internally clogged, or externally clogged. A leaking radiator is among the simplest concerns to diagnose, by pressure testing the cooling system. When a leak is located in the radiator, it should be removed and sent out to a specialty shop for repair or replaced.

External clogging can be caused by external visible debris (see **Figure 33**). This debris should be carefully cleaned off without forcing the debris farther into the tubes or damaging the fins. Often, debris will actually collect between the condenser and the radiator. Many times, this type of restriction cannot be seen unless the condenser and radiator are separated. When this occurs, both heat exchangers should be properly cleaned.

Internal restrictions are slightly more difficult to detect. A good place to start is by observing the amount of mineral deposits that are located on the ends of the radiator tubes (see **Figure 34**). A large amount of mineral buildup is an indication that the radiator should be disassembled and cleaned. The simplest method of pinpoint radiator restriction

Figure 33. Debris can seriously restrict airflow and cause an overheating condition.

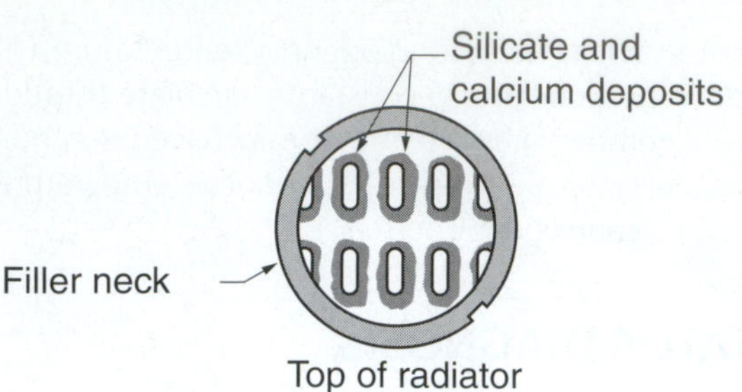

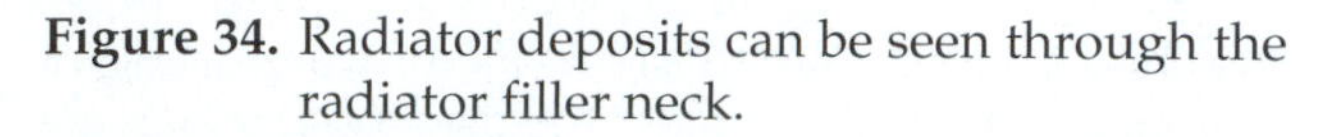
Figure 34. Radiator deposits can be seen through the radiator filler neck.

is by "scanning" the radiator surface with an infrared surface thermometer. With the engine at operating temperature, the surface of the radiator should be warmest near the radiator inlet and gradually cool toward the outlet. If you detect areas of the radiator that are significantly cooler than the inlet, there is a strong probability that a restriction exists.

All major services require the radiator to be removed for service. The following steps outline the basic steps for radiator removal.

1. Drain the coolant from the radiator. If the vehicle is equipped with a belt-driven fan, proceed to step 2. If the vehicle is equipped with an electric fan, proceed to step 3.
2. If the fan is equipped with a shroud, remove the attachments and slide the shroud toward the engine. Proceed to step 3.
3. Remove the electrical connector from the fan motor. Remove the fasteners from the fan assembly brackets and lift the fan assembly out. Care must be taken not to damage the radiator cooling fins.
4. Carefully remove the upper and lower coolant hoses from the radiator.
5. Remove the transmission cooler lines (if the vehicle has an automatic transmission) from the radiator. Plug the lines to prevent transmission fluid loss.
6. Remove the radiator attaching bolts and brackets.
7. Carefully lift out the radiator.
8. Reverse steps 1–6 for reinstallation.
9. Refill the radiator with the proper 50-50 mixture of coolant.

THERMOSTAT SERVICE AND DIAGNOSIS

The thermostat is often associated with both engine overheating as well as overcooling. The following steps outline the removal and testing of a typical thermostat.

A typical procedure for removing a thermostat follows.

Thermostat Removal

1. Reduce the engine coolant to a level below the thermostat.
2. Remove the bolts holding the thermostat housing onto the engine (see **Figure 35**). It is not necessary to remove the radiator hose from the housing.
3. Lift off the thermostat housing. Observe the pellet-side down position of the thermostat to ensure proper replacement. Do not install the thermostat backward.
4. Lift out the thermostat (see **Figure 36**).
5. With the thermostat cooled down to room temperature, make sure the thermostat is fully closed. If the thermostat is not fully closed at room temperature, it should be replaced with a new one.

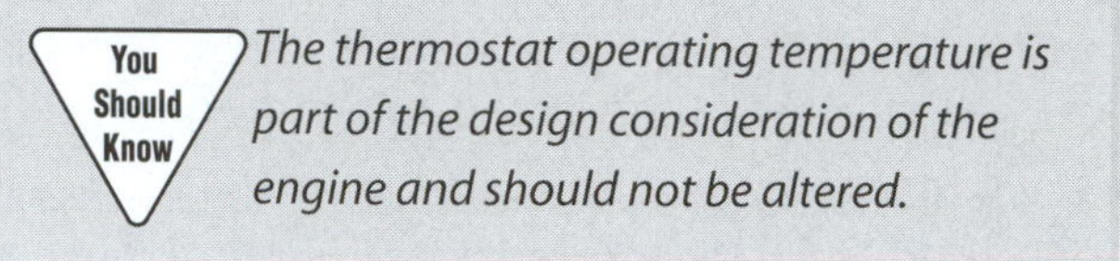

The thermostat operating temperature is part of the design consideration of the engine and should not be altered.

Figure 35. Removal of the thermostat housing bolts.

Figure 36. Note the position in which the thermostat is installed before removal. For proper operation the thermostat must be installed in the correct position.

6. Clean all the old gasket material from the thermostat housing and engine-mating surface. Many thermostats today use O-ring seals, whereas others may still use a gasket. Be sure the replacement thermostat physically matches the one being removed.
7. Install the new thermostat in the same position as the one that was removed, and torque it to manufacturer's specifications.
8. Refill with coolant as needed.

Thermostat Testing

There are several ways to test a thermostat. However, many technicians believe that if a thermostat is suspect, it should be replaced. The labor cost for the time required to test a thermostat often outweighs the cost of a thermostat. Although this is a good practice, one also should know how to test a thermostat. Following is a procedure for testing a thermostat that has been removed from the engine.

1. Note the condition of the thermostat.
2. Is the thermostat corroded or open? If no, proceed with step 3. If yes, replace the thermostat.
3. Note the temperature range of the thermostat.
4. Suspend the thermostat in a heatproof glass container filled with water (H_2O) (see **Figure 37**).
5. Suspend a thermometer in the container. Neither the thermostat nor the thermometer should touch the container or touch each other.
6. Place the container and contents on a stove burner and turn on the burner.
7. Observe the thermometer. The thermostat valve should begin to open at its rated temperature value. If it begins to open more than ±3 degrees F of its rated value, replace the thermostat.
8. Is the thermostat fully opened at approximately 25 degrees F (11 degrees C) above its rated value? If yes, the thermostat is all right. If not, replace the thermostat.

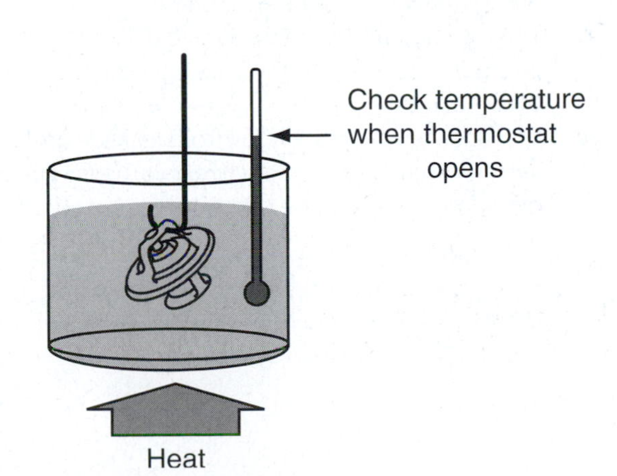

Figure 37. Testing thermostat operation.

There is generally no service interval stated for the replacement of the cooling system thermostat. But, as a preventative maintenance item, thermostat replacement should be suggested when the engine coolant is changed.

COOLANT PUMP DIAGNOSIS

Coolant pump failure will traditionally fall into two categories: noise and leakage. Water pump leakage typically starts at the shaft seal located at the front of the water pump. Leaks from this area are typically traced to coolant leaking from a hole that is located underneath the shaft on most water pumps (see **Figure 38**). When a coolant leak is indicated at this point, the water pump should be replaced. These types of leaks can be tricky to locate. In the best scenario, the leak can be located by pressure testing. However, many times these seals will only leak when they are cold with a small amount of pressure applied to them. As the engine cools down, the seal material shrinks slightly. The pressure that remains on the cooling system pushes coolant out from between the shaft and the seal, creating a leak. When the engine is started and heat is transferred to the seal area, the material expands and seals off the leak.

The difficulty comes when you try to diagnose the problem. The customer notices a leak on the floor and brings the car in for service. However, as the vehicle is being driven, the seal expands to seal off the leak. Consequently, when you pressure test the system there is no leak present. The solution is to pressure test the system and allow the engine to cool off when the pressure is still applied. As the engine cools, the seal retracts and will begin to leak.

Noise concerns that are water pump related emanate from the front bearing assembly. The tricky part about this diagnosis is that accessory drive noises can be easily transferred and may sound as if they are coming from a different location. The way to diagnosis this problem is to

Figure 38. The water pump will usually leak from the vent hole.

remove the belt and grasp the coolant pump pulley at the top and bottom and rock the pulley to and fro. There should be very little, if any, detectable movement. Next, slowly turn the pulley by hand. The pulley should turn smoothly. If there is the slightest hint of binding, grabbing, or noise, the pump should be replaced. The following steps outline coolant pump replacement:

1. Drain the cooling system as needed.
2. Loosen and remove all belts. If there is an engine-mounted fan, proceed with step 3.
3. If there is an electric fan, proceed with step 4.
4. Remove the fan and fan/clutch assembly.
5. Remove the coolant pump pulley.
6. Remove accessories as necessary to gain access to the water pump bolts.
7. Remove the lower radiator hose from the water pump.
8. Remove the bypass hose, if equipped.
9. Remove the bolts securing the water pump to the engine.
10. Tap the water pump lightly, if necessary, to remove it from the engine.
11. Clean the old gasket material from all surfaces. Take care not to scratch the mating surfaces.
12. Clean the mounting bolts of any rust, scale, or debris.
13. Install new gaskets and seals and coat bolts with thread sealant (liquid Teflon).
14. Reverse the removal steps to install a new water pump assembly and torque fasteners to manufacturer's specifications.
15. Fill the coolant system as needed with clean 50-50 mixture of the proper coolant.

Summary

- The best coolant replacement solution is to use exactly what is recommended by the manufacturer. A properly stocked shop should have an IAT, Dex-Cool replacement, OAT, and HOAT.
- A 50-50 mixture provides the best balance of freeze protection, boilover protection, and heat transfer. Distilled water is recommended for coolant mixtures.
- The cooling system should be inspected each time the vehicle is presented for service. This should include inspection of the coolant level, coolant condition, accessory belt condition, hose condition, and brief inspection of the cooling fan and associated shrouds and air dams.
- Coolant should be inspected for overall condition and freeze protection. Coolant protection can be measured using a hydrometer, a refractometer, or reactive test strips. Testing the pH level gives the technician an indication of the overall coolant condition. If coolant becomes acidic, electrolysis can develop.
- Coolant should be replaced when any one of these conditions exists: manufacturer recommended service interval, inadequate freeze protection, incorrect pH levels, or when obvious contamination has occurred. Typical procedures include drain and refill, manual flush, and automated flush.
- Accessory drive belt and pulleys should be inspected each time the vehicle is serviced. The serpentine belt should be replaced when large cracks appear or large chunks are missing from the belt ribs.
- Proper hose maintenance can help avoid more costly repairs. To ease the process of hose removal, the hose can be split with a knife. Constant tension clamps should be used whenever possible.
- Leaks can be located by pressure testing the cooling system. Some leaks may not appear until the cooling system has cooled down.
- Two common temperature-related concerns are overheating and overcooling. Overheating can be caused by low coolant, defective thermostat, inoperative fans, or a clogged radiator. Overcooling is typically caused by a stuck-open thermostat.
- The radiator should be tested for leaks, external restrictions, and internal restrictions. External restrictions can be visually identified. Internal restrictions can be identified with an infrared thermometer.
- The thermostat should be replaced if a problem is suspected. A stuck-open thermostat is a typical cause of overcooling.
- Coolant pumps typically fail from leakage and bearing failure. It may be necessary to allow the engine to cool off before a leak can be properly diagnosed.

Review Questions

1. Technician A says that proper coolant concentration should contain 50 percent antifreeze and 50 percent distilled water. Technician B says that the use of tap water can cause the formation of mineral deposits on the surfaces of cooling system components. Who is correct?
   A. Technician A
   B. Technician B
   C. Both Technician A and Technician B
   D. Neither Technician A nor Technician B
2. List the basic steps that a basic cooling system inspection should include.
3. List three things that engine coolant should be inspected for.
4. Technician A says that as long as the freeze protection is okay, there is no need to replace the coolant. Technician B says that over a period of time coolant can become acidic and cause cooling system damage. Who is correct?
   A. Technician A
   B. Technician B
   C. Both Technician A and Technician B
   D. Neither Technician A nor Technician B
5. Explain when a serpentine belt drive system should be replaced.
6. Technician A says that white streaks located at a hose connection indicate a coolant leak. Technician B says that coolant hoses should be replaced as they fail. Who is correct?
   A. Technician A
   B. Technician B
   C. Both Technician A and Technician B
   D. Neither Technician A nor Technician B
7. Technician A says that new clamps should always be installed when replacing cooling system hoses. Technician B says that constant tension clamps alleviate the need for periodic tightening and can help prevent outlet damage. Who is correct?
   A. Technician A
   B. Technician B
   C. Both Technician A and Technician B
   D. Neither Technician A nor Technician B
8. Technician A says that engine overheating can be caused by a thermostat that is partially stuck in open position. Technician B says that engine overheating can be caused by a missing air deflector. Who is correct?
   A. Technician A
   B. Technician B
   C. Both Technician A and Technician B
   D. Neither Technician A nor Technician B
9. Technician A says that accessory noises can be easily transferred to other components. Technician B says that water pump noises can be isolated by rotating the pulley by hand. Who is correct?
   A. Technician A
   B. Technician B
   C. Both Technician A and Technician B
   D. Neither Technician A nor Technician B
10. Technician A says that when pressure testing a cooling system, the pressure applied should not exceed the rating of the pressure cap. Technician B says that most pressure caps are rated at 14–17psi (97–117 kPa). Who is correct?
    A. Technician A
    B. Technician B
    C. Both Technician A and Technician B
    D. Neither Technician A nor Technician B

Chapter 12 Heater System Diagnosis and Service

Introduction

Proper operation of the heater system is vital to both consumer comfort and safety. This chapter will review heater system diagnostic and repair techniques.

POOR HEATING DIAGNOSIS

The amount of heat that is produced by the heater core is directly related to both the amount of coolant and the temperature of the coolant that is moving through the core. The most common concerns that cause a lack of heat condition are low coolant level, faulty thermostat, and heater core restriction. Other causes of poor heating may be related to the operation of the temperature control doors in the air distribution system. The operation and diagnosis of this system will be covered in Section 6 of this textbook.

Leaking Heater Core

Low coolant level is probably the most common cause of a poor heating condition. When coolant level drops below a given point, there is no longer enough coolant to circulate through the heater core. Because the heater core is also a component of the cooling system, any leaks that are present within the cooling system also will affect the operation of the heater. A cooling system inspection and pressure test should be among the first diagnostic procedures performed when a low coolant condition exists.

In some cases, however, the heater core can be the actual cause of the coolant leak. Heater core leaks are typically recognized by wet carpet, oily fog being blown against the windshield in the defrost mode, or pungent sweet or mildew type odors in the passenger compartment (see **Figure 1**). In some models, the air distribution case is designed in such a manner that when a heater

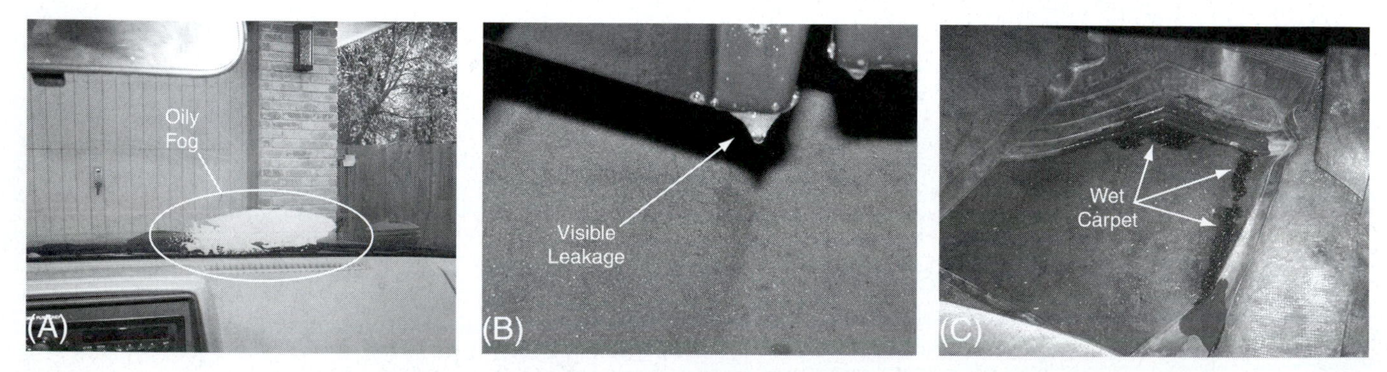

Figure 1. Any one of these concerns could indicate a leaking heater core: foggy windshield (A), coolant visibly leaking into the interior of the car (B), or wet carpet (C).

core leak occurs, the lost fluid will drain outside the vehicle through the condensation drain. However, in many cases, the coolant will overflow into the passenger compartment of the vehicle. Wet carpet is an indication of this. If the leak is small and slow, the coolant can drain in between the air distribution case and the firewall. The coolant will end up seeping under the carpet, which will cause the carpet to soak from the bottom toward the top. This process takes a long time and may go unnoticed until the carpet becomes saturated or odors begin to develop.

You Should Know *A wet floor carpet also can be caused by a leaking seal around the windshield.*

A vehicle with a leaking heater core will most likely arrive at the repair facility with fog, odors, or wet carpets. However, in some cases, a leaking heater core can appear as a slow undetectable coolant loss. When presented with this concern, begin with a cooling system pressure test. Be sure to look for key indicators by pulling back the passenger side carpet to see if the underside is wet and inspect the firewall near the air distribution case for signs of leakage and note any particularly bad odors (see **Figure 2**). If your investigation turns up any one of these indicators, the heater core should be visually inspected.

Figure 2. It may be necessary to check the underside of the carpet to verify the presence of a leaking heater core.

Restriction

Internal heater core restriction typically occurs when scale and debris collect in the heater core. This can restrict part or all of the coolant flow through the core. A quick analysis of heater core flow can be made by monitoring the temperature of both the inlet and outlet hoses. With the engine at operating temperature, measure the temperature of both heater inlet and outlet hoses and the upper radiator hose using an infrared thermometer (see **Figure 3**). With the blower in the off position, all of the hoses should be very close to the same temperature. Slight variations are normal because of the loss of heat within the heater core.

Figure 3. Heater core operation can be judged by comparing the temperature of both the inlet and outlet hoses.

- Inlet and outlet are close to the same temperature: This is an indication that coolant is properly flowing through the heater core.
- Inlet is hotter than the outlet: If the temperature between the inlet and the outlet hoses is significant, an internal heater core restriction is indicated.
- Both the inlet and outlet are cooler than the radiator hose: This may be an indication of an extremely low coolant level, a restriction in one of the hoses or connections, or an inoperative heater control valve.
- All three hoses are the same temperature but are cooler than normal: This is an indication that the coolant level is extremely low or that the thermostat is stuck in the open position.
- If a heater core restriction is suspected, the core can sometimes be cleaned by back-flushing. Disconnect both the inlet and outlet hose from the core. Using a water hose, force clean water into the outlet of the core and flush toward the inlet side of the core (see **Figure 4**). In some cases, this will dislodge any trapped debris. If the obstruction cannot be dislodged, the heater core must be replaced. In any case, when a heater core restriction occurs, the technician must be sure to find and correct the cause of the obstruction.

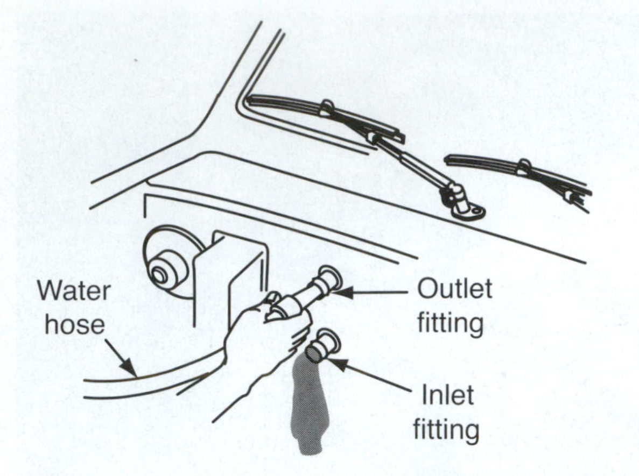

Figure 4. Some heater obstructions can be cleared by back-flushing the heater core.

This may require a complete flushing of the entire cooling system. Failure to do so will result in a reoccurrence of the restriction.

HEATER CONTROL VALVE

The heater control valve operates by restricting the flow of coolant to the heater core. There are three problems that you are likely to encounter with a heater control valve: leakage, stuck open, or stuck closed. Begin heater valve diagnostics by inspecting the control valve for external leaks. Heater valve leaks will usually be very slight in nature and will often "grow" significant amounts of calcium and mineral deposits on the exterior of the valve. On externally controlled valves, these deposits can be large enough to keep the valve from opening or closing. When inspecting the valve, look for mineral buildup on the exterior of the valve and obvious drips (see **Figure 5**). With the engine at operating temperature and the heater valve opened, both the valve inlet and outlet should be the same temperature. When closed, the inlet should remain hot, but the outlet should become significantly cooler.

The control mechanism of the vacuum valve can be tested using a vacuum pump. Simply disconnect the hose, connect an external vacuum pump, and apply a vacuum. The diaphragm should be capable of holding a vacuum and the valve should move from one position to the other (see **Figure 6**). If the diaphragm holds a vacuum but the valve position does not change, the valve should be replaced.

The solenoid valve can be tested by measuring resistance and supplying artificial power and ground connections. Simply measure the resistance between the two terminals of the solenoid. The resistance should not indicate an open circuit; refer to the service manual for specifications. To test the operation of the valve, use a fused jumper to momentarily supply an artificial power and ground connection to the solenoid. When this is done, the valve position should move from the open to the closed position.

Figure 5. A corroded heater control valve is a sure sign of a coolant leak. Excessive amounts of corrosion can keep the valve from operating.

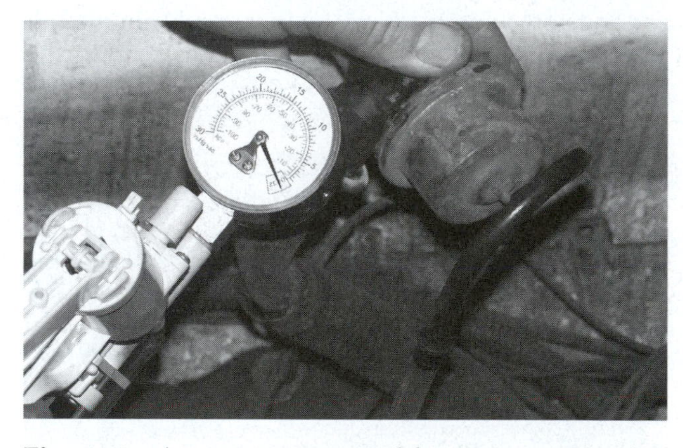

Figure 6. A vacuum actuated heater control valve should open and close as vacuum is applied and released.

Once the valve has been tested, continue by testing the control mechanisms to make sure that they are capable of operating. This mechanism can be cable, vacuum, or electrical. To test the cable, simply disconnect the cable from the valve and move the lever or knob through its full range of motion and note the action of the cable. For vacuum systems, connect a vacuum gauge in place of the valve and move the lever to the full heat position. The gauge should respond accordingly. Electrical valves are operated using an electrical solenoid. Because of the many different control variations, refer to the appropriate service manual for testing the control circuit of this valve.

STUCK-OPEN THERMOSTAT

When a thermostat sticks in the open position, coolant is allowed to flow through the engine and radiator at all times. As a result, the coolant moves through the cooling system very rapidly. The effect of this is that the coolant absorbs very little heat from the engine and the heat that is absorbed is dissipated by the radiator very easily. This results in relatively low temperature coolant being introduced into the heater core, causing low heater outlet temperatures. This problem also can be associated with low engine operating temperatures. Customers may also notice that engine or heater temperatures may be higher when the vehicle is sitting still, but tend to decrease when the vehicle starts moving.

HEATER CORE REPLACEMENT

There are no repair procedures for a leaking or restricted heater core. If the core develops a leak or a severe restriction, it must be replaced. The following is a basic list of procedures to remove and replace a heater core.

1. Drain the coolant as needed.
2. Remove the access panel(s) as needed to access the heater case (see **Figure 7**).
3. Loosen the hose clamps and remove the heater coolant hoses.
4. Remove the cable or vacuum control lines (if equipped).
5. Remove the heater core, securing brackets, or clamps.
6. Lift the core from the case.
7. Do not use force. Take care not to damage the fins of the heater core when removing and replacing it.
8. Reverse steps 2–6 to reinstall the core.
9. Refill the cooling system with proper amount of clean 50-50 mixture of coolant.

CONTROL VALVE REPLACEMENT

There are no repair procedures for a damaged heater control valve. To replace the valve:

Figure 7. The heater core is located within a confined distribution case that directs passenger compartment air both to and from the heater core.

1. Drain the coolant to a level below that of the control valve.
2. Remove the cable linkage, vacuum hose(s), or electrical connector from the control valve.
3. Loosen the hose clamps and remove the inlet hose from the control valve.
4. Remove the heater control valve, as applicable. Remove the outlet hose from the heater core. Remove the attaching brackets or fasteners from the control valve.
5. Inspect the hose ends. If they are hard or split, cut 0.5–1.0 in. (12.7–25.4 mm) from the damaged ends.
6. Reverse steps 2–4.
7. Fill the system as needed with a 50-50 mixture of coolant.

HEATER HOSE REPLACEMENT

Heater hose replacement procedures are virtually the same as those for other cooling system hoses. What must be remembered is that the heater core is more delicate than any of the other cooling system components, so extreme care must be taken when removing hoses from the core outlets. Heater core outlets are also more susceptible to being crushed, so the use of constant tension clamps is encouraged (see **Figure 8**).

Because heater core outlets are sometimes located in inaccessible places, as shown in Figure 8, it is a practice of some technicians to use a hose that is too large for the application. When this procedure is used, the hose clamp must be overtightened to stop a potential leak. This allows the hose to be installed relatively easily; however, a hose clamp that is too large for the hose is often distorted when tightened enough to secure the hose, and the extra clamp

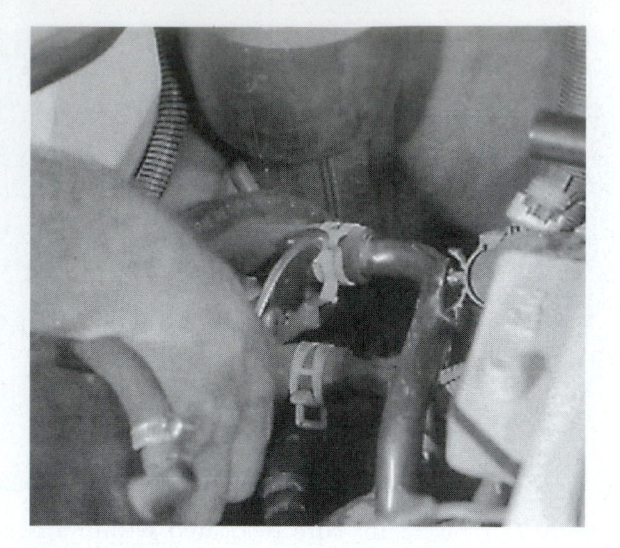

Figure 8. Heater hose clamps can often be difficult to access.

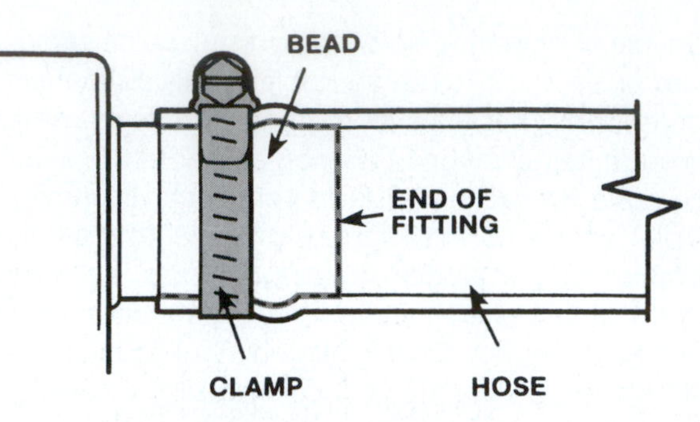

Figure 9. Use of the proper size hoses is critical to proper sealing.

tension can damage the heater core. Although it is sometimes difficult for the correct hose to slide over the various fittings, it is imperative that the proper size hose be used (see **Figure 9**). It is sometimes helpful to lubricate hose ends before attempting installation.

Summary

- Most heater system concerns are caused by a low coolant level, faulty thermostat, or heater core restrictions. Heater core leaks can be recognized by wet carpet, the formation of oily fog on the windshield, and pungent odors.
- Heater core restriction occurs when debris clogs the internal passages of the heater core. Heater core flow can be detected by measuring heater core inlet and outlet temperatures. In some cases, a restricted heater core can be cleaned by back-flushing the system.
- Heater control valve concerns typically revolve around leakage, stuck in the closed position, or stuck in the open position.
- A thermostat stuck in the open position may be recognized as a poor heating condition. A stuck-open thermostat can additionally cause an engine overcooling condition.
- Great care must be exercised when replacing cooling system hoses. Proper hoses and clamps should be used to avoid heater core damage and various leaks.

Review Questions

1. List thee conditions that can cause poor heater performance.
2. Technician A says that low coolant level can keep the heater from working properly. Technician B says that a thermostat that is stuck in the closed position will block coolant flow to the heater core. Who is correct?
   A. Technician A
   B. Technician B
   C. Both Technician A and Technician B
   D. Neither Technician A nor Technician B
3. List three indicators of a leaking heater core.
4. Technician A says that if the heater core inlet, outlet, and radiator hoses are the same temperature, the heater core is working properly. Technician B says that if the inlet and outlet are cooler than the radiator hose, the heater core is working properly. Who is correct?
   A. Technician A
   B. Technician B
   C. Both Technician A and Technician B
   D. Neither Technician A nor Technician B
5. Which of the following will not cause a poor heating condition?
   A. Low coolant level
   B. Stuck-open thermostat
   C. Kinked inlet hose
   D. Stuck-open control valve

Section 4

Refrigeration System

Air conditioning system repair is among the most lucrative repair areas for an automotive technician.

SECTION OBJECTIVES

At the conclusion of this section you should be able to:

- Describe the construction and operation of the individual refrigeration system components.
- Discuss construction variations of individual refrigeration system components.
- Describe how each of the individual components works in conjunction to form the complete refrigeration system.
- Describe the operation of the complete refrigeration cycle.
- Discuss the construction of a fixed orifice tube.
- Describe how refrigerant is metered in a fixed orifice tube refrigeration system.
- Discuss the operation of a variable orifice valve.
- Discuss how evaporator temperature is controlled in a fixed orifice tube refrigeration system.
- Discuss the operation and construction of a thermostatic expansion valve.
- Discuss how refrigerant is metered in a thermostatic expansion valve.
- Discuss various refrigerant contaminates and how they affect refrigeration system operation.
- Discuss the importance of refrigeration system identification.
- Explain the methods and procedures used to recover, recycle, and evacuate a refrigeration system.
- Explain how oil balance is maintained within the refrigeration system.
- Explain the external factors that effect refrigeration system operation.
- Perform refrigeration system quick tests and performance tests.
- Perform refrigeration system diagnosis by observing suction and discharge pressures.
- Diagnose refrigeration system noises and odors.
- Use various methods to test a refrigeration system for leaks.
- Properly replace various refrigeration system components.
- Know the procedures required to retrofit an R-12 refrigeration system.
- Perform a refrigeration system retrofit.

Chapter 13 System Components

Introduction

The automotive air conditioning system is made up of many components that work together to provide effective climate control in the vehicle passenger compartment. Although most of the components within the air conditioning system are simple in construction, their tasks are somewhat complex. In this chapter, we will examine the construction and operation of the individual components and how each component contributes to the operation of the entire system.

COMPRESSOR

The basis of air conditioning system operation is the circulation, evaporation, and condensation of a refrigerant. The compressor is the heart of the air conditioning system, which makes these actions possible. The compressor has three specific tasks: to promote the circulation of refrigerant through the air conditioning system, to raise the pressure of the vaporized refrigerant, and to increase the temperature of the refrigerant.

The compressor is essentially an air pump that is used to facilitate the movement of refrigerant through the air conditioning system by creating a low-pressure condition in one area of the system while creating a high-pressure area in another. The pressure differential between these two areas creates the conditions that are required for refrigerant to easily move through the system. In operation, the compressor increases both the pressure and temperature of the refrigerant. In general, this is accomplished by drawing low-temperature low-pressure vaporized refrigerant from the evaporator through the suction hose into the suction port of the compressor. The refrigerant is then compressed and expelled into the compressor discharge port and into the system, where it moves through the discharge line and into the condenser. The refrigerant leaves the compressor as a high-pressure, high-temperature vapor (see **Figure 1**).

> **Interesting Fact**
> *According to the laws of physics, when the pressure of a liquid or vapor is increased, so is its temperature. When the temperature of gaseous refrigerant is increased, its ability to readily condense is enhanced. This is a critical process in the operation of the air conditioning system.*

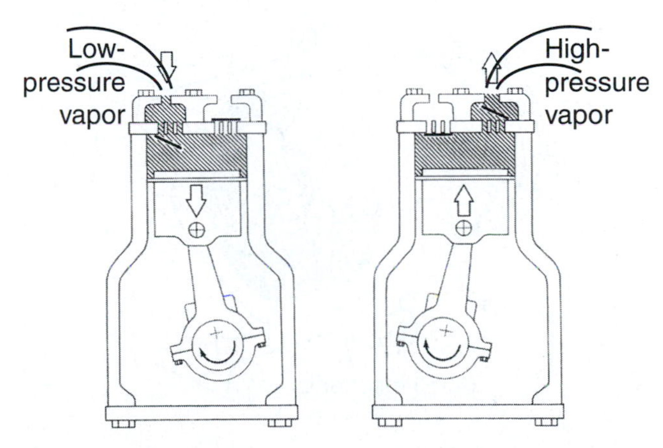

Figure 1. Refrigerant enters the compressors as a low-pressure vapor and leaves as a high-pressure vapor.

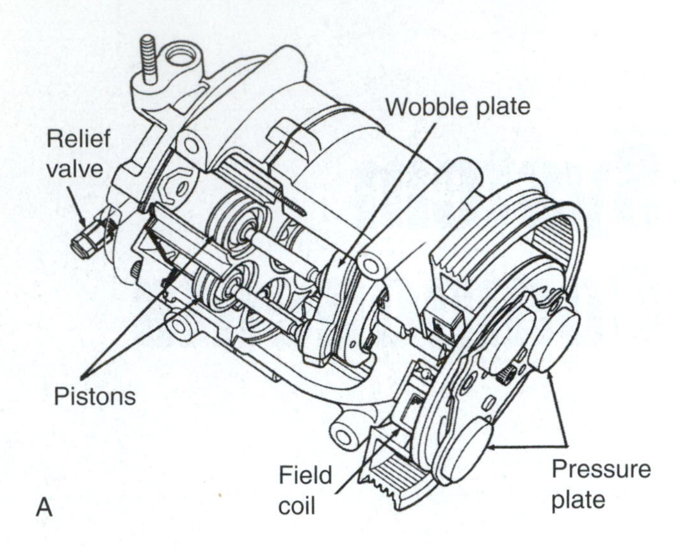

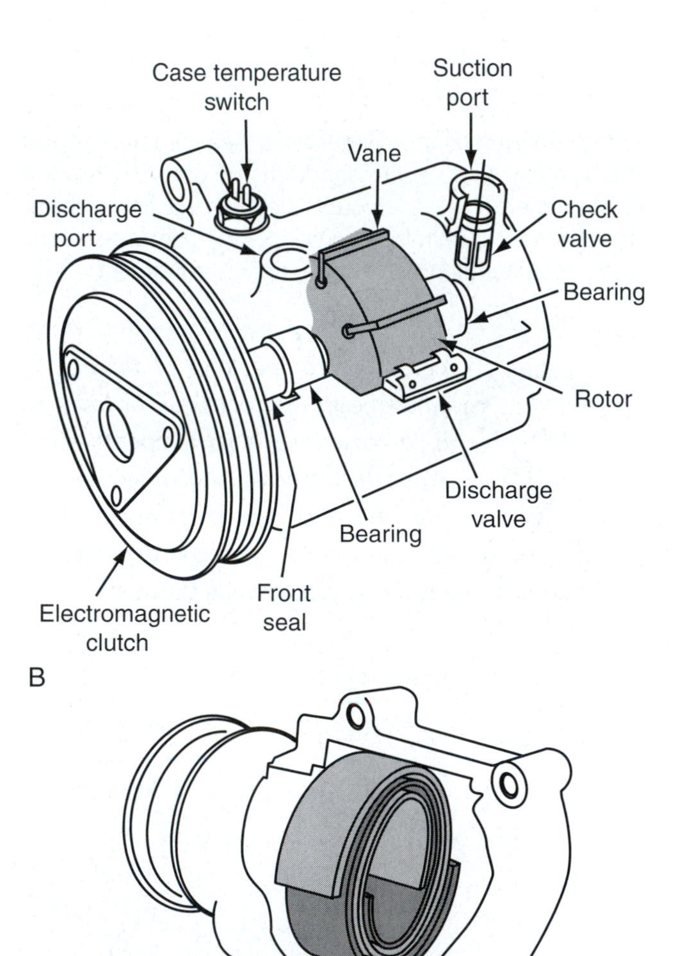

Figure 2. A. Reciprocating piston B. Rotary vane C. Moveable scroll.

There are three primary types of compressors used in a modern automobile: reciprocating piston, rotary vane, and moveable scroll (see **Figure 2**). Within these types, the compressor can be one of two styles: **fixed displacement** or **variable displacement**. Displacement is the amount of refrigerant that is displaced and expelled into the air conditioning system by the compressor in one revolution of the compressor shaft. Fixed displacement means that the amount of refrigerant displaced and expelled into the air conditioning system during each revolution of the compressor shaft remains the same, regardless of system operating conditions. Variable displacement compressors have the ability to vary the amount of refrigerant that is displaced and expelled into the system in response to system conditions. The type and style of compressor are determined when the air conditioning system is designed and cannot be altered during service. Most of the compressors in use today are of the reciprocating piston design.

RECIPROCATING COMPRESSOR

All reciprocating compressors use pistons sealed within a cylinder to create suction and compression. Like an internal combustion engine, each cylinder of the compressor uses an intake and exhaust valve. However, rather than using a poppet valve, an air conditioning compressor uses a special steel plate called a **reed valve**. The reed valve is a thin strip of steel that covers either a suction (inlet) port or a discharge (outlet) port (see **Figure 3**). Operation of the reed valve relies on the deflection of the reed to control refrigerant flow in and out of the compressor. Both the inlet reed and the outlet reed are exposed to the piston. As the piston moves down in its bore, a low-pressure area is created. When this occurs, the suction reed is deflected and the port is exposed to the cylinder, allowing refrigerant to enter the cylinder. When the piston begins its upward movement in the cylinder, pressure is created above the piston. The pressure within the cylinder forces the suction reed to close the port; however, at the same time, the discharge reed is deflected, allowing refrigerant to exit the cylinder (see **Figure 4**). If a reed valve should become

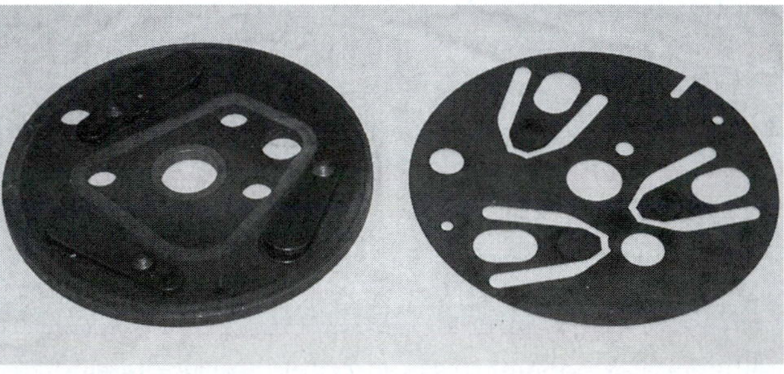

Figure 3. A typical reed plate and valve plate.

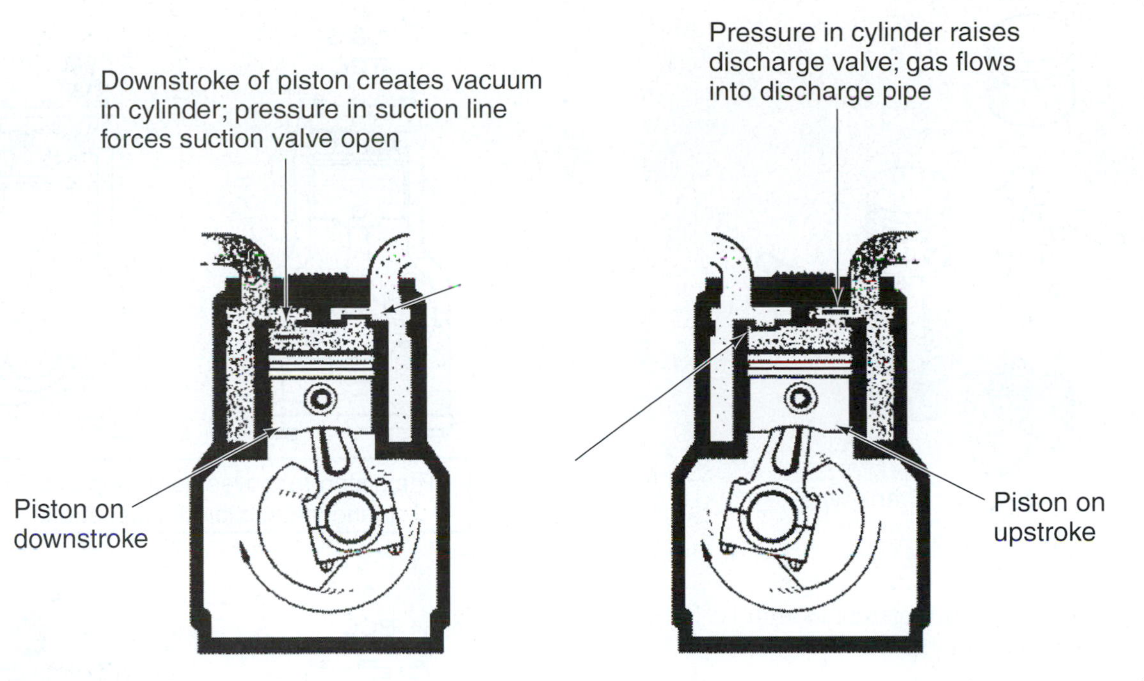

Figure 4. Operation of a compressor using reed valves. Note how the movement of the piston affects the deflection of the reed valve.

weakened or break the cylinder in which the valve is located, it will become inoperative, reducing the overall capacity of the compressor.

Reciprocating compressors may have configurations of two, four, five, six, seven, and even ten cylinders. The cylinders of the reciprocating compressor can be arranged in a variety of fashions: upright inline, v-type, radial, and axial (see **Figure 5**). Although hundreds of thousands of inline and radial style compressors have been used throughout the years, the axial design is the most prominent design found today. There are three basic methods of driving the piston of a reciprocating compressor: by **crankshaft**, by **axial plate**, and by **scotch yoke**.

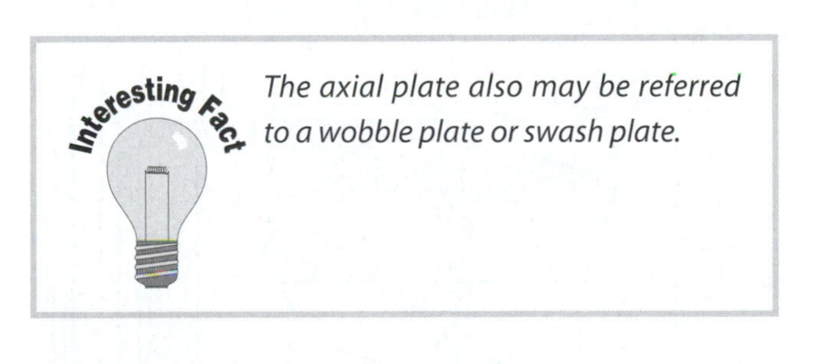

The axial plate also may be referred to a wobble plate or swash plate.

Crankshaft

The rotating assemblies of the inline and v-type compressors operate similarly to an engine. These style compressors use a crankshaft, connecting rods, and pistons (see **Figure 6**). The **crankshaft** is the part of a reciprocating compressor on which connecting rods and pistons are attached. The crankshaft is connected to the compressor clutch and is driven by the engine accessory drive system.

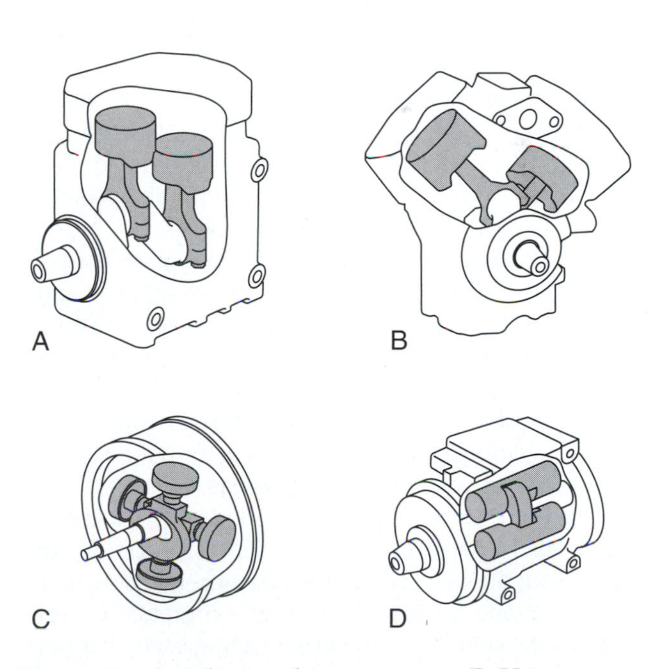

Figure 5. A. Inline-style compressor, B. V-type compressor, C. radial compressor, and D. axial compressor.

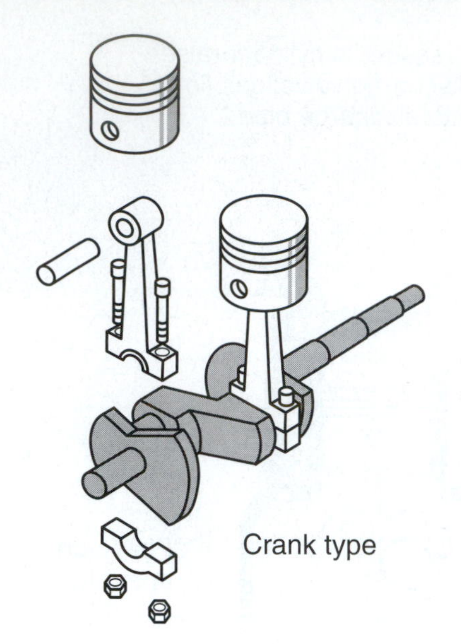

Figure 6. Crankshaft and piston assembly for an inline compressor.

The connecting rods are connected to offset areas of the crankshaft called **throws.** When the crankshaft spins, the connecting rods are forced to move to and fro, causing the pistons to do the same within the cylinders (see **Figure 7**). These style compressors are relatively heavy and require a significant amount of mounting space.

Axial Plate Compressors

The pistons of an axial compressor are arranged around the centerline and are parallel to the driveshaft. This style of compressor may have one or two banks of cylinders that are located at the front or the rear of the compressor. The axial plate compressor is driven through a drive shaft that is attached to the A/C clutch and driven by the accessory belt system. Inside the compressor, an axial plate is attached to the driveshaft. The axial plate is an offset concentric plate attached to the driveshaft at an angle (see **Figure 8**). Within the axial compressor family, there are two major designs: a double-ended piston assembly and a single-ended piston.

The double-ended piston axial compressor uses two banks of cylinders located at each end of the drive shaft. Essentially, this style of compressor has two sets of horizontally opposed cylinders. The axial plate is located between each bank of cylinders. Each piston in this compressor operates two of the opposed cylinders simultaneously; one piston is on the compression stroke whereas the other is on the intake stroke. Each piston is a long cylindrical-shaped slug. A notch is located between each one, directly in the center between each piston head. The axial plate contacts the piston in this notch. As the drive shaft spins, the axial plate moves to and fro. When

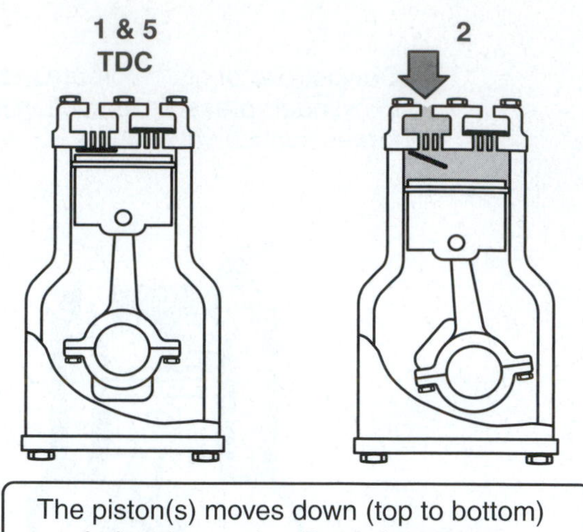

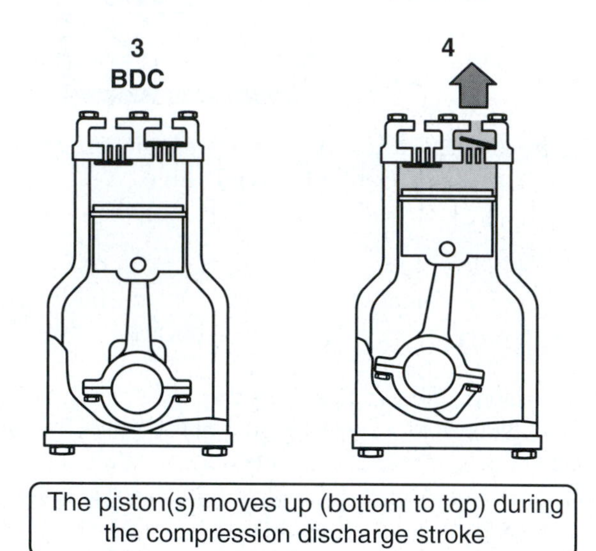

Figure 7. One complete cycle of an inline compressor.

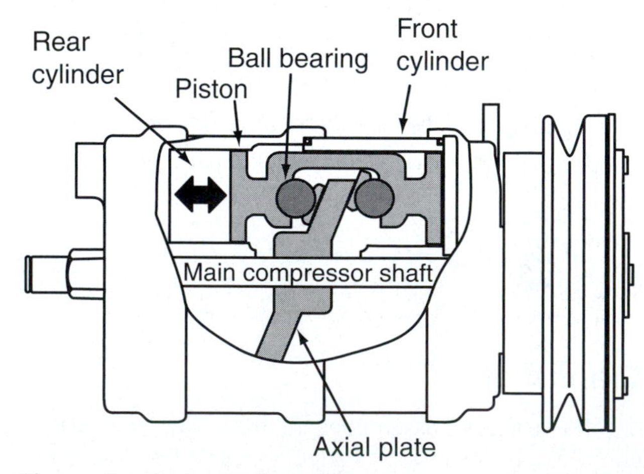

Figure 8. A typical axial compressor with two opposed cylinder banks.

this occurs, it forces the piston to move in the same direction as the plate (see **Figure 9**).

A second design uses only one bank of cylinders located at the rear of the compressor (see **Figure 10**). In this design, the pistons are connected to the axial plate using a

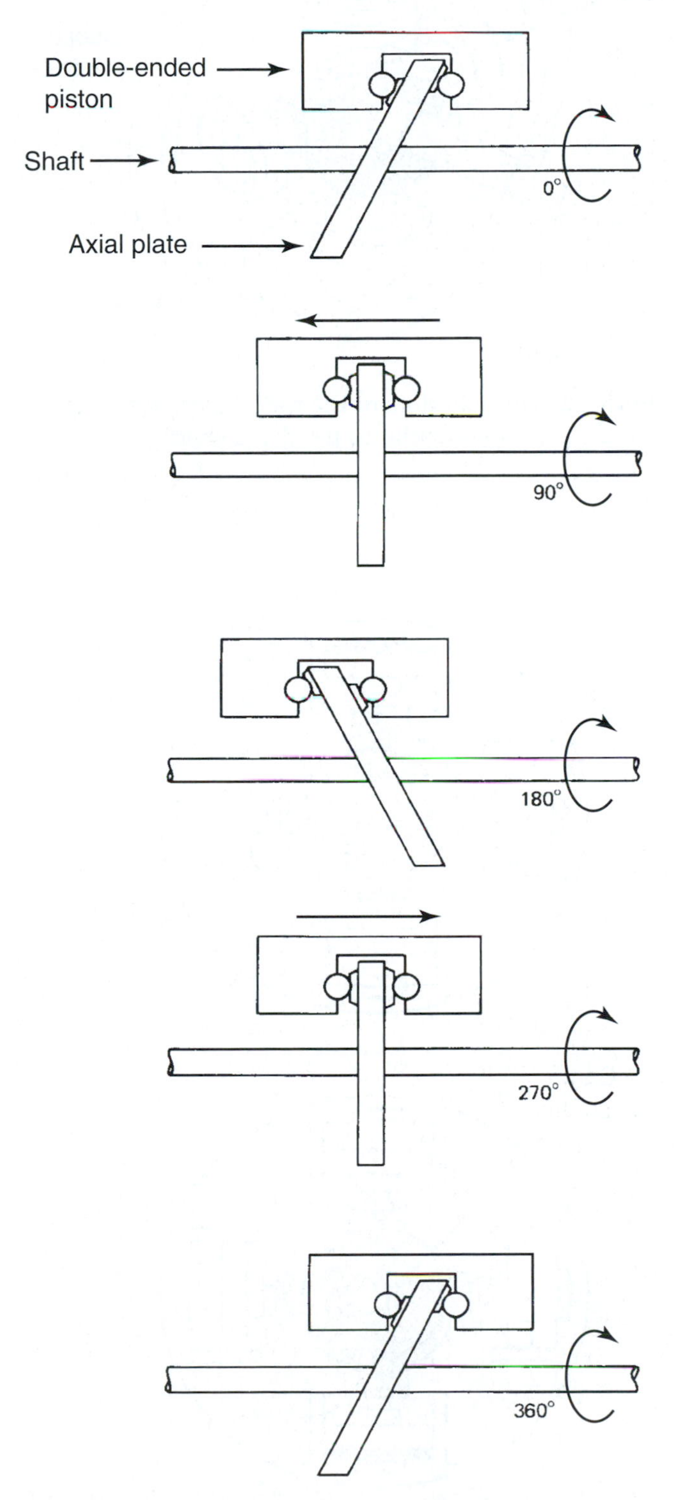

Figure 9. Operation of the an axial plate.

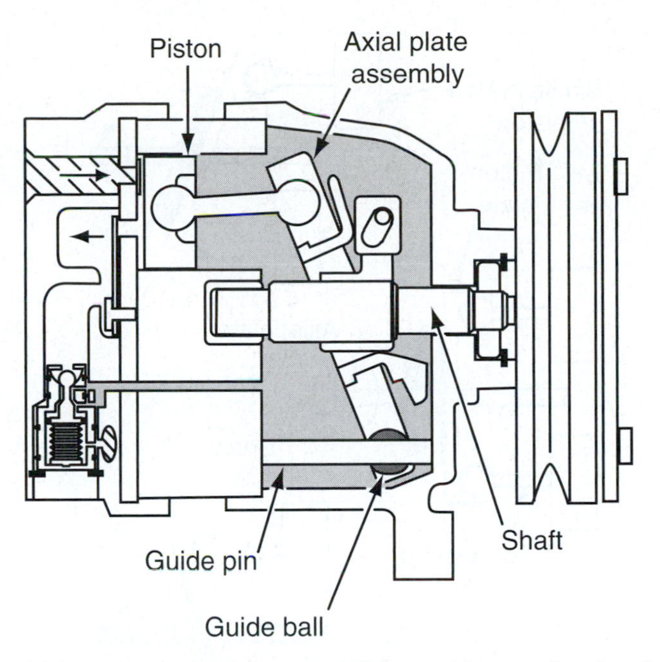

Figure 10. An axial compressor with one bank of cylinders.

short connecting rod. The axial plate in this design is slightly different in that there are three components to the axial plate assembly. An axial plate is connected to and spins with the compressor driveshaft. A second plate, which we will call a piston plate, is attached to the piston rods and held in place by a guide rod; the piston plate is allowed to move to and fro but does not rotate. A needle bearing separates the two plates. As the driveshaft spins, the axial plate moves to and fro against the bearing and the piston plate. Because the piston plate cannot rotate, its only action is to move to and fro, transferring this motion to the pistons (see **Figure 11**). This design is exceptionally unique in the fact that it can be equipped so that the pitch of the axial plate can be changed according to A/C operating conditions. Changing the pitch of the axial plate effectively changes the piston stroke, either increasing or decreasing the displacement of each cylinder. The design of the axial compressor is extremely compact in nature and is relatively light in weight.

Radial Compressors

A radial compressor also uses cylinders that are arranged around the centerline of the driveshaft; however, rather than being in parallel arrangement, the cylinders of the radial compressor are arranged perpendicular to the driveshaft (see **Figure 12**). The pistons of the radial compressor are attached to a common yoke. The driveshaft has an eccentric that rotates in the center of the yoke. As the driveshaft rotates, the eccentric forces the yoke to move in a perpendicular fashion away from the crankshaft. This forces the pistons to move up down within their respective bores (see **Figure 13**). Radial compressors are smaller and

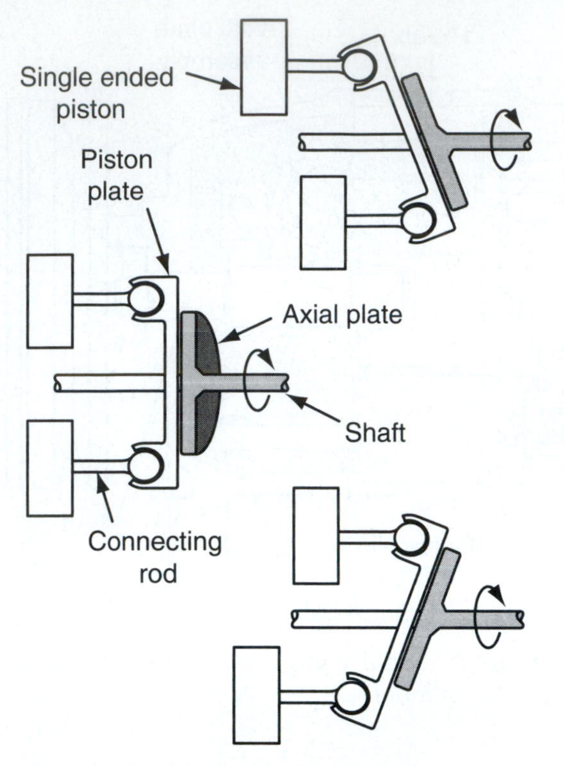

Figure 11. Operation of axial compressor using single-ended pistons.

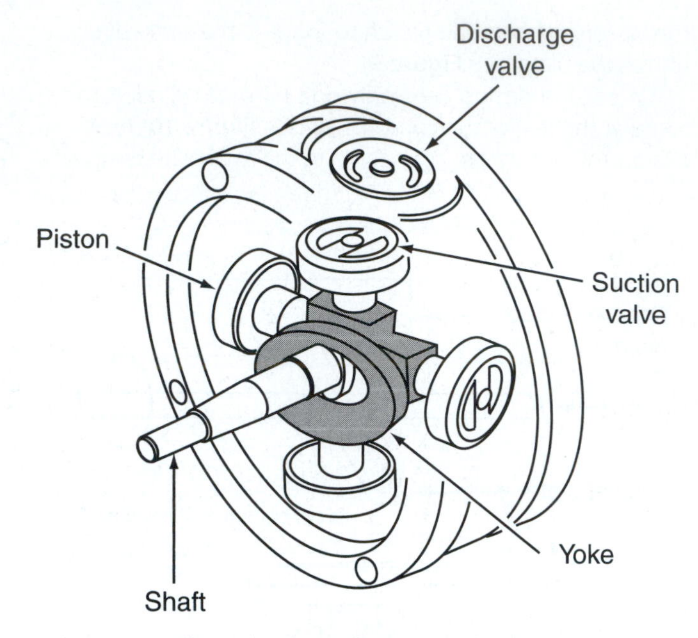

Figure 12. The cylinders of the radial compressor are perpendicular to the drive shaft.

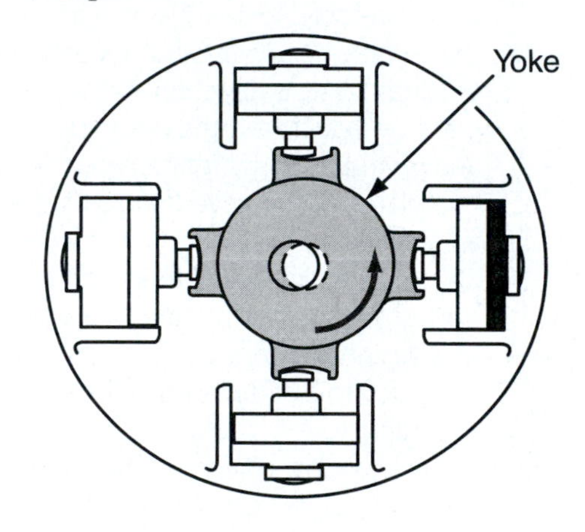

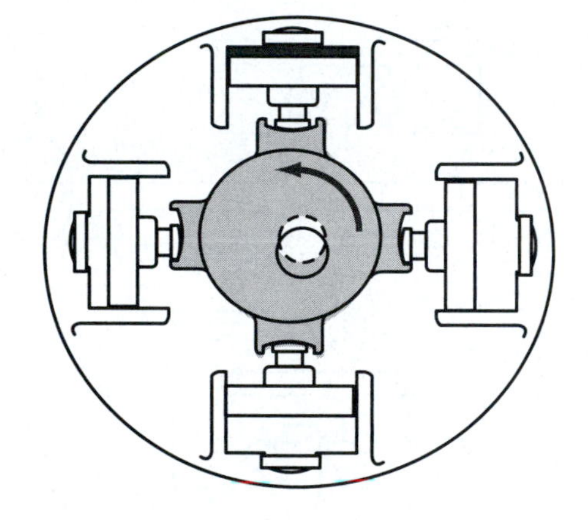

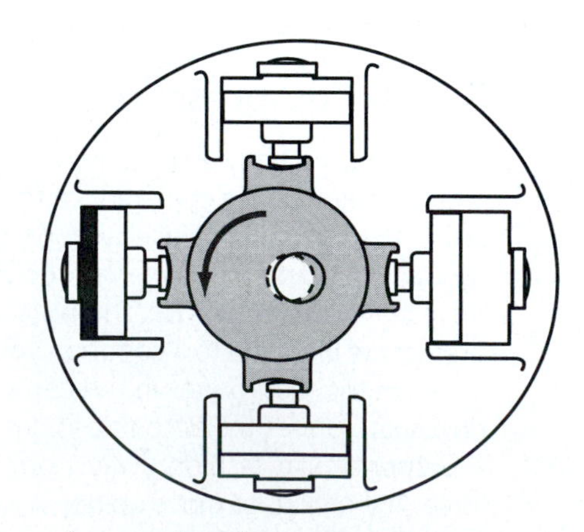

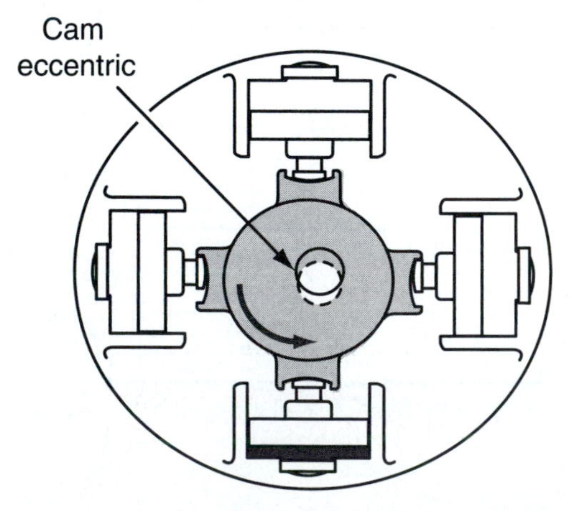

Figure 13. Operation of a scotch yoke compressor.

lighter than v-type or inline designs, but are heavier and larger than axial compressors.

ROTARY VANE COMPRESSORS

The rotary vane compressor is significantly different than a reciprocating compressor in that it uses no pistons. Suction and compression are created using a rotor and a set of vanes (see **Figure 14**). The rotor is attached to a driveshaft that is attached to the clutch and spun by the accessory drive system. The rotor is slotted, creating a space for the vanes to operate. The slots in the rotor are deep enough so that the operating depth of the vanes can change when the compressor is operating. The rotor and vanes operate inside of a closed cavity. The cavity is exposed to the inlet and outlet ports of the compressor. In this configuration, the space between each rotor becomes its own compression chamber. The rotor is installed in such a position that there is an unequal amount of space between the inlet and outlet sides of the rotor and vane assembly. As the rotor spins, centrifugal force pushes the vanes out to contact the walls of the cavity, creating a tight seal. As the individual compression chambers pass by the inlet port, refrigerant enters the cavity and individual chamber. When the chamber moves across the outlet port, the compressed refrigerant is expelled into the outlet of the system. Refer to **Figure 15** for a complete description of operation.

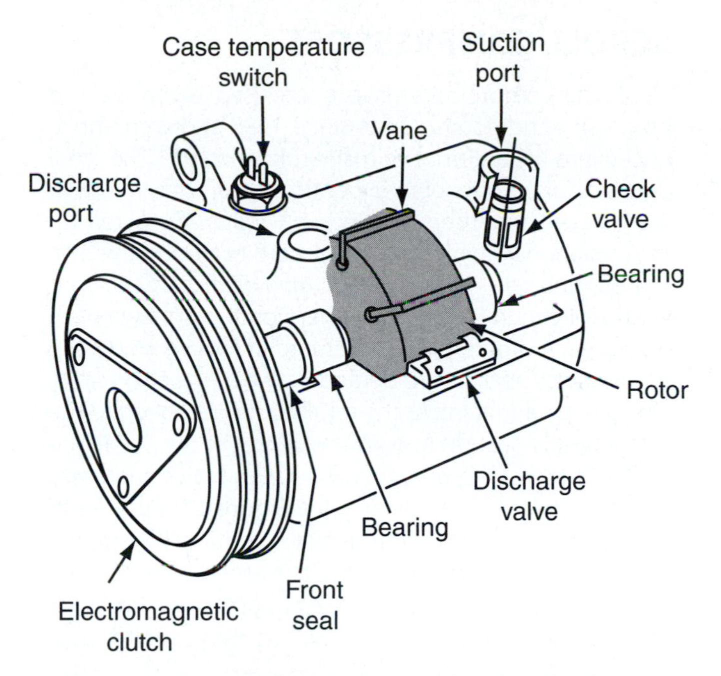

Figure 14. A typical rotary vane compressor.

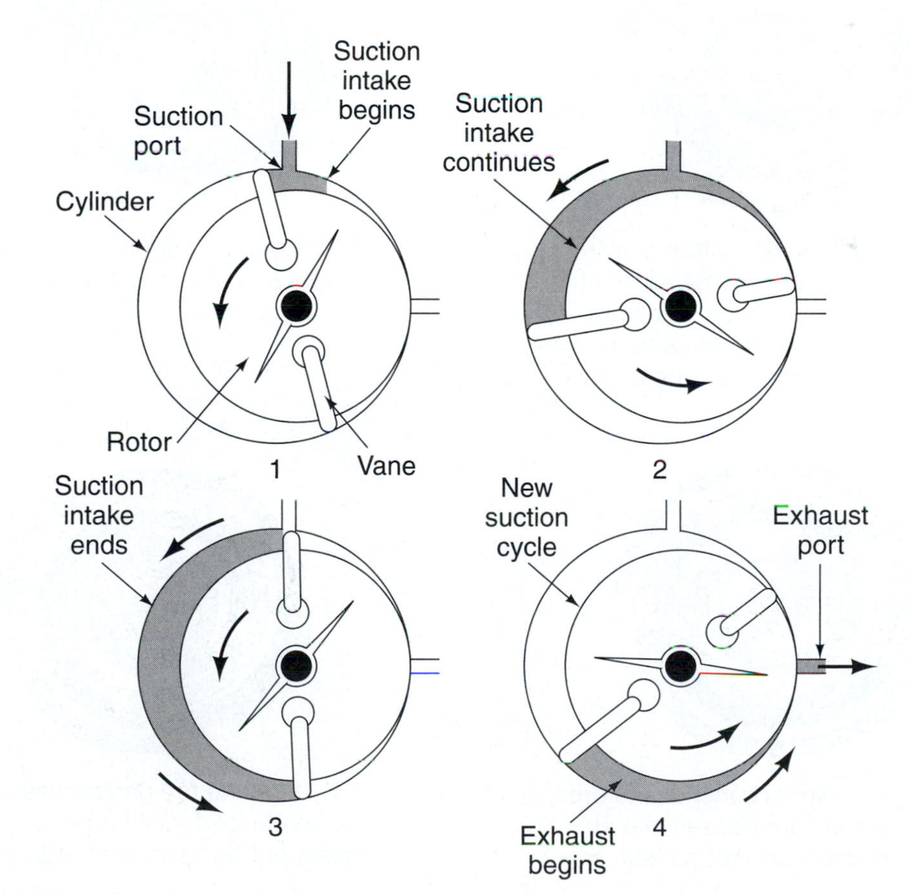

Figure 15. Basic operation of a rotary vane compressor.

SCROLL COMPRESSORS

Scroll compressors have been used extensively in home air conditioning units since 1988 and were introduced into the automotive marketplace in 1993. The scroll compressor consists of two identical spirals or scrolls that are meshed together 180 degrees apart. One scroll remains stationary while the second orbits around the first. The orbiting scroll is driven through a driveshaft and conventional clutch, just as the other compressors we have discussed. A simple scroll is pictured in **Figure 16.** As the scrolls mesh, pockets are formed between the two scrolls, which trap refrigerant. As the scroll continues to rotate, the refrigerant is pushed toward the center of the scroll. The pocket becomes progressively smaller, which increases the pressure of the refrigerant (see **Figure 17**). Because of the design of the scrolls, refrigerant is continually entering and moving through the pockets formed between the two scrolls. Because the scrolls stay continually "loaded," there are no pressure pulses such as those found in all reciprocating compressors. This makes for very smooth compressor operation and creates continuous suction and discharge pressures.

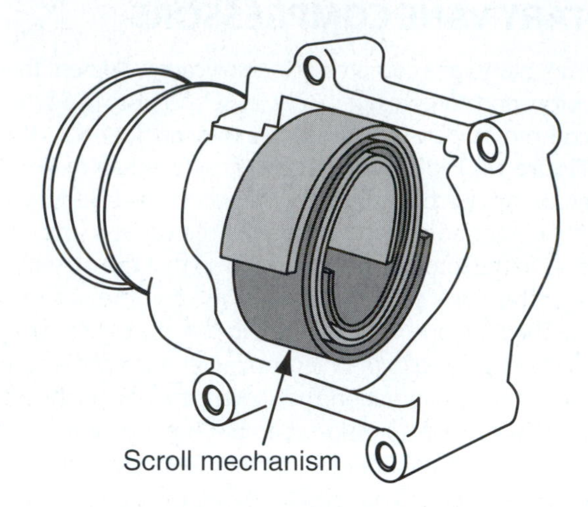

Figure 16. A simplified scroll compressor.

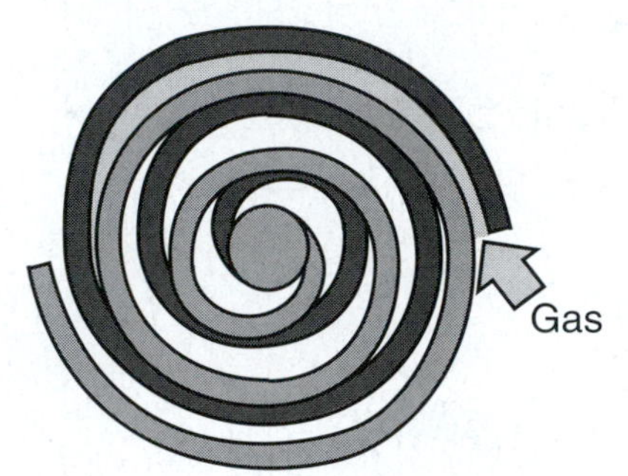

1. Compression in the scroll is created by the interaction of an orbiting spiral and a stationary spiral. Gas enters an outer opening as one of the spirals orbits.

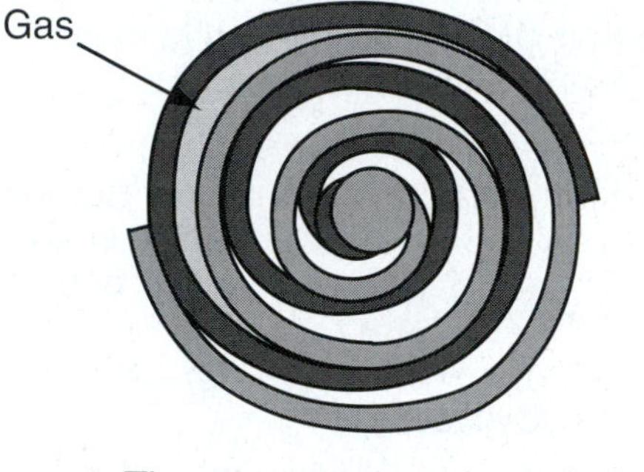

2. The open passage is sealed off as gas is drawn into the spiral.

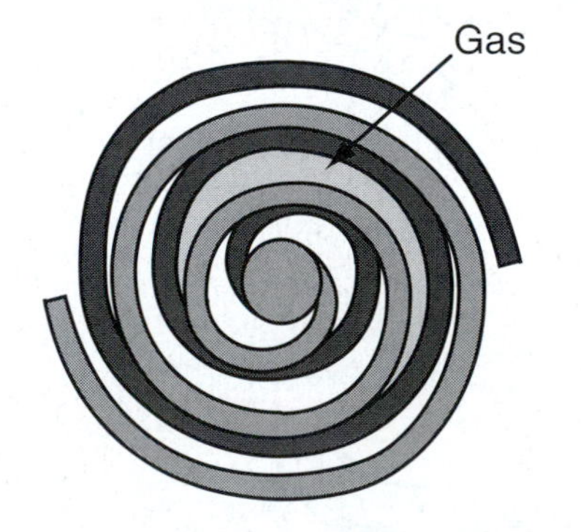

3. As the spiral continues to orbit, the gas is compressed into an increasingly smaller pocket.

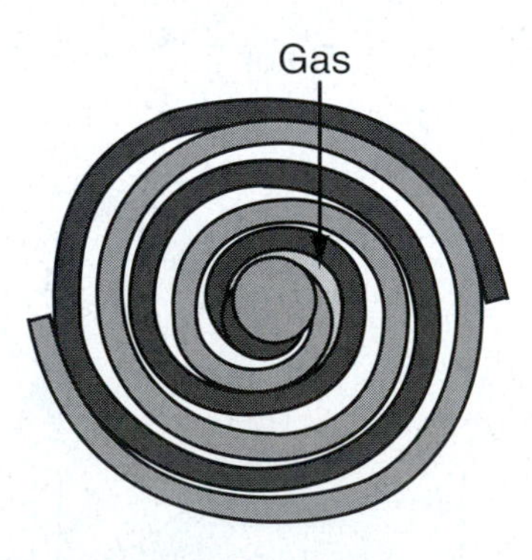

4. By the time the gas arrives at the center port, discharge pressure has been reached.

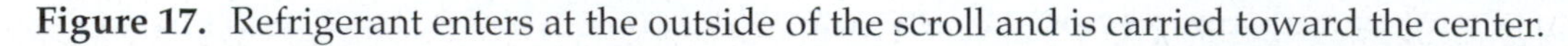
Figure 17. Refrigerant enters at the outside of the scroll and is carried toward the center.

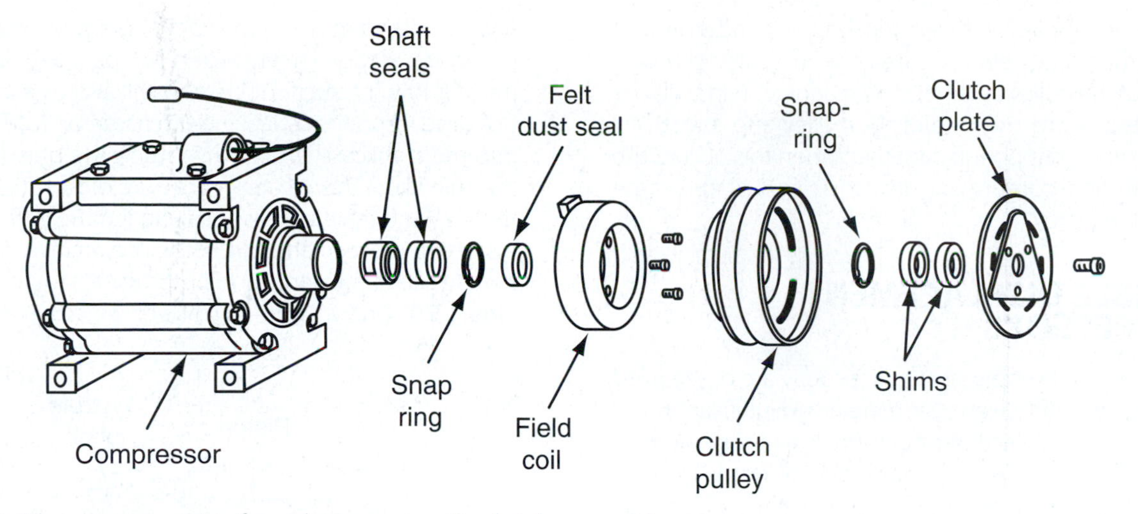

Figure 18. The components of an electromagnetic clutch assembly.

ELECTROMAGNETIC CLUTCH

All automotive air conditioning compressors use an electromagnetic clutch to engage and disengage the A/C compressor. The electromagnetic clutch assembly is mounted to the front of the compressor. Typically, the front compressor housing has a bearing support in which the clutch coil and pulley assembly mount. The clutch assembly consists of a pulley, a drive plate, and a magnetic coil (see **Figure 18**). The electromagnetic coil or "clutch coil" is the basis for the operation of the A/C clutch. The clutch coil is mounted stationary to the front bearing support of the compressor directly behind the pulley assembly.

The pulley assembly consists of a pulley and a sealed bearing. The bearing is pressed into the pulley assembly and held in place by a snapring or by **staking.** Staking is a process in which the metal surrounding the bearing is disturbed, creating a lip that holds the bearing in place. The bearing and pulley assembly is pressed onto the front bearing support of the compressor, directly on top of the clutch coil. When mounted on the compressor, the pulley covers the clutch coil. The pulley is unique in that the face of the pulley has a flat machined surface. This acts as a friction surface for the next component, called the clutch drive plate, to clamp to.

The clutch drive plate is located opposite of the pulley assembly and connected to the compressor shaft, which extends through the center of the clutch coil/pulley assembly. The drive plate consists of mounting hub attached to the drive plate that fits over the top of the compressor shaft. The clutch plate is mounted to the drive hub using flat springs that allow the clutch plate to be pulled toward the pulley without moving the mounting hub. Some clutch drive plate assemblies additionally may make use of rubber damper across the outer face of the plate. The damper absorbs some of the shock that is associated with the engagement of the compressor. The face of the clutch plate is the same diameter as the face of the pulley. It is located in a specific position on the compressor shaft, either by the use of splines or a square-cut steel key, and held onto the shaft by an interference press fit or a mechanical connection such as a bolt or nut (see **Figure 19**). When the

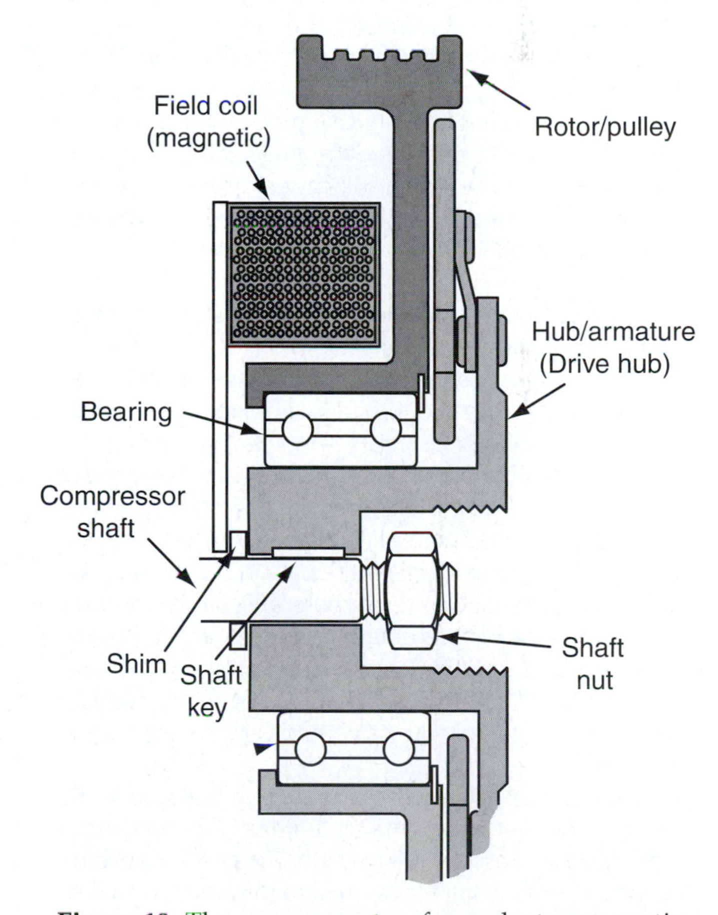

Figure 19. The components of an electromagnetic clutch assembly installed on the compressor.

clutch coil is supplied with electrical current and a ground source, a magnetic field is created and projected through the front of the pulley assembly. The magnetic field pulls the drive plate toward the face of the pulley and effectively locks the two components together. When this occurs, the drive plate and compressor driveshaft spin at the same speed as the pulley.

VARIABLE DISPLACEMENT COMPRESSORS

Most of the compressors in use today are of the fixed displacement variety. However, there are a significant number of variable displacement compressors in use as well. The variable displacement compressor has the ability to alter the volume of refrigerant that is circulated through the air conditioning system based on the demands of the system. By changing the compressor output, the need to cycle the clutch on and off is eliminated, reducing potential noise and engine performance concerns that are often associated with the cycling and parasitic load caused by the compressor. This is accomplished by varying the displacement of the compressor by means of altering the stroke of the pistons. This is done automatically by a control mechanism contained within the compressor.

This style of compressor uses a single bank of cylinders located at the back of the compressor. The pistons are connected to the variable pitch axial plate using short connecting rods. The more severe the angle of the axial plate, the more stroke the piston will have. Conversely, the less the angle or closer to perpendicular the plate is, the less stroke the piston will have. The angle of the axial plate is controlled by the pressure differential that exists between the compressor suction inlet and the crankcase, or more simply stated, the pressure that exists on the top of the piston versus the pressure on the back of the piston. When the suction pressure and the crankcase pressure are equal, no pressure difference exists and the pistons can be moved in and out of their bores relatively easy. The angle of the axial plate remains relatively high. As crankcase pressure increases, a pressure differential exists between the compressor suction inlet and the back of the pistons. This pressure differential makes it more difficult for the axial plate to pull the pistons out of their bores and causes the plate to stand upright, lessening the angle of the axial plate, causing the piston stroke to decrease. The pressure differential is controlled by a control valve located in the rear head of the compressor.

The control valve senses the suction pressure at the compressor inlet. When the air conditioning demand is high, suction pressure will be high. The control valve responds by leaking inlet pressure into the crankcase, effectively equalizing the pressures between the inlet and the crankcase. This means that there is no pressure differential, and the compressor will have maximum displacement. The wobble plate is at maximum angle, providing greatest piston travel. Conversely, when the air conditioning demand is low, the suction pressure will also be low. The control valve responds by leaking discharge or outlet pressure into the crankcase, effectively raising the pressure within the crankcase. This creates a pressure differential between the inlet and the crankcase, making it more difficult for the pistons to move in and out of their respective bores, causing the axial plate angle and piston stroke to decrease (see **Figure 20**). Only a slight increase of the crankcase-suction

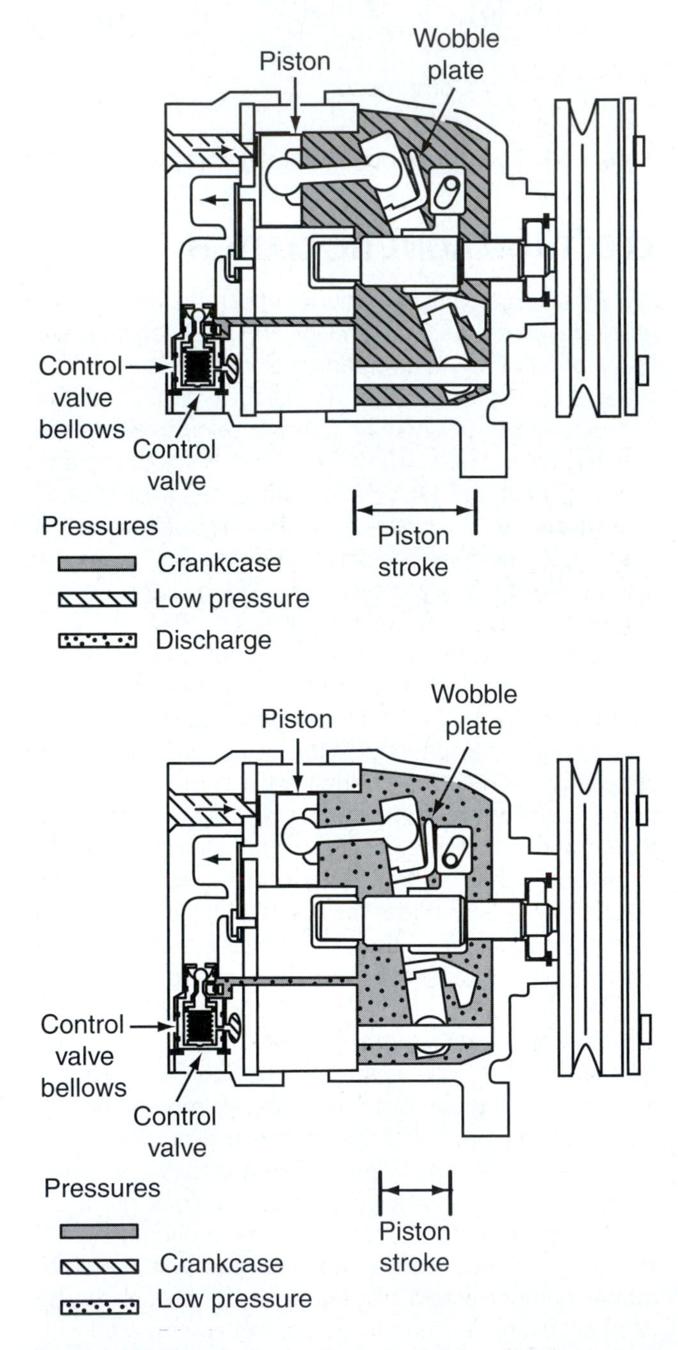

Figure 20. Operation of a variable displacement compressor.

pressure differential is required to create a force on the pistons sufficient to result in a movement of the wobble plate.

CONDENSER

The condenser is the next destination for the refrigerant after it leaves the compressor. When the refrigerant leaves the compressor and enters the condenser, it does so as a high-pressure, high-temperature vapor. The purpose of the condenser is to remove enough heat from the gaseous refrigerant to cause a change of state into a high-pressure liquid.

The condenser is a heat exchanger that is physically located at the front of the vehicle and installed in front of the radiator (see **Figure 21**). The placement of the condenser allows for maximum airflow to pass across the condenser to provide maximum heat transfer.

The condenser is constructed of a series of tubes that transport refrigerant. Between each tube there is a set of fins that provide surface area in which heat can be dissipated. The basic design and operation is similar to that of a radiator (see **Figure 22**). Refrigerant enters at the top of the condenser and flows toward the outlet located at the bottom of the core. As high-pressure refrigerant is forced through the condenser, the heat that is present within the refrigerant is transferred to the tubes and fins through the process of conduction. As air is forced across the surfaces of the condenser, heat from the tubes and fins is transferred into the atmosphere by convection. Sufficient heat is removed from the refrigerant to cause the refrigerant to condense into a liquid refrigerant; hence the name condenser. There are several different condenser core and tube designs, but each design accomplishes the same task. These designs are shown in **Figure 23**.

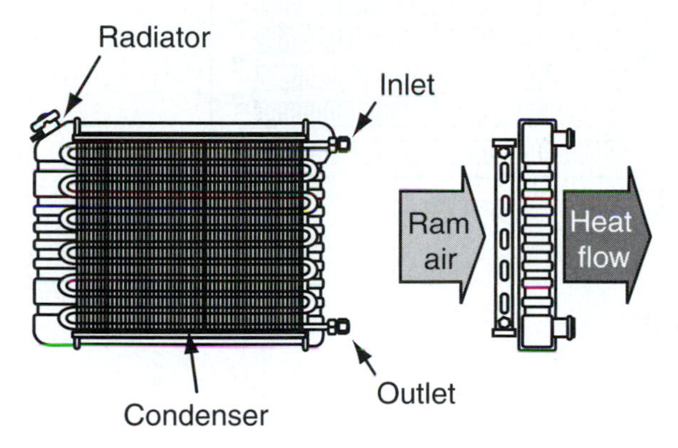

Figure 21. The condenser is located in front of the radiator.

EVAPORATOR

The evaporator, like the condenser, is a heat exchanger. This heat exchanger is located within the air distribution ductwork inside the vehicle, where it can be exposed to the ambient air of the passenger compartment. In this location, it can remove both heat and moisture from the air entering the passenger compartment or from the air that is recirculating through the cabin.

Basic construction of the evaporator is similar to that of a condenser in which refrigerant flows through aluminum tubes. The tubes, like the condenser, are surrounded by fins that are used to dissipate heat. Even though the condenser and evaporator are similar in construction, they operate in opposite ways.

Refrigerant enters the inlet at the bottom of the evaporator as a cool, low-pressure liquid. As refrigerant flows upward through the tubes toward the outlet, the surface of the evaporator becomes cool. A blower fan is used to

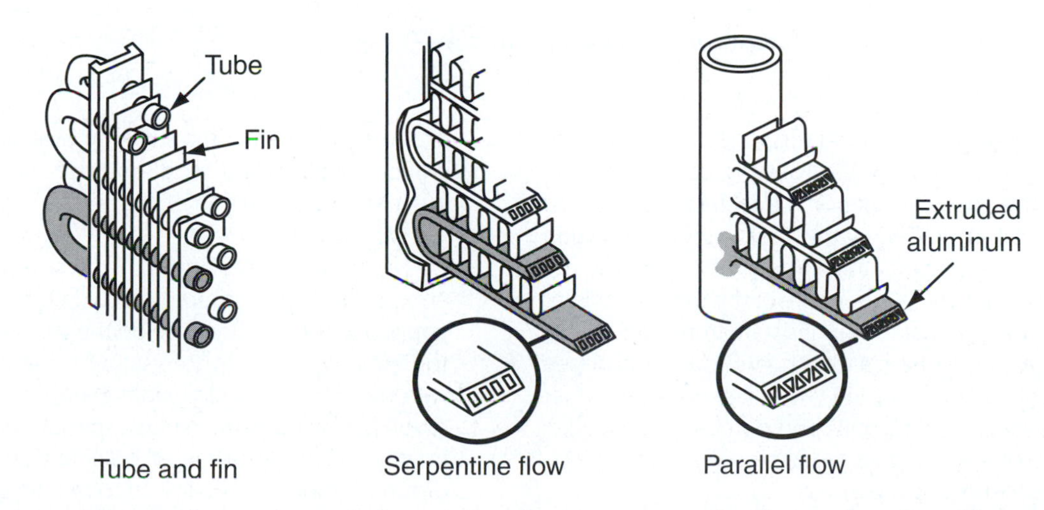

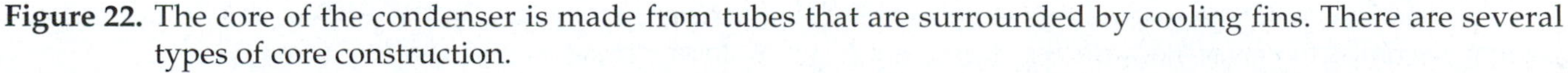

Figure 22. The core of the condenser is made from tubes that are surrounded by cooling fins. There are several types of core construction.

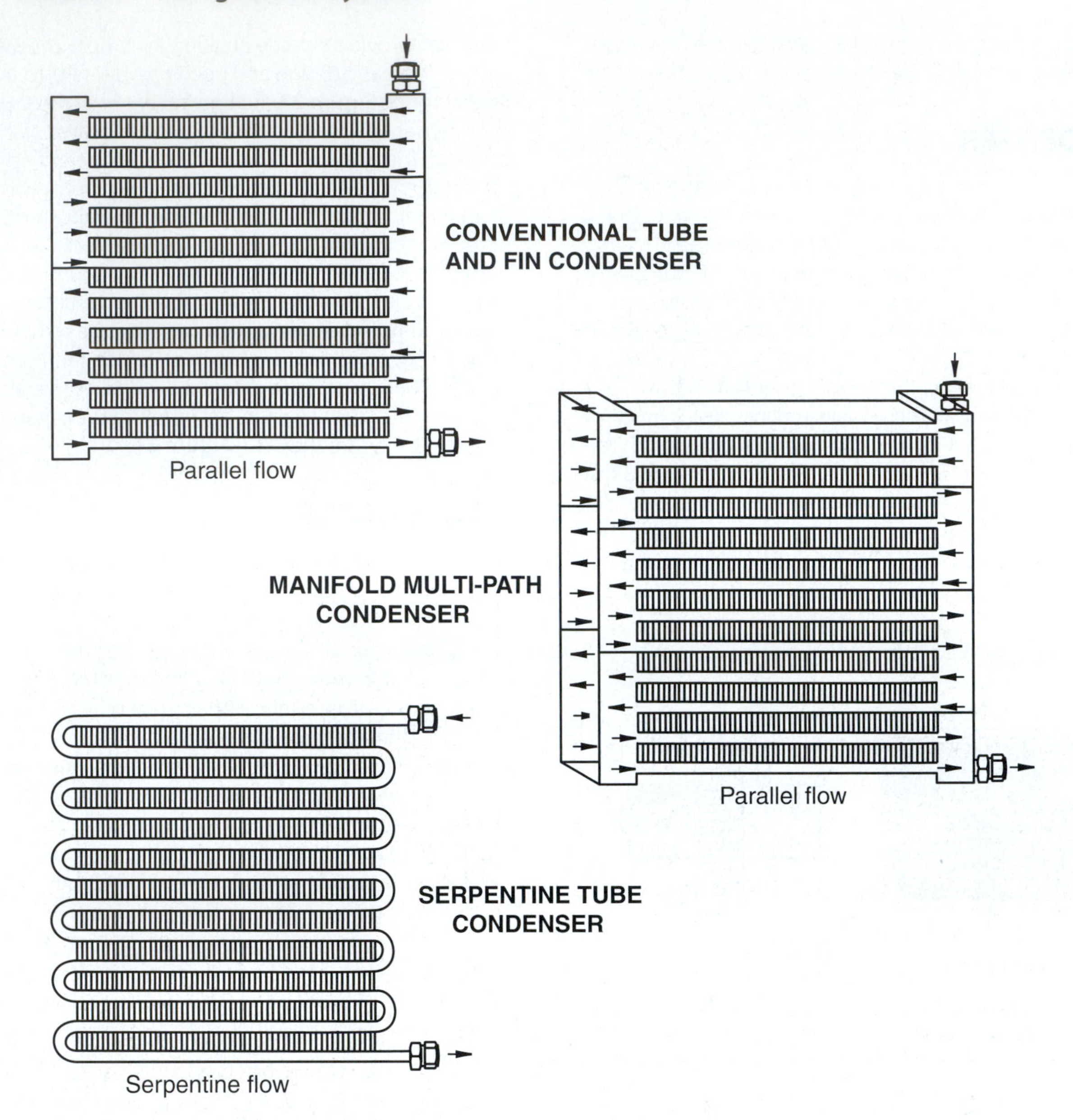

Figure 23. There are several different condenser configurations.

circulate warmer cabin air across the surface of the evaporator. As the warm air passes across the cool fins of the evaporator, the heat from the air is transferred to the fins and is eventually absorbed by the liquid refrigerant within the tubes. As the refrigerant absorbs the heat energy, the air becomes cooler. The temperature of the refrigerant is increased to the point that it changes state from a liquid to a vapor and expands; hence the name evaporator (see **Figure 24**). On inspection, you will notice that the outlet tube of the evaporator is significantly larger than the inlet; the difference in size provides the necessary area for the refrigerant to expand as it vaporizes.

A second function of the evaporator is to dehumidify passenger compartment air. Because the evaporator is extremely efficient at moisture removal, air conditioner operation is also called for during defroster operation. This prevents the formation of fog and frost on the interior of the windshield.

As the air passes across the surface of the evaporator, enough latent heat is removed so that the water molecules trapped in the air condense on the surface of the evaporator. The amount of condensation present naturally varies with the relative humidity in the ambient air. The condensation is allowed to drain into a special area of the evaporator case where it is drained outside the vehicle onto the ground. If moisture is allowed to remain on the surface of the evaporator and in the evaporator case, bacteria can form and cause a variety of foul odors.

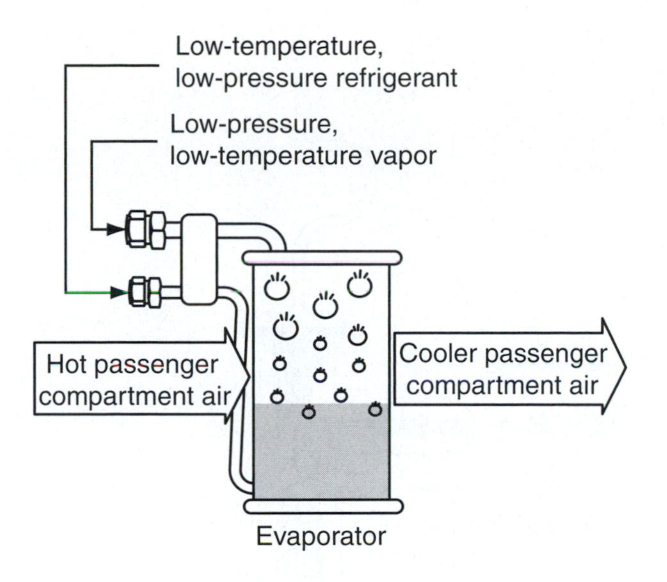

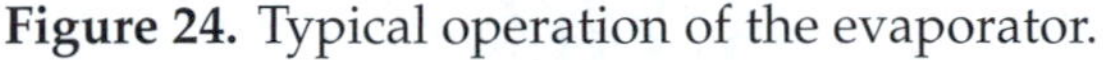

Figure 24. Typical operation of the evaporator.

METERING DEVICE

All air conditioning systems use a metering device. The metering device is a calibrated restriction that is located in the liquid line between the condenser and the evaporator. The job of the metering device is to change hot high-pressure liquid refrigerant from the condenser, to the cool low-pressure liquid that enters the evaporator. This is accomplished by restricting the amount of refrigerant that is allowed to flow into the evaporator, which decreases both the pressure and the volume of the refrigerant entering the evaporator.

There are two basic styles of metering devices in use today: the fixed orifice tube and the thermostatic expansion valve. There have been a number of other designs used in years past, but for the most part they are obsolete.

Interesting Fact

Each evaporator located in the vehicle will use only one metering device, either a fixed orifice tube or a thermostatic expansion valve. However, a vehicle that uses multiple evaporators, typical of many modern SUVs with separate air conditioning systems for the rear passenger compartment, has multiple metering devices. These vehicles may not use the same type of metering device on each evaporator. For example, it is common to use a fixed orifice tube on the front evaporator but a thermostatic expansion valve on the rear evaporator.

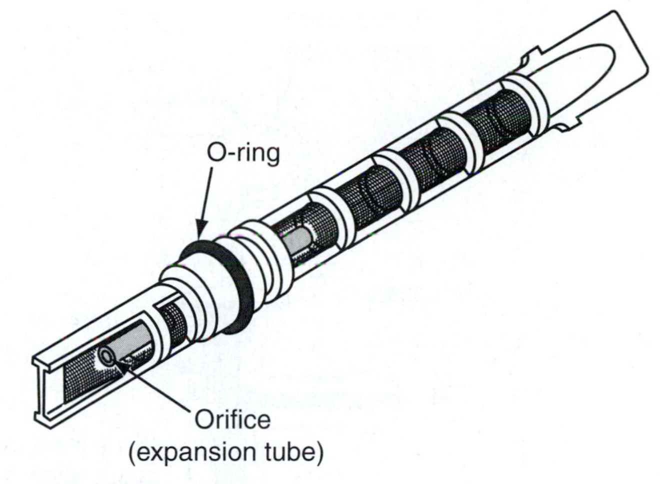

Figure 25. An example of a typical orifice tube.

Fixed Orifice

The **fixed orifice tube (FOT)** is the simplest design; it consists of a small brass tube encased in a plastic housing with a mesh filter (see **Figure 25**). The orifice tube is equipped with a small mesh screen on each side of the tube to protect the tube from debris. The brass tube is considerably smaller than the tubes leading into and away from the orifice. As the liquid refrigerant is passed through the brass tube, its volume and therefore its pressure is reduced. The size of the tube is fixed; therefore, the amount of refrigerant that is allowed to pass through the orifice is relatively consistent, changing only in relation to the difference in system pressures.

You Should Know

Although most orifice tubes have a similar external appearance, the size of the internal orifice is designed to match the needs required by the type of refrigerant and the capacity of the evaporator and condenser.

Thermostatic Expansion Valve

The **thermostatic expansion valve (TXV)** is also located in the refrigerant liquid line between the condenser and the evaporator. However, operation of the thermostatic expansion valve is entirely different. The thermostatic expansion valve is variable in nature, allowing the volume of refrigerant that is let into the evaporator to be changed relative to the temperature of the evaporator. This is accomplished most often by placing a sensing bulb in a location where it can sense the temperature of the evaporator outlet tube. As the temperature of the evaporator increases above a desired temperature, the valve opening is increased and the amount of refrigerant allowed to flow into the

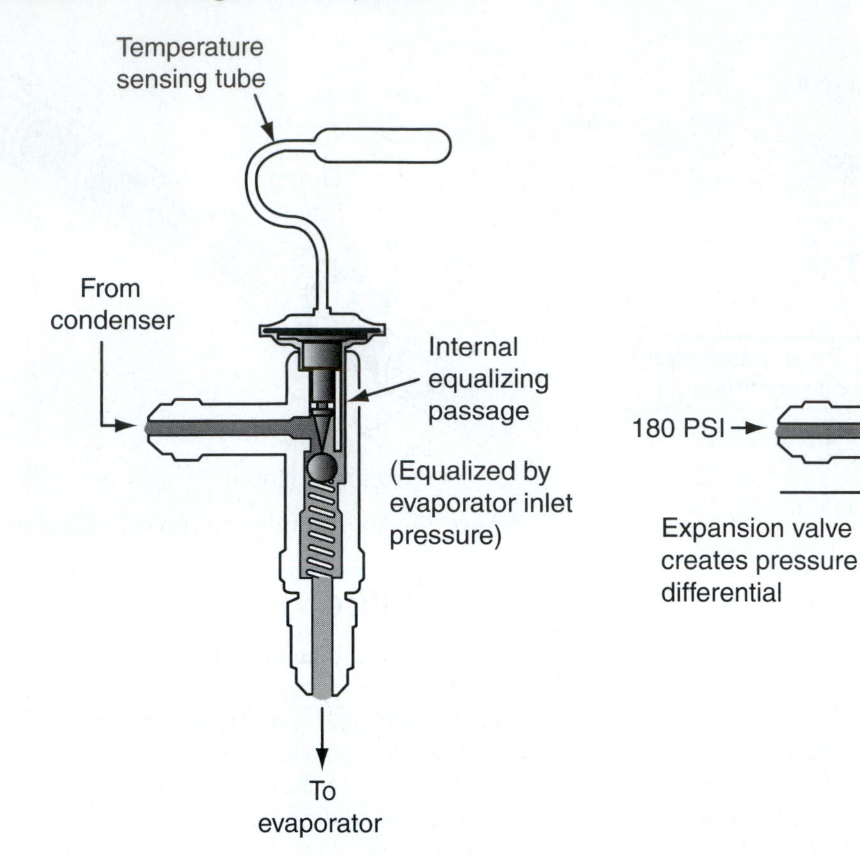

Figure 26. A common expansion valve designs.

evaporator is increased accordingly. As the temperature of the evaporator drops, the valve opening is reduced and refrigerant flow is decreased accordingly (see **Figure 26**).

STORAGE AND DRYING

All systems use some method to remove any transient moisture from the refrigerant and provide refrigerant storage. The type of device used is completely dependent of the type of metering device used. The two components used are the accumulator and the receiver dryer.

The accumulator is used exclusively with fixed orifice tube systems. The accumulator is a large canister that is located on the outlet of the evaporator. The accumulator serves two purposes. The first is to capture any liquid refrigerant that might leave the evaporator. This stops any liquid refrigerant from entering the compressor and causing serious damage. Second, the accumulator is equipped with a desiccant bag. The desiccant bag is charged with collecting any transient moisture that might be present in the refrigerant.

The receiver dryer is similar in appearance to the accumulator but is used exclusively with the thermostatic expansion valve. The receiver dryer is placed in the line between the condenser and the expansion valve. Like the accumulator, it is equipped with a desiccant to trap any moisture that might remain in the refrigerant.

The most important job of the receiver dryer is as a storage device. Because the expansion valve typically will not pass the full amount of liquid refrigerant that is available to it, the receiver dryer stores the excess refrigerant; this ensures that the expansion valve can always be supplied with liquid refrigerant and will operate efficiently. Both the accumulator and receiver dryer are illustrated in **Figure 27**.

HOSES AND FITTINGS

The air conditioning system uses a variety of specialized hoses, pipes, and fittings to connect all of the components of the system. Ridged aluminum lines are used in areas where flexing and abrasion are not likely. In areas where flexing and unusual bends occur, specially designed rubber hoses are used. Any rubber hoses that are used within the air conditioning system must be of a barrier type construction. These hoses consist of several layers of materials. The hose itself is made from a synthetic rubber product that is surrounded by a nonpermeable nylon layer to guard against leakage. This layer is covered by another rubber layer, and then a braided nylon layer is added to provide strength. This combination is lastly covered by a layer of butyl rubber to provide abrasion resistance to the entire hose (see **Figure 28**). Recent laws

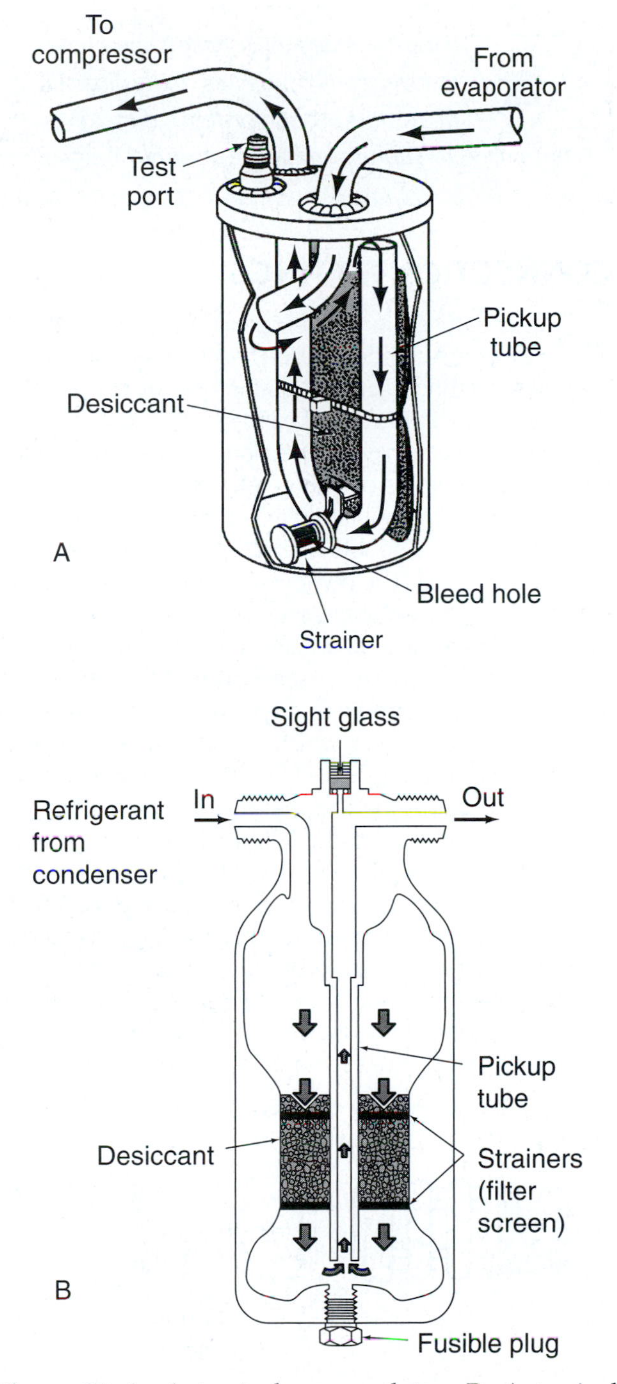

Figure 27. A. A typical accumulator. B. A typical receiver dryer.

concerning refrigerant requirements require that all vehicles manufactured for use with R-134a refrigerant and replacement hoses must be a barrier type. Nonbarrier hoses that were previously used with R-12 refrigerant were somewhat porous and allowed some amount of refrigerant to escape. Because R-134a molecules are smaller than those of R-12, the use of a nonbarrier hose in a R-134a application will exhibit a significant amount of leakage.

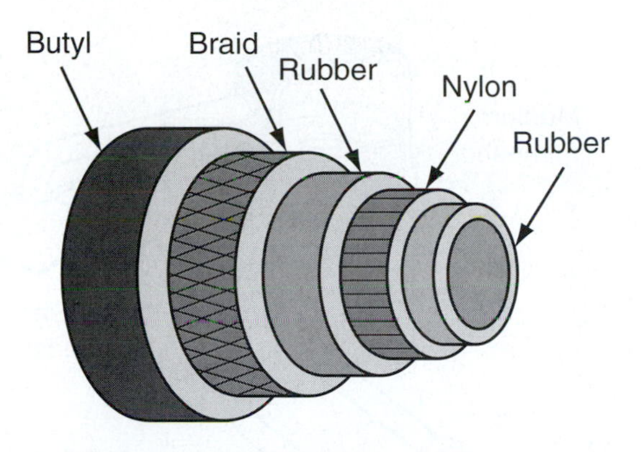

Figure 28. Barrier hose details.

The lines used in the air conditioning system have the following specific names:

- Discharge hose: connects the compressor discharge outlet to the condenser (typically made from barrier hose with aluminum fittings).
- Liquid line: connects the condenser to the evaporator inlet. This also may contain a location for the orifice tube or may connect directly to the receiver dryer in an expansion valve type system. This line is typically constructed from rigid aluminum.
- Suction hose: connects the evaporator/accumulator outlet to the compressor inlet. This hose is significantly larger than those lines used between the compressor and the condenser inlet. The added size allows the hose to accommodate expansion of the refrigerant as it exits the evaporator. This hose is constructed of barrier hose with aluminum fittings. The suction hose also can be attached to a common manifold with the discharge hose. This assembly is called the suction/discharge hose assembly.
- In many applications, a muffler is installed in the suction or discharge hose to dampen the pulses created by the compressor. When left unchecked, these vibrations can travel through various lines and can eventually be heard within the passenger compartment (see **Figure 29**).

SERVICE FITTINGS

Refrigeration systems use service fittings located in both the high side and the low side of the refrigeration system. The fittings are necessary so that service hoses can be connected to service the system. There are several different styles of service fittings used on motor vehicles. Each different style of service fitting requires the use of the corresponding service hose fitting. The type of fitting used is determined by the type of refrigerant that is used in the system. Each style of fitting additionally will have a unique high- and low-side fitting to prevent personal injury and equipment damage. Although there may be many different styles of service fittings, the operation of the fitting is virtually identical.

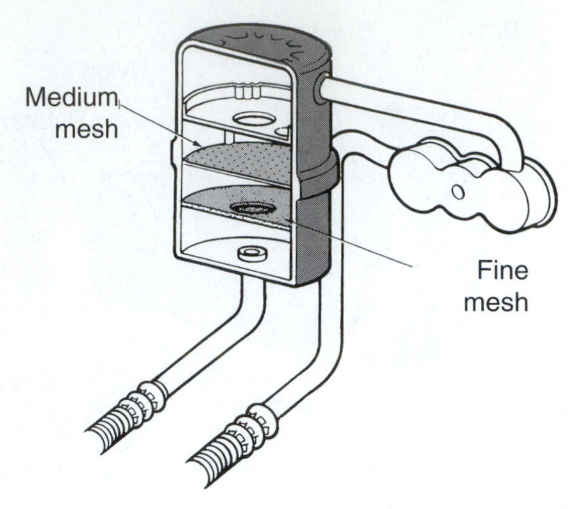

Figure 29. Some vehicles use a muffler to dampen compressor noise.

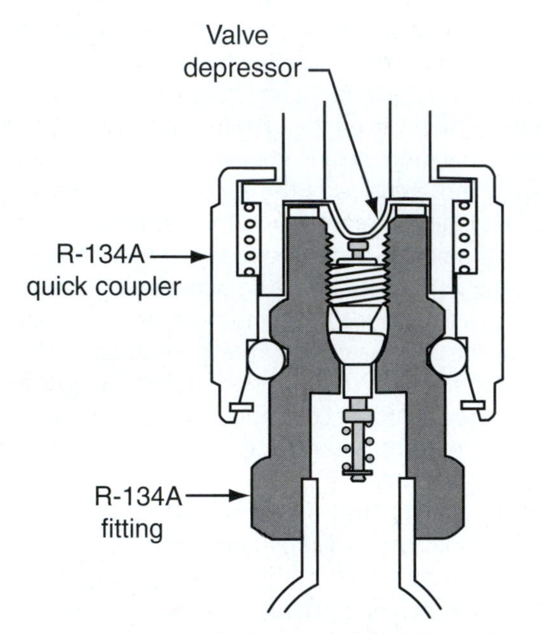

Figure 30. Service fittings use Schrader valves to seal the system and control the flow of refrigerant between the refrigeration system and the service equipment.

The fitting consists of an outer housing to which the service hose connects. Inside the outer housing there is a Schrader valve. The valve consists of a valve seat and spring. The spring acts on the valve to keep it closed (see **Figure 30**). The service hose contains a device called a **depressor**, which pushes the valve against spring pressure and off its seat to connect the refrigeration system with the service hoses. It is very important for the technician to pay special attention to the location of the service fittings. System pressures can vary depending on the location of the fittings. It is important to take this into consideration when performing system diagnoses using a pressure gauge.

You Should Know *If the low-side service hose is somehow inadvertently connected to the high side of the refrigeration system, refrigerant containers can rupture and cause personal injury.*

CONNECTION FITTINGS

The lines used in the refrigeration system are connected to system components and each other using special O-ring style fittings. The fittings themselves consist of an O-ring seal and a mechanical connection to keep the lines together. The mechanical connections used can be a typical thread and nut design or a quick-release spring lock fitting.

The thread and nut fitting may consist of either a compressed O-ring or a double O-ring that uses an interference fit to form the seal. The compressed O-ring design uses a specially flared tube that provides a collar to hold the O-ring as well as flange to equally distribute force against the O-ring and help it hold its form (see **Figure 31**). The flared tube and O-ring are inserted into a specially formed tube that provides a smooth seat which the O-ring seats against. When the nut is tightened, the O-ring is compressed and seals tightly against the two lines. In this style of fitting, the O-rings can be distorted or cut if the nut is overtightened. The double O-ring design uses two O-rings that are located in two machined grooves in the fitting (see **Figure 32**). The fitting is inserted into a specially designed fitting that provides the seat for the O-ring. The sealing ability of this style fitting is totally dependant on the interference fit between the O-ring and the two tubes. As the fitting is tightened, the O-rings remain the exact same shape and size.

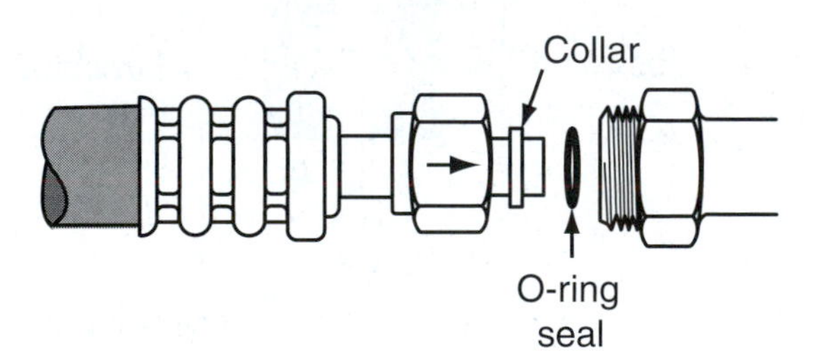

Figure 31. Single O-ring style service fittings.

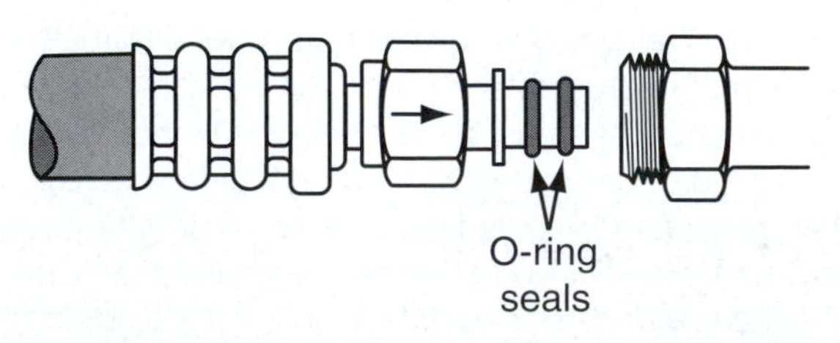

Figure 32. Double O-ring style service fitting.

The spring lock connector uses only the double O-ring style fitting. This style of fitting is only held together using a captured spring that is stretched around a flange of the mating line (see **Figure 33**). Because of its design, it is unable to apply any pressure to compress the O-ring. A special tool must be used to retract this spring when service is required.

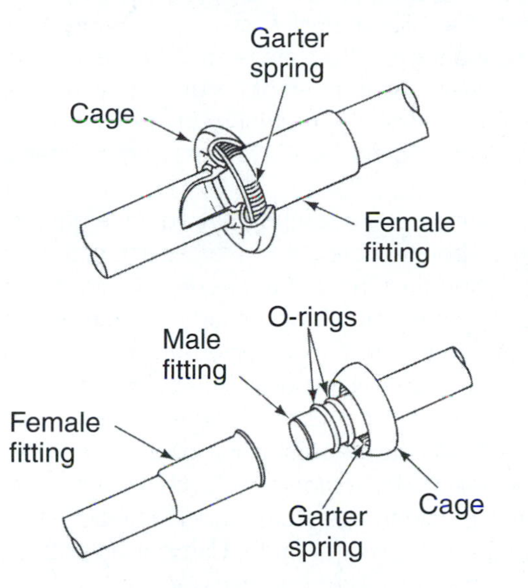

Figure 33. Spring lock style refrigeration fittings.

REFRIGERATION CYCLE

The refrigeration cycle is the series of events that occur to allow the refrigerant to both absorb and release heat energy. (Use **Figure 34** as a reference as you study this section.) The refrigeration system is divided into a low-pressure side and a high-pressure side. There are two specific points within the refrigeration system where the system divides: the compressor and the metering device. The high side of the system begins with the compressor outlet and includes the condenser and associated lines and ends at the metering device inlet. The low side of the system begins at the outlet of the metering device and includes the evaporator and associated lines and ends at the compressor inlet.

The compressor draws in low-pressure vapor from the suction line. In the compressor, the refrigerant is pressurized, causing the temperature to increase relative to the pressure applied. High-pressure, high-temperature refrigerant vapor exits the compressor and travels through the discharge line to the condenser.

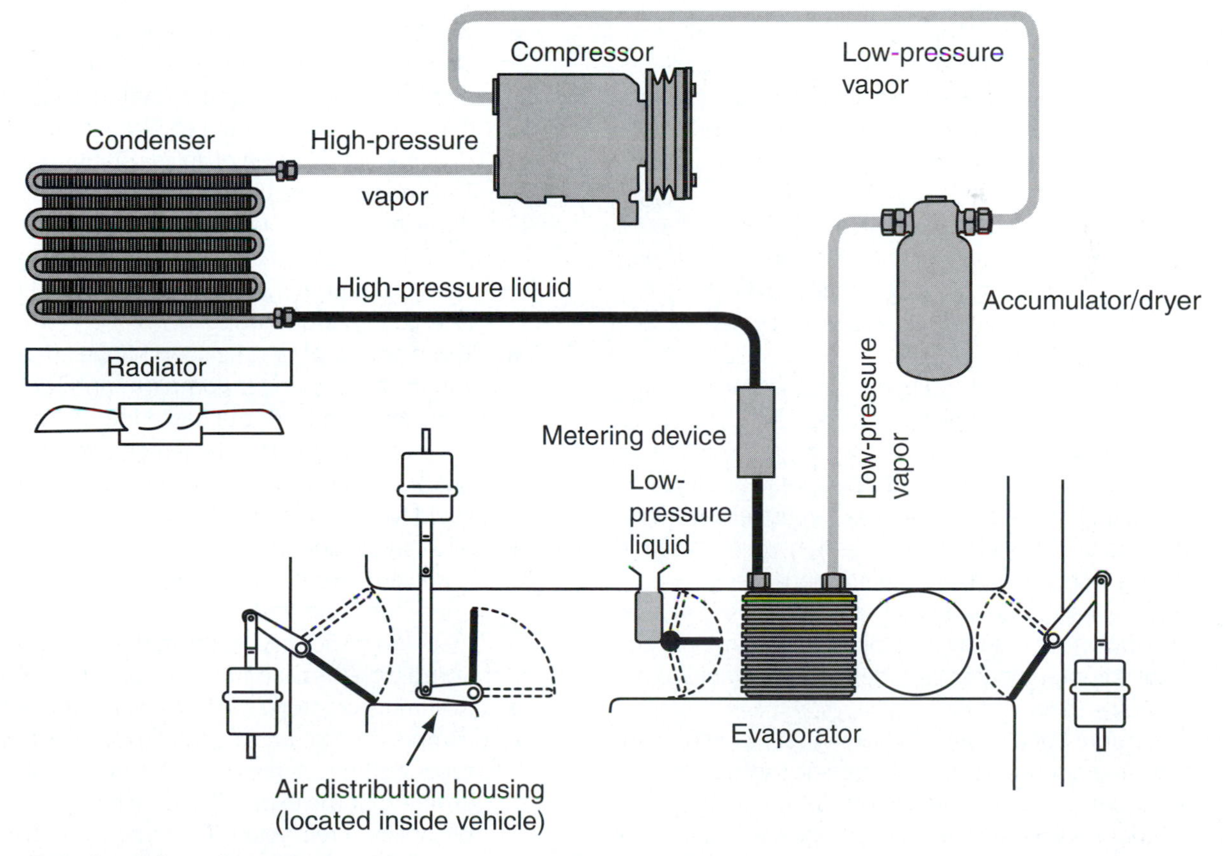

Figure 34. Details of the refrigeration cycle.

> **Interesting Fact**
>
> *Within the industry, the low-pressure side of the refrigeration system is usually called the "low" side or "suction side," whereas the high-pressure side can be called the "high side," "pressure side," or "discharge side."*

When the refrigerant reaches the condenser inlet, it is almost 100 percent vapor. A small amount of liquid may remain. As the refrigerant moves from the top of the condenser, it passes through the tubes. Much of the heat that is present within the vapor is transferred to the tubes and fins of the condenser. As air moves across the surface of the fins and tubes, heat is dissipated into the atmosphere. This process is aided by a ram air effect provided by the movement of the vehicle and the operation of the cooling fan. As the refrigerant moves down through the tubes of the condenser, much of the latent heat stored in the vaporized refrigerant is released, causing the vapor to condense into a liquid. The refrigerant exits the condenser as a high-pressure liquid.

High-pressure liquid refrigerant exits the condenser into the liquid line, where it is transported to the metering device. The metering device acts as a restriction, reducing the amount of refrigerant pressure and volume. As the liquid passes through the metering device, the pressure is reduced by approximately 75 percent or more. Because pressure and temperature are relative to one another, the temperature of the refrigerant is significantly reduced as the refrigerant exits the metering device. The refrigerant exits the metering device as a low-pressure, low-temperature liquid.

Low-pressure, low-temperature liquid refrigerant enters the evaporator core. As air is forced across the surface of the evaporator, heat that resides within the air is absorbed by the evaporator core and into the refrigerant, causing the refrigerant to vaporize, thus reducing the air temperature. As the air is cooled, moisture present in the air molecules condenses on the surface of the evaporator. Refrigerant exits the evaporator as a low-pressure, low-temperature vapor. The refrigeration cycle continues as the low-pressure, low-temperature vapor refrigerant enters the compressor suction hose and is transported to the compressor. At this point, the cycle starts over.

Summary

- The compressor has three specific tasks: to promote the circulation of refrigerant through the air conditioning system, to raise the pressure of the vaporized refrigerant, and to increase the temperature of the refrigerant. Refrigerant enters the compressor as a low-pressure, low-temperature vapor and exits as a high-pressure, high-temperature vapor.
- There are three primary types of compressors: reciprocating piston, rotary vane, and moveable scroll. Compressors can be either fixed displacement or variable displacement.
- The pistons in a reciprocating compressor can be driven by a crankshaft, axial plate, or scotch yoke. The crankshaft is used with inline and v-type compressors. The axial plate is used in compressors that use an axial cylinder arrangement. These compressors can use either single- or double-ended pistons. The scotch yoke is used in compressors with radial cylinder arrangement.
- Rotary vane compressors use a rotor and a set of vanes to form several small compression chambers. The scroll compressor uses two identical moveable spirals to compress refrigerant.
- All compressors use an electromagnetic clutch to engage and disengage the compressor. The electromagnetic clutch consists of a pulley, a drive plate, and a magnetic coil. When supplied with power and ground, the clutch is engaged and the compressor operates.
- Variable displacement compressors have the ability to alter the amount of refrigerant that is circulated throughout the refrigeration system based on system demands. Many variable displacement compressors depend on the action of an adjustable axial plate.
- The condenser is a heat exchanger that is located at the front of the vehicle in front of the radiator. The condenser removes heat from the refrigerant and changes refrigerant state from a high-pressure vapor to a high-pressure liquid.
- The evaporator is a heat exchanger that is located within the passenger compartment. The evaporator removes heat from the passenger compartment air. The heat absorbed by the refrigerant within the evaporator causes it to change state from a low-pressure liquid to a low-pressure vapor.
- All air conditioning systems use one metering device per evaporator. Types of metering devices include the fixed orifice tube and the thermostatic expansion valve. The metering device causes the refrigerant to change state from a high-pressure liquid to a low-pressure liquid.
- Accumulators stop liquid refrigerant from entering the compressor. Accumulators are used only in fixed orifice tube systems. A receiver dryer stores refrigerant to ensure that the expansion valve has a consistent supply of liquid refrigerant. Receiver dryers are only used with thermostatic expansion valves. Both accumulators

and receiver dryers remove transient moisture from the refrigerant.

- Barrier hoses are multilayered hoses that are used to prevent the leakage of refrigerant. Rigid aluminum lines may be used in some applications.
- Refrigerant changes state two times through the course of the refrigerant cycle, either absorbing or releasing heat each time a change of state occurs.

Review Questions

1. What are three specific tasks of the compressor?
2. Describe how the refrigerant changes as it moves through the compressor.
3. Technician A says that compressors have an intake and a compression stroke similar to an internal combustion engine. Technician B says that refrigerant enters the compressor cylinder when the reed valve deflects toward the cylinder. Who is correct?
   A. Technician A
   B. Technician B
   C. Both Technician A and Technician B
   D. Neither Technician A nor Technician B
4. Technician A says that a broken reed valve can cause the corresponding cylinder to lose compression. Technician B says that a broken inlet reed will allow high-pressure refrigerant to be forced back into the compressor inlet and suction hose. Who is correct?
   A. Technician A
   B. Technician B
   C. Both Technician A and Technician B
   D. Neither Technician A nor Technician B
5. Review the design and operation of the crankshaft, axial, and radial reciprocating compressors. Describe how each of these compressors are alike and how they differ.
6. Describe the operation of the rotary vane compressor.
7. Technician A says that the electromagnetic clutch uses friction to engage the compressor. Technician B says that the electromagnetic clutch uses the principles of magnetic attraction to engage the compressor. Who is correct?
   A. Technician A
   B. Technician B
   C. Both Technician A and Technician B
   D. Neither Technician A nor Technician B
8. All of the following statements about the variable displacement reciprocating compressor are true *except*:
   A. The angle or pitch of the axial plate can be changed when the compressor is operating.
   B. The variable compressor uses a double-ended piston.
   C. When suction pressure is high, crankcase pressure is equal.
   D. When suction pressure is low, outlet pressure is leaked into the crankcase.
9. Explain how refrigerant changes state within the condenser.
10. Why is the evaporator outlet tube larger than the inlet?
11. Describe how the action of the evaporator dehumidifies the air within the passenger compartment.
12. Technician A says that refrigerant enters the metering device as a high-temperature vapor. Technician B says that the refrigerant leaves the metering device as a low-pressure liquid. Who is correct?
   A. Technician A
   B. Technician B
   C. Both Technician A and Technician B
   D. Neither Technician A nor Technician B
13. Which of the following statements about an accumulator is incorrect?
   A. It is used in conjunction with a fixed orifice tube.
   B. It is located in between the evaporator and compressor.
   C. It ensures that the orifice tube has a constant supply of liquid refrigerant.
   D. It uses a desiccant to remove moisture from the refrigerant.
14. Describe which components are connected together using the following hoses and lines: discharge hose, liquid line, and suction hose.
15. Explain what occurs during the refrigeration cycle beginning at the compressor inlet and ending at the compressor inlet.

Chapter 14

Orifice Tube Refrigeration Systems

Introduction

All automotive refrigeration systems, regardless of manufacturer, use basically the same components that operate in the same fashion. And although the component itself may look different depending on the manufacturer, the operation remains almost identical. The most obvious difference in operation of refrigeration systems is the type of metering device that is used. The metering device serves as restriction to the flow of liquid refrigerant between the condenser and the evaporator (see **Figure 1**). This restriction effectively meters the amount of liquid refrigerant that is allowed to enter the evaporator. In this chapter, we will take a close look at those systems that employ the use of an orifice tube.

REFRIGERANT METERING

In the automotive industry, the efficiency of the air conditioning system is represented by the temperature of the air that is exiting the dash registers. That air temperature is a direct representation of the temperature of the evaporator. Consequently, the temperature of the evaporator is directly proportional to the amount of refrigerant that is moving through the evaporator tubes. The proper operation of the air conditioning system is dependent on the flow of the refrigerant through the system. In simple terms, the metering device provides a separation between the high and low sides of the refrigeration system and provides low-pressure liquid refrigerant into the evaporator (see **Figure 2**). However, to a large extent, the metering device actually controls the flow of refrigerant throughout the entire refrigeration system.

FIXED ORIFICE TUBE SYSTEMS

Fixed orifice tube (**FOT**) systems make use of a "flooded" evaporator. In the basic description of refrigeration system operation, we made the distinction that refrigerant leaves the evaporator as a low-pressure vapor. This is only partially true. In order for the refrigerant to exit the evaporator as 100 percent vapor, the top portion of the evaporator would only contain vapor. Because vaporized refrigerant is unable to absorb as much heat as a liquid refrigerant, the top portion of the evaporator would be warmer than the bottom. This would create uneven cooling across the surface of the evaporator and effectively reduce its capacity. By flooding the evaporator, enough liquefied refrigerant is allowed into the evaporator to virtually fill it, making the entire surface of the evaporator available for cooling. This increases its overall capacity (see **Figure 3**).

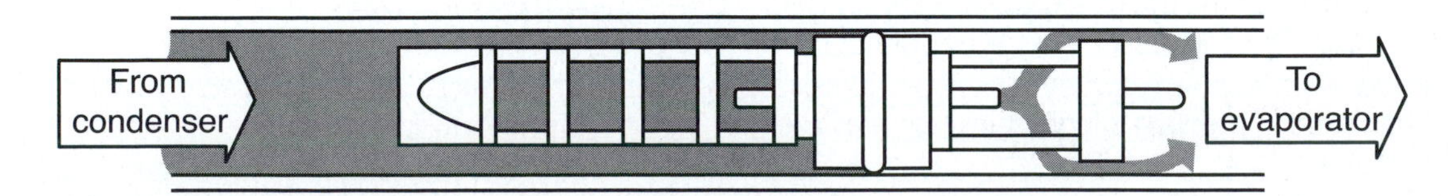

Figure 1. The orifice tube meters the amount of refrigerant that is allowed to enter the evaporator.

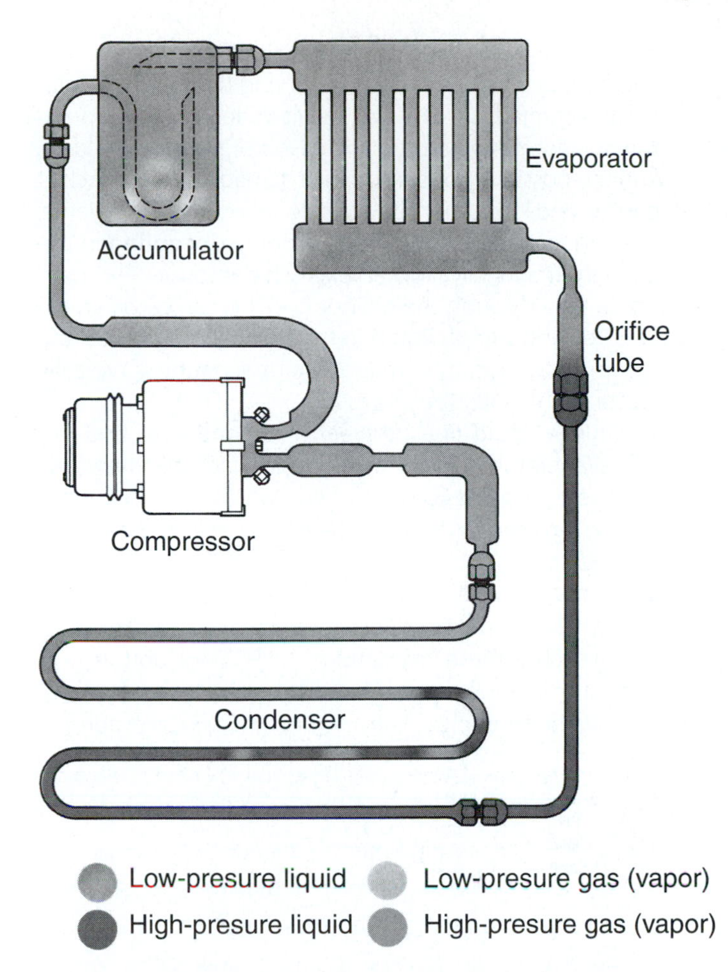

Figure 2. The orifice tube separates the high-pressure and low-pressure sides of the system and meters liquid refrigerant into the evaporator.

> **You Should Know** *When the air conditioning system is turned off, the pressure within the system will immediately begin to equalize. This can be recognized by a hissing noise in the engine compartment. This is a normal condition and will usually last up to about 1 minute.*

The negative side to evaporator flooding is that some liquid refrigerant will leave the evaporator. An accumulator is installed on the evaporator outlet to collect the liquid refrigerant that escapes the evaporator and to provide a place for the refrigerant to vaporize before returning to the compressor (see **Figure 4**).

ORIFICE TUBE

The fixed orifice tube is the simplest design of a metering device. It consists of a small brass tube encased in a plastic housing with a mesh filter. The orifice tube is equipped with a small mesh screen on each side of the tube to protect the tube from debris (see **Figure 5**). The brass tube is considerably smaller than the tubes leading into and away from the orifice. The size of the tube is fixed; therefore, the amount of refrigerant that is allowed to pass through the orifice is relatively consistent, changing only in relation to the difference in system pressures and refrigerant volume. Orifice tube sizes range from .047 in. (1.19 mm) to .067 in. (1.7 mm) and are usually color-coded for easy identification (see **Figure 6**). The selection of the FOT by the manufacturer is based on a compromise of the flow needs of the refrigeration system during all possible

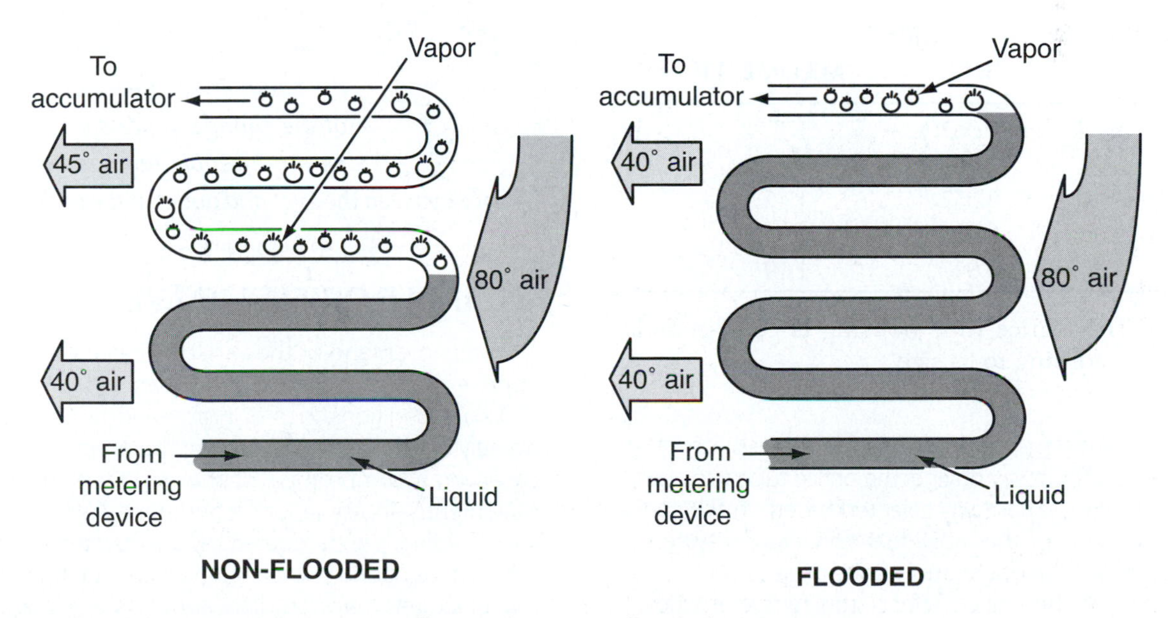

Figure 3. A flooded evaporator is more efficient than a nonflooded evaporator.

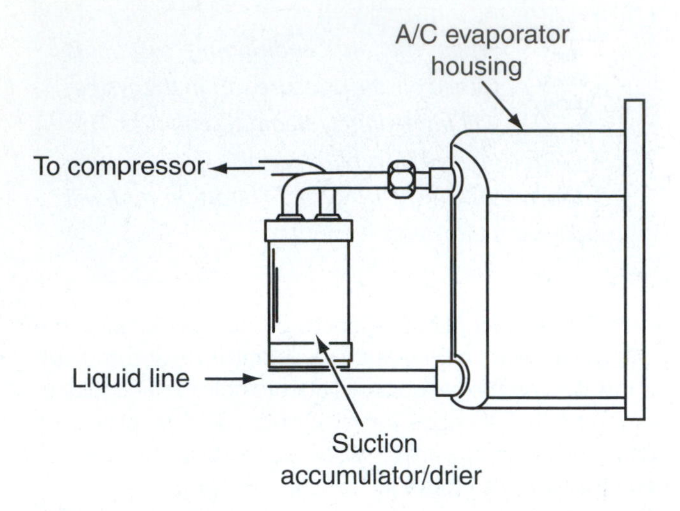

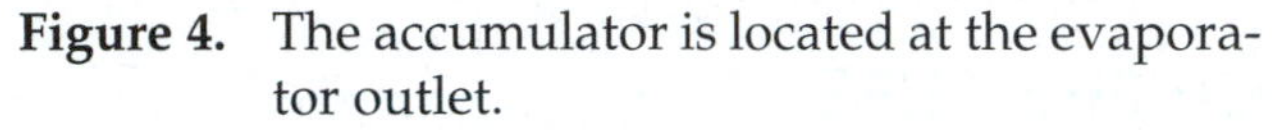
Figure 4. The accumulator is located at the evaporator outlet.

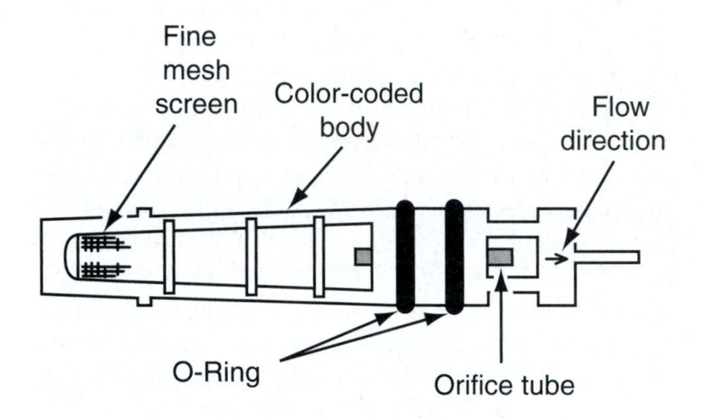

Figure 5. A cutaway view of an orifice tube.

COLOR	ORIFICE	
	INCH	MILLIMETER
Blue or black	0.067	1.7
Red	0.062	1.57
Orange	0.057	1.45
Brown	0.053	1.35
Green	0.047	1.19

Figure 6. The orifice tube housing is color-coded according to its size.

operating conditions, with primary emphasis on the requirements of highway driving. The orifice tube size used in any application is carefully selected based on the specific requirements of the individual refrigeration system; therefore, orifice tube size should not be altered.

The size of the orifice affects the refrigerant flow both before and after the orifice. If the size of the orifice is intentionally decreased or becomes restricted, the flow of liquid refrigerant is decreased, starving both the evaporator and compressor. This can result in decreased evaporator efficiency and cooling, increased high-side pressures, and the possibility of compressor damage due to a lack of cooling and lubrication. Conversely, if the orifice is too large, the refrigerant will not receive proper **subcooling** in the condenser and will allow an excessive amount of refrigerant to flow through the evaporator. This can cause an excessive amount of liquid refrigerant to return to the compressor, causing damage. The orifice tube is typically located in one of three places:

- In the liquid line between the condenser and the evaporator. Most of these are replaceable; however, there are some Ford vehicles in which the tube cannot be replaced. In this case, the entire liquid line has to be replaced.
- Installed in the evaporator inlet.
- Installed in the condenser outlet.

In each of these installations, the FOT will "bottom out" against indentions in the line. This stops the tube from moving too far into the line and becoming irretrievable.

You Should Know *Subcooling is an important term that is basically a measurement of refrigerant temperature in relation to its vaporization point. It is expressed as the number of degrees that a refrigerant is below its vaporization point. The more a refrigerant has been subcooled, the more heat it is able to absorb within the evaporator.*

Interesting Fact *If you are unsure of the orifice tube location, it can be located by feeling the liquid line temperature. Because of the change in pressure, there will be a noticeable temperature difference between the inlet and outlet of the orifice tube.*

VARIABLE ORIFICE VALVE

The advantage of the variable orifice valve or VOV is improved air conditioning performance during high heat and load conditions, such as in stop-and-go traffic or on extremely hot days. The selection of the proper size of FOT is based on a compromise of the refrigeration system requirements under various conditions such as highway speeds, idling, high temperatures, and low temperatures. To obtain better performance during all operating conditions, manufacturers have studied and, in some applications, used variable orifice valves, or VOV. These valves began to

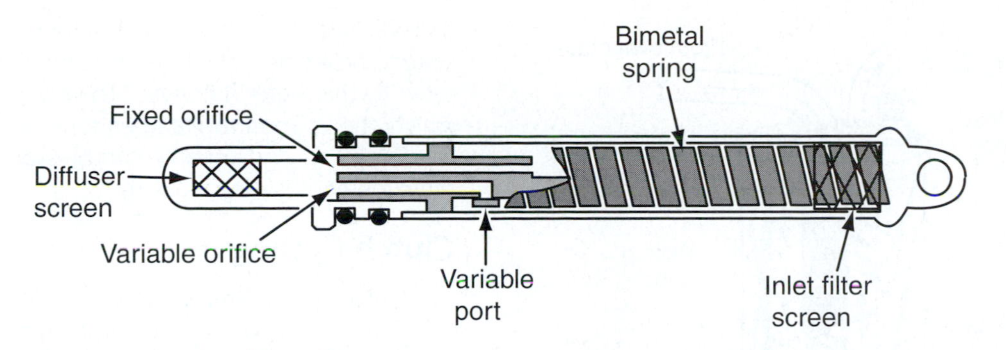

Figure 7. Cutaway view of a variable orifice tube.

appear in some 1999 model vehicles and are currently available for retrofit for some applications. Externally, the VOV is similar in appearance to the FOT but internally the VOV provides two parallel refrigerant paths. One is a fixed orifice opening similar to an FOT, and the other is a variable orifice opening. The opening through the fixed orifice tube is normally 0.047 in. (1.1938 mm), and the variable orifice tube opening ranges from 0 to 0.015 in. (0 to 0.381 mm). A VOV is illustrated in **Figure 7**.

The variable orifice tube operates in relation to the temperature of the refrigerant exiting the condenser. In an FOT system, when the vehicle is operating at low speeds or is idling, the refrigerant retains a great deal of heat because of the decreased airflow through the condenser. The volume of refrigerant is also low as a result of the low engine speed. Under these conditions, the refrigerant is able to move through the condenser very rapidly—so rapidly, in fact, that it does not fully condense as it passes through, and sends a very high-temperature vapor and liquid refrigerant mixture to the orifice tube and evaporator. This creates poor cooling conditions in the evaporator. When a VOV is used, the bimetal control spring acts to close off one of the ports within the valve assembly. This slows down the flow of refrigerant through the system—specifically, through the condenser—allowing more heat to be removed and the availability of more liquid refrigerant at the evaporator.

When the vehicle again begins to move, the amount of refrigerant circulated through the system is increased at the same time as the airflow is increased, effectively lowering the pressure and temperature of the refrigerant. When this occurs, the bimetal spring again moves, opening the auxiliary passage in the VOV and providing maximum refrigerant flow to the evaporator.

ACCUMULATOR

The accumulator is a tanklike vessel that is located at the outlet of the evaporator (see **Figure 8**). It is an essential part of an orifice tube air conditioning system. In a flooded evaporator system, some liquid refrigerant can flow from the evaporator outlet. Although the compressor can

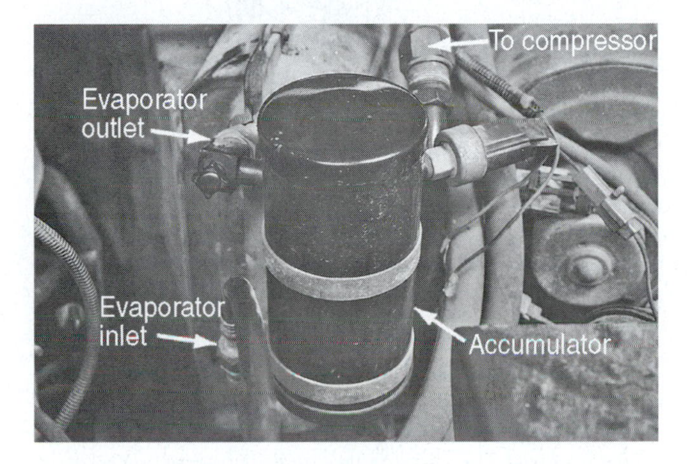

Figure 8. The accumulator is installed on the evaporator outlet.

tolerate a small amount of liquid refrigerant, large amounts of refrigerant can cause **slugging**. Slugging is a condition in which the compressor pumps excessive amounts of liquid. Because liquid cannot be compressed, compressor damage can occur. To prevent this problem, all refrigerant and oil leaving the evaporator must enter the accumulator. The accumulator allows the refrigerant vapor to pass on to the compressor and traps the liquid refrigerant and oil.

The refrigerant enters at the top of the accumulator tank and the heavier liquid refrigerant and oil fall to the bottom of the tank, whereas the vapor rises to the top. The accumulator outlet is connected to a specially designed "u"-shaped pickup tube. The inlet is located near the top of the tank. This prevents large amounts of liquid refrigerant from entering the compressor. At the bottom of the pickup tube there is a calibrated orifice, known as an "oil bleed hole," not to be confused with the orifice tube. It allows small amounts of liquid refrigerant and oil to enter the vapor path and return to the compressor with the vapor, providing both lubrication and cooling (see **Figure 9**).

A secondary task of the accumulator is to remove any transient moisture that may be present in the refrigerant. The accumulator contains **desiccant** that is sealed within the accumulator tank. A desiccant is a chemical drying agent. The

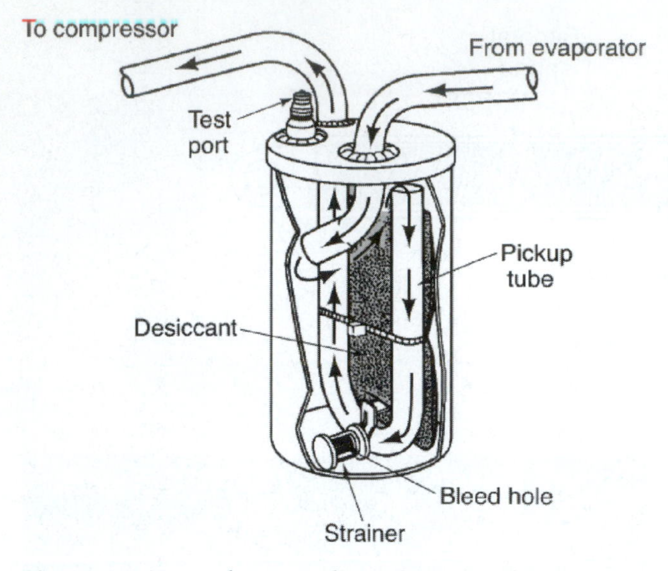

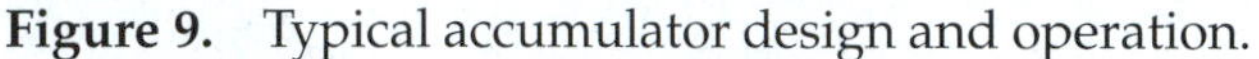

Figure 9. Typical accumulator design and operation.

desiccant attracts, absorbs, and holds moisture that may have entered the system during service. The desiccant is sealed within the accumulator and is not serviced separately from the accumulator. If desiccant replacement is indicated, the accumulator must be replaced as an assembly.

There are several desiccant types that have been used through the years. Not all desiccants are compatible with all refrigerants. The desiccants are classified XH5, XH7, and XH9. Only XH7 and XH9 are acceptable for use on R-134a systems. To be sure of system compatibility, use only replacement components specifically designated for a particular application.

You Should Know *Because the compressor bleed hole is essential to proper compressor oiling, any restriction of the orifice can result in compressor damage. The accumulator should be replaced any time the compressor is replaced.*

EVAPORATOR TEMPERATURE CONTROL

Evaporator temperature is of critical importance to the air conditioning technician. If the surface of the evaporator is too warm, then the air within the passenger compartment will not be properly cooled. Conversely, if the evaporator temperature becomes too cold, the condensation that is removed from the passenger compartment air will freeze and block the airflow across the core. This will cause a reduction in airflow and an increase in outlet temperatures. In a TXV system, the temperature of the evaporator can be controlled by altering the amount of refrigerant that is allowed to enter the evaporator. However, in the FOT system, the amount of refrigerant is fixed and cannot be altered by the metering device. There are two methods in which the evaporator temperature can be controlled: clutch cycling and variable compressor output. Each of these methods is used regularly.

Clutch Cycling

Clutch cycling simply involves turning the compressor off when evaporator temperature drops to a specific temperature. When surface temperature increases, the compressor is turned back on. Clutch cycling can be accomplished using two methods: temperature clutch cycling and pressure cycling. These methods will be covered extensively in section 5. Very basic descriptions of these methods are presented here.

- Temperature cycling switch. This is a temperature sensitive switch that uses a sensing bulb or line to measure evaporator temperature. When temperatures reach preset high and low thresholds, the switch turns the clutch on and off.
- Pressure cycling switch. Because pressure and temperature are relative to one another, a pressure switch that monitors low-side pressure is used to turn the clutch off and on. Like the temperature cycling switch, the pressure cycling switch is calibrated to cycle at specific pressures. This switch is typically located on the accumulator.

Variable Compressor Output

A variable displacement compressor is used in some FOT models. The variable compressor controls evaporator temperature by adjusting the volume of refrigerant that is introduced into the system. This is accomplished by varying the stroke of the pistons within the compressor (see Chapter 13). When the operator of the vehicle requests the operation of the A/C system, the compressor clutch engages and remains engaged until the vehicle is turned off or the A/C is no longer requested.

The control valve assembly located within the compressor monitors suction pressure to determine refrigeration system load. When refrigeration system demands are high, it causes an increase in suction pressure. When the pressure is above the **control point** of the variable displacement control valve, the compressor will operate at full capacity. The control point is simply the point at which the valve activates or deactivates to make a change in displacement. As the demand on the system decreases, the suction pressure will drop accordingly. When the suction pressure drops below the control point, the valve acts by decreasing the stroke of the pistons and reducing the capacity of the compressor, thereby increasing the temperature of the evaporator (see **Figure 10**).

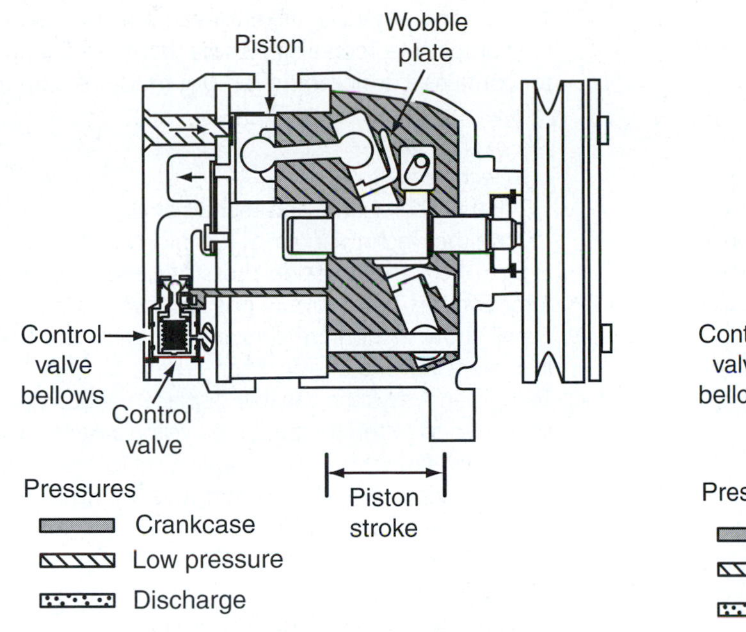

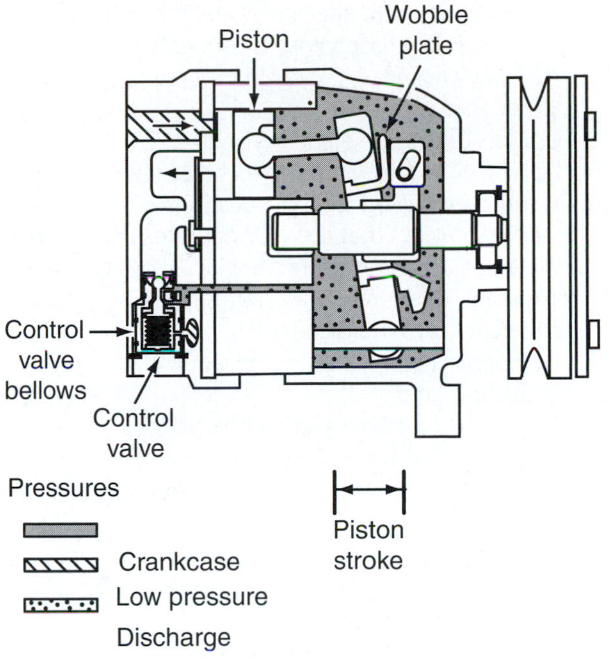

Figure 10. Operation of a variable displacement compressor.

Summary

- Evaporator temperature is directly related to the amount of refrigerant that is passing through the tubes. The metering device separates the high side and low side of the refrigeration system.
- The metering device controls the flow of refrigerant through the refrigeration system and meters the amount of liquid refrigerant allowed into the evaporator.
- The FOT system makes use of a flooded evaporator. Evaporator efficiency is increased by increasing the amount of liquid refrigerant allowed to enter the evaporator. Flooded evaporator systems must use an accumulator to capture any liquid that exits the evaporator.
- The FOT is among the simplest metering devices. The FOT uses small brass tube of a fixed diameter to provide the necessary refrigerant restriction. Orifice tube size is carefully selected, based on the specific requirements of the refrigeration system.
- Orifice tube size affects the flow of refrigerant both before and after the restriction. The orifice can be located in any one of three locations: the liquid line, the condenser outlet, and the evaporator inlet.
- The VOV provides both a fixed restriction and a variable restriction to refrigerant flow. The VOV operates in relation to the temperature of the liquid refrigerant exiting the condenser.
- The accumulator is located at the evaporator outlet and is designed to capture liquid refrigerant and lubricant and to remove any transient moisture.
- The evaporator temperature may be controlled by compressor cycling and variable compressor output.

Review Questions

1. Technician A says that the metering device controls the amount of liquid refrigerant that enters the evaporator. Technician B says that the metering device controls the flow of refrigerant in the entire system. Who is correct?
   A. Technician A
   B. Technician B
   C. Both Technician A and Technician B
   D. Neither Technician A nor Technician B
2. Explain how a flooded evaporator is more efficient than a nonflooded evaporator.
3. Technician A says that altering the size of the orifice can cause damage to the compressor. Technician B

says that it is okay to alter the size of the orifice as long as the size is decreased. Who is correct?
   A. Technician A
   B. Technician B
   C. Both Technician A and Technician B
   D. Neither Technician A nor Technician B
4. List three specific orifice tube locations.
5. Technician A says that the VOV operates in relation to refrigerant temperature. Technician B says that when refrigerant temperature is high, the VOV increases refrigerant flow. Who is correct?
   A. Technician A
   B. Technician B
   C. Both Technician A and Technician B
   D. Neither Technician A nor Technician B
6. Which of the following statements about an accumulator is correct?
   A. It is located between the condenser and evaporator.
   B. It ensures a constant supply of liquid refrigerant to the orifice tube.
   C. It separates the oil and refrigerant.
   D. It meters small amounts of oil and refrigerant to the compressor.
7. Technician A says that all refrigerants and desiccants are compatible. Technician B says that the accumulator contains a desiccant to remove moisture from the refrigerant. Who is correct?
   A. Technician A
   B. Technician B
   C. Both Technician A and Technician B
   D. Neither Technician A nor Technician B
8. Explain how the low-side pressures and evaporator temperatures are relative to one another.
9. Explain how suction pressure and refrigeration system load are relative to one another.
10. Technician A says the variable displacement compressor cycles off when the pressure cycling switch opens. Technician B says that the variable displacement compressor uses a temperature sensitive switch to cycle the compressor. Who is correct?
   A. Technician A
   B. Technician B
   C. Both Technician A and Technician B
   D. Neither Technician A nor Technician B

Chapter 15 Thermostatic Expansion Valve Refrigeration Systems

Introduction

The thermostatic expansion valve (TXV) is the second type of metering valve that is used in automotive applications. Like the orifice tube, the thermostatic expansion valve provides a restriction to refrigerant flow between the condenser and the evaporator, thus providing a separation between the high and low side of the refrigeration system. However, the orifice tube provides a fixed restriction, while the expansion valve provides a constantly variable restriction that actively adjusts the flow of refrigerant based on system requirements. In this chapter, we will examine the construction and operation of those systems that use a thermostatic expansion valve.

REFRIGERANT METERING

Efficiency of the automotive refrigeration system is represented by the temperature of the air exiting the dash registers, and that air is a direct representation of evaporator temperature. Consequently, the evaporator temperature is directly related to the amount of refrigerant that is flowing through the evaporator core. It is the job of the TXV to meter the proper amount of refrigerant into the evaporator based on evaporator temperatures. When evaporator temperatures become too high, more refrigerant is admitted into the evaporator. As temperatures decrease, the amount of refrigerant entering the evaporator is restricted. Like the FOT, the TXV manages the flow of refrigerant throughout the system (see **Figure 1**). Factors that affect refrigerant flow through the TXV are evaporator temperature, refrigerant volume, and pressure. However, because the TXV is constantly variable, it is better equipped to handle these conditions. Unlike the FOT, the TXV system does not use a flooded evaporator. Therefore, to prevent compressor damage, the refrigerant must exit the evaporator as almost pure vapor. The TXV has three main functions:

- **Throttling:** to provide a pressure drop across the system between the high side and low side of the system by providing a restriction.
- **Modulating:** the valve can move in a range from a wide-open position to a fully closed position to provide the proper amount of refrigerant to the evaporator.
- **Metering:** the active process by which the valve senses evaporator load conditions and adjusts refrigerant flow accordingly.

THERMOSTATIC EXPANSION VALVE

The thermostatic expansion valve is located at the inlet side of the evaporator and receives high-pressure liquid refrigerant from the receiver dryer (see **Figure 1**). The TXV may be physically located in the engine compartment or within the evaporator case inside the vehicle; typical installations are shown in **Figure 2**. The TXV consists of a valve body that contains various internal passages and provisions for refrigerant to both enter and exit. Between the inlet and the outlet is a valve and seat assembly. These components comprise the adjustable orifice that controls the amount of refrigerant entering the evaporator. The control components of the valve consist of a calibrated spring and a sealed diaphragm assembly. The calibrated spring is installed in such a manner that it acts to close the valve, blocking off the flow of refrigerant. The diaphragm is installed in a position and it acts to open the valve. TXV construction is illustrated in **Figure 3**.

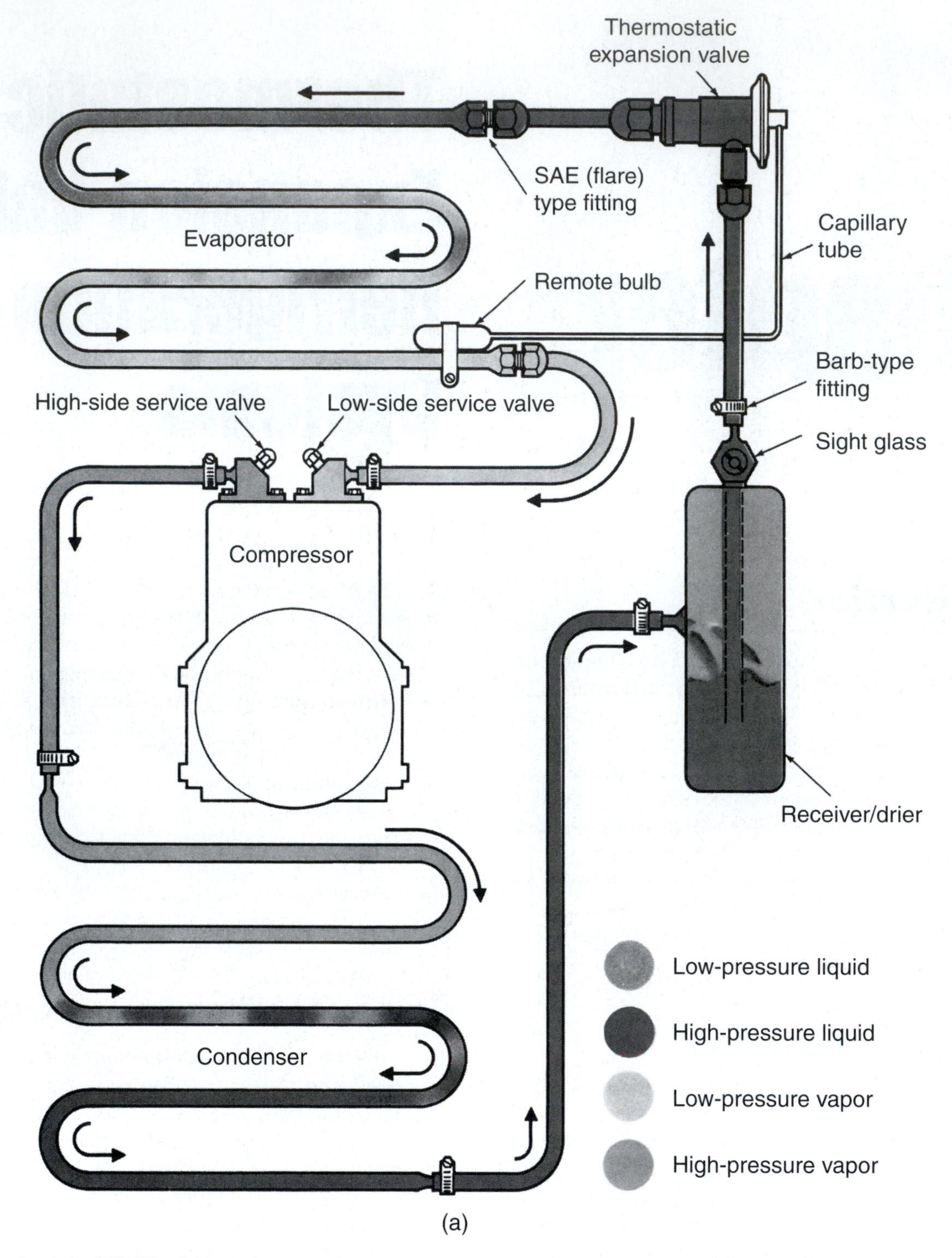

Figure 1. A typical TXV refrigeration system.

The diaphragm is a sealed dual-sided control mechanism. The diaphragm is located at the top of the TXV. One side of the diaphragm acts against the internal valve and is exposed to the same pressure as the evaporator. This is called **balance pressure**. The balance pressure is used to help balance the actions between the valve and the diaphragm. The balance pressure helps the valve make very fine and smooth adjustments to the refrigerant flow rate; this assists the valve in reducing broad evaporator temperature fluctuations. The TXV may be **internally balanced**, meaning that balance pressure is supplied through a passage drilled inside the valve, or **externally balanced**, meaning that an external tube is used to supply evaporator pressure to the diaphragm.

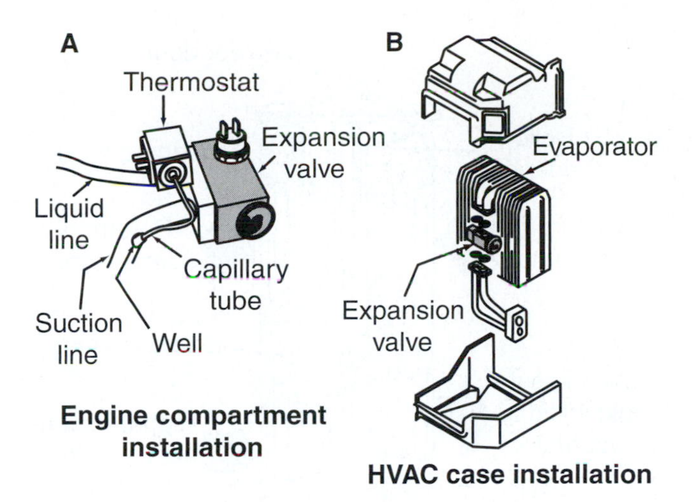

Figure 2. The TXV can be located in the engine compartment (A) or in the evaporator case (B).

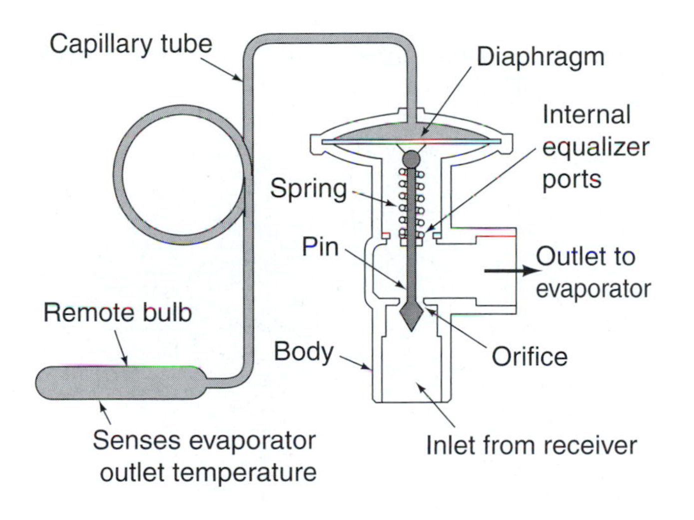

Figure 3. Typical construction of a TXV.

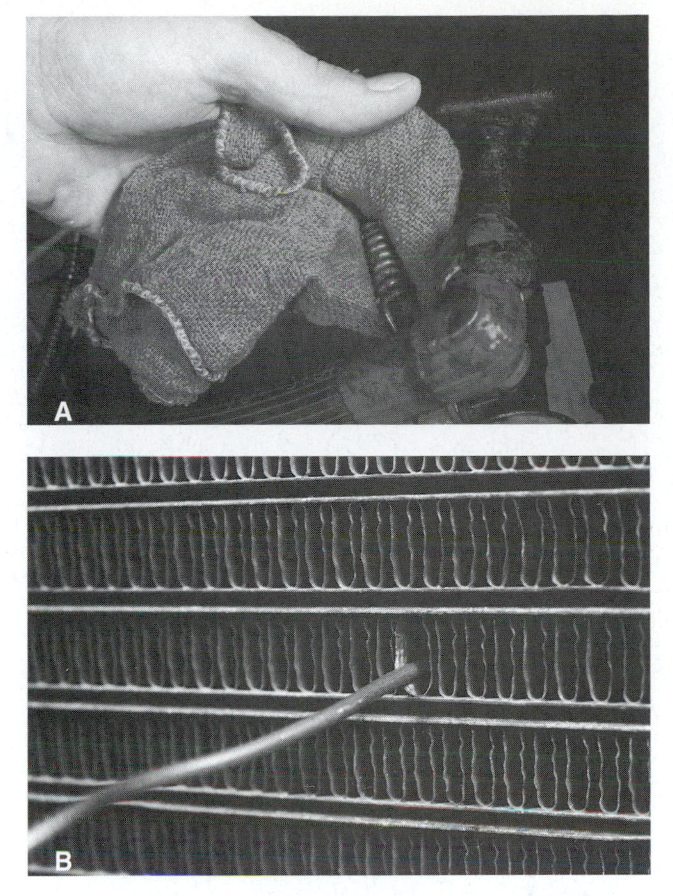

Figure 4. The capillary tube can be attached to the evaporator outlet (A) or can be inserted into the evaporator fins (B).

The top of the diaphragm forms a sealed chamber. Attached to the chamber is a copper tube that is used as a remote temperature sensing device. This is called a **capillary tube.** Some capillary tube designs have a large sensing bulb attached to the end of the tube; this is called the **remote sensing bulb**. The diaphragm, tube, and bulb are filled with a temperature sensitive gas such as refrigerant or carbon dioxide. The tube or bulb is attached to the evaporator so that it is able to sense the evaporator temperature near the outlet. The tube can be clamped against the outlet pipe (see **Figure 4A**), or inserted into the fins of the evaporator core (see **Figure 4B**). As the temperature of the evaporator increases, it causes the gas within the bulb, tube, and diaphragm to expand. This action causes the valve to open and allows more refrigerant into the evaporator.

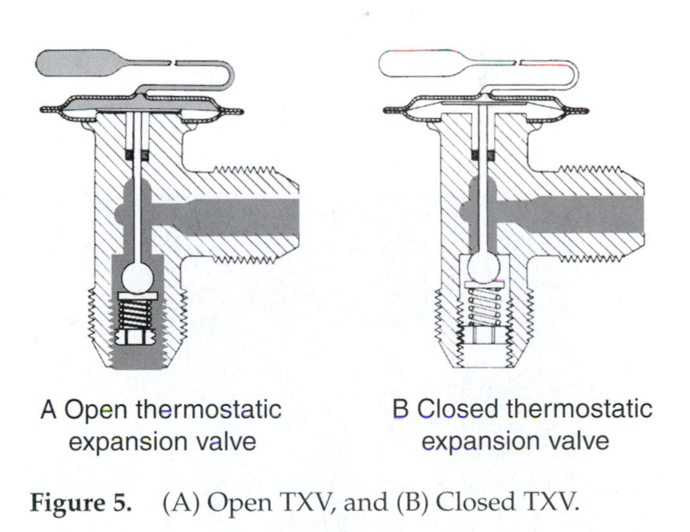

Figure 5. (A) Open TXV, and (B) Closed TXV.

When evaporator temperatures drop, the gas within the tube and diaphragm contract and the valve is allowed to close. Opening and closing actions of the TXV are illustrated in **Figure 5**.

You Should Know *To operate properly, the remote sensing bulb and sensing end of the capillary tube cannot be exposed to ambient air temperatures. When the remote sensing bulb or capillary tube is clamped to the evaporator outlet line, it is sealed and insulated with a specialized tape that prevents exposure to the atmosphere.*

H-Valve

A second style of TXV used in a large number of Chrysler products through the years and in a more limited capacity by other manufacturers is the H-valve. The H-valve is a completely self-contained version of the TXV. The H-valve has four ports rather than two and mounts to both the inlet and outlet of the evaporator (see **Figure 6**). Although the H-valve still uses a sealed gas-filled control diaphragm, called a **power dome**, it does not use a capillary tube. The control mechanism for the power dome is located in the upper portion of the valve, where it is exposed to the refrigerant vapors leaving the evaporator (see **Figure 7**). Other than these design variations, the operation of the valve is virtually the same as the TXV. Various styles of thermostatic expansion valves are shown in **Figure 8**.

RECEIVER DRIER

The receiver drier is used exclusively with TXV systems. The receiver drier is installed in the high-pressure side of the system between the condenser and the metering device. The receiver drier may attach directly to the condenser or can reside in the liquid line (see **Figure 9**). The receiver drier separates liquid refrigerant from any vapors that may

Figure 6. The H-valve is attached to both the evaporator inlet and outlet.

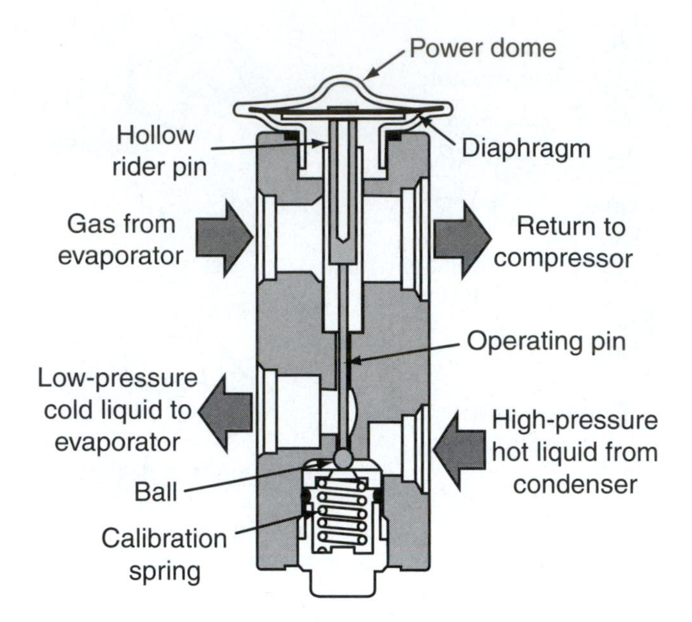

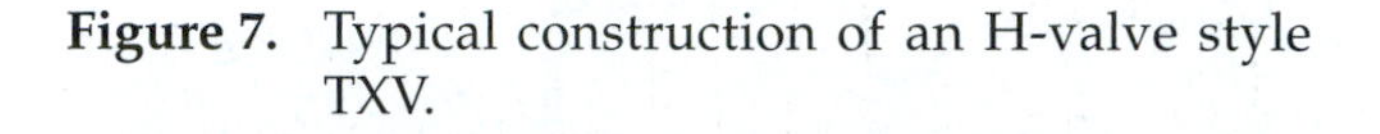

Figure 7. Typical construction of an H-valve style TXV.

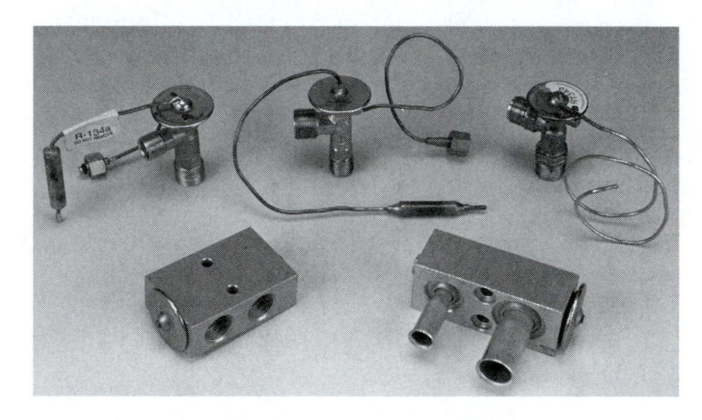

Figure 8. Various styles of thermostatic expansion valves.

have entered the liquid line. This helps ensure that 100 percent liquid is supplied to TXV.

In appearance, the receiver drier is very similar to the accumulator. The receiver section of the receiver drier is a tanklike storage compartment. This section holds the proper amount of reserve refrigerant required to ensure acceptable air conditioning system performance under variable operating conditions. The refrigerant from the condenser enters the receiver drier at the inlet; heavy liquid refrigerant falls to the bottom of the tank, whereas the vapor rises toward the top. Because the TXV requires a steady supply of liquid refrigerant, liquid is drawn from the bottom of the receiver drier rather than the top, as in an accumulator. An outlet tube extends to the bottom of the tank and is submerged in liquid refrigerant. The pickup tube is fitted with a strainer that is used to block the circulation of debris out of the receiver drier. Typical receiver drier construction is illustrated in **Figure 10**.

Figure 9. Typical receiver drier installation.

A desiccant, located within the tank of the receiver drier, is used to remove transient moisture that may be present within the refrigerant. The desiccants used in the receiver drier are classified exactly as those used in the accumulator and must be compatible with the intended refrigerant.

> **You Should Know** *Some receiver driers may have a sight glass installed. The sight glass can actually allow the technician to see the refrigerant as it flows within the liquid line. This can sometimes be useful in performing system diagnosis.*

> **You Should Know** *Because the receiver drier will inevitably collect any debris resulting from a compressor failure, the component should be replaced any time that the compressor is replaced.*

EVAPORATOR TEMPERATURE CONTROL

We want to keep the temperature of the evaporator surface as near to 32 degrees F (0 degrees C) as possible without causing the condensation that is present on the surface to freeze. The temperature of the evaporator in a TXV system is controlled by the opening and closing of the TXV. In most TXV systems, the compressor will stay engaged at all times under normal circumstances. Some manufacturers may make use of a variable displacement compressor to enhance the operation of the TXV.

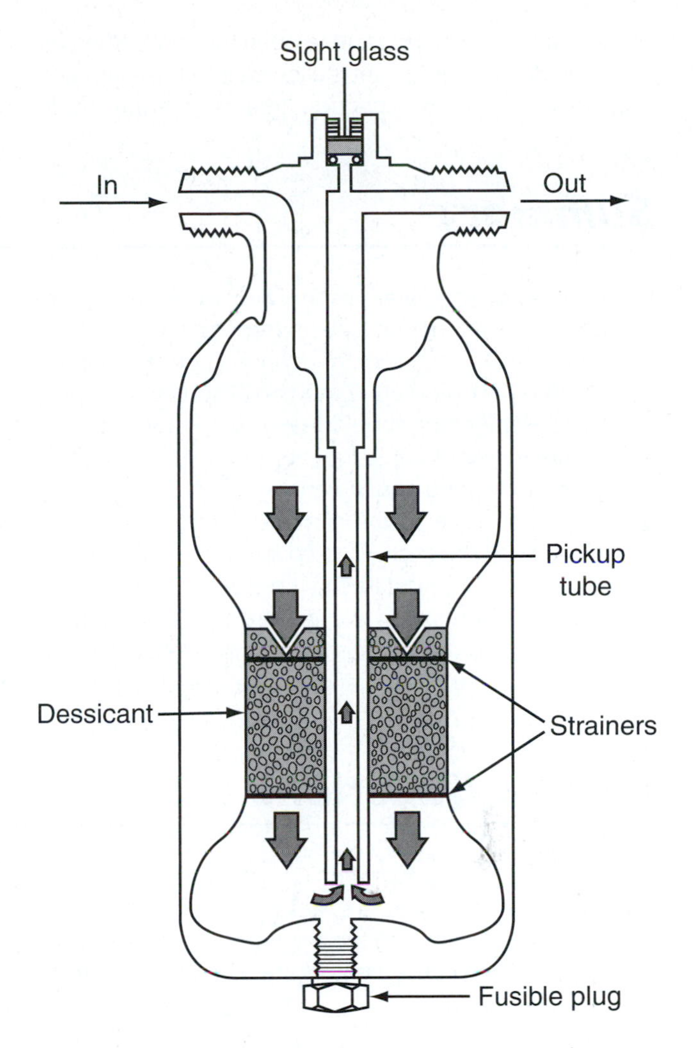

Figure 10. Typical construction of a receiver drier.

In the FOT system, a constant amount of refrigerant is allowed into the evaporator. When evaporator temperatures become low enough, the compressor is turned off for a short time until the TXV has detected that the surface temperatures have increased. The control of the TXV is much more precise. As the system operates, the sensing bulb responds to minute changes in surface temperature. When evaporator temperatures are low, the control diaphragm retracts and the spring is allowed to close the valve; this reduces the amount of refrigerant allowed to enter the evaporator. When the refrigerant supply is restricted, the evaporator temperatures will again increase. As temperatures increase, the gas within the capillary tube expands and the diaphragm moves against both spring pressure and balance pressure to open the valve. When this occurs, additional refrigerant is allowed into the evaporator and again causes a reduction in temperature. Under normal operating conditions, once the system becomes stabilized the valve will only experience small incremental movements in valve position (see Figure 5).

Summary

- Evaporator temperatures are directly related to the amount of refrigerant flowing through the evaporator core. The TXV meters in the proper amount of refrigerant into the evaporator based on evaporator temperatures. Refrigerant should leave the evaporator as a complete vapor. Three functions of the TXV are throttling, modulating, and metering.
- The TXV is located in the inlet side of the evaporator. The TXV can be physically located in the engine compartment or it can be installed in the evaporator case.
- The TXV consists of a valve and seat that form a variable orifice. A sealed diaphragm and spring are used to control the operation of the valve. The sealed diaphragm, capillary tube, and sensing bulb are filled with a temperature-sensitive gas. The remote sensing bulb is located in a position in which it can sense evaporator temperature.
- The receiver drier is used only in TXV systems. The receiver drier separates liquid and vapor to ensure 100 percent liquid refrigerant reaches the metering device. The receiver drier contains a desiccant that is used to remove transient moisture.
- The TXV controls evaporator temperature by restricting refrigerant flow when temperatures become cold and increasing refrigerant flow when temperatures increase.

Review Questions

1. Technician A says that the TXV separates the high side and the low side of the system. Technician B says that the TXV is installed on the evaporator outlet to control how much refrigerant enters the core. Who is correct?
   A. Technician A
   B. Technician B
   C. Both Technician A and Technician B
   D. Neither Technician A nor Technician B
2. Explain how the metering valve controls the flow of refrigerant into the evaporator.
3. Technician A says that all TXVs use a sealed diaphragm to control their operation. Technician B says that all TXVs must use a capillary tube. Who is correct?
   A. Technician A
   B. Technician B
   C. Both Technician A and Technician B
   D. Neither Technician A nor Technician B
4. Technician A says that if the gas leaked out of the capillary tube, the valve would always be open. Technician B says that if the gas leaked out of the capillary tube, the balance pressure would be able to control the valve. Who is correct?
   A. Technician A
   B. Technician B
   C. Both Technician A and Technician B
   D. Neither Technician A nor Technician B
5. Technician A says that receiver dryer converts high-pressure liquid into low-pressure liquid. Technician B says that the receiver drier supplies the TXV with high-pressure liquid refrigerant. Who is correct?
   A. Technician A
   B. Technician B
   C. Both Technician A and Technician B
   D. Neither Technician A nor Technician B

Chapter 16 Refrigeration System Service

Introduction

Almost any time a refrigeration system repair is performed, the refrigerant must be removed and the proper amount reinstalled. In the process of doing this, any moisture must be removed and system oil must be restored. These procedures are among the most commonly performed and are among the most important. This chapter will outline the procedures required for the recovery, evacuation, recycling, and recharging of the refrigeration system. The performance of these procedures will revolve around the use of a commercially available recovery, evacuation, recycling, and recharging station. This will be referred to as a recovery station (see **Figure 1**).

Figure 1. A typical recovery, recycling, and recharging unit.

SYSTEM CONTAMINATES

A major function of the recovery, evacuation, and recharging process is to remove contaminates from the refrigeration system. System contamination can come from a variety of sources; however, each of these is usually introduced through inefficient or incorrect service procedures. Any amount of contamination is detrimental to the operation of the refrigeration system. The refrigeration system is considered to be contaminated if it contains any substance for which the system is not designed. This can include, but is not limited to, air, moisture, a refrigerant of the incorrect composition, or a combination of refrigerants. Other sources of contamination can be from the introduction of the incorrect refrigeration system lubricant, a system sealer, or a chemical leak detector. Contamination can affect the operation of the refrigeration system in several different ways.

Air

Because air cannot transfer heat as efficiently as refrigerant, any amount of air that is allowed into the system has the potential to alter the operation of the refrigeration system significantly. Air within the system will alter heat transfer properties as well as vaporization and condensation qualities. In addition to the reduction in efficiency, the introduction of air also allows moisture into the system, which can cause the formation of acids and ice. Ice can intermittently block the metering device, causing a loss of refrigerant flow through the system. Air is routinely introduced into the refrigeration system as a result of improper service procedures or because of an extremely low refrigerant level.

Moisture

Any time a refrigeration system is extremely low on refrigerant or is opened for service, the potential for the entry of moisture exists. This is compounded by the fact that the synthetic oils used with R-134a are hygroscopic, meaning that they easily attract moisture. When moisture is present within a refrigeration system, it reacts with both the refrigerant and the refrigeration oil; when these are combined under the heat and pressure found within an operating refrigeration system, hydrochloric acid and sludge are formed, which can damage system components and cause clogging of mesh screens found in the system. As little as one drop of moisture is enough to cause system damage. Although the desiccant in the accumulator or receiver drier is designed to attract any stray moisture, these components can be quickly saturated if the system is left open for extended periods of time.

Refrigerant

System contamination often results from the use of an improper refrigerant to "top off" a refrigeration system. This results in a blended refrigerant. Blended refrigerant compounds can be very unpredictable in nature and can cause unusually high operating pressures. Not only can this cause system damage; it also can cause significant difficulty in performing system diagnostics.

Lubricants

Although the introduction of the incorrect system lubricants is not necessarily considered contamination, it is a critical factor in system operation, as the refrigerant and oil must be mixed in order to be effective. In the event that lubricant and refrigerant are not compatible, eventual compressor damage will occur. It is very difficult to detect the composition of the lubricant when a system service is being performed. Because of this, great care must be exercised when replacing refrigerants and lubricants.

Sealers

System sealers are often presented as a cure-all for leaking air conditioning systems. There are two different varieties of sealers found in the market today: seal swellers and stop-leak formulas. These chemicals are installed in the refrigeration system and are circulated with the refrigerant.

Seal swellers cause seals within the system to swell. The chemicals react with rubber seals, causing them to swell, theoretically forming a tighter seal. Leak sealers work in a different way. When refrigerant levels are low, air and moisture are introduced into the system. The sealant reacts with moisture at the point of entry to form a seal over the hole.

If this type of chemical is introduced into the recovery equipment, it can react with moisture that is present within the machine and act to stop up filters. In some cases, it can cause more serious damage. To protect your equipment from contamination, specialized sealant identification equipment is available. If a system is known or suspected to have a sealer compound installed, it is recommended that the refrigerant not be recovered using a recovery unit, as damage may result. If any of these conditions are suspected, these systems should be considered contaminated and handled accordingly.

IDENTIFICATION

With the availability of alternative and blended refrigerants on the market today, the possibility of air conditioning contamination is a significant problem that must be properly identified and corrected. Even though R-12 and R-134a refrigeration systems are required to have different fittings, it is still relatively common to find refrigeration systems that contain entirely the wrong refrigerant or a blend of refrigerants. Refrigerant is considered to be pure and safe for recovery when it is at least 98 percent by weight of a single type of refrigerant. If contaminated refrigerant is introduced into any type of recovery unit, the entire storage of refrigerant also will become contaminated and must be commercially reclaimed or destroyed.

To prevent potential system and equipment contamination, refrigerant quality should be identified before connecting any recovery equipment to a refrigeration system. It is also good practice to test the quality of the refrigerant that is used to refill a recovery station; this is particularly important when using recycled refrigerant.

The refrigerant composition can be detected using a refrigerant identifier, sometimes called an analyzer (see **Figure 2**). The identifier is connected to the refrigerant source and, when activated, takes a refrigerant sample and analyzes the contents. The refrigerant identifier as described in Chapter 4 has the ability to detect R-12, R-134, R-22, air, and hydrocarbons. The identifiers may display each substance by a percentage of quantity or may simply be displayed as a "pass" if it has a 98 percent or higher purity, or "fail" if the quantity is less than 98 percent. The identifier is also equipped with an audible alarm to warn the technician

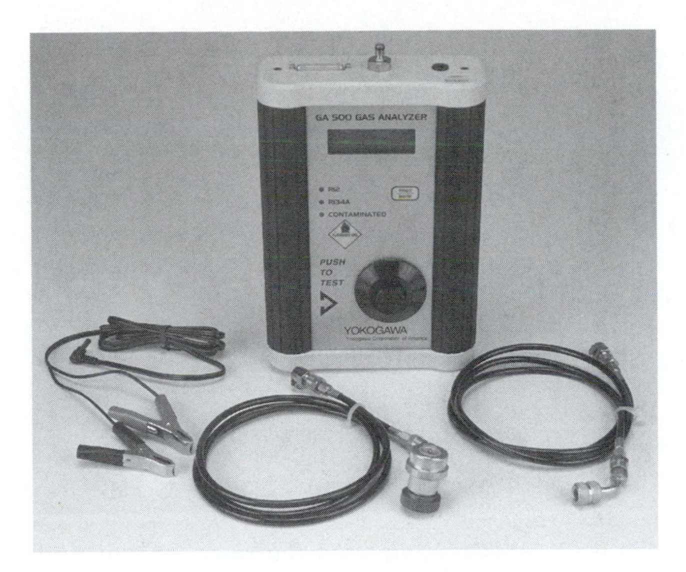

Figure 2. A typical refrigerant identifier.

of the presence of hydrocarbons in the refrigerant sample. If it is determined that the refrigerant is contaminated, it must be recovered using a recovery unit dedicated to the recovery of contaminated refrigerant. The refrigerant must then be sent to a commercial recycler to either be reclaimed or destroyed.

Operation

Because of the differences in the construction and operation of refrigerant identifiers of different makes and models, you should refer to the documentation supplied with your device for specific operating instructions. There are some general operating procedures for using a refrigerant identifier.

1. Before attempting to use a refrigerant identifier, a thorough visual inspection should be performed. The condition of the filter should be inspected first. When contaminated, the filter will have red spots visible on the filter surface. When this condition exists, the filter must be changed or the identifier can be damaged. Additionally, the sampling hoses should be checked for cracks and restrictions and any intake or vent ports should be checked for restrictions.
2. Connect the identifier to the appropriate electrical source.
3. Allow the system to warm up. Calibrate and purge any existing refrigerant as needed (see **Figure 3**). It also may be necessary to enter the local elevation as part of the calibration requirements. This will vary between manufacturers.
4. Select the proper adapter in order to connect the refrigerant identifier to the low side of the refrigeration system or refrigerant tank (see **Figure 4**).
5. Follow the instructions on the display and activate the testing process. The identifier will obtain a refrigerant sample and analyze the contents and display the results.
6. Disconnect the identifier as soon as the identifier has completed its analysis.
7. Refrigerant concentrations that are less than 98 percent pure should be considered contaminated and should be recovered in a dedicated recovery machine so that they can be properly reclaimed or destroyed. This rule does not apply to a mixture that contains a large amount of air. If an excessive amount of air is present but the refrigerant concentration is otherwise pure, the mixture can be recovered. The air will be removed during the recycling process.

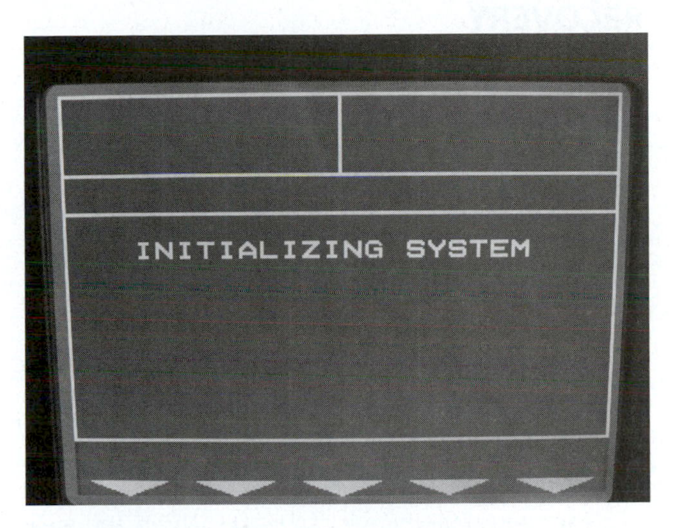

Figure 3. The refrigerant identifier must warm up and calibrate before operation.

Figure 4. Select the proper adapters for the refrigerant being used.

You Should Know *It is a poor practice to determine the type of refrigerant used in a system by the type of service fittings installed. It is not uncommon to encounter a refrigeration system in which someone has changed the refrigerant type without changing the fittings or in which someone has installed a blended refrigerant. When this situation is encountered, you must remember that the refrigerant must be recovered with a unit of the same type of refrigerant, regardless of the fitting type.*

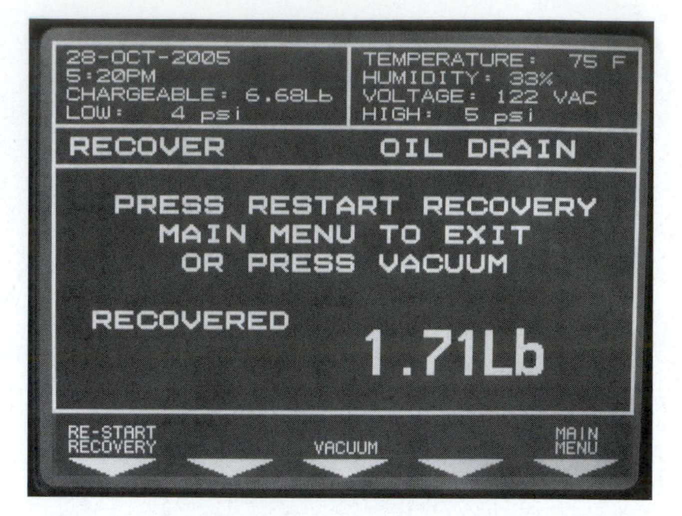

Figure 5. When completed the recovery unit will display the amount of refrigerant that was removed from the system. This information can be very useful in the diagnosis of a refrigeration system.

RECOVERY

Recovery is the process of removing all available refrigerant from the refrigeration system when service is required. This is a mandatory process used to prevent the escape of refrigerant into the atmosphere. Recovery should be performed any time that the system has to be opened for service. The ability to know exactly how much refrigerant has been removed from the system has become a secondary benefit to aid in the diagnostic process whenever a low refrigerant condition is suspected.

As the recovery process is started, a compressor within the recovery unit pulls refrigerant from the vehicle and into the unit. Once inside, it is cycled through the recovery unit, where the refrigerant, oil, and air can be separated. The refrigerant is sent to a storage tank, whereas the oil is directed to an accumulator where it can be dispensed and measured once recovery is completed. When the recovery is completed, the recovery unit will display the amount of refrigerant that was removed from the system and will either prompt you to drain the accumulated refrigerant oil or will drain the accumulator automatically (see **Figure 5**). The oil is dispensed into a graduated container where the quantity can be measured. Maintaining the oil balance within the refrigeration system is of critical importance. For this reason only, the exact amount of oil removed should be replaced.

Operation

The following general steps are provided for the recovery of refrigerant.

1. Select the correct recovery unit for the type of refrigerant being removed.
2. Perform a visual inspection of the equipment including hose connections, hoses, and electrical connections. Correct any concerns as needed.
3. Connect the recovery unit to the appropriate electrical outlet.
4. Turn the unit to the ON position.
5. Observe the oil level in the drain container and empty as needed.

Figure 6. After the fitting is connected turn the knob clockwise.

6. Remove the protective caps from the fittings and connect the color-coded hoses to the refrigeration system. The blue hose connects to the low side, whereas the red hose connects to the high side. The fittings used in an R-134a system are a quick-connect design. When using this type of fitting, the knob on the top must be turned all of the way down in a clockwise direction (see **Figure 6**). This activates a built-in depressor that opens the valve in the system fitting (see **Figure 7**).
7. Open both low- and high-side valves (this may vary between units).
8. Activate the recovery process. The recovery unit is equipped with an on-board scale that will weigh the amount of refrigerant that is being removed from the system.

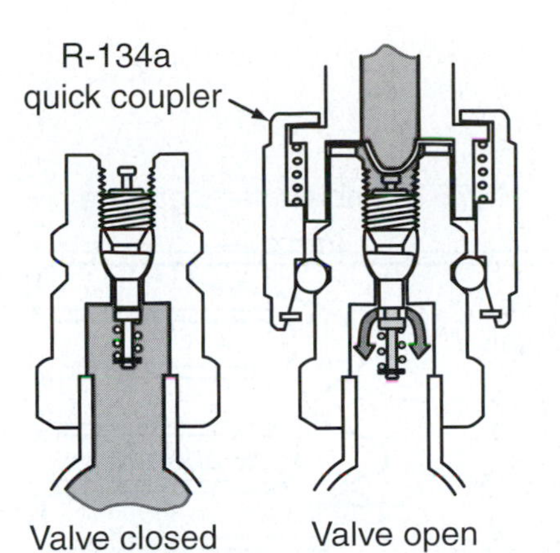

Figure 7. The fitting opens the Schrader valve to connect the service equipment to the refrigeration system.

9. The machine will continue the recovery process until system pressure has reached approximately 13 in-Hg; at this point, the recovery unit will automatically stop the process. The amount of refrigerant that was removed from the system will be displayed (see Figure 5).
10. When recovery has stopped, the oil that was recovered from the system should be drained. Depending on the particular unit being used, this may be accomplished by turning a knob on the machine or may be completed automatically.
11. Observe the amount of oil that was drained into the oil container (see **Figure 8**); this is the amount of clean refrigeration oil that will be returned to the system. This amount will be added in addition to any amount that might be necessary as a result of other service procedures.
12. To ensure that a complete recovery has occurred, wait 5 minutes and observe the pressure on the low-side gauge. If the pressure has risen above 0 in-Hg, refrigerant is still present in the system and the recovery process must be repeated. Repeat the process until the system maintains 0 psi or lower for 2 minutes. Once the pressure stays below 0 psi or can maintain a vacuum, the system can then be opened for service.
13. When completed, all valves should be turned to the OFF position. The knobs on the quick-connect fittings need to be turned counterclockwise and the machine must be disconnected from the vehicle.

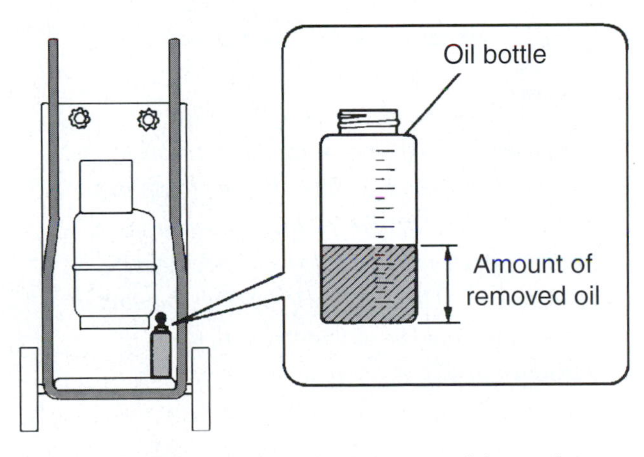

Figure 8. The amount of oil removed from the system can be observed in the drain bottle.

You Should Know *Because of the low system pressure experienced when refrigerant recovery is activated, it is normal for some of the refrigerant to freeze in the refrigeration system. When this occurs, it appears that the system has reached a proper vacuum and the recovery unit will turn off. However, once the recovery process has stopped, the frozen refrigerant expands and pressure builds up within the system. When this occurs, the system pressure observed on the gauges rises above a vacuum and indicates that the system must be evacuated again.*

Alternate Recovery Method

This method of recovery uses an empty refrigerant cylinder and a manifold gauge set and is presented for information only. It should only be accomplished by, or under the direct supervision of, an experienced technician.

In some situations, it may not be practical to have a recovery unit dedicated to the recovery of contaminated refrigerant. In this case, there is an alternative method available. This recovery method works on the basis that the refrigerant will travel from a hot area to a cold area. In this case, a refrigerant drum is cooled to a very low temperature in an effort to have the refrigerant migrate from the warmer refrigeration system to the cooler storage drum. The most important consideration is that the recovery cylinder will not have been filled to more than 80 percent of capacity when the temperature is increased. Refer to **Figure 9** and follow these instructions:

1. Place a properly identified recovery cylinder into a tub of ice on the floor beside the vehicle. The recovery cylinder should be below the level of the air conditioning system.
2. Add water and ice cream salt to the ice. This will lower the temperature to about 0 degrees F (–17.7 degrees C).
3. Connect a service hose from the high-side fitting of the system to the gas valve of the recovery cylinder.
4. Open all valves.
5. Cover the recovery cylinder and tub with a blanket to insulate them from the ambient air.
6. Place the shop light(s) or other heat source near the accumulator or receiver.

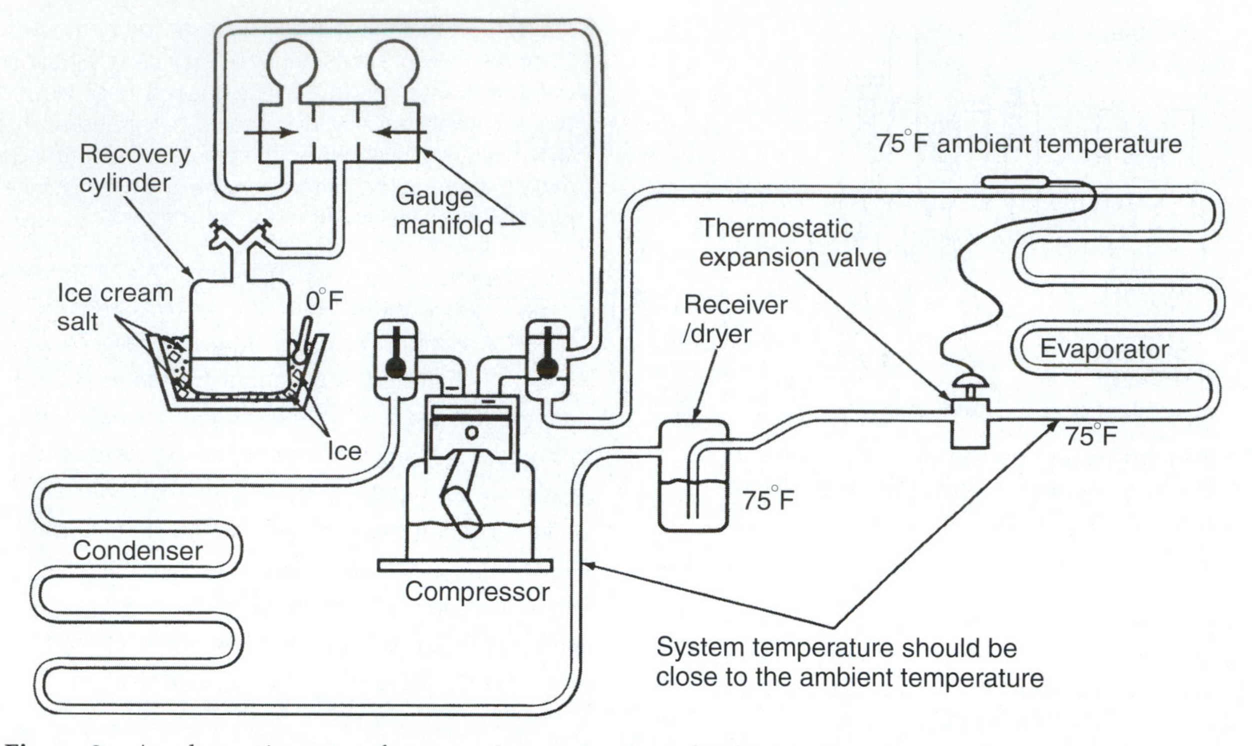

Figure 9. An alternative setup for removing contaminated refrigerant.

7. Allow 1 to 2 hours for recovery. The actual time that is required will depend on the ambient temperature and the amount of refrigerant to be recovered.

Recycling

There are two terms with which the technician should be familiar: **recycling** and **reclamation**. Recycling is a filtering process in which air, oil, and moisture are removed from refrigerant. This process can be done in-house with most recovery units. On most equipment, this is done automatically when the refrigerant is being recovered. When refrigerant is being recovered, but before it reaches the internal storage tank, the refrigerant is passed though a filter system in which moisture, oil, and air are removed. Recycling equipment cannot separate two types of refrigerants or other contaminates from the refrigerant.

Contaminated refrigerant must be recovered and sent to a commercial reclaiming facility, where it can be reclaimed or destroyed. Reclamation is a process in which all impurities have been removed and the refrigerant returned to a nearly pure state. The service facility that was responsible for the recovery of the refrigerant is required by law to keep records indicating the facility and the physical address to which the refrigerant was sent for reclamation.

EVACUATION

When the system is opened for service, it fills with both air and moisture. The amount of moisture that enters the system is directly related to how humid the air is and how long the system is left open. Both air and moisture are removed from the refrigeration system by the process of evacuation. Once repairs have been completed and the system has again been resealed, a vacuum pump is used to maintain a constant vacuum of at least 29.76 in-Hg (.81 kPa) on the refrigeration system for a predetermined amount of time. This is called "pulling a vacuum" or "pumping down." The amount of time required for evacuation is determined by the humidity of the ambient air and the amount of time that the system remains open. It is relatively obvious how the vacuum pump is able to remove air from the system; what is not as obvious is how moisture is removed from the system.

> **You Should Know** *When pressure is applied to a liquid in a sealed system, the internal pressure increases, as well as its boiling point. Conversely, when the pressure within a sealed system is reduced, the boiling point is also reduced. When water in a sealed system is exposed to a vacuum, the boiling point can be as low as 40 degrees F (4.44 degrees C) at 29.76 in-Hg (.8 kPa). This amount of vacuum should be obtainable with a quality vacuum pump.*

As the vacuum pump runs, it creates a vacuum within the system. The presence of vacuum causes the moisture

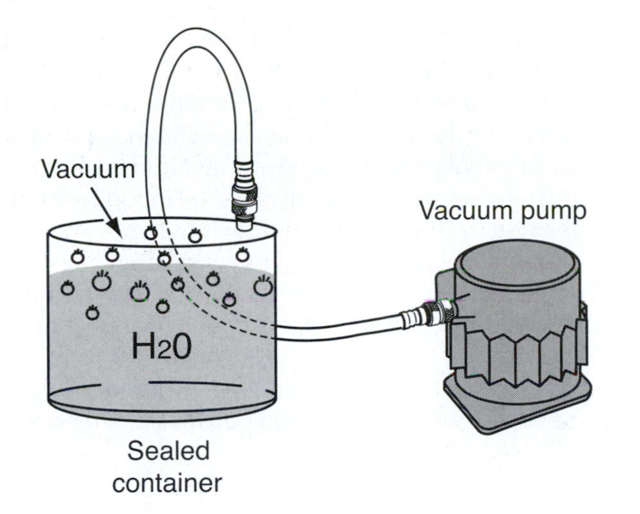

Figure 10. When a closed system is placed under a vacuum, moisture will boil and can be pulled out of the system.

that is present within the system to boil. The vapors are drawn out and expelled through the vacuum pump (see **Figure 10**). The longer the pump runs, the more moisture can be expelled. Before recharging the refrigeration system with refrigerant, a vacuum should be pulled for a minimum of 15 minutes if the system has not been opened, or for 30 minutes if the system has been opened for a short time during a repair. If, however, the system has remained open for long periods of time or has been exposed to exceptionally high humidity, longer evacuation times may be appropriate. For maximum efficiency, it is a good idea to allow the system to "pump down" as long as possible beyond these minimums.

You Should Know

If the refrigeration system has been open for long periods of time or has been exposed to exceptionally high humidity, the desiccant in the accumulator or receiver drier most likely will become saturated with moisture. In these cases, the accumulator or receiver drier should be replaced before evacuating and recharging takes place.

Operation

Here are the basic steps used to evacuate a refrigeration system using a typical recovery station:

1. Plug the recovery unit into the proper power source.
2. Connect the hoses and open the hose fittings.
3. Turn the recovery unit to the ON position.
4. Turn both valves on the recovery unit to the OPEN position.

Figure 11. Select the desired length of time the system should be evacuated.

5. Set the timer on the recovery unit to the desired amount of time (see **Figure 11**).
6. Activate the vacuum pump. When the timer expires, the vacuum pump will turn off automatically.
7. Leave the system connected to the vehicle and prepare for recharging.
8. Once completed, the system should be capable of holding a vacuum for several minutes.

OIL BALANCE

The oil level of the refrigeration system is critical to proper operation. If the system is low on oil, the compressor can starve and compressor damage will occur. If too much oil is in the system, heat transfer efficiency of the heat exchangers will suffer. Because of this, lubricant should only be added to the system when lubricant was removed during recovery or when a component is replaced. Although the compressor is the only component that needs lubrication, refrigeration oil is distributed among every component of the refrigeration system. Because there is no way to measure the amount of oil within the refrigeration system, the oil must be replaced in the quantity that it was removed. The vehicle service manual will usually list the oil distribution by component; this is the amount of refrigerant that should be added to the component when it is replaced. When replacing a component, the oil should be added directly to the component. If a component is not being replaced, the lubricant can be added through the recovery unit before refrigerant recharging takes place. If the unit is not equipped to inject oil, oil can be introduced using an oil injector.

You will often encounter situations in which a refrigerant leak is present and a noticeable lubricant loss has occurred. Many times, these conditions only require the replacement of a seal or minor component that retains no lubricant. In these cases, it is impossible to calculate the amount of lubricant that has been lost. In this case add 2 ounces (59 ml) of lubricant in addition to the amount that was recovered with the refrigerant. Under no circumstances should oil be added for "good measure."

RECHARGING

Recharging is the process of filling the system with the proper amount of refrigerant. It is absolutely imperative that the proper amount of refrigerant be restored to the system. Too little refrigerant will cause poor cooling and rapid cycling of the A/C clutch and, in some cases, no engagement at all. Excessive refrigerant can cause system pressures to be too high, and also can cause compressor slugging as well as poor cooling. In some systems, refrigerant capacity variations as small as 4 oz (.113 kg) can cause poor system performance. It should not be assumed that the amount of refrigerant removed from the systems is the correct amount. Proper capacity can be found in the service manual and is also usually located on a decal under the hood.

Operation

1. At this point, the system should be under a vacuum and the recovery machine should be attached to the vehicle and the hose fittings opened. The vehicle and A/C system must be turned off.
2. The refrigerant oil should be injected. If your recovery unit is equipped with an oil injector, you may be prompted to inject oil at this point (see **Figure 12**). Otherwise, it should be injected with an oil injector (see **Figure 13**).
3. When equipped with an on-board injector, check to make sure that the reservoir has a sufficient amount of clean refrigeration oil present (see **Figure 14**). If the reservoir should run low, air could be introduced into the system. The on-board injector is used by either depressing a button or twisting a knob. When doing so, observe the amount of oil that is draining from the reservoir and stop when the proper amount has been introduced.
4. Once the oil has been introduced, the refrigerant can be added. Set the recovery unit up for recharging and

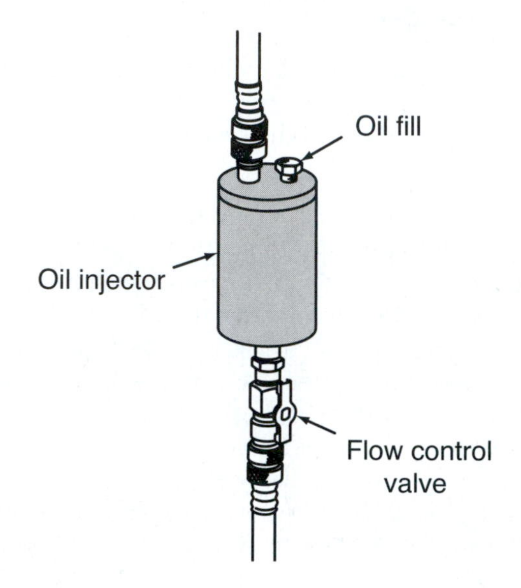

Figure 13. A forced injector can be used to inject oil into a sealed system.

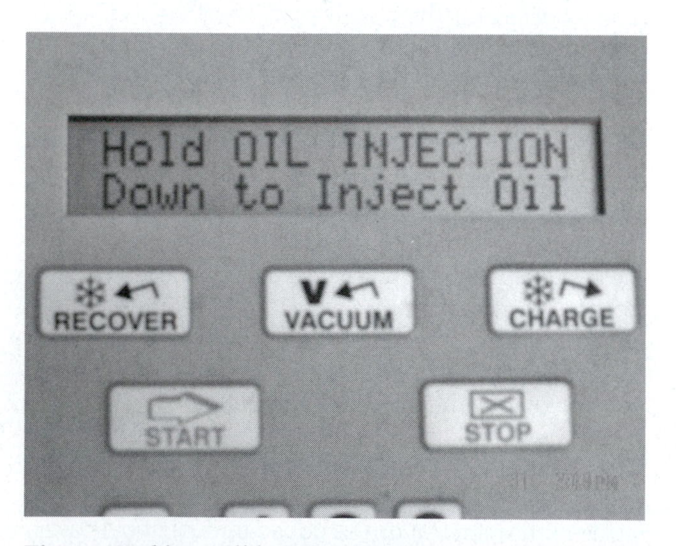

Figure 12. You will be prompted to inject oil into the system.

Figure 14. The oil level in the oil fill bottle should be checked before opening the oil fill valve.

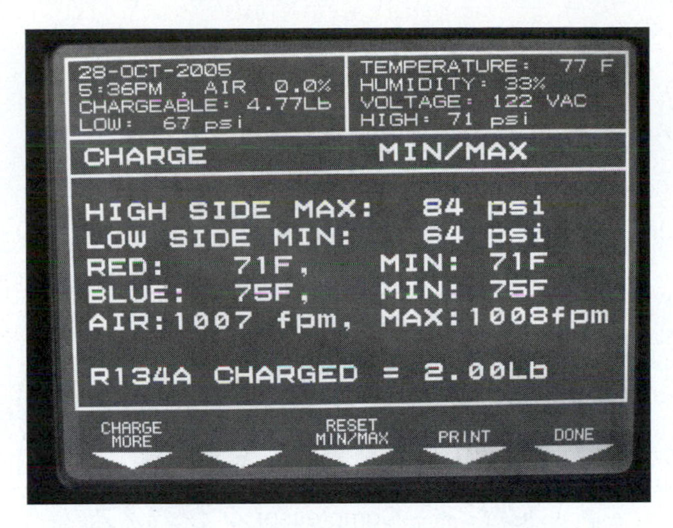

Figure 15. The unit will alert the operator when charging has been completed.

enter the amount of refrigerant to be injected into the system.

5. In most cases, the system will be filled through the high side of the system. Turn the high-side valve to the ON position and the low side to the OFF position.
6. Activate the charge cycle. The on-board compressor will pump the proper amount of refrigerant into the system. When the proper amount has been injected, the compressor will turn off (see **Figure 15**).
7. Turn the high-side valve to the OFF position. The refrigeration system is now full and ready to be returned to operation. It may be desirable to monitor system pressure using the gauges of the recovery unit. Once any testing is done, the unit can be disconnected.

> **You Should Know** *The recovery unit uses a scale to determine the amount of refrigerant that is both removed and injected into the system. Because of the sensitivity of the scale, it is a good habit not to touch or disturb the recovery unit when it is in the recovery or recharge modes.*

ALTERNATIVE SERVICE METHODS

Before recovery stations became the norm, refrigerant service was performed with a manifold gauge set, a vacuum pump, and a scale. Refrigerant was added using a large cylinder or using individual cans. With great care, these methods can still be used today. Refer to **Figure 16** through **Figure 20**.

EQUIPMENT MAINTENANCE

Service of the refrigeration service equipment is crucial to the proper operation of the equipment and the quality of service that it provides. This section will deal with the service of a recovery, evacuation, recycling, and recharging station. However, service for the individual components is very similar in nature and frequency. Most recovery units have the added benefit of using the on-board microprocessor to track maintenance intervals accurately and provide feedback to the operator about when service should be completed. When these units are serviced as part of a recovery station, it is imperative that the maintenance indicators be reset.

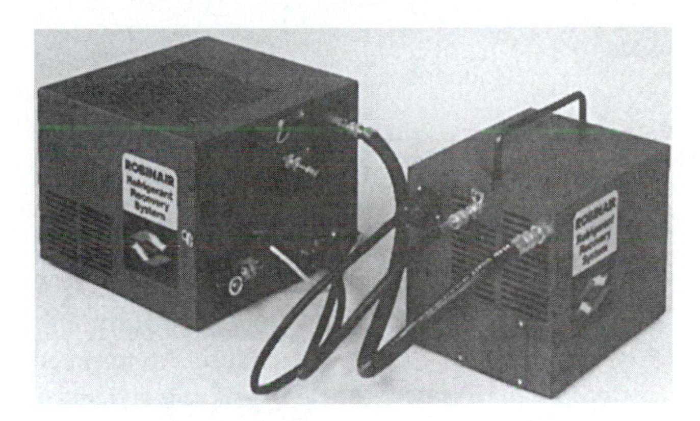

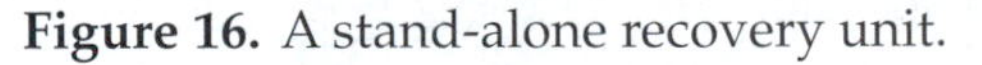

Figure 16. A stand-alone recovery unit.

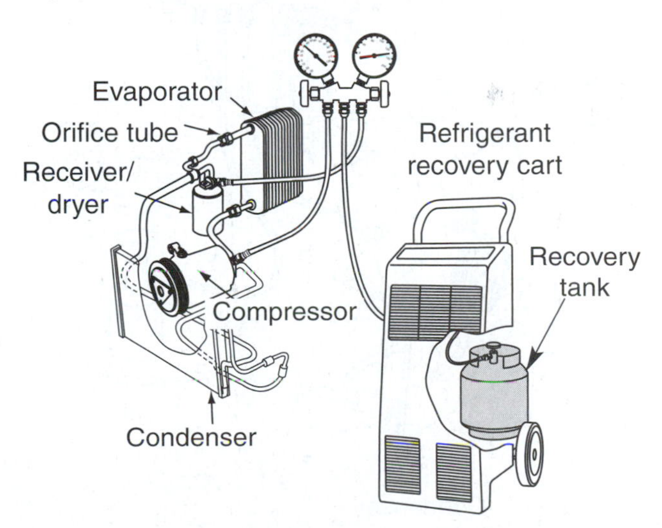

Figure 17. The recovery unit is used in conjunction with a manifold gauge set. Refrigerant is pumped from the refrigeration system into a storage container.

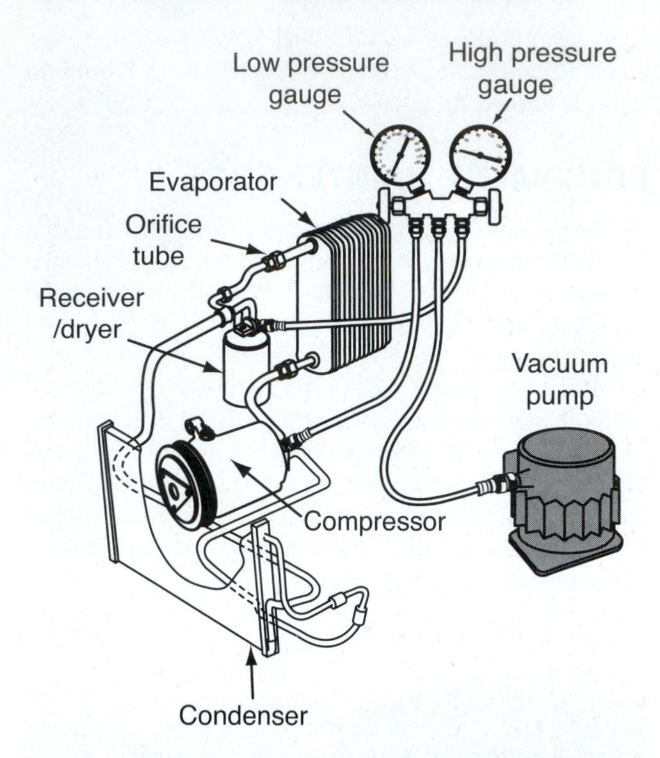

Figure 18. Proper procedure for connecting a vacuum pump when a recovery unit is not used.

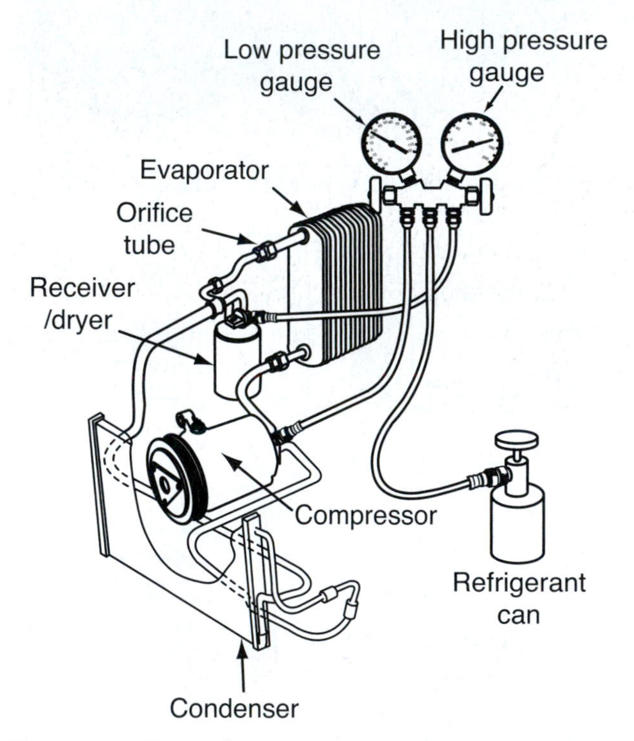

Figure 19. The refrigerant source is connected in place of the vacuum pump. The proper number of cans is determined by the amount of system refrigerant required.

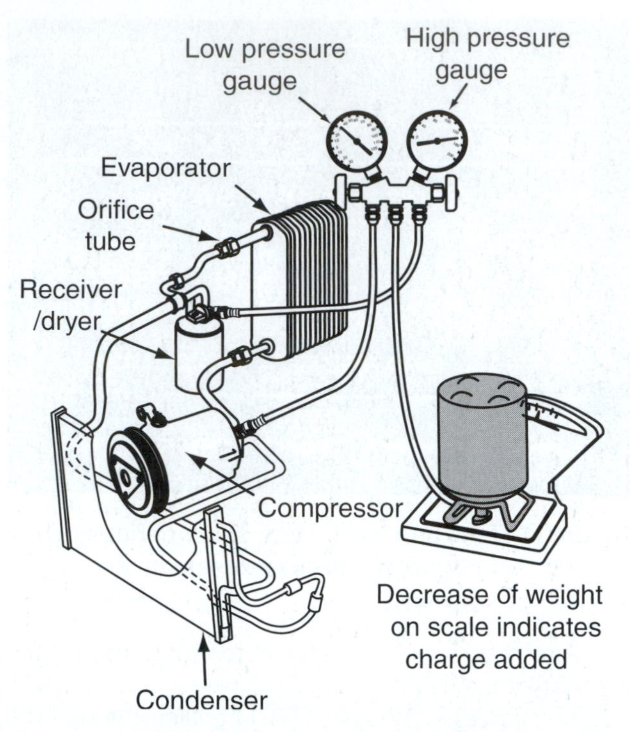

Figure 20. When using a large cylinders a scale is used to measure the amount of refrigerant introduced into the system.

Vacuum Pump

The vacuum pump uses a specially formulated lubricant designated as "vacuum pump oil" to lubricate the mechanical components of the pump and to help form a tight seal within the pump itself. The performance of the vacuum pump is directly related to the condition of the oil. As moisture is removed from a refrigeration system, it is inevitable that some of that moisture will mix with the oil. This causes the oil to dilute and will reduce its sealing ability. To ensure proper performance, the refrigeration oil should be changed (see **Figure 21**) when any of the following conditions are met:

- If the pump has operated for 10 hours since the oil was changed
- If oil becomes cloudy or milky
- If the pump will not produce sufficient vacuum
- If a system with known contamination or high moisture content has been evacuated

Recovery Unit

All recovery units use some type of filtering system to remove moisture and other impurities from the refrigerant. The filter should be changed (see **Figure 22**) after 150 lbs (68 kg) of refrigerant has been recycled. Recovery of

Figure 21. For maximum efficiency the vacuum pump should be changed according to manufacturer's specifications.

contaminated refrigerant will decrease the life expectancy of this filter significantly.

Refrigerant Identifier

The refrigerant identifier is equipped with a filter that is used to protect the identifier from oil and moisture contamination. In most identifiers, the filter is located in a clear filter housing located at the top of the identifier (see **Figure 23**). This filter should be bright white in color. When contaminated, the red spots will be visible on the filter's surface. When this occurs, the filter should be changed immediately.

Figure 22. Recovery unit filters should be changed according to manufacturer's specifications.

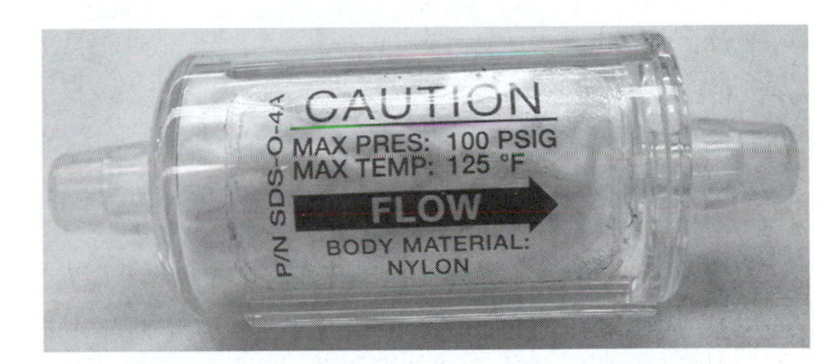

Figure 23. The identifier filter should be changed when red spots appear on the filter's surface.

Summary

- System contamination can come from any substance not originally intended for that system. Contamination is detrimental to the operation of the refrigeration system. Potential contaminates include air, moisture, improper refrigerants, improper lubricants, and sealers.
- Moisture reacts with refrigerant and lubricant to form hydrochloric acid and sludge. As little as one drop of water is enough to cause system damage.
- The introduction of the wrong refrigerant or a blended refrigerant will contaminate the system and can cause unusually high operating pressures.
- Introduction of the wrong refrigerant oil can act as a contaminant and impede proper mixing of the refrigerant and lubricant.
- System sealers may be useful when in the refrigeration system; however, these same sealers may cause damage

to recovery equipment if they are inadvertently ingested. Refrigeration systems that are known to have system sealers should be classified as contaminated.

- Because of the availability of a wide variety of refrigerant products on the market, it is imperative that the refrigerant content of a system be accurately identified before any recovery procedures take place. Refrigerant that is determined to be less than 98 percent pure by weight should be considered contaminated.
- Recovery is a mandatory process that is used to remove all refrigerant from the refrigeration system before service takes place. Refrigerant and oil are separated during the recovery process. The amount of refrigerant removed is displayed to the operator. The recovery procedure may have to be repeated two or three times to remove all refrigerant.
- Recycling is a filtering process in which moisture, oil, and air are removed from a refrigerant. Reclamation is a process in which impurities have been removed from the refrigerant, returning it to a nearly pure state.
- Air and moisture enter the refrigeration system any time that it is opened for service. System evacuation is the process used to remove both air and moisture from the refrigeration system. By creating a vacuum in the system, any moisture that is in the system will boil off. Evacuation times will vary.
- Refrigeration oil is distributed among all system components. There is no way to determine the exact amount of oil within a refrigeration system. For this reason, oil should be replaced only in the quantities in which it was removed.
- The refrigeration system requires the exact amount of refrigerant to operate properly. Amounts above or below recommendations can cause a decrease in system efficiency.
- The alternative recovery process may be used to recover contaminated refrigerant if a dedicated recovery unit is not available.
- Equipment maintenance is crucial to proper refrigeration system service. Some recovery units will give the operator feedback on when service needs to occur.

Review Questions

1. List three possible system contaminants.
2. Technician A says that when water and refrigerant combine, they can form hydrochloric acid. Technician B says that one drop of water is enough to cause system damage. Who is correct?
   A. Technician A
   B. Technician B
   C. Both Technician A and Technician B
   D. Neither Technician A nor Technician B
3. Technician A says that most commercial refrigerant identifiers have the ability to detect the presence of system sealers. Technician B says that an audible alarm will sound when sealers are present in the system. Who is correct?
   A. Technician A
   B. Technician B
   C. Both Technician A and Technician B
   D. Neither Technician A nor Technician B
4. Technician A says that refrigerant in a refrigeration system must be at least 98 percent pure in order to be recovered. Technician B says that refrigerant mixtures that are contaminated with large amounts of air should be commercially reclaimed or destroyed. Who is correct?
   A. Technician A
   B. Technician B
   C. Both Technician A and Technician B
   D. Neither Technician A nor Technician B
5. A commercial recovery, evacuation, recycling, and recharging unit can do all of the following tasks, *except*:
   A. remove moisture from a refrigeration system
   B. separate oil and refrigerant from one another
   C. display the amount of refrigerant removed from a refrigeration system
   D. separate two types of refrigerant
6. Technician A says that recovery of contaminated refrigerant does not require the use of dedicated equipment. Technician B says that contaminated refrigerant can be recovered using a manifold gauge set and empty refrigerant storage drum. Who is correct?
   A. Technician A
   B. Technician B
   C. Both Technician A and Technician B
   D. Neither Technician A nor Technician B
7. All of the following statements about system evacuation are true, *except*:
   A. A vacuum pump should be able to "pull" a vacuum of at least 29.76 in-Hg.
   B. The evacuation process should last only until the system has reached a vacuum of 29.76 in-Hg.
   C. When a sealed system is pulled into a vacuum, the water molecules can begin to boil.
   D. The length of time in which a system is "pumped down" is determined by the amount of time the system was opened.

8. Technician A says a small amount of additional oil should be added to the system any time refrigerant is added to the system. Technician B says that, when possible, oil should be poured directly into a replacement component during installation. Who is correct?
   A. Technician A
   B. Technician B
   C. Both Technician A and Technician B
   D. Neither Technician A nor Technician B
9. Technician A says that the refrigeration oil should be added just before the refrigerant is introduced into the system. Technician B says that refrigerant should be added to the system just before evacuation takes place. Who is correct?
   A. Technician A
   B. Technician B
   C. Both Technician A and Technician B
   D. Neither Technician A nor Technician B
10. Technician A says the performance of quality refrigeration service, among other things, is dependant on the use of properly maintained service equipment. Technician B says that the required service intervals of the vacuum pump and recovery unit can vary depending on the condition of the refrigeration systems that have been serviced. Who is correct?
    A. Technician A
    B. Technician B
    C. Both Technician A and Technician B
    D. Neither Technician A nor Technician B

Chapter 17 Refrigeration System Diagnosis

Introduction

This chapter will focus on the systematic diagnosis of the refrigeration system. Customer concerns resulting from problems in the refrigeration system can usually be limited to poor cooling, noise, and odors. Poor cooling is the most common concern that is faced by an air conditioning technician, but it is also one of the most difficult problems to diagnose. Poor cooling can be caused by a wide range of component or system malfunctions, ranging from an electrical problem to a mechanical system failure. Conversely, noises are usually caused by a compressor failure, whereas odors are typically caused by the growth of fungus on the surface of the evaporator. In this chapter, we will explore customer concerns that are related to the mechanical operation of the refrigeration system. The conditions in this chapter will focus solely on mechanical conditions and will assume that all electrical systems, including the compressor clutch, are in proper working order. Diagnosis of the electrical systems will be covered in Section 5.

EXTERNAL FACTORS

In the automotive industry, the performance of the air conditioning system will be judged by the customer's perceived comfort level—not how cool the outlet air temperature actually is, but how cool it "feels." This is called the **comfort factor**. The human body is comfortable at temperatures of 75 degrees F to 80 degrees F (22.2 degrees C to 26.6 degrees C) and at humidity levels of 45 to 50 percent. Conditions outside of these norms feel uncomfortable.

At this point, you know how the refrigeration system operates. You also know that as air is circulated across the evaporator, both heat and moisture are removed from the air. But how cold should the air be? How dry should the air be? There is no one correct answer to either of these questions. As air is circulated across the surface of the evaporator, heat is removed. As the air moves farther away from the evaporator, the temperature will increase. The operation of the refrigeration system and the comfort factor are heavily influenced by external environmental factors, including:

- Ambient temperature
- Humidity
- Sunload
- Air volume
- Customer perception

Ambient Temperature

Ambient air temperature greatly influences the comfort factor and system operation. Air temperature is a numerical indicator of how much heat energy is present within the air. The more heat that is present within the air, the less the temperature can be lowered as it passes across the evaporator. Therefore, the human comfort level is decreased.

Humidity

Humidity is a significant factor in both human comfort level and actual refrigeration system operation. The human body relies heavily on evaporation to provide cooling. The higher the humidity, the more difficult it is for moisture from the skin to evaporate into the atmosphere, thereby reducing the human comfort level.

Humidity also directly affects the operation of the refrigeration system. In order for moisture to evaporate into the atmosphere, it must absorb enough heat to change from a liquid to a vapor. This is called latent or hidden heat (see Chapter 5). As the air circulates across the surface of the evaporator, both sensible and latent heat are removed from the air. As latent heat is removed, moisture condenses on the surface of the evaporator, effectively dehumidifying the air. As sensible heat is removed, air temperature is reduced. Even though the latent heat cannot be measured, much of the evaporator's capacity is used to remove latent heat, reducing the amount of sensible heat that can be removed. In simple terms, the higher the humidity, the less efficiently the evaporator operates.

Sunload

Sunload is directly related to how bright the sun is on a given day. The brighter the sun, the more radiant energy is projected across the earth's surface. As modern vehicles have become sleeker and more aerodynamic, the glass panels of the passenger compartment have increased in size and angles have increased, significantly increasing the exposure to the sun and the effects of sunloading on the interior of the automobile (see **Figure 1**). This can cause the air temperature inside of a parked vehicle to increase by ½ or more than that of the ambient air. This requires the air conditioner to work much harder on initial startup and increases the amount of time it takes for passenger compartment air to become "comfortable." Effects of sunloading can be decreased with window shields and window tinting.

Air Volume

The larger the air volume within the passenger compartment of a vehicle, the longer it will take for the interior of the vehicle to become "comfortable." When sunload is factored in, the length of time for the passenger compartment to become comfortable may seem excessive to the consumer. This is of particular concern if the vehicle is used for very short trips in which the vehicle is driven for a short time and parked for a long time. Under these conditions, the air conditioner may never actually run long enough to cool the interior of the vehicle. This is of particular concern if the vehicle has a large amount of cabin air space. The customer may never feel that the air conditioner is cooling satisfactorily, regardless of how efficient the refrigeration system actually operates. It is important for you to be able to communicate these points to the customer in a respectful and effective manner.

Figure 1. Sunload can significantly increase the interior temperature of a vehicle. This increases the time it takes to cool off and the load placed on the air conditioner system.

Customer Perception

Customer perception is often the most difficult repair to make. However, this can usually be accomplished through thoughtful education. Many times, the customer is not aware of the complexity of a vehicle system or the factors that affect its operation, and can only make comparisons to other vehicles to establish what they may consider normal for their vehicle. This is of particular concern for new vehicle owners. For example, an owner who has just moved from a midsize car to a large SUV may expect the SUV interior temperature to cool in the same amount of time as the car. When it doesn't, they perceive this to be a problem even though the system is working to specifications. In this or similar situations, it is critical that the technician carefully and thoughtfully explain the operating characteristics and reasonable expectations of the vehicle to the customer. In some cases, it may be necessary to compare the vehicle with another of similar design. It is important that, before any attempts are made in this direction, the customer's concern is clearly understood. The system should be thoroughly tested to ensure that no system malfunctions exist.

ESTABLISHING NORMS

The diagnosis of any air conditioning system concern will almost always begin with a measurement of outlet vent temperature. But what is the proper outlet temperature? There is no one right or wrong answer. More than any other automotive system, the proper operation of the air conditioning system is determined by external conditions, particularly ambient temperature and humidity. Because these external factors can vary greatly from day to day and from region to region, it is necessary for the technician to establish what the "normal" outlet temperatures and pressure variations should be. **Figure 2** provides a range of normal operating parameters for an orifice tube system. This chart should be used only as a reference, as you will likely find that the actual performance of a properly operating air conditioning system exceeds these performance expectations. It is generally accepted that the refrigeration system should be able to provide a 20 degree F (11 degree C) reduction in temperature between the air as it enters the inlet and as it exits at the outlet.

Relative Humidity	Ambient Air Temp		Max Low Side Pressure		Max High Side Pressure		Max Right CTR Air Outlet Temp	
%	F	C	PSIG	kPaG	PSIG	kPaG	F	C
20	70 80 90 100	21 27 32 38	23 30 35 37	159 207 242 225	190 250 300 330	1313 1728 2074 2281	42 50 54 57	6 10 12 14
30	70 80 90 100	21 27 32 38	23 30 35 37	159 207 242 255	200 280 310 335	1383 1936 2143 2310	42 51 55 58	6 11 13 15
40	70 80 90 100	21 27 32 38	23 31 36 43	159 214 248 296	200 285 320 370	1379 1965 2206 2551	43 52 56 64	6 11 14 18
50	70 80 90 100	21 27 32 38	23 33 39 47	159 228 269 324	200 300 340 385	1379 2068 2344 2654	43 54 60 69	6 13 16 21
60	70 80 90 100	21 27 32 38	23 36 43 55	159 248 296 379	200 315 365 375	1379 2171 2516 2585	43 56 64 78	6 14 18 26
70	70 80 90	21 27 32	26 38 47	179 262 324	240 325 380	1655 2240 2620	46 58 67	8 15 20
80	70 80 90	21 27 32	30 40 49	207 276 338	260 340 380	1792 2344 2620	49 60 72	10 16 22
90	70 80	21 27	32 41	221 283	275 345	1896 2378	52 62	11 17
NOTE: PERFORMANCE DATA OBTAINED AT ENGINE SPEED OF 2000 RPM.								

Figure 2. This chart represents the relationship of temperature, pressure, humidity, and outlet temperature of an orifice tube refrigeration system.

DIAGNOSIS OF POOR COOLING CONCERNS

Any successful diagnosis is dependent on the acquisition of information. First, obtain a clear description of the customer concern. This gives the technician a clear picture of what he is trying to fix. An accurate description can significantly reduce diagnostic time and improve accuracy. The description should include:

- What the customer is experiencing.
- When the customer experiences the concern.
- How long the condition has been present, and when it was first noticed.
- How often the problem is experienced, and if it is consistent or intermittent in nature.
- Any additional information that can help the technician duplicate the concern.

You Should Know *Because most diagnostic procedures must be done with the air conditioning system turned on, you will be required to perform many tasks with the engine running. To avoid personal injury, extreme caution must be exercised at all times when working in or around the engine compartment when the engine is running.*

After obtaining any customer information, it is recommended that you refer to the vehicle service manual to familiarize yourself with the normal function of the particular system that is being diagnosed. This will include how and when the compressor operates as well as the normal operating pressures and vent temperatures. Being aware of this information will often lead you to small clues that may appear during the diagnostic process. The methodical diagnosis of a refrigeration system should begin with a quick system check, a visual inspection, and a performance test.

QUICK CHECK

The quick check is a process of verifying what part(s) of the system operate and which ones don't. This can be done very quickly and without the aid of any test equipment. The information gathered in the quick check will ultimately guide the rest of your diagnosis. Many steps of this quick test can take place during a short test drive and often can be accomplished when you are pulling the vehicle into the service bay. Once you have completed the quick test, you will have a much better idea of where to focus your attention during the visual inspection and the performance test. The quick test should include the following steps:

- Clutch engagement: turn the A/C from OFF to ON several times and determine if you can hear the clutch engage. This is often recognized by an audible click or by a slight decrease in engine rpm when the compressor engages. If the clutch engages, this will give you a solid indication that the compressor's electrical control system is operational.
- Vent temperature: the vent temperature is a direct indication of how the refrigeration system is operating. With the engine near normal operating temperature, the air conditioner in the ON position, and the temperature range moved to full cold, simply determine the outlet vent temperature with your hand. Cold or slightly cooled vent temperatures most likely mean that the compressor is engaging and that the system has some cooling capacity. Temperatures that are close to ambient temperatures most likely indicate that the compressor is not turning on or that the refrigeration system is not operating properly. If temperatures are above ambient, this may be an indication that heated air may be entering the passenger compartment and that an air distribution problem exists.
- Air mixing: to achieve comfortable vent temperatures, heated air and cooled air are often mixed within the air distribution case; refer to Section 6 for more information. If the door fails in a position that allows refrigerated and heated air to mix, the vent temperatures will never be desirable, no matter how well the refrigeration system operates. This can be tested by slowly moving the temperature control from full cold to full hot. The result should be that air temperatures get increasingly warmer as the control is moved toward hot, and decreasingly cool as temperatures are moved back toward the cold. Many times you will be able to hear an audible "knock" as the door reaches the end of its travel in either direction.
- Blower speed: many times a poor cooling concern is actually related to the amount of available airflow. Manually activate the blower to move through all of its ranges. Make note of the air temperature and blower volume at each position. The air temperature should remain relatively consistent when the airflow inclemently increases or decreases. Any results other than these may indicate a blower control circuit problem or an air distribution concern. Refer to the appropriate section for diagnosis.
- Compressor operation: once the vehicle is in the service bay, place the transmission in park and set the parking brake. Raise the hood and visually verify that the clutch is engaged and that the compressor is turning.
- Cooling fan: observe the operation of the cooling fan. Most vehicles equipped with electric fans engage the fan any time that the air conditioner is turned on.
- Refrigerant level: with the system operating, feel the temperature of the evaporator inlet after the metering device and compare it to the temperature of the evaporator outlet (see **Figure 3**). Both tubes should be significantly colder than ambient air and close to the same temperature. If the inlet is significantly colder than the outlet, a low refrigerant charge is indicated. If both tubes are near ambient temperature with the compressor engaged, a refrigeration system problem exists.

Figure 3. The refrigerant charge can be quickly analyzed by comparing the temperature of the evaporator inlet and outlet.

VISUAL INSPECTION

The completion of a visual inspection is intended to eliminate any obvious system concerns that may be present in the air conditioning system as well as provide you with clues that can be used as part of a broader diagnostic strategy. The items that you check as part of the visual inspection are often influenced by what was observed during the quick check. A thorough visual inspection should at the very least include the following steps:

- Check the condition of all related electrical connections.
- Check hoses and service fittings for leaks and the presence of service caps. Leaks often will appear as an oily film near the leak area (see **Figure 4**). Make a note of any problems.
- The belt should be inspected for overall condition and signs of slippage. Correct problems as needed.
- The clutch mechanism should be inspected for signs of oil leakage and signs of slippage. Slippage can often be identified by large amounts of black dust around the clutch plate or clutch components that have obvious exposure to excessive heat (see **Figure 5**).

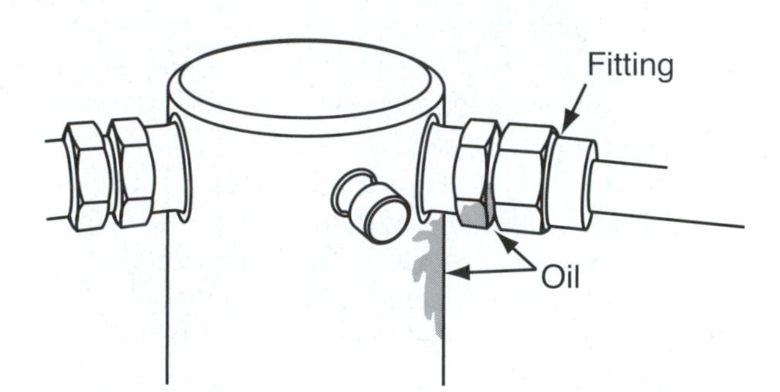

Figure 4. Oil film is a sign of a potential refrigerant leak.

Figure 5. Inspect the compressor clutch for signs of obvious damage.

Figure 6. The condenser should be inspected for debris and damaged fins.

- Check the condenser for obvious signs of obstruction or damaged fins (see **Figure 6**). Make corrections as needed.

PERFORMANCE TEST

The performance test is a procedure that allows the technician to assess air conditioning system performance. The performance test is conducted under controlled conditions while information is gathered about the current suction and discharge pressures as well as vent temperatures. This information can then be compared to a pressure/temperature chart, similar to the one shown in Figure 2, to determine if the system performance is acceptable. This information, along with the information gathered from the customer, can then be used with charts in this chapter to make an accurate system diagnosis. By making this procedure part of your regular diagnostic routine, you will quickly establish what the normal operating parameters are for a particular system and your geographic area.

Preparation

1. Connect the manifold gauge set or recycling unit. All valves should remain closed during the test.
2. All windows and doors should remain closed during the test. The hood should remain open to allow adequate airflow into the engine compartment.
3. Place a thermometer in the center air outlet. This location provides an accurate representation of average outlet temperature (see **Figure 7**).
4. Measure and record the temperature and humidity near the condenser (see **Figure 8**).
5. Place a large fan in front of the condenser to produce sufficient airflow for proper operation (see **Figure 9**).

Figure 7. A thermometer placed in the outlet vents is used to gauge air conditioner performance.

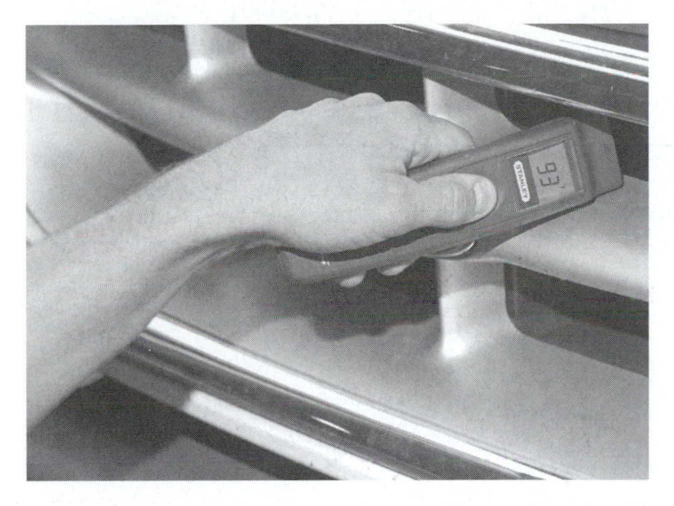

Figure 8. Ambient temperature and humidity should be measured near the condenser.

Figure 9. A large fan placed in front of the condenser is used to simulate the airflow that would normally be created while the vehicle is being driven.

Figure 10. Gauge readings should be observed and recorded to analyze test results.

Testing

1. Start the engine.
2. Set the air conditioning controls to maximum cooling and the blower speed to high.
3. Allow the engine to idle for 5 minutes. This will allow the refrigeration system pressures to stabilize.
4. Increase the engine speed and maintain 2,000 rpm.
5. Observe and record the outlet temperature and both high- and low-side gauge readings (see **Figure 10**).
6. If you are working on a system that has a cycling clutch system, note the pressures at which the clutch cycles and the frequency in which cycling occurs.
7. Compare the readings to those found in Figure 2.

PRESSURE GAUGE DIAGNOSIS

Poor cooling is one of the most common customer concerns encountered by the air conditioning technician. However, there are many different things that can cause this concern. Because refrigerant has a direct pressure-to-temperature relationship, a pressure gauge is used to monitor the pressures within an operating refrigeration system. This allows the technician to monitor exactly what is occurring within the system. As you have learned, the refrigeration system is divided into a high-pressure discharge side and a low-pressure suction side (see **Figure 11**). The two sides are separated at the compressor and the metering device. Each side of the system will exhibit specific pressure characteristics based on system operation and environmental conditions. When the system is operating properly, the relationship between the suction and discharge pressure is highly predictable.

Base pressure readings used for diagnosis can be obtained by completing the performance test described earlier. After you have obtained the pressure readings and determined what the normal readings should look like for the vehicle that is being serviced, take some time to think about what is happening at each point in the system. This

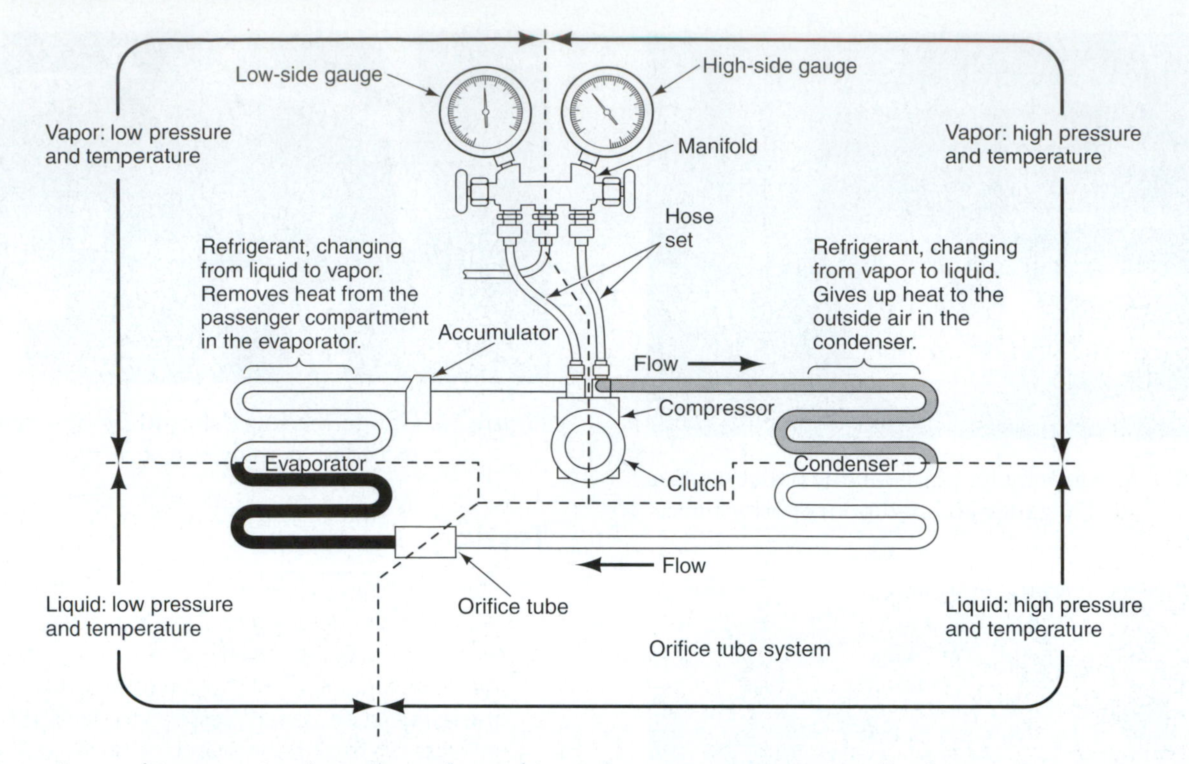

Figure 11. As a refrigerant circulates through a refrigeration system, it continually changes in pressure temperature and state.

will help you better understand the pressure readings that have been obtained. The pressure readings for the suction and discharge sides of the system can exist in one of three states: normal, higher than normal, and lower than normal. Pressure readings obtained from a malfunctioning system can appear in any combination (for example, the low side may be low, whereas the high side is normal). It is, however, the occurrence of various combinations that helps you decipher exactly what is occurring within the system. **Figure 12** and **Figure 13** outline some of the most common air conditioning system symptoms, pressures, and solutions. These charts are a helpful resource when repairing an inoperative air conditioning system. Let's take a look at why each of the following pressure readings occur and how they affect air conditioning performance.

Service Fitting Location

Service fittings can be located in various positions throughout the refrigeration system (see **Figure 14**). It is worth noting that the location of the service fittings can significantly affect the pressure readings obtained with the gauge. For example, a high-side fitting that is located on a vapor side of the condenser can exhibit pressures that are significantly higher than those obtained from a fitting that is located on the liquid side of the condenser. This is of particular importance when diagnosing a system malfunction. Although the service valve location should not change how you approach the system diagnosis, it should be factored in when analyzing the results of testing.

High-Side Readings

Normal high-side readings can range from 150 to 375 psi (1,034 to 2,586 kPa); however, these readings are fully dependent on ambient conditions (see **Figure 15**). Readings that are out of range—either too high or too low—have the potential to reduce the cooling capacity of the air conditioning system.

Readings that are higher than these are most often caused by poor condenser airflow, engine overheating, or a refrigerant overcharge. A restriction in the discharge line between the compressor and the condenser also could be a possible cause for high discharge pressures. However, this condition would most likely be caused by a pinched line, which should be readily visible through visual inspection.

- Poor condenser airflow will limit the amount of heat that can be removed from the refrigerant. This will cause refrigerant temperatures to remain higher than normal as they pass through the condenser, also resulting in higher pressure. Poor condenser airflow is

Orifice Tube Systems

System Operating Normally—Fully Charged				
Low Side	**High Side**	**Sight Glass**	**Evaporator Outlet**	**Duct Temperature**
15–30 psi	150–300 psi	None	Cold	40°–55°F
23–48 psi	150–375 psi	None	Cold	40°–55°F

Diagnostic Chart				
Low Side	**High Side**	**Symptoms**	**Diagnosis**	**Solutions**
Low to Normal 10–46 psi	Low 120–170 psi	Poor cooling. Warm evaporator outlet line. Compressor clutch cycling rapidly	Low refrigerant charge.	Check and repair any leaks in the system. Recharge system as needed.
Low to Vacuum −10 to 10 psi	Low to Normal 90–170 psi	Poor cooling. Warm evaporator outlet line. Compressor clutch cycling rapidly.	High-side restriction. Bad-orifice tube. Gauge reading may be higher if the restriction is directly past the service fitting.	Check for a clogged orifice tube. Evacuate and recharge the system.
Normal to Low 5–48 psi	Normal 185–375 psi	No air or warm air from the ducts. Evaporator lines cold or iced.	Evaporator freeze-up. Bad thermostatic switch or cycling switch. Evaporator freeze-up at low blower speeds or during long drives.	Replace cycling switch or thermostatic switch. Make sure you reinstall capillary tube in the original location.
High 60–100 psi	Low 70–120 psi	No cooling. Warm evaporator outlet pipe.	Bad compressor	Repair or replace compressor. Replace accumulator. Evacuate and recharge the system.
High 40–60 psi	High 200–400+ psi	Fair to poor cooling. Evaporator outlet cool to warm. Compressor doesn't cycle.	System overcharged	Bleed excess R-12 or R-134a until the A/C system operation returns to normal.
Normal to High 15–55 psi	High 200–300+ psi	Fair to poor cooling. Evaporator outlet cool to warm.	Engine overheating Restricted airflow past condenser.	Check cooling system operation. Check cooling fan operation Clear radiator or condenser restriction.

Figure 12. This chart represents the relationship of temperature, pressure, humidity, and outlet temperature of an orifice tube refrigeration system.

most often caused by debris blocking the flow of air through condenser fins, damaged condenser fins (see **Figure 16**), or an inoperative cooling fan. Systems that suffer from poor condenser airflow will often cool normally when the vehicle is moving but cool very poorly when the vehicle is stopped. Additionally, these vehicles will often exhibit higher than normal engine coolant temperatures.

- Engine overheating adds to the amount of heat that is present in the radiator condenser area, effectively reducing the efficiency of the condenser. This eventually leads to elevated refrigerant temperatures and pressures.
- A system that has a refrigerant overcharge condition will have an excess amount of refrigerant in both the evaporator and condenser. This results in poor vaporization within the evaporator, which can allow some liquid refrigerant to return to the compressor, causing excessive compressor discharge pressures and noisy operation. This condition is caused by improper service procedures.

Expansion Valve Systems

System Operating Normally—Fully Charged				
Low Side	**High Side**	**Sight Glass**	**Evaporator Outlet**	**Duct Temperature**
15–30 psi Pressure will be higher at higher blower speeds	150–285 psi Low airflow past the condenser increases high-side pressures.	Clear—any color other than white or clear indicates system contamination.	Cold—lines sweating heavily, no frost	40°–50°F

Diagnostic Chart

Low Side	High Side	Symptoms	Diagnosis	Solutions
Low 15–30 psi	Low 110–150 psi	Poor or no cooling. Foamy bubbles in the sight glass. Compressor cycles rapidly. Warm evaporator outlet line.	Low or improper refrigerant charge.	Check and repair any leaks in the system. Recharge system as needed.
Low or Vacuum −20 −10 psi	Low to Normal 115–160 psi	No cooling. Sight glass is clear. Warm evaporator outlet line.	High-side restriction. Bad expansion valve. Gauge reading may be higher if the restriction is directly past the service fitting.	Check the expansion valve and screen. Look for icing on the high-side lines. Clear the restriction or replace necessary components. Evacuate and recharge the system.
Normal to Low 0–25 psi	Normal 150–285 psi	Unit works fine for a while, then begins to blow warm air. Evaporator pipes frozen. Compressor doesn't cycle.	Evaporator freeze-up. Bad thermostatic switch or cycling switch.	Replace cycling switch or thermostatic switch.
High or Equal to High-Side Gauge 70–90 psi	Low or Equal to Low-Side Gauge 90–110 psi	No cooling. Warm evaporator outlet pipe. Compressor won't cycle.	Expansion valve stuck open. Bad compressor.	Repair or replace compressor. Replace expansion valve. Evacuate and recharge the system.
Normal to High 30–60 psi	High 200–300 psi	Fair to poor cooling. Sight glass clear to foamy. Evaporator outlet cool to warm.	System overcharged	Bleed excess R-12 or R-134a until the A/C system operation returns to normal.
Normal to High 30–60 psi	High 200–300+ psi	Fair to poor cooling. Sight glass clear. Evaporator outlet cool to warm.	Engine overheating Restricted airflow past condenser.	Check cooling system operation. Check cooling fan operation. Clear radiator or condenser restriction.

Figure 13. This chart represents the relationship of temperature, pressure, humidity, and outlet temperature of an expansion valve equipped refrigeration system.

Lower than normal high-side readings often result from a low refrigerant charge, a high-side restriction, or a faulty compressor.

- The development of pressure in the refrigeration system is dependent on the ability of the compressor to pressurize against a restriction, in this case, the metering device. When the refrigeration system is low on refrigerant, both the condenser and evaporator are starved. When this occurs, the refrigerant flows through the system with little or no restriction. This results in proportionately low discharge pressures. A low refrigerant condition can be caused by a system leak or improper service procedures.
- The metering device provides an engineered restriction between the condenser and the evaporator. When the amount of restriction is increased, refrigerant flow through the device is decreased. This can cause slightly lower than normal high-side pressures

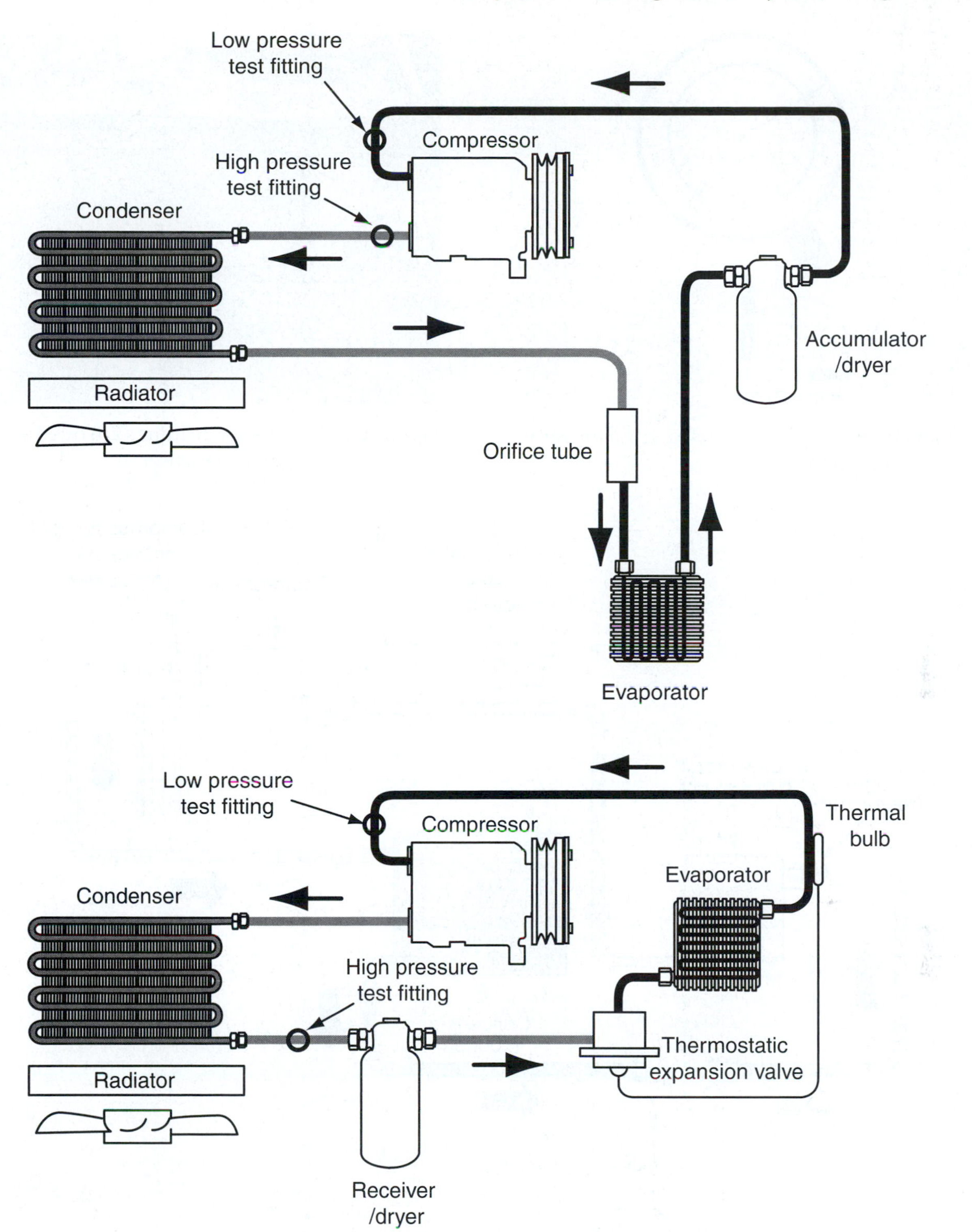

Figure 14. Discharge service fitting locations often vary between vehicle models.

because the refrigerant will spend more time in the condenser, resulting in slightly cooler refrigerant temperatures and pressures (see **Figure 17**). The visibility of these differences is relative to the location of the high-side refrigerant service fitting. Restriction in this location can be caused by debris clogging the metering device inlet screen (see **Figure 18**) or physical damage to the liquid line. A restriction in this location also can affect the low-side readings as well.

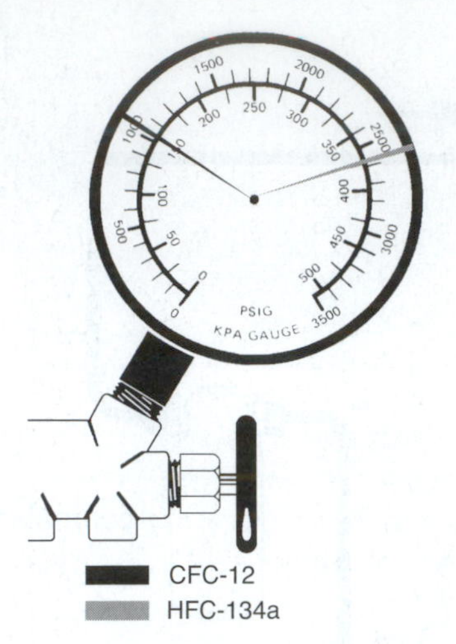

Figure 15. Gauge readings caused by moisture within the system.

Figure 16. These gauge readings are often a result of a clogged condenser.

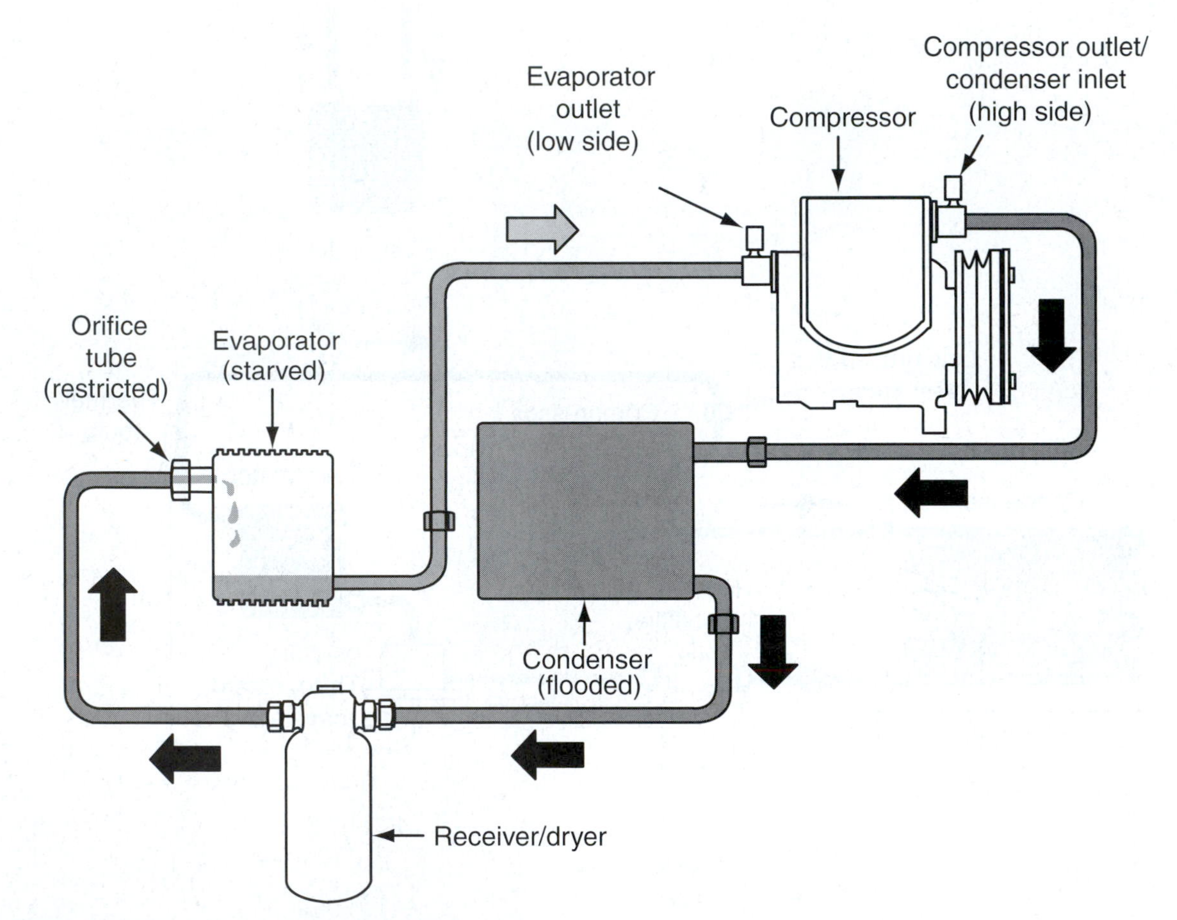

Figure 17. A restriction between the condenser and evaporator can result in a starved evaporator and or a flooded condenser.

- The efficiency of the compressor is relevant to its ability to seal. When the compressor's cylinders fail to seal, its ability to create both adequate suction and compression is hindered. When this problem becomes severe enough, suction pressures become higher than normal, whereas discharge pressures will be reduced. This condition often can be detected by near-equal suction and discharge pressures (see **Figure 19**) or sharp fluctuations noted on the high and low pressure gauges.

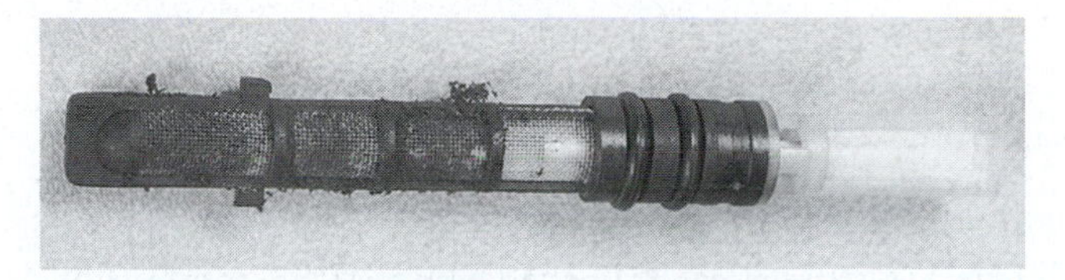

Figure 18. A clogged metering device will cause an unwanted restriction between the condenser and evaporator.

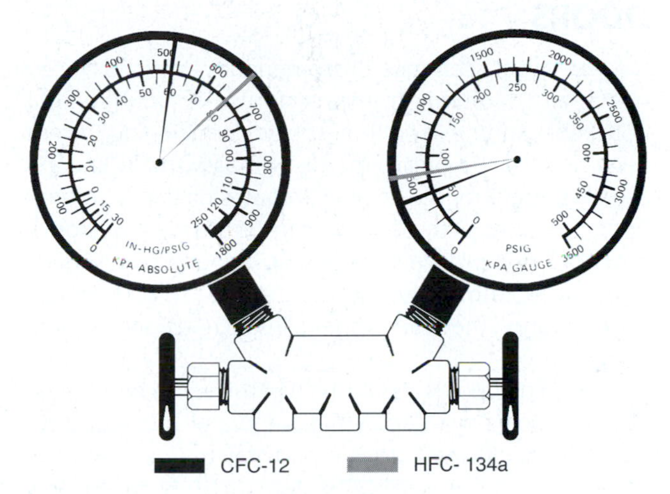

Figure 19. Gauge readings caused by a restriction in the high side of the system.

Low-Side Readings

Normal low-side readings are usually in a range between 30 and 48 psi (1,034 and 2,586 kPa) (see **Figure 20**). When readings exceed or drop below these norms, refrigeration system efficiency will suffer. Like the high-side readings, these pressures can vary greatly in relation to ambient conditions. Readings that are higher than normal are often a result of a refrigerant overcharge or a faulty compressor.

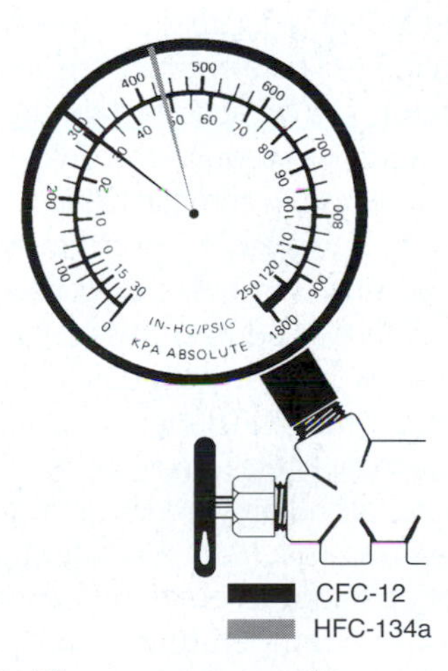

Figure 20. These gauge readings are caused by an inoperative or faulty compressor.

Lower than normal low-side readings are usually caused by low refrigerant, high-side restriction, moisture in the system, or defective cycling switch.

- A low refrigerant level will not provide an adequate amount of refrigerant. When this occurs, the evaporator becomes starved and the refrigerant that does enter the evaporator is able to vaporize very easily. Under these conditions, the compressor can rapidly evacuate all of the refrigerant vapor from the evaporator, causing very low suction pressures and, in some instances, the suction pressures may even pull into a vacuum. The technician also may notice very rapid compressor clutch cycling. This condition can be caused either by a system leak or through improper service procedures.
- A high-side restriction between the evaporator and the condenser also will result in a starved evaporator (see Figure 15). When this condition is severe, the compressor can quickly evacuate the refrigerant from the evaporator, causing the low-side pressure to rapidly pull into a vacuum. This condition is most often caused by a restriction in the metering device inlet screen or by an expansion valve that is stuck in the closed position.
- An excessive amount of moisture in the system can intermittently cause lower than normal low-side readings. As trapped moisture circulates through the system, it can freeze and cause a restriction at the metering device, blocking the flow of refrigerant. Pressures will initially appear normal but, as flow decreases, low-side pressures may become very low. When this condition is present, the customer may notice that the air conditioner is able to produce cold air initially, but outlet temperatures increase as the compressor continues to operate. This condition can be caused by a system leak, a saturated desiccant bag, or improper service procedures.
- A faulty pressure cycling switch will allow the compressor to run for long periods of time at very low pressures. This in turn creates lower than normal suction pressures as well as low evaporator surface temperatures. When this condition exists, the surface of the evaporator can remain below freezing, causing the condensation present on the evaporator's surface to freeze. This creates an obstruction to airflow across the evaporator. The customer will notice this condition as a reduction in outlet airflow. This condition can be caused by a faulty cycling switch.

NOISES

There are several noises that emanate from the refrigeration system that cause customer concern. Most of these noises will be related in some manner to the engagement of the compressor. In this section, we will examine the most common noises that are often associated with refrigeration systems.

Hissing or Whistling

One common concern that air conditioning technicians are confronted with is a hissing noise after the air conditioner has been turned off; this also may present itself as a high-pitched whistle. The customer will often hear these noises after the car has been turned off. These noises are most often caused by high-side pressure passing through the metering device to equalize system pressures when the system has been turned off. This is a normal condition in most cases.

Clutch Noises

The customer also may be alerted to a random clicking noise when the air conditioner is operating. This is often caused by the clutch plate engaging when the compressor initially engages. This noise may become more apparent in vehicles that have some clutch wear. This is usually a normal condition. However, in the event that the clutch has severe wear, a squealing noise or screeching may be noticed when engaged. This can be a sign that the clutch is severely worn or is contaminated with engine or refrigerant oil. In some severe cases, this may be caused by a locked compressor.

Suction/Discharge Hose Noise

Any time that the compressor is engaged, various vibrations are created. These vibrations are transmitted through the suction/discharge hose and compressor mounting brackets. Groaning noises produced by compressor vibrations often may appear under very specific engine loading conditions. These types of noises are easily transferred to the other refrigeration system components, which can give the impression that the noise originates from inside the vehicle. When faced with this type of noise, the technician should first inspect the suction/discharge hose for proper routing and signs of abrasion. If abrasion is noticed, the hose should be adjusted so that it is not touching any other engine or body component. In some cases, a "tuned" hose may be available for some applications that address this type of noise; this will often be addressed in a technical service bulletin.

Compressor Noises

Compressor noise is a very common cause of customer dissatisfaction. Many compressor noises, however, are a result of loose mounting brackets. Loose mounting brackets can produce noises ranging from rattling to groaning. In many cases, these noises will only occur under specific vehicle loading conditions and are usually fairly random in nature. Furthermore, any noise that is caused by the compressor or its mounting brackets will be transferred through the accessory belt drive system. Often this can lead to the unnecessary replacement of other accessory components.

Although loose brackets can produce a multitude of noises, internal compressor failure will often surface as excessive noise before a loss of cooling is noticed. Compressor failure is often recognized by a knocking or rattling noise that is relative to engine speed. These types of noises can usually be accurately diagnosed with a stethoscope. Further indications can be detected by observing pressure gauge readings, giving special attention to gauge needle action.

ODORS

A relatively common customer concern that is faced by the air conditioning technician is the occurrence of noticeable odors exiting from the air conditioning system vents. These are typically moldy or musty smells and are usually present for the first few minutes of air conditioner operation. Most of these odors can be traced to microbial growth of mold, mildew, and various fungi on the surface of the evaporator. Although most of these odors are just an annoyance, they can trigger allergic reactions in some people.

These microorganisms thrive in the warm moist environment that is provided by the evaporator. The natural accumulations of small organic particles that are naturally drawn in by the ventilation system, such as pollen, only serve to feed the growing organisms. Although this growth is primarily a natural occurrence, it can be significantly compounded by the accumulation of water in the evaporator case. This most often occurs when the evaporator drain is clogged. Another contributing factor may be dirty cabin air filters.

Once microbial growth occurs, it can only be removed through the use of a disinfectant applied to the surface of the evaporator. There are many products on the market that can be used to combat microbial growth. You should be aware that some of these products are merely deodorizers applied through the instrument panel vents that will mask the odors but will not kill the growth. Products that kill the growth must be applied directly to the evaporator's surface. The application will greatly vary from vehicle to vehicle; some evaporators can be accessed through fresh air vents, other applications may require the drilling of access holes in the evaporator case. In some instances, the evaporator may have to be removed to perform a sufficient cleaning. Many major vehicle manufacturers have provided technical service bulletins to address this very issue. These should be consulted whenever possible. Some steps to help prevent microbial growth include:

- Make sure all cabin air filters remain clean.
- Ensure that evaporator drains are able to drain freely.
- Run the blower motor on high speed for the last several minutes of operation, preferably with the air conditioner off and the temperature control moved to hot. This will aid in drying the evaporator. Many manufacturers have developed systems called **after blow**

modules that turn the blower motor on for several seconds after the air conditioner has been turned off.

LEAK TESTING

Refrigeration system leaks are among the most common causes of poor air conditioning system performance. As a result, the diagnosis of these leaks is the most common procedure performed by an air conditioning technician. Leak testing should be performed any time that refrigerant leak is suspected. Leaks in the refrigeration system can occur at any point within the system, but the most common locations can be found at various connections, crimped hose connections, various compressor seals, or a damaged heat exchanger (see **Figure 21**).

Before leak testing is performed, there are a few preliminary checks that should be performed to aid in the location of potential system leaks. Among the first steps in leak testing is the visual inspection of all refrigeration system components for signs of lubricant loss. Traces of lubricant can often indicate a refrigeration system leak. Whenever refrigerant oil is deposited on a system component, the area should be heavily scrutinized as a potential leak point. Once a visual inspection has been performed, the technician needs to verify that at least 50 psi of pressure exists within the system. If the system has less than 50 psi, the system should be recovered, evacuated, and properly recharged.

Refrigeration systems that use mineral-based oils tend to leave traces of oil at or near the leak point. PAG oils, however, tend to evaporate as they escape the refrigeration system and often leave no visual indication of a leak.

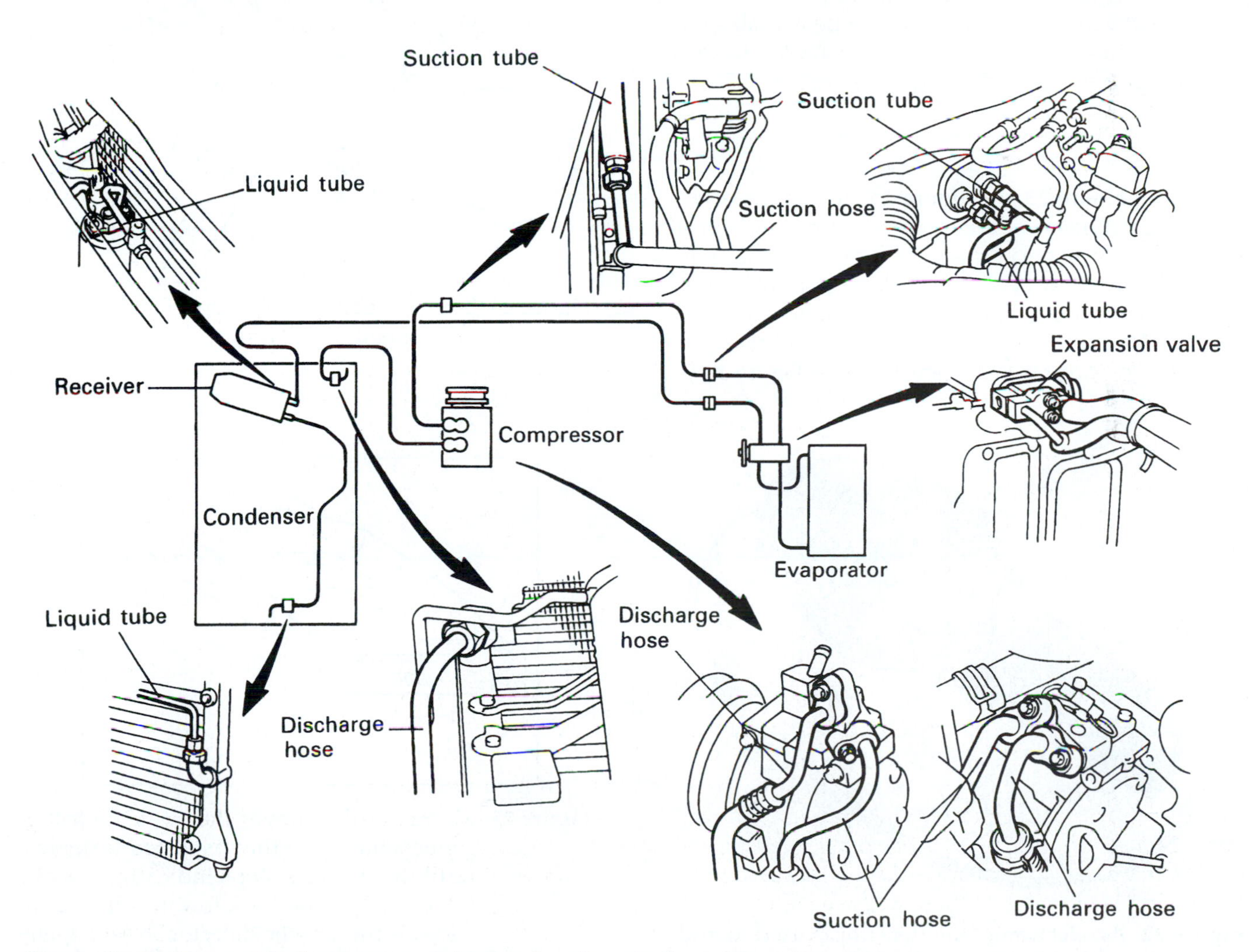

Figure 21. Refrigerant leaks can occur at many locations in the refrigeration system.

Electronic Leak Testers

The electronic leak detector is the most common method of leak detection and can provide extremely accurate results in the hands of an experienced technician (see **Figure 22**). In order to achieve accurate results, there are some steps that must be followed.

- Ensure that the instrument has clean tips and has been properly maintained and calibrated as necessary. Refer to the maintenance and calibration section of the owner's manual for specific instructions about your instrument.
- Most electronic leak detectors have a sensitivity setting. It is imperative that the sensitivity be properly adjusted. Refer to the owner's manual for proper sensitivity settings.
- The electronic leak detector can be very sensitive to many types of gases. In order to obtain accurate results, use compressed air to clear the testing area of any significant concentration of gases.
- To prevent contamination of the sensing tip with oil or grease, the tip should never contact any surface.

Once the instrument has been properly calibrated, leak testing can begin. Testing should begin with the various hoses and component connections. Because refrigerant is heavier than air, start by moving the probe on the underside of the hose or connection being tested, eventually moving the probe to encircle the entire area being tested (see **Figure 23** and **Figure 24**). To improve accuracy, only test a small area at a time. When thoroughly tested, move on to the next area. Once the tester has indicated a leak, move away from the suspect area until the instrument stops indicating a leak. Then slowly move back toward the leak area. If a leak exists, the instrument should consistently "sound off" in the same area (see **Figure 25**).

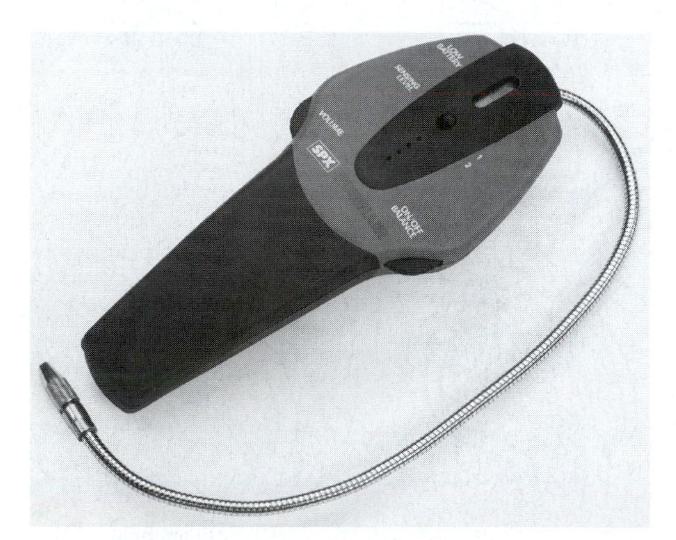

Figure 22. An electronic leak detector is used to find refrigerant leaks.

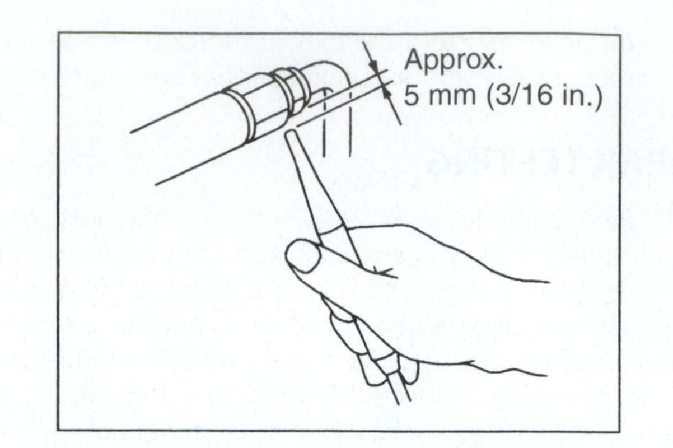

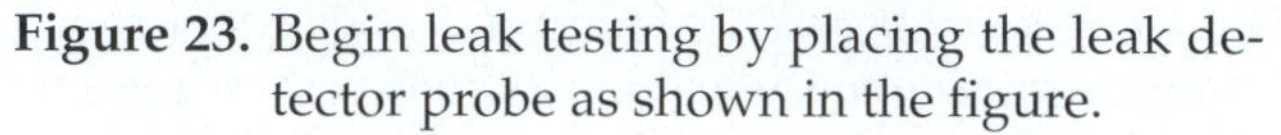

Figure 23. Begin leak testing by placing the leak detector probe as shown in the figure.

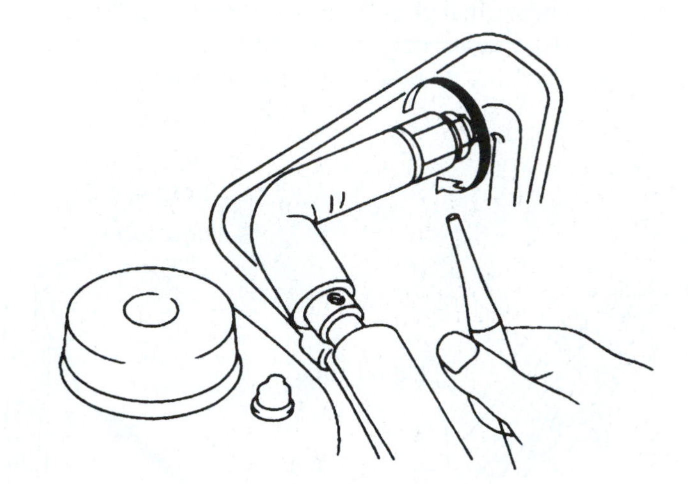

Figure 24. Slowly encircle the suspected leak area with leak detector probe.

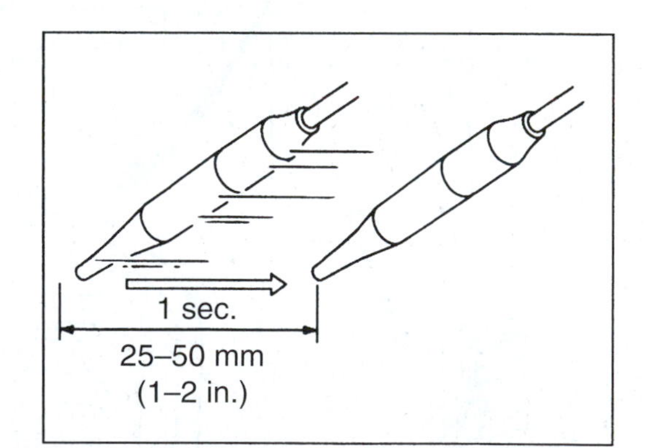

Figure 25. Momentarily move the leak detector probe away from the suspected leak area until the detector stops indicating a leak, then slowly move back toward the area. If a leak is present the detector should again indicate a leak.

Two components that require special attention when performing a leak test are the compressor and the evaporator. The compressor has several different locations in which a leak could develop, and each one of these locations must be thoroughly inspected.

- The front seal is a common leak location that should be inspected for oil residue as well as tested with an electronic tester. It may be necessary to insert the instrument probe through the front of the pulley where the clutch is pressed on (see **Figure 26**).
- The crankcases of most compressors are constructed of multiple pieces. Those pieces are connected by a slip fit and are sealed with O-rings. These connecting joints are typically visible on inspection. Because the compressor crankcase retains a relatively large amount of refrigerant oil, large amounts of oil can often leak from a crankcase seal. When severe enough, drips may emerge (see **Figure 27**).

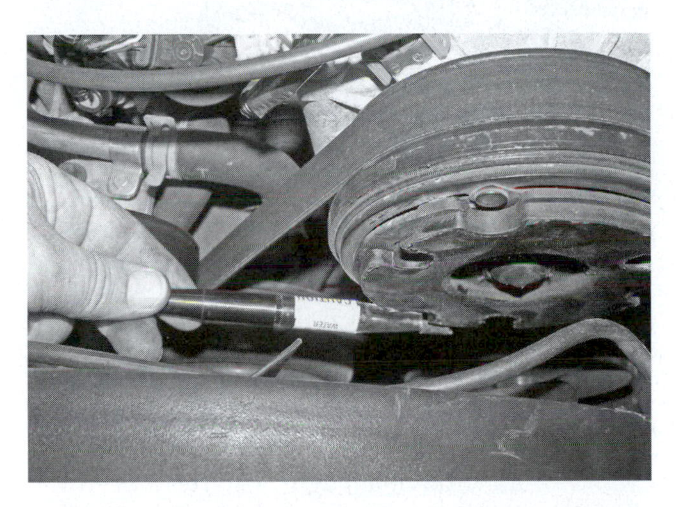

Figure 26. Checking for a leak at the front compressor seal.

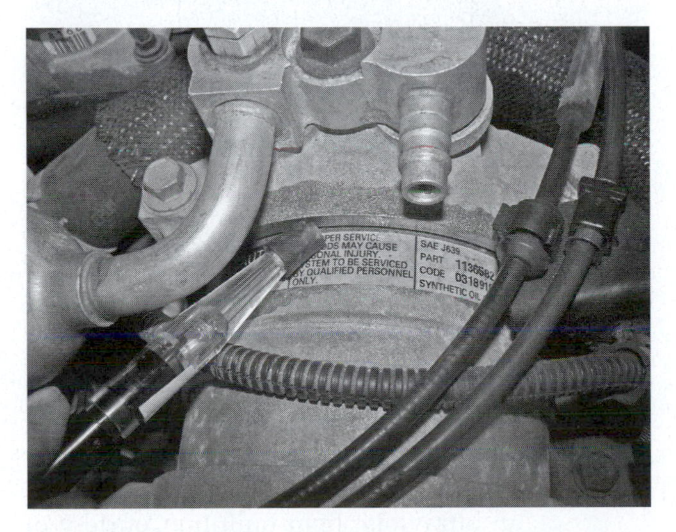

Figure 27. Compressors can leak where the case joins together.

- Some compressors are equipped with high-pressure relief valves designed to relieve refrigeration system pressure in the event that the pressure approaches a dangerous level. A detectable amount of lubricant or refrigerant may further indicate that the valve has been forced to relieve pressure. If this is suspected, the refrigeration system should be thoroughly tested to ensure that a high-pressure condition does not exist. This condition is particularly prevalent when a loss of condenser airflow has occurred, usually from an inoperative engine cooling fan.

> **Interesting Fact**
>
> *Compressor crankcase leaks often can be caused by improper tightening procedures, causing crankcase distortion. To reduce the chance of crankcase distortion, torque limits and proper tightening sequences should be observed.*

Because the evaporator is located within a sealed case in the passenger compartment, it can be extremely difficult to pinpoint a leak in the evaporator. One method of leak testing the evaporator can be accomplished by inserting the leak detector probe in the condensation drain immediately after the air conditioner has been turned off (see **Figure 28**). When this test is performed, great care must be taken not to contaminate the sensing probe with moisture. On some vehicles it is possible to access the evaporator by removing the blower motor resistor module (see **Figure 29**). This will often allow direct access to the evaporator's surface, and the probe can be placed near the core. In extreme cases, if a refrigerant loss is well documented but a leak cannot be located, it may be necessary to remove the evaporator for further inspection. Although this

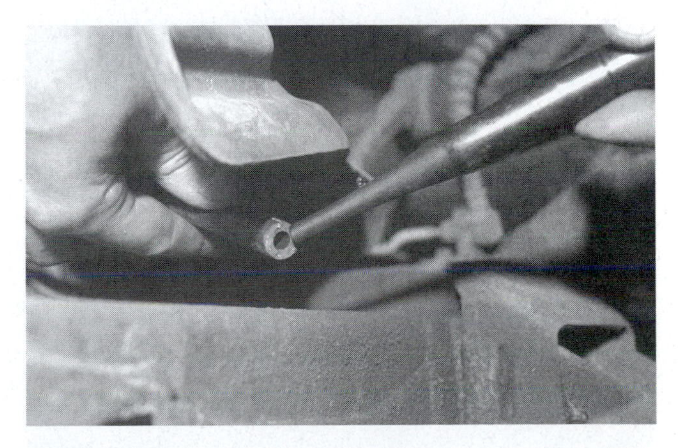

Figure 28. Some evaporator leaks can be found by placing the detector probe near the evaporator case drain tube.

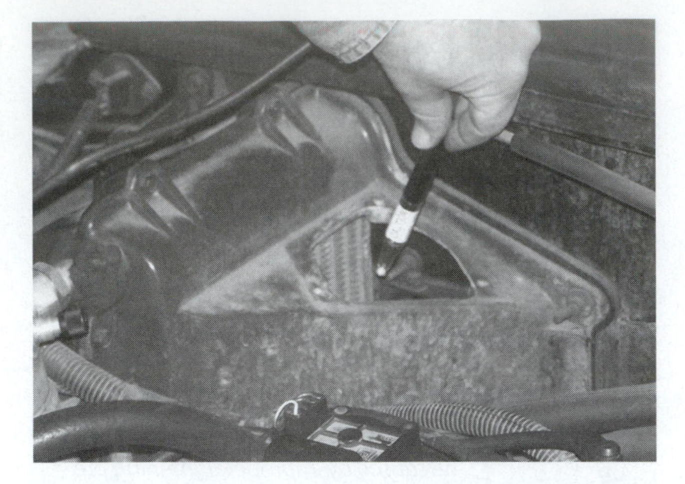

Figure 29. The evaporator can be accessed in some vehicles by removing the blower motor module or other case components.

is a common occurrence, you should be positive that no other external system leaks exist.

Ultraviolet Tracer Dye

An increasingly popular method of detecting leaks is the addition of ultraviolet dye to the refrigeration system. This method is particularly useful in locating those leaks that are not detectable using an electronic tester. The dye is introduced into the system and combines with the refrigeration oil. In order for the dye to work, it must be circulated throughout the system. Therefore, the refrigerant levels must be adequate for the refrigeration system to operate. To begin, the prescribed amount of dye is injected into the system (see **Figure 30**). This may be accomplished by directly injecting the dye, or, in some instances, it is premixed with the refrigeration system lubricant. Once the dye has been introduced, the system

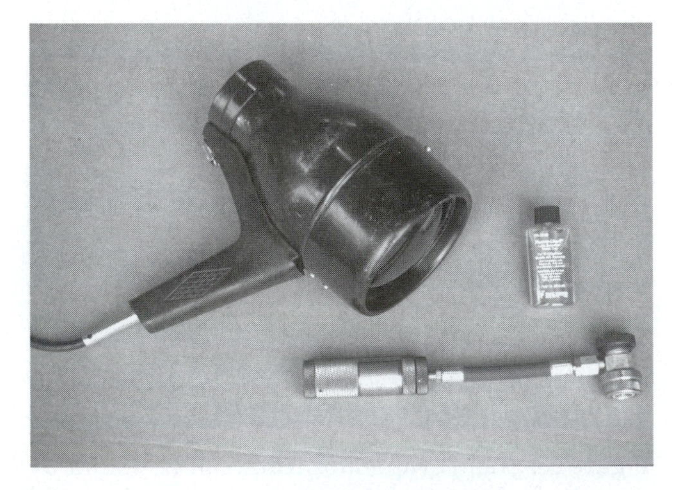

Figure 30. Ultraviolet light, dye injector and dye.

There are some electronic leak detectors that are equipped with an ultraviolet light on the end of the wand. This makes it possible to reach many inaccessible locations.

should be run for several minutes. After turning the engine off, an ultraviolet light is used to inspect each of the refrigeration system components (see **Figure 31**). Leaks will appear as a very bright yellow/green glow. A mirror often may be useful to find those leaks that are not readily accessible (see **Figure 32**). Inspection of the

Figure 31. Testing for refrigeration system leaks with an ultraviolet light. Courtesy of Tracer Products

Figure 32. A mirror may helpful in locating some refrigeration system leaks. This mirror has an ultraviolet light built in. Courtesy of Tracer Products

Figure 33. A bubble solution is helpful in the diagnosis of some refrigeration leaks. When a leak is present bubbles form in the immediate area.

evaporator using this method can be quite difficult because visual contact is required.

Soap Solution

A third method of leak detection is the use of a soap solution. This method requires that the suspected leak areas be soaked with the soap solution. If a leak is present, bubbles will appear at the location of the leak (see **Figure 33**). The drawback of this method is that the liquid solution must come into direct contact with the leak area. This can be quite difficult in many situations. However, it can be an effective method for locating larger refrigerant leaks, particularly if electronic or dye test equipment is not available. Although a homemade soap solution can be made by mixing a common dishwashing liquid with water, purpose-manufactured leak testing solutions tend to work better.

Summary

- Automotive air conditioning system performance will be judged by the customer's perceived comfort level. Human comfort level exists at temperatures of 75 degrees F to 80 degrees F (22.2 degrees C to 26.6 degrees C) and humidity levels of 45 percent to 50 percent.
- Air conditioning system performance is heavily influenced by ambient temperature, humidity, customer perception, sunload, and vehicle air volume. Technicians must be able to determine normal operating temperatures and pressures for the conditions that exist in their geographic regions.
- Diagnosis of the air conditioning system should begin with a clear concise description of the customer concern. The technicians should familiarize themselves with the operating characteristics of a particular system.
- Performing a quick system check will give the technician quick feedback on the operation of the entire air conditioning system. The visual inspection is used to identify any obvious air conditioning system faults. Performance testing allows the technician to test the refrigeration under a controlled set of circumstances.
- Diagnosis of the air conditioning system should begin by comparing all of the data that has been accumulated, including customer description, quick check, visual inspection, and performance test. The actual temperatures and pressures produced by the refrigeration system are often indicative of the condition and operation of specific components.
- Most noises related to the refrigeration system are generally connected to the engagement of the compressor. These noises can be projected as hissing noises, clicking noises produced by the clutch, groaning, knocking, and rattling noises.
- Odors within the refrigeration system most often are produced by microbial growth on the surface of the evaporator. Warm moist conditions that are present within the evaporator housing produce ideal conditions for the growth of various microorganisms.
- The diagnosis of refrigeration system leaks is one of the most common procedures performed by an automotive technician. Leaks can be present at almost any location within the system. The three most common methods of leak detection are the use of an electronic tester, a dye solution, and a bubble solution.

Review Questions

1. Technician A says that humidity affects how cool the air feels to the human skin. Technician B says that humidity affects the actual operation of the refrigeration system. Who is correct?
   A. Technician A
   B. Technician B
   C. Both Technician A and Technician B
   D. Neither Technician A nor Technician B
2. Technician A says that normal refrigeration system pressures and temperatures should be the same, regardless of geographic location. Technician B says that the outlet air temperature should be at least 20 degrees F (11 degrees C) cooler than the air at the inlet. Who is correct?
   A. Technician A
   B. Technician B
   C. Both Technician A and Technician B
   D. Neither Technician A nor Technician B
3. Describe how the quick check, visual inspection, and the performance test can be used to help the technician make an accurate diagnosis.
4. A technician is diagnosing an air conditioning system with a poor cooling condition. The customer states that the air conditioner works fine when it is first turned on, but after the vehicle has been driven for an hour at highway speeds, the airflow from the vents significantly decreases. However, after the air conditioner is turned off for a while, normal air conditioner system operation is resumed. Technician A says that the refrigeration system is contaminated with moisture. Technician B says that a faulty pressure cycling switch could be the cause of the concern. Who is correct?
   A. Technician A
   B. Technician B
   C. Both Technician A and Technician B
   D. Neither Technician A nor Technician B
5. A technician is diagnosing an air conditioning system with a poor cooling condition. The customer states that the air conditioner works fine when the vehicle is moving, but when the vehicle is stopped, the temperature of the outlet air noticeably increases. A second complaint on the repair order states that the engine temperature seems to be warmer than normal.
   Technician A says that a faulty cooling fan relay could be the cause. Technician B says that a low refrigerant charge could be the cause. Who is correct?
   A. Technician A
   B. Technician B
   C. Both Technician A and Technician B
   D. Neither Technician A nor Technician B
6. Noises within the air conditioning system can be caused by which of the following:
   A. Metering device (pressure equalization)
   B. Loose compressor mounts
   C. Improper hose routing
   D. All of the above
7. Technician A says that running the blower motor on low speed for the first few minutes of air conditioner operation can help alleviate many odor concerns. Technician B says that odors are usually noticed after the air conditioning system has run for several minutes. Who is correct?
   A. Technician A
   B. Technician B
   C. Both Technician A and Technician B
   D. Neither Technician A nor Technician B
8. Which of the following is not a method of refrigeration system leak detection?
   A. Electronic leak detector
   B. Compressed air method
   C. Soap/bubble method
   D. Ultraviolet dye method
9. Technician A says that a contaminated sensing tip can cause an electronic leak detector to indicate a false reading. Technician B says that to obtain correct readings, the sensing tip should be cleared of trace gases using compressed air. Who is correct?
   A. Technician A
   B. Technician B
   C. Both Technician A and Technician B
   D. Neither Technician A nor Technician B
10. All of the following statements about the use of ultraviolet refrigerant dye are true, *except:*
    A. Dye can be added directly to the system.
    B. The refrigeration system must contain enough refrigerant to operate.
    C. The dye can be premixed with refrigerant oil.
    D. The dye is exposed using fluorescent light.

Chapter 18 Refrigeration System Procedures

Introduction

Once a customer concern has been duplicated and diagnosed, it is time to perform the necessary repairs. Like any other automotive system, it is critical that proper procedures be closely followed. Failure to do so will likely lead to a customer comeback. This chapter will detail the most common service procedures and provide helpful tips for performing service.

SERVICE VALVE REPAIR

From time to time, the service valves will develop leaks. Often these leaks develop immediately after system service has been performed. In most instances, the Schrader valve core can be replaced, whereas in other cases the entire valve may need to be replaced. In extreme cases, the component in which the service valve is attached may have to be replaced as well. Service of the Schrader valve or fitting requires that the refrigerant be removed from the system.

1. Recover the refrigerant as needed.
2. Engage the slots of the core remover (see **Figure 1**) over the shoulders of the valve core and turn counterclockwise until the core is removed (see **Figure 2**).
3. Lubricate the replacement valve core using mineral oil and reinstall. The core should be carefully tightened using only finger pressure.
4. Evacuate and recharge the system to the appropriate level.

Figure 1. Valve core removal tool.

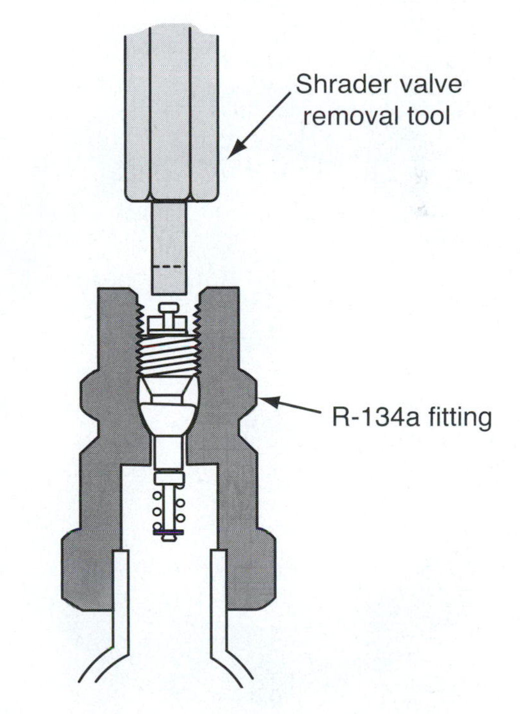

Figure 2. The valve core removal tool engages the shoulders of the valve core.

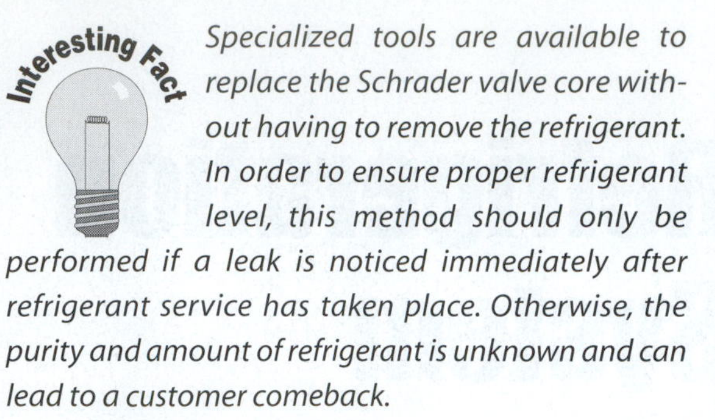

Specialized tools are available to replace the Schrader valve core without having to remove the refrigerant. In order to ensure proper refrigerant level, this method should only be performed if a leak is noticed immediately after refrigerant service has taken place. Otherwise, the purity and amount of refrigerant is unknown and can lead to a customer comeback.

O-RING REPLACEMENT

O-ring replacement is one of the most frequent repairs performed to a refrigeration system. O-rings may be replaced to repair a refrigerant leak and should always be replaced any time that they are removed or disturbed. The procedures outlined also should be followed when replacing lines or fittings.

O-rings fail and begin to leak when they are physically damaged. Damage typically occurs as a result of being cut, pinched, cracking from old age, or exposure to excessive heat (see **Figure 3**). Connections that are sealed by O-rings require very little pressure to seal, and therefore leaks are rarely a result of loose fittings. When fittings are overtightened, O-rings often can be overcompressed, pinched, or cut, resulting in a leak (see **Figure 4**).

Figure 3. Inspect O-rings for signs of damage.

Figure 4. O-rings can be damaged during installation if proper procedures are not observed.

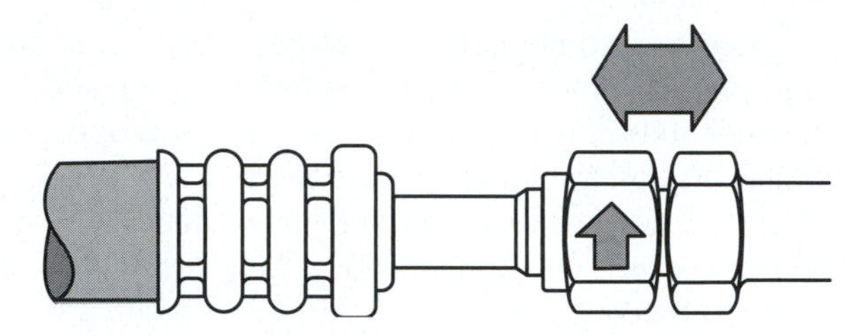

Figure 5. Disassembly of a threaded fitting.

O-ring service is relatively straightforward and can be successfully completed following just a few steps. The biggest challenge in servicing O-rings is separating the mechanical connection. Some use large threaded fittings, others use quick-connect spring fittings, and yet another type uses a flange fitting held together with a stud and nut. Refer to **Figure 5**, **Figure 6**, and **Figure 7** for specific fitting disassembly procedures. Once the fittings are disassembled, the following steps should be followed:

1. Inspect the O-ring for any damage. The damage sustained to an O-ring will give the technician a strong sense of why the O-ring failed.
2. Inspect both tubes for nicks or signs of distortion.
3. Clean all of the O-ring mating surfaces.
4. Inspect threaded fittings for damaged or corroded threads; clean or replace as needed. Spring-lock connectors should be examined for distortion and spring condition.
5. Used O-rings should never be reinstalled.

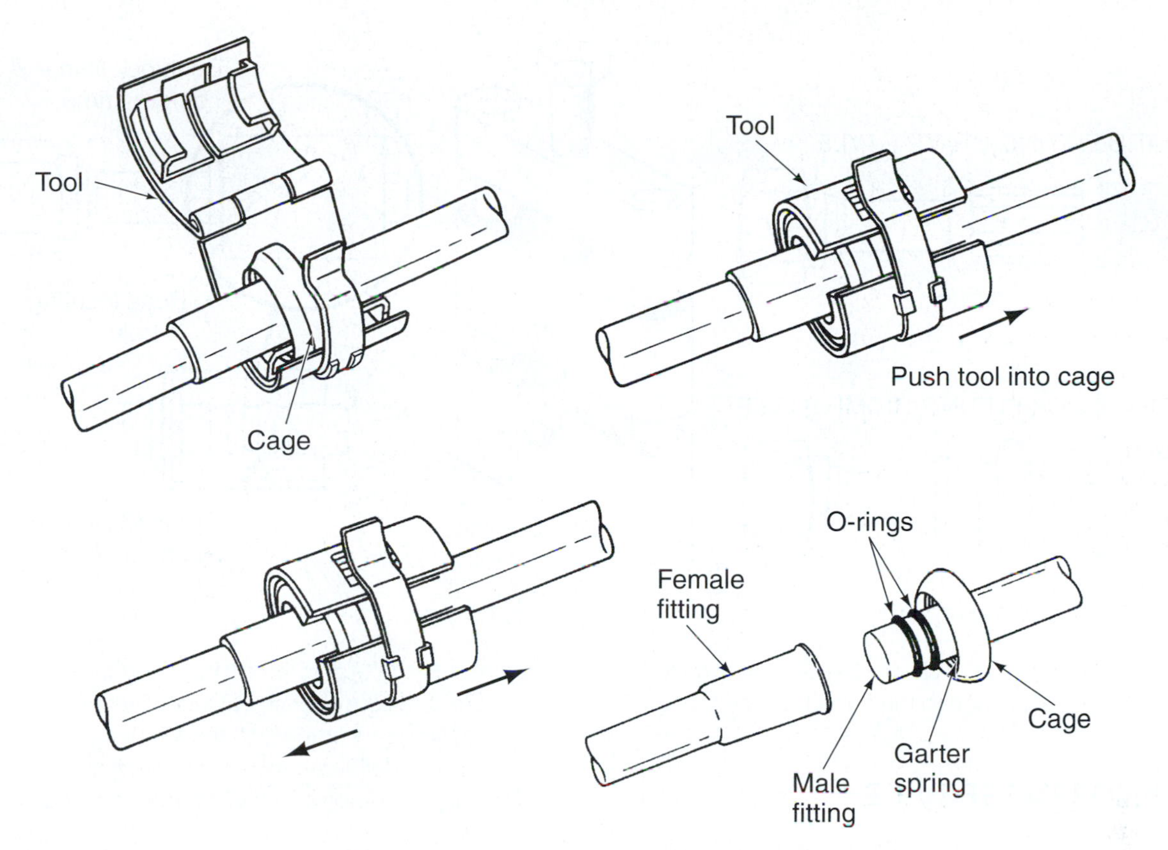

Figure 6. Procedures for disassembly of a spring lock connector.

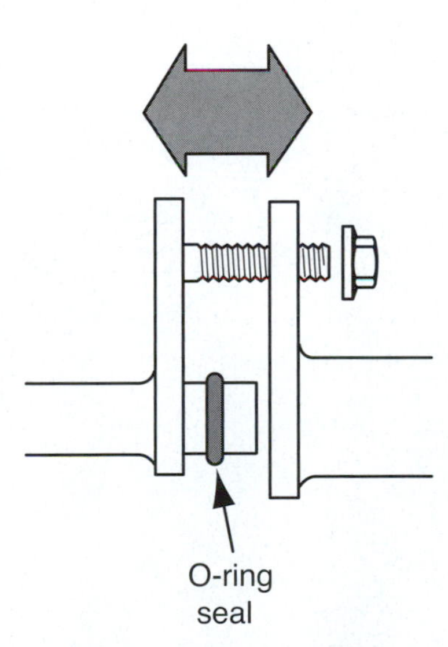

Figure 7. Disassembly of a stud and flange type fitting.

Reassembly

6. Compare the new O-ring to the old one. The new O-ring should be the same size in outer diameter, inner diameter, and the O-ring cross section should be the same. It also should be verified that the new O-ring is compatible with the type of refrigerant being used.
7. Lubricate new O-rings using mineral oil and slide them into position.
8. Carefully join the two tubes back together.
9. Reconnect the mechanical connections. Do not lubricate any component of the mechanical connector. Do not overtighten threaded fittings. Overtightening

You Should Know *Only mineral oil should be used to lubricate refrigeration system O-rings. Synthetic oils used with R-134a are hygroscopic in nature, attracting moisture into the fitting. This can lead to the formation of corrosion, making subsequent separation of fittings difficult and, in some instances, impossible.*

You Should Know *Threaded hose connections do not require a great deal of torque to properly compress and hold the O-ring seal. Do not overtighten threaded fittings; doing so can cause damage to the fitting and the O-ring.*

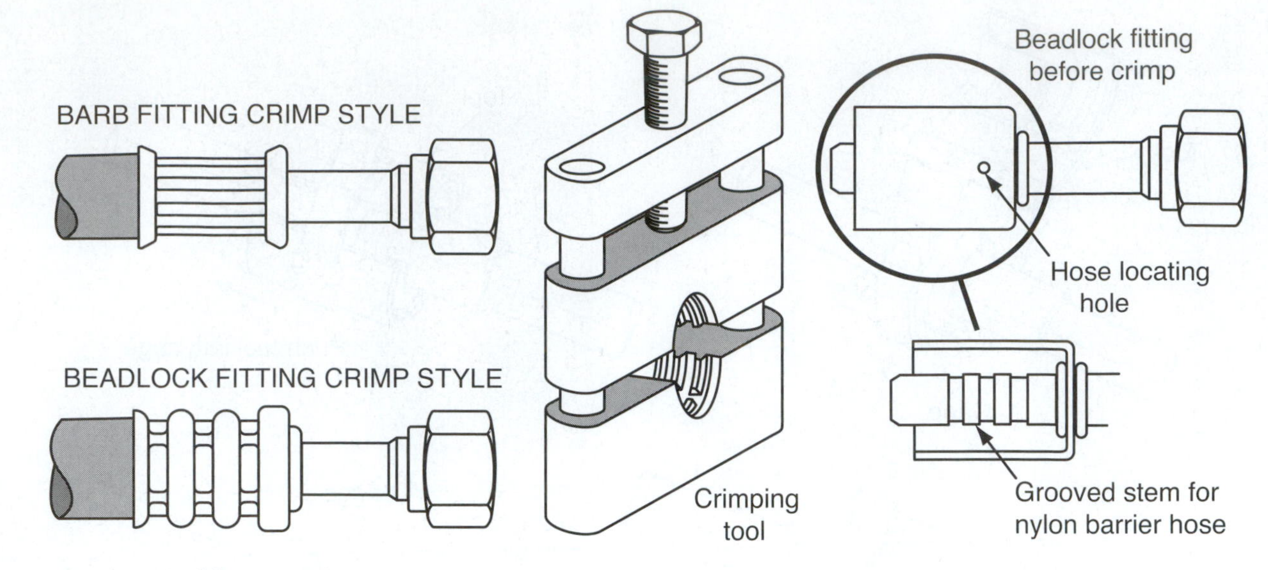

Figure 8. A typical hose repair system.

can cause serious damage to the newly installed O-ring, as well as damage to the threads of the fitting.

HOSE AND LINE SERVICE

In most situations, if a hose or line is damaged, the entire assembly will be replaced. However, some specialty shops and dealers have the equipment with which to repair damaged hoses. In most cases, this equipment will be part of an overall "system" in which the equipment is designed to work with specific fittings (see **Figure 8**). Most repair systems consist of various size and shape fittings and **ferrules** that are specifically designed for a particular type of hose. The ferrule is a sleeve that is placed around the fitting and the hose. When compressed, the ferrule becomes a permanent clamp to hold the fitting and the hose together. The fittings are usually equipped with external barbs that help the fitting grip the hose. The information given here is intended to be generic in nature to provide the student with some valuable insight as to how these repairs are completed. To ensure a quality repair to the hose or line, strictly adhere to the directions included with the equipment. These procedures can be used when repairing an existing hose or in the construction of a new assembly.

1. Measure and mark the required length of replacement hose. If repairs are being made to an existing hose, determine how much hose must be cut ahead of the damaged fitting. Every effort should be made to remove the least amount of hose possible.
2. Using a hose cutter cut the hose to the proper length. To ensure a proper seal, the end of the hose must be cut perfectly square.
3. Using the proper system oil, apply a liberal amount of clean refrigeration oil to the inside of the hose.
4. Install the proper ferrule onto the end of the hose.
5. Inspect the fitting to ensure that is free of all nicks or burrs and coat liberally with clean refrigeration oil.
6. Slip the insert fitting into the refrigeration hose in one constant, deliberate twisting motion.
7. Using the crimping tool, crimp the ferrule (see **Figure 9**).

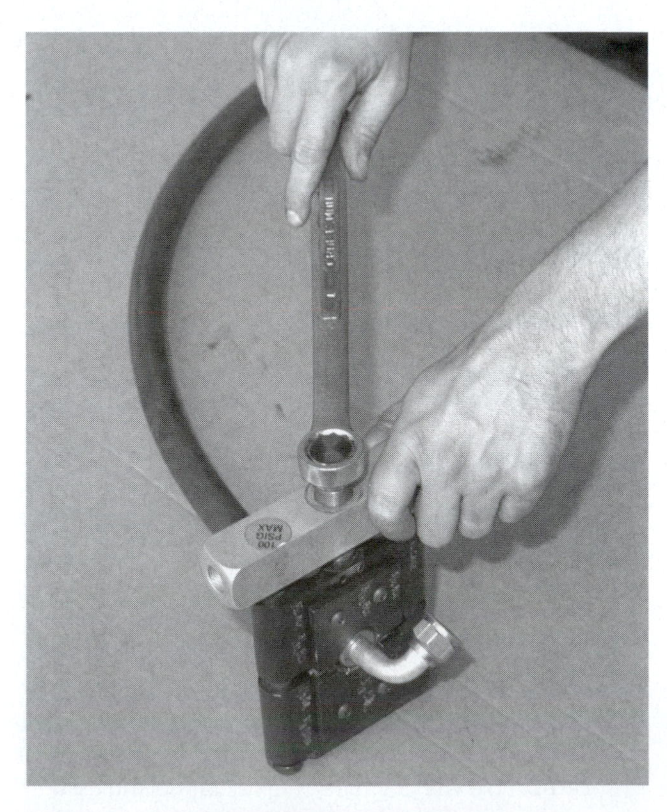

Figure 9. Crimping the ferrule to the hose.

Repairing Rigid Lines

In the past, when a rigid refrigerant line became damaged, it was necessary to replace the entire line. However, in recent years, repair systems have become available that make the repair of damaged rigid lines practical. The system uses special collars to repair damaged lines (see **Figure 10**). Lines ranging in size from 5/16 in. (8 mm) up to 3/4 in. (19 mm) can be repaired using this system.

1. Recover the system refrigerant as needed.
2. Locate the damaged area of the tube.
3. Remove the tube as needed.
4. Cut the tube on each side of the damaged section.
5. If a large section of tube had to be removed, it may be necessary to install a section of replacement tube.
6. Clean any burrs or grease from the repair area of the tube.
7. Apply one drop of required sealant.
8. Insert each end of the tube into the proper size fitting (see **Figure 11**).
9. Rotate the fitting to distribute the sealant.
10. Install the crimping tool over the fitting.
11. Tighten the forcing screw until the jaws bottom against each other (see **Figure 12**).
12. Remove the crimping tool and verify that the fitting is evenly crimped and tight.
13. Evacuate and recharge the system as required.

ORIFICE TUBE REPLACEMENT

The fixed orifice tube (FOT) can be located anywhere between the condenser and the evaporator. The most common locations are in the condenser outlet, the evaporator inlet, or the liquid line. The FOT will be located in a position so that it can be extracted from the end of the tube in which it is installed. The FOT location can be located by three dimples in the line. These dimples or a ridge will stop the tube from going too far into the tube (see **Figure 13**).

1. Recover the refrigerant charge as needed.
2. Using the proper size and type wrenches, remove the liquid line connection at the FOT location to expose

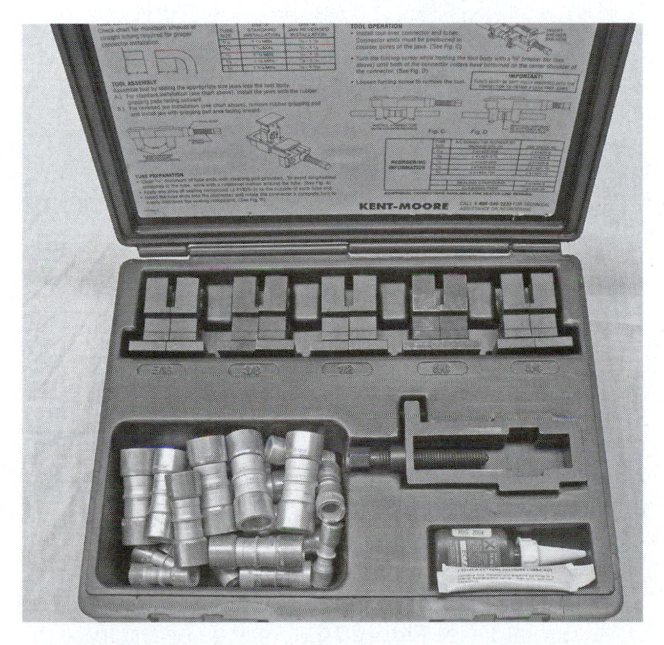

Figure 10. This kit is used to repair rigid refrigerant lines.

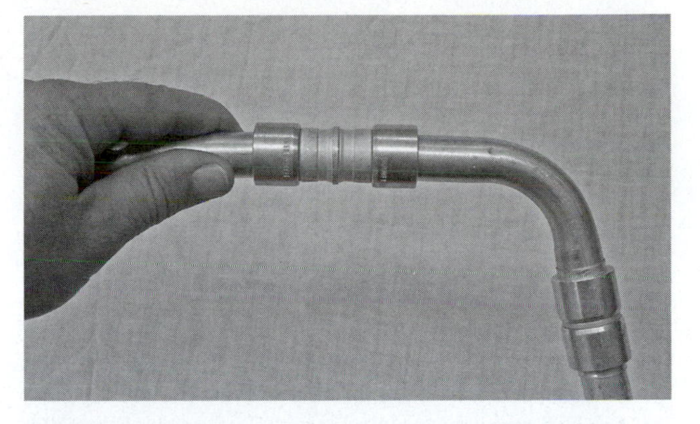

Figure 11. The cleaned and prepped tubes are inserted into the repair sleeve.

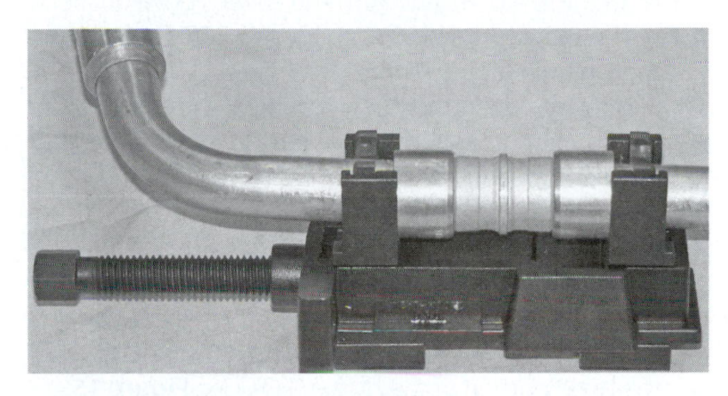

Figure 12. The crimp permanently connects the two tubes.

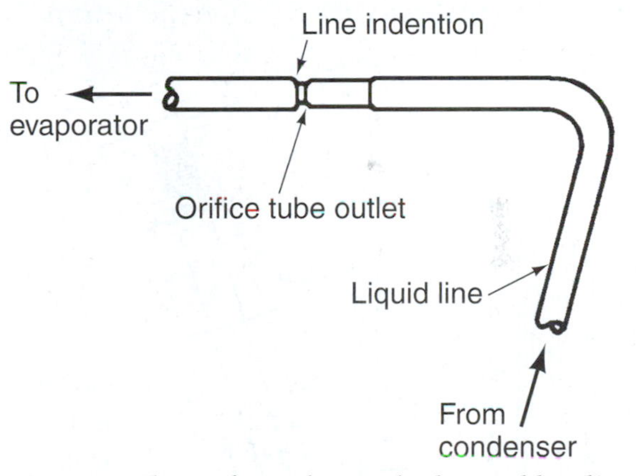
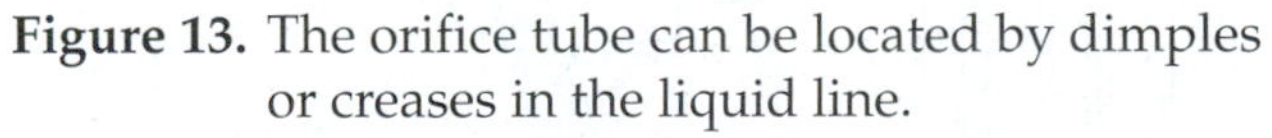

Figure 13. The orifice tube can be located by dimples or creases in the liquid line.

the FOT. The FOT should appear very close to the end of the tube.

3. Remove and discard the O-ring(s) from the liquid line fitting, if equipped.
4. Pour a very small quantity of clean refrigeration oil into the FOT well. This step will help loosen debris and stuck O-rings. To prevent contamination, the oil used should be of the same type used in the system: mineral oil for R-12 and PAG for R-134a.
5. An orifice tube removal tool (see **Figure 14**) should be used to remove the FOT from the tube in which it is located; this will help avoid damage to the tube. Insert the FOT removal tool onto the FOT. Turn the T-handle of the tool slightly clockwise only enough to engage the tool onto the tabs of the FOT.
6. Hold the T-handle and turn the outer sleeve or spool of the tool clockwise to remove the FOT. Do not turn the T-handle; doing so will break the end of the FOT.
7. It is not an uncommon occurrence for the FOT to seize within the tube and break during removal. If this occurs, proceed to step 8. If the tube comes out cleanly, proceed to step 11.
8. A broken tube can be removed with a threaded extractor. To remove the tube, insert the extractor into the well and turn the T-handle clockwise until the threaded portion of the tool is securely inserted into the brass portion of the broken FOT (see **Figure 15**).
9. Pull the tool. The broken FOT should slide out.
10. NOTE: The brass tube may pull out of the plastic body. If this happens, remove the brass tube from the puller and reinsert the puller into the plastic body. Repeat steps 8 and 9.
11. Clear the tube of any debris.
12. Liberally coat the new FOT with clean refrigeration oil.
13. Note the direction that old tube was removed and re-install in the same direction. Place the FOT into the evaporator cavity and push it in until it stops against the evaporator tube inlet dimples.
14. Install a new O-ring, if equipped.
15. Replace the liquid line and tighten it to the recommended torque.
16. Evacuate and recharge the system according to manufacturers specifications.

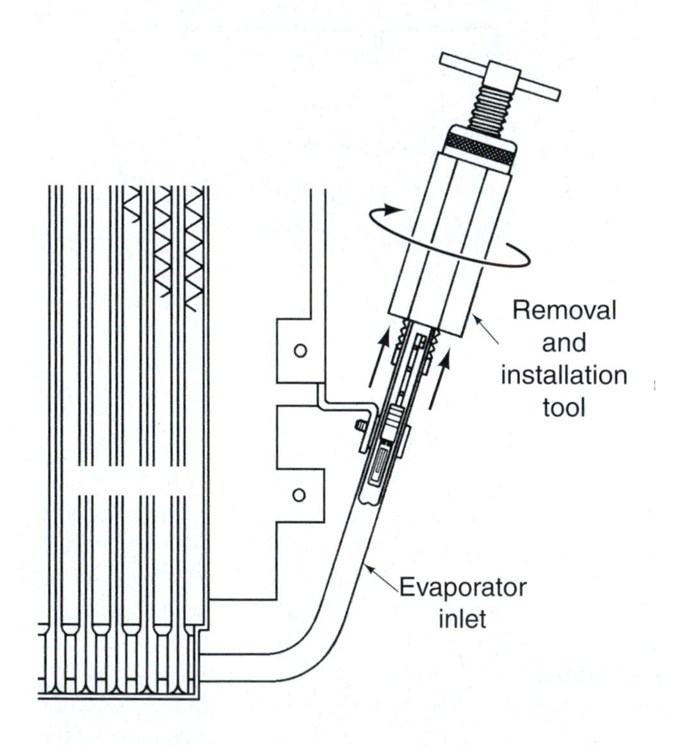

Figure 14. Proper orifice tube removal.

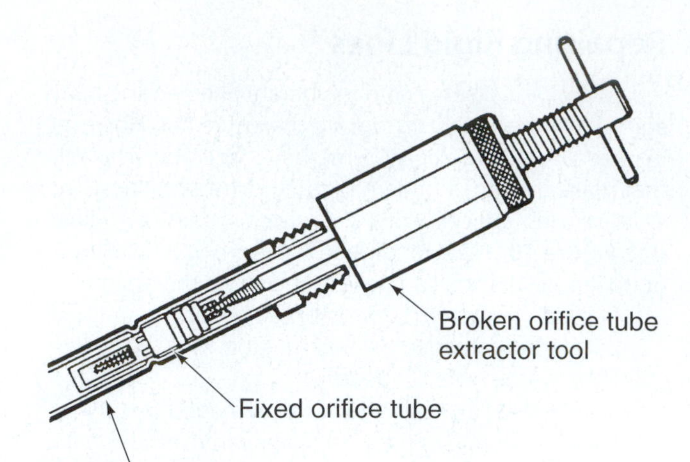

Figure 15. A broken orifice tube can be removed using a threaded removal tool.

> **You Should Know** *Some vehicle models use a nonserviceable FOT located in the liquid line. On these vehicles, the entire line must be replaced.*

EXPANSION VALVE REPLACEMENT

Locations for the expansion valve can vary from application to application; some expansion valves may be accessible from under the hood of the vehicle, whereas others may require disassembly of the vehicle instrument panel. The following procedures are general procedures used during the replacement of most any TXV.

1. Recover the refrigerant as necessary.
2. Locate the capillary tube and remote sensing bulb. This can be found at the evaporator inlet or in the fins of the evaporator (see **Figure 16**). It may be necessary to remove part or all of the instrument panel to access the TXV. Refer to the service manual for the required steps.

Figure 16. This capillary tube is located between the evaporator fins.

3. If the sensing bulb is attached to the evaporator inlet, it will be most likely wrapped in an insulating tape; remove the insulation tape from the remote bulb. If the bulb is inserted into the evaporator fins, no insulation will be present.
4. Loosen any clamps or fasteners that may be holding the sensing bulb.
5. Remove the liquid line from the inlet of the TXV.
6. Remove the evaporator inlet fitting from the outlet of the TXV.
7. Remove and discard all used O-rings, if equipped.
8. Inspect the inlet screen. If the screen is clogged, the source of the debris should be located and the valve should be replaced with a new one.
9. Remove the holding clamp (if provided on the TXV), and carefully lift the TXV from the evaporator.
10. Carefully locate the new TXV in its proper location.
11. Lubricate and install new O-ring(s) on the lines.
12. Attach the evaporator inlet to TXV outlet. Tighten to the proper torque.
13. Attach the liquid line to the TXV inlet. Tighten to the proper torque.
14. Position the remote bulb and secure it with a clamp.
15. Using purpose-made insulating tape, wrap the remote bulb to insulate it from ambient air.
16. Replace any access panels that were previously removed.
17. Evacuate and recharge the refrigeration system as required.

RECEIVER DRYER/ACCUMULATOR REPLACEMENT

The receiver dryer and accumulator components are typically only replaced as part of a compressor replacement or if a system has been severely contaminated with moisture. To prevent contamination of the new accumulator or receiver dryer unit, make sure it remains sealed until you are ready to install it. To further prevent contamination, the unit should be the last component installed before the system is evacuated. The receiver dryer can be located in almost any position between the condenser and the metering device (see **Figure 17**), whereas the accumulator will always be located between the evaporator outlet and the compressor inlet (see **Figure 18**). Refer to the specific service manual for exact locations.

1. Recover the air conditioning system refrigerant as needed.

Figure 17. Typical receiver dryer location.

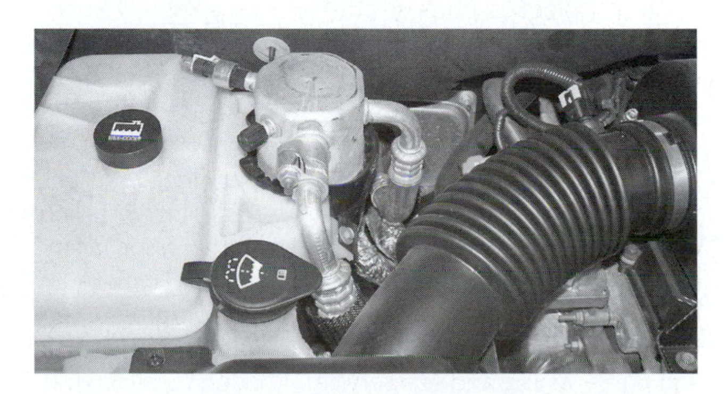

Figure 18. Typical accumulator location.

2. Remove any electrical connections that may be present.
3. Remove the inlet and outlet hoses or lines. Be sure to use the proper size wrenches and backup fittings as needed to avoid damage to the lines, the evaporator, or receiver dryer. Plug any remaining lines to avoid moisture contamination.
4. Remove and discard O-rings.
5. Loosen or remove the mounting hardware as needed.
6. Remove the unit from the vehicle.
7. Remove the pressure switch from the dryer, if applicable. Discard the gasket or O-ring.
8. Using new O-rings install any switches removed in step 7 on the new unit.
9. Install the proper amount of clean refrigerant oil to the new unit. Refer to the service manual for the proper amount.
10. Lubricate and install new O-rings. The new receiver dryer or accumulator can be installed in the reverse order.
11. Evacuate and recharge the refrigeration system to the proper amount.

You Should Know *Most receiver dryers are marked with an arrow or the word(s) IN and OUT to denote the direction of refrigerant flow. Remember, flow is away from the condenser and toward the metering device.*

You Should Know *Almost all compressor manufacturers and rebuilders require the replacement of the accumulator or receiver dryer when a replacement compressor is installed. Failure to do so will often result in the voiding of any warranties.*

COMPRESSOR SERVICE

For all practical purposes, the A/C compressor is a nonserviceable component in most repair facilities. The large number of required specialty tools, lack of parts availability, the desire for longer warranties on repairs, and the required commitment of time to perform the job properly have made compressor service a job for the specialized rebuilder. This leaves all but the most dedicated A/C facilities installing new or rebuilt compressors when a compressor failure occurs. Although internal compressor repair is very unusual, in most service facilities there are several compressor services that are still performed on-site. These include clutch and pulley service and seal service. Although these services often can be performed with the compressor still installed on the vehicle, it is likely that a more efficient repair can be accomplished with the compressor removed. Although it is practical to perform these services in-house, one must be very careful to analyze the cost to repair versus cost of replacement. Although replacement is usually more expensive initially, it is more cost-effective over a longer period of time. These are issues that must be considered and discussed with the customer. Factors that should be considered include:

- Compressor/vehicle age: Is the compressor likely to have another unrelated problem in a short amount of time?
- Why the component failed: Was the failure that is being repaired caused by another compressor fault? Could this repair lead to another fault in a short amount of time?
- Repair costs versus replacement versus warranty: Do the repairs, including parts and labor, come close to the cost of replacement? One also must consider the length time the repair will be warrantied versus the warranty of a replacement compressor.

Clutch Service

Having the proper compressor clutch air gap is essential to the longevity of the clutch and its components. There are several compressor service procedures that require the removal of one or more compressor clutch components. Because of this, it is important that the service technician know and understand how clutch components are removed, installed, and properly adjusted.

When servicing any part of the clutch assembly, the clutch plate is the first component that must be removed. Depending on the repair being performed, a technician may use one or several of the procedures outlined here. The service of various compressor models can be slightly different from the steps described here; for that reason, the proper service manual information should always be referenced to obtain proper specifications and procedures.

1. If necessary, recover the system refrigerant and remove the compressor from the vehicle.
2. Using a clutch plate holding tool, remove the nut or bolt that is used to retain the clutch plate to the compressor shaft (see **Figure 19**).
3. Insert the push bolt into the end of the puller shaft.
4. Screw the hub of the puller into the inside threads of the clutch plate. Hold the hub using the proper sized wrench and turn the push bolt until the clutch plate has released from the shaft (see **Figure 20**). Some models are not press-fit and can be removed by hand; this style of clutch plate uses small washers to set the clutch plate to pulley clearance. Do not discard these washers.
5. Remove the retaining ring holding the pulley on the front of the compressor.

Figure 19. A holding tool is used to keep the compressor shaft from spinning while the nut is being removed.

Figure 20. The clutch removal tool is threaded into the clutch plate.

6. Insert the puller pilot on the compressor shaft to protect the threads on the compressor shaft.
7. Lock the teeth of the pulley remover into the slots of the puller. Tighten the center bolt against the compressor shaft (see **Figure 21**). The pulley will pull free from the compressor.

Figure 21. A puller is used to remove the pulley from the compressor.

8. Mark the location of the clutch coil connector. Remove the retaining ring holding the coil place on the front of the compressor.
9. Align the puller jaws behind the coil. Turn the center screw against the compressor shaft. The coil will separate from the compressor. The clutch coil on some compressor models is held in place with an external snapring and a puller is not required.

Reassembly

10. Remove any scale from the coil or pulley mounting surfaces.
11. Align the clutch coil with the index mark that was created earlier.
12. Place the coil installer over the internal opening of the clutch coil. Install the puller and the proper adapters used to apply force to the front of the puller. Turn the center bolt against the installer. The clutch coil will be forced over the nose of the compressor. Continue to apply pressure until the coil is seated. Pay special attention to make sure that the coil is properly aligned and is being installed straight (see **Figure 22**).

Figure 22. A puller and a special attachment are used to reinstall the clutch coil onto the compressor.

13. Align the pulley by hand on the compressor snout. Install the pulley installer. Install the puller using the same arrangement described in step 10. Tighten the center bolt to force the pulley onto the compressor snout. Continue forcing the pulley until the edge of the bearing completely clears the snapring groove. Install the snapring.
14. Splined clutch plate:
    a. The splined clutch plate uses a slip fit between the plate and the compressor shaft. The clearance of this type plate is set using washers between the clutch plate hub and the compressor shaft. Start by installing the washers removed in step 4. Install the plate and tighten the bolt to specifications.
    b. Using a feeler gauge, measure the amount of clearance between the clutch plate and the pulley (see **Figure 23**). The clearance should typically measure .015–.025 in. (.381–.635 mm). If the clearance is too tight, remove the plate and add the appropriate combination of washers to achieve the proper clearance. If an excessive amount of clearance is present, remove washer as needed to obtain the proper clearance.
15. Press-fit clutch:
    a. The press-fit clutch uses the same tool that was used to remove the plate to install the plate. Following the manufacturer's instructions, prepare the removal tool to install the clutch.
    b. Align the plate to the key on the end of the compressor shaft by hand.

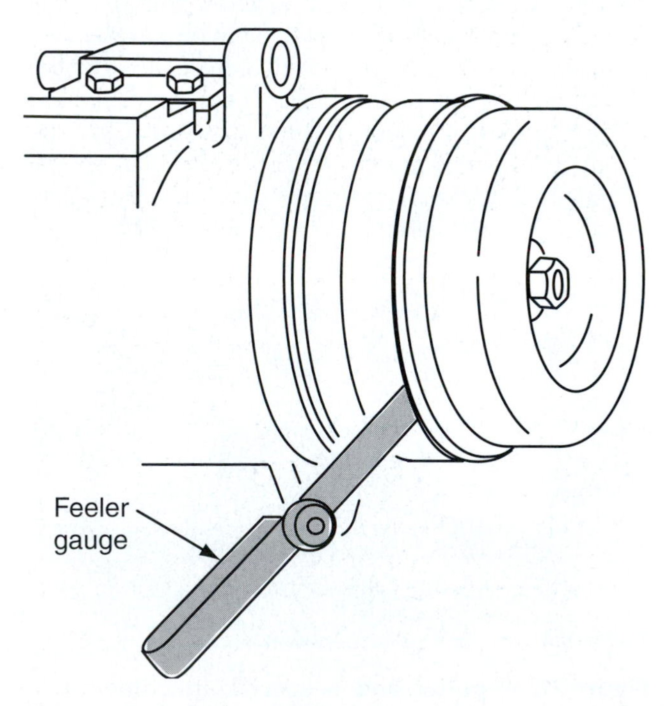

Figure 23. A feeler gauge is used to measure the clearance of the clutch plate.

    c. Screw the center installer bolt to the compressor shaft by hand.
    d. Using the proper-sized wrench hold the center bolt while turning the outer hub to force the bearing against the clutch plate.
    e. Once the clutch plate nears the pulley, start gauging the clearance between the pulley and clutch plate. Use the same procedure used in step 14b. The clearance should be between .015 and .025 in (.381 and .635 mm). Do not force the plate against the feeler gauge. If the plate is forced against the feeler gauge, the plate clearance will snap closed when the feeler gauge is removed, leaving too little clearance. When this occurs, repeat step 4 to pull the clutch plate away from the pulley, and repeat steps 15d and 15e to reset the clearance.
16. Reinstall the compressor and evacuate and recharge the refrigeration system as needed.

You Should Know *Some compressors use only a press-fit to retain the clutch plate. Although the shaft may be threaded, a nut may not be present or required.*

Pulley Bearing Replacement

The clutch pulley bearing can be held in place using a variety of methods. The most popular are the use of a snapring and an upset stake. Use steps 1 through 6 to remove to pulley from the compressor. Once the pulley is removed, continue as follows:

1. Inspect the pulley and determine if the bearing is retained using a snapring or an upset stake. If a snapring is present, remove it at this time.
2. Note the direction in which the bearing will be removed from the pulley. Rest the pulley in upright position against wooden blocks in a direction so that the bearing can be forced downward out of the pulley.
3. Using a bearing driver or press, force the bearing out the front side of the pulley (see **Figure 24**). If the bearing is retained using the upset stake method, the upturned metal will flatten out as the bearing passes through the pulley.

Installation

4. Turn the pulley over and rest the pulley on wooden blocks in a direction that allows the bearing to be forced down into the pulley.
5. Align the pulley bearing by hand and use the bearing driver or press to seat the bearing against the back of the pulley.

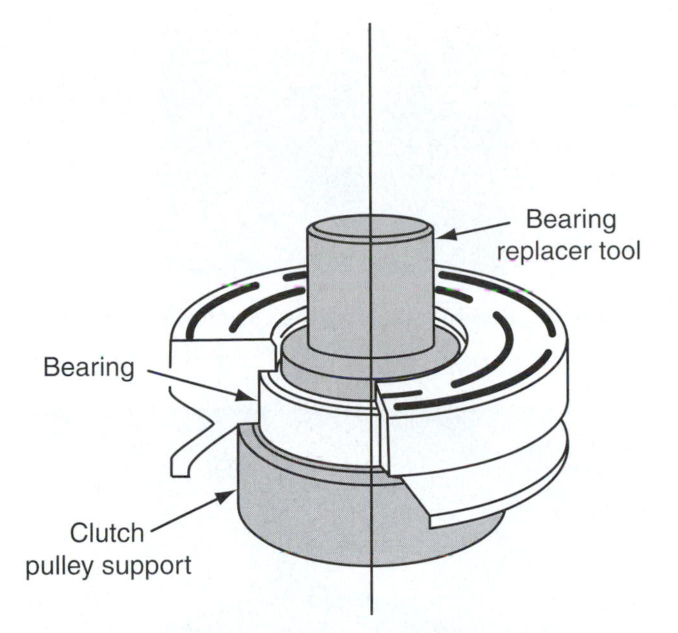

Figure 24. A bearing driver is used to remove the pulley bearing from the pulley.

6. If a snapring is used, reinstall at this time and continue to step 12 of the previous procedure. If an upset stake is used to retain the bearing, proceed to step 7.
7. Place the staking guide and pin against the bearing in the pulley bearing bore.
8. Using a hammer, sharply rap the staking pin. The staking pin will push a layer of metal from the pulley against the bearing (see **Figure 25**). Rotate the guide and pin assembly and repeat the staking process until the original number of staking positions has been evenly placed around the bearing, typically three to six (see Figure 25).
9. Spin the bearing by hand to ensure that it spins freely.
10. Continue to step 12 of the previous procedure.

Shaft Seal Replacement

In order to replace the compressor shaft seal, the refrigeration system must be evacuated of all refrigerant. As with the other procedures detailed in this section, it may be beneficial to remove the compressor to perform compressor seal service.

1. Follow steps 1–4 of the clutch service section for the steps to properly remove the clutch plate.
2. Thoroughly clean the area around the compressor shaft.
3. Remove the internal snapring that is used to hold the shaft seal in place.
4. Install the seal remover/installer against the face of the seal. Twist to expand the jaws of the tool against the seal.
5. Use a gentle twisting/pulling motion to remove the seal from the front compressor housing (see **Figure 26**).

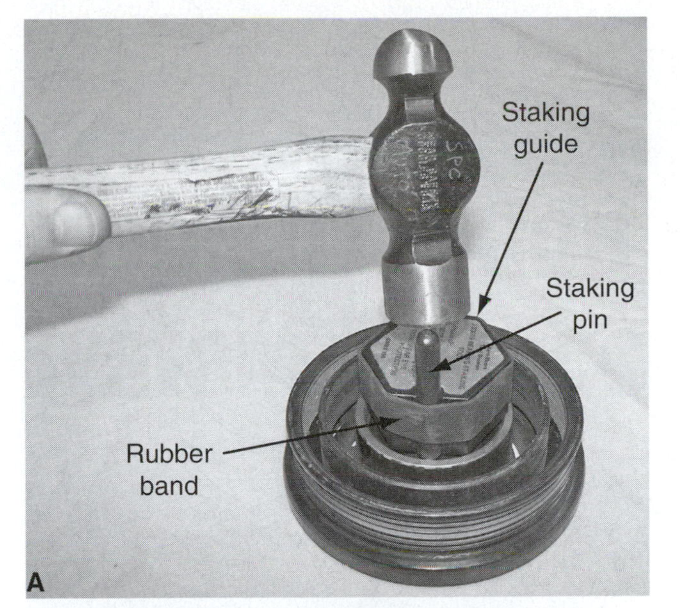

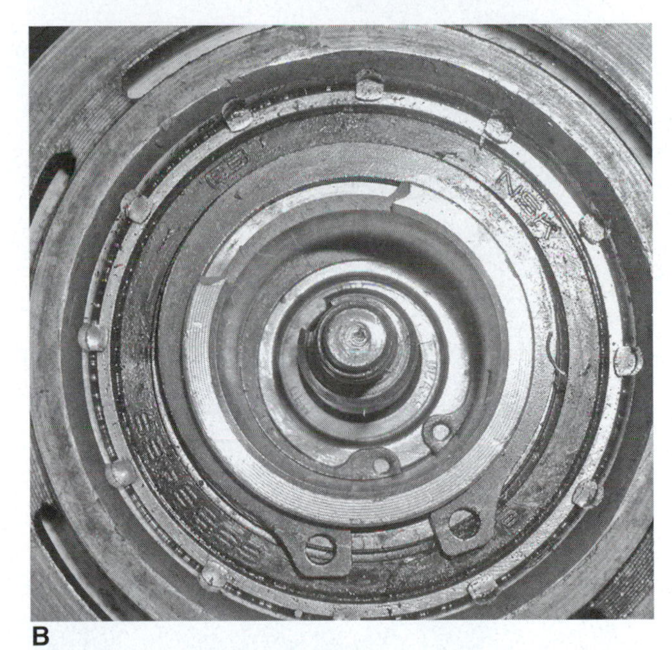

Figure 25. (A) This tool is used to stake the bearing in place. (B) The staking tool upset the pulley material to hold the bearing in place.

6. Remove and discard the O-ring seal that is used between the compressor housing and the seal (see **Figure 27**).
7. Lubricate and install the new O-ring into the proper groove in the compressor housing.
8. Place the new seal into the jaws of the seal remover/installer.
9. Lubricate the surface of the seal with clean refrigerant oil.
10. Install the cone-shaped seal protector over the threads of the compressor shaft.

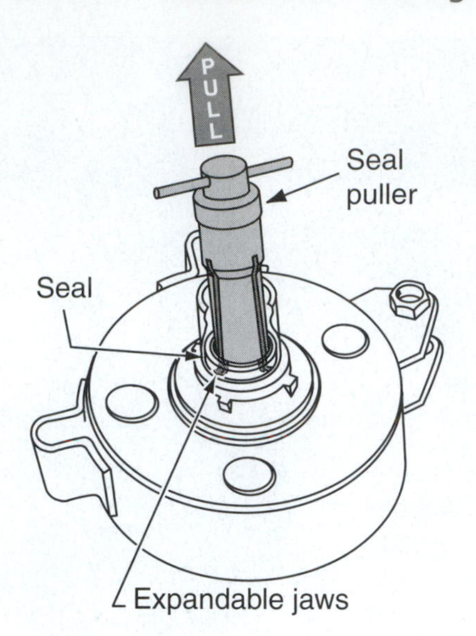

Figure 26. (A) A detailed look at a compressor seal removal tool. (B) Twist to lock the seal tool into the seal. Gently pull using a slight twisting motion to remove the seal.

11. With a gentle twisting motion, slide the seal over the installer and into the groove in the front of the compressor housing (see **Figure 28**). Release the installer from the seal.

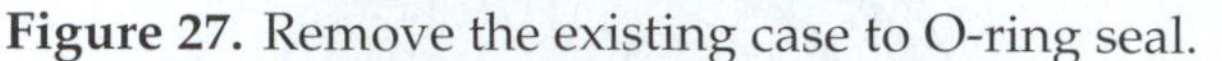

Figure 27. Remove the existing case to O-ring seal.

Figure 28. Install the thread protector over the threads of the compressor shaft. Lock the seal onto the installer, lubricate and carefully install with a slight twisting motion.

12. Reinstall the snapring.
13. Follow steps 14 or 15 as required to reinstall and adjust the clutch plate clearance.
14. Evacuate the refrigeration system and recharge.

Compressor Replacement

In today's automotive market, with the availability of quality, competitively priced rebuilt compressors and long warranty periods, it has become more feasible to replace the compressor in many situations rather than attempt even the simplest repairs. Because compressor failures usually flood the refrigeration system with some amount of debris, it is also good practice to replace the metering device or, at the very least, inspect and clean metering device inlet screens. Some compressor manufacturers and rebuilders also require the replacement of the receiver drier or accumulator. The following steps outline the procedures required for successful compressor replacement.

1. Recover the system refrigerant as needed.
2. Remove the compressor drive belt.
3. Disconnect any electrical connectors. Typical connector locations are at the clutch coil; some compressors

have additional pressure switches located at the back of the compressor.

4. Disconnect the suction/discharge hose assembly. Discard the seals.
5. Remove the bolts and brackets that hold the compressor to the engine (see **Figure 29**).
6. Drain and measure the lubricant from the compressor (see **Figure 30**). Depending on the compressor style, oil may be drained from the suction/discharge ports or, in some cases, the compressor can be equipped with a drain plug.
7. Remove any pressure switches that may be attached to the compressor.

Reinstallation

8. Some manufacturers and rebuilders choose to ship their compressors with a full charge of refrigerant oil. This oil is only used to preserve the compressor during shipment and storage. If your replacement compressor is equipped with an oil charge, drain and discard the oil at this time, unless the included directions indicate otherwise.

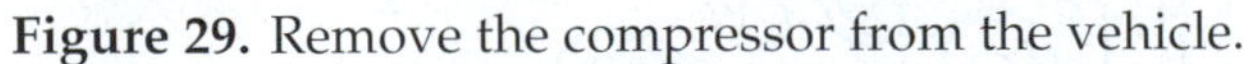

Figure 29. Remove the compressor from the vehicle.

Figure 30. Drain and measure the lubricant from both the old and new compressors.

9. Remove any applicable switch plugs and install pressure switches as needed.

You Should Know *The compressor typically holds a large amount of the refrigeration system oil. Oil fill procedures vary from manufacturer to manufacturer. In order to maintain a proper system oil balance, oil fill procedures must be explicitly followed.*

10. If the amount of oil drained from the original compressor is more than 3 oz (88.72 ml), add the same amount of refrigerant oil to the new compressor. If the amount is less than 3 oz (88.72 ml), add 3 oz (88.72 ml) to the new compressor. Turning the compressor shaft when adding the oil may help the oil enter the compressor.
11. Lubricate and install new suction discharge hose seals (see **Figure 31**).
12. Reverse steps 2–5 to reinstall the compressor.
13. Evacuate and recharge the refrigeration system to the proper specifications.

DEBRIS REMOVAL

Often when a compressor failure occurs, the entire system becomes contaminated with debris. Much of the debris will be forced into the discharge side of the system; however, some of the debris can be deposited in the suction side, often finding its way into the accumulator. If not contained, this debris will eventually collect in the metering device filter and, in some cases, return to the compressor, causing serious damage. The only way to ensure that all of the debris has been removed is to replace all of the system's components. For obvious reasons, this is

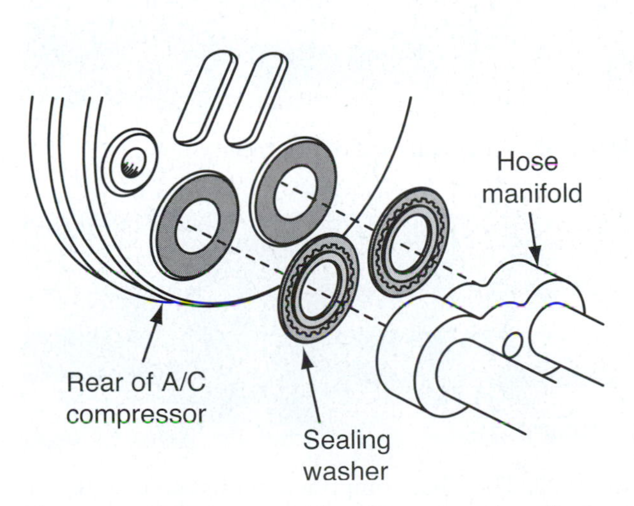

Figure 31. Lubricate and install new suction discharge seals.

an impractical solution. There are several methods that can be employed to help avoid system damage from debris. These methods are installing a liquid line filter, flushing the system, and installing a compressor inlet screen.

Liquid Line Filter

The most effective method of containing system debris is through the use of a liquid line filter (see **Figure 32**). Liquid line filters are installed between the condenser and the metering device. In most cases, the line has to be cut and the filter installed. The steps for filter installation are as follows.

1. If you have not already done so, recover the refrigerant from the system as needed.
2. Examine the filter to determine its physical size. Inspect the area of the vehicle in which the filter will be installed. Select a location that provides enough straight line so that at least 1 in. (25.4 mm) of straight line is present on each side of the filter and so that the filter will not be in the way of any other components.
3. Using a tubing cutter, cut the line in the selected location.
4. Use a small file to clean up any nicks or burrs that may have been created when cutting.
5. Slide the nuts onto the lines with the threads facing each other.
6. Slide the ferrule onto the line with the cone facing the nut (see **Figure 33**).
7. Slide the tube into the filter until it bottoms out.
8. Slide the nut and ferrule up the line and tighten the nut onto the filter and torque to specifications. Repeat with the opposite tube. This procedure will swell and lock the ferrule onto the tube.

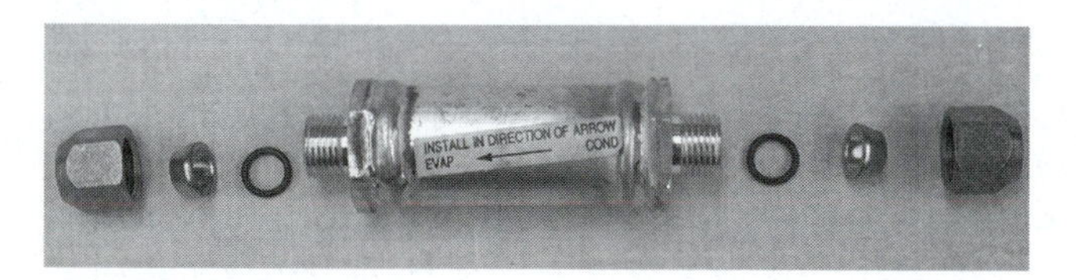

Figure 32. A typical liquid line filter.

Figure 33. The filter components must be installed in a specific order.

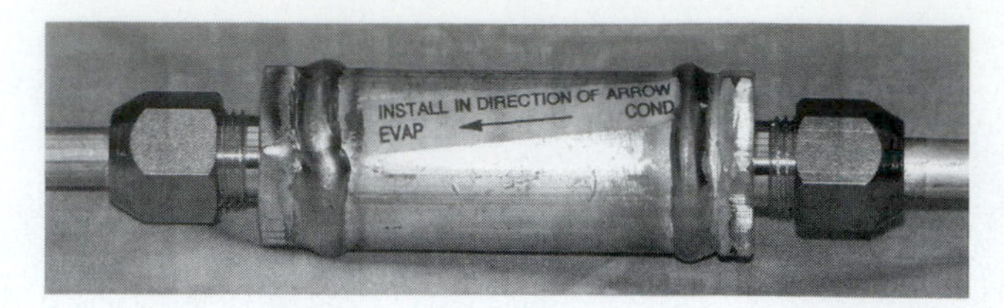

Figure 34. The completed filter installation.

9. Unscrew both fittings and remove the tube from the filter.
10. Lubricate and install the seals onto the lines.
11. Reassemble the lines to the filter (see **Figure 34**).
12. Evacuate, leak test, and recharge the system to specifications.

You Should Know *Some filters have built-in orifice tubes for use in those applications in which the orifice tube was installed in the condenser outlet. When using one of these filters, the original orifice tube should be discarded.*

Inlet Screen

Compressor inlet screens serve to block debris from entering the suction port of the compressor. The inlet screen is lightly pressed into the suction hose at the compressor connection. If an inlet screen is to be used, it should be installed at the same time that the replacement compressor is installed (see **Figure 35**). When

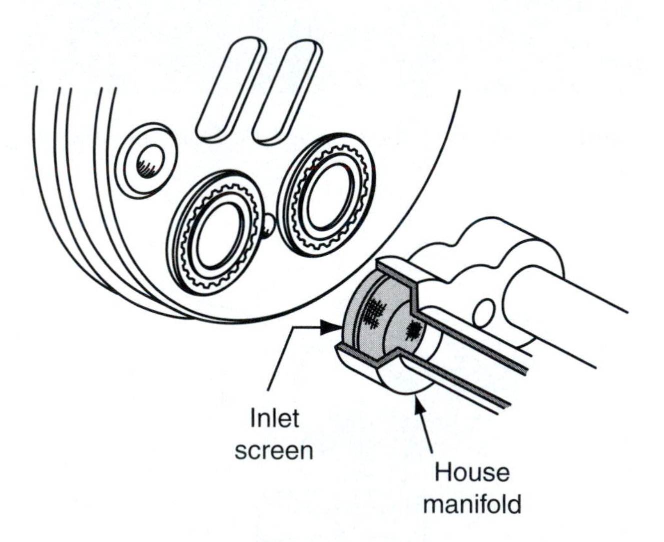

Figure 35. Inlet filters are used to protect the compressor from debris entering from the suction side of the system.

coupled with the installation of the liquid line filter, the metering valve and compressor are now protected from the circulation of residual debris. Replacing the receiver drier or accumulator serves as an additional layer of protection.

System Flushing

System flushing is the third method of removing system debris. Flushing involves sending pressurized solvent or refrigerant through the system to push out any existing debris. Although this method can be effective, it does remove lubricant from the system. If not properly cleaned, it will leave solvent residue in the system. Even under the best circumstances, this method provides no guarantee that all of the debris will be removed. For this reason, many manufacturers do not endorse the practice of system flushing unless it is done using the specified system refrigerant. However, using refrigerant to flush the system provides its own set of problems in terms of the capture and disposal of the used material. It is more practical and environmentally friendly to install liquid line filters and suction inlet screens.

EVAPORATOR REPLACEMENT

Without a doubt, the evaporator is the most difficult refrigeration system component to replace. The difficulty of evaporator replacement will differ significantly between vehicles. The evaporator may be readily accessible from under the hood in some vehicles, whereas in other vehicles extensive disassembly of the instrument panel and dash may be required (see **Figure 36**). Rear air conditioning units pose their own unique set of conditions and circumstances. Many of the panels that have to be removed may have screws that are at best difficult to see and in some cases hidden. Failure to locate and remove these hidden fasteners often can result in damage to the vehicle. The following steps are intended as a general guide to the procedures required to removing the evaporator. It is imperative that the service manual be strictly adhered to in order to avoid damage to the vehicle.

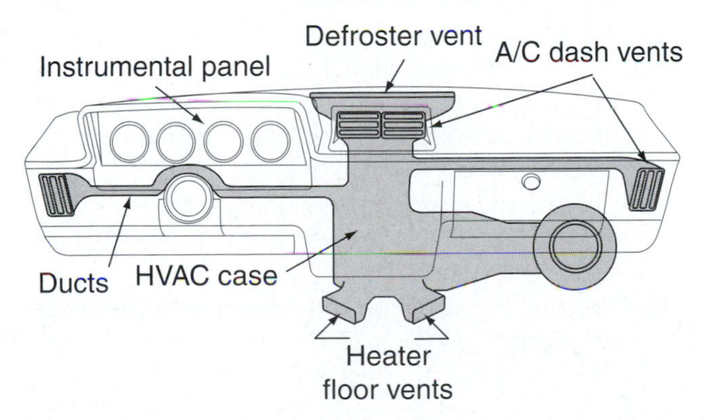

Figure 36. In some vehicles the entire dashboard may have to be removed to access the evaporator.

1. Recover the system refrigerant as needed.
2. Following service manual procedures, gain the necessary access to the evaporator.
3. Remove the liquid and suction lines from the evaporator. Plug any remaining lines to prevent moisture contamination.
   a. If the vehicle is equipped with H-valve style expansion valves, both lines will be removed together.
   b. If the vehicle is equipped with an accumulator, the suction hose must be removed from the accumulator and the accumulator from the evaporator.
4. Remove and discard used O-rings.
5. Remove mounting bolts and hardware, as applicable, from the evaporator housing.
6. Separate the housing to gain access to the evaporator core (see **Figure 37**).
7. Carefully lift the evaporator assembly from the vehicle, being careful not to damage evaporator fins.
8. Install the proper amount of refrigerant oil into the new evaporator.
9. If the refrigeration system uses an expansion valve, a new valve should be installed at this time by placing the capillary tube and remote sensing bulb in the same locations as the old one. Install insulating tape as needed.
10. Install a replacement evaporator by reversing steps 2 through 7.
11. Evacuate and recharge the system to the proper capacity.

EVAPORATOR CLEANING

Most odor concerns related to the air conditioner system are a direct result of fungal growth on the surface of the evaporator. The first step in odor diagnosis is to ensure that the evaporator drain is opened so that condensation can exit to prevent moisture from collecting in the housing. Second, the housing must be free from debris and cabin air filters must be clean. If a large amount of debris is present, it must be removed or no amount of disinfectant is going to make a difference.

There are several different products that can be used to clean the evaporator. In order to be effective, both sides of the evaporator must be thoroughly coated. This can be difficult at best, unless the evaporator is removed from the vehicle. If evaporator removal is not an option, the evaporator can be accessed by removing the blower motor or control

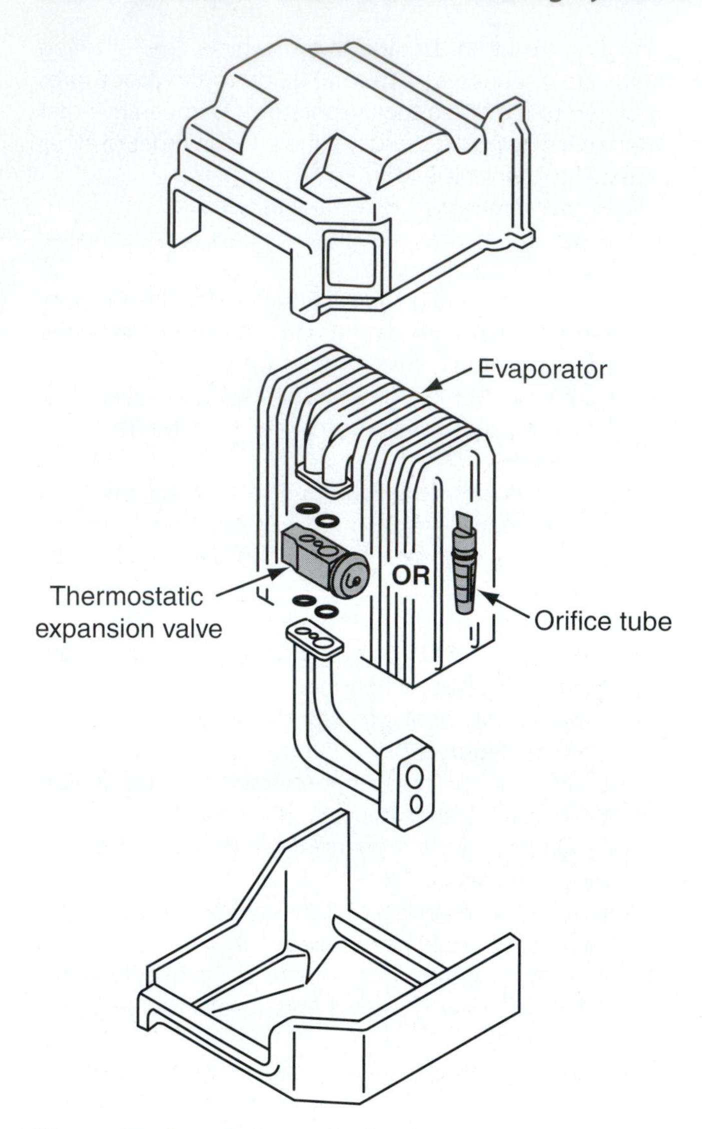

Figure 37. Separation of the evaporator case is required to gain access the evaporator.

module. (see **Figure 38**) Some evaporators can be accessed through the cabin air filter inlet. If access is not readily available, a hole must be drilled in the evaporator housing large enough for the wand to enter the evaporator area. When finished, the hole must be resealed. Although this is an option, it is only recommended as a last resort. Additionally, great care must be exercised when drilling in order to avoid damage to the evaporator or other components.

Once completed, the customer should be advised that reoccurrence of problems can be reduced by running the blower motor on high for a few minutes with the compressor turned off. This will help remove moisture from the evaporator and stifle the growth of microorganisms.

Figure 38. Evaporator disinfectants can sometimes be introduced into the evaporator by removing the blower motor, resistor block, or control module from the evaporator housing.

CONDENSER REPLACEMENT

Replacement of the condenser is generally a straightforward procedure. Because the condenser is located in front of the radiator grill, however, core support for the radiator may have to be removed or relocated in order to complete the procedure. Extreme care should be taken to prevent damage to the fins of the radiator and condenser. The following are general procedures required to replace the condenser.

1. Recover the system refrigerant as needed.
2. Remove the grill, hood latch mechanism, and cooling fans as needed to gain access to the condenser. Refer to the vehicle service manual for specific details.
3. Remove the discharge and liquid lines from the condenser. Plug the remaining lines to prevent moisture contamination when the system is open.
4. Remove and discard any O-rings.
5. Remove and retain any attaching bolts or nuts holding the condenser in place.
6. Lift the condenser from the car.
7. Add the proper amount of the proper refrigerant oil to the replacement condenser. Information on this step can be found in the service manual.
8. Lubricate and install the new O-rings.
9. Install the condenser by reversing the procedure given in steps 2, 3, 5, and 6.
10. Evacuate and recharge the system with the proper refrigerant.

Summary

- Failure to follow established service procedures can often lead to a customer comeback.
- Leaking Schrader valves can usually be repaired by replacement of the valve or, in some cases, the entire fitting.
- Used O-rings should never be reused. O-rings usually fail as result of physical damage. Overtightening fittings can lead to O-ring damage. Replacement O-rings should be exactly the same size as the originals. Only mineral oil should be used to lubricate O-rings.
- Refrigeration system hoses can be repaired using specialized fittings. A hose repair system consists of various styles of fittings and the equipment to install those fittings properly.
- Lubricating a stuck orifice tube provides lubrication to stuck O-rings and flushes away debris. An orifice tube removal tool should be used to remove the orifice. It is a common occurrence for an orifice tube to break during removal.
- Some TXVs can be accessed from under the hood, whereas others require removal of the evaporator. The remote sensing bulb can be found in the fins of the evaporator or clamped to the evaporator inlet line. To work properly, the remote sensing bulb should be insulated.
- Common compressor repair procedures include clutch and pulley service and shaft seal replacement. In many cases, it may be economical over a long period of time to replace a compressor rather than repair it.
- An air gap between the clutch plate and the pulley is essential to the longevity of each component. Splined clutch plates use washers to compensate for plate to pulley clearance. Pressed clutch plates depend on an interference fit to maintain clearances.
- Compressor pulley bearings can be held in place using a snapring or may be retained using upset staking.
- Shaft seal replacement requires the removal of all refrigerant.
- When a compressor fails, it often floods the refrigeration system with debris. When replacing a compressor, it is good practice to replace the metering device or at least clean the inlet screens. To maintain the warranty, it may be necessary to install a new receiver dryer or accumulator.
- The amount of oil drained from the old compressor is a critical part of determining how much oil should be installed in the replacement compressor. Three methods of debris removal are liquid line filters, suction inlet screens, and system flushing. Liquid line installation requires the liquid line to be cut. Inlet screens should be installed when the compressor is removed. Flushing is not recommended by most manufacturers.
- Evaporator replacement is one of the most difficult procedures performed to a refrigeration system. Extensive disassembly is often required to access the evaporator. Many of the fasteners that must be removed to access the evaporator are often hidden.
- In order for evaporator cleaning to be effective, both sides of the evaporator must be accessible and thoroughly coated with the cleaning agent. The evaporator can be accessed through the removal of the evaporator case components or by drilling holes in the evaporator housing.
- Minor disassembly of the front of the vehicle may be required to remove the condenser.

Review Questions

1. Technician A says that a leaking Schrader valve can usually be replaced. Technician B says that in some cases the Schrader valve can be replaced without removing the refrigerant from the system. Who is correct?
   A. Technician A
   B. Technician B
   C. Both Technician A and Technician B
   D. Neither Technician A nor Technician B
2. All of the following statements are true, *except*:
   A. O-rings should be replaced if they are ever disturbed.
   B. O-rings should be lubricated with the type of lubricant that is used in the system.
   C. O-rings require only a small amount pressure to properly seal.
   D. All O-rings should be lubricated with mineral oil.

3. Technician A says that when replacing a compressor, the old compressor should be drained. Technician B says that some compressors are shipped with proper amount of system lubricant already installed. Who is correct?
   A. Technician A
   B. Technician B
   C. Both Technician A and Technician B
   D. Neither Technician A nor Technician B
4. Technician A says that flushing is the desired method of removing debris from a refrigeration system. Technician B says that flushing can remove all of the debris from the refrigeration system. Who is correct?
   A. Technician A
   B. Technician B
   C. Both Technician A and Technician B
   D. Neither Technician A nor Technician B
5. Which of the following is NOT a method of removing refrigeration system debris?
   A. Liquid line filter
   B. Chemical disinfectant
   C. Suction inlet screen
   D. Chemical flushing
6. Technician A says that the surface of the evaporator can be most thoroughly cleaned with the evaporator out of the vehicle. Technician B says that the evaporator can be accessed for cleaning by drilling a small hole in the evaporator housing. Who is correct?
   A. Technician A
   B. Technician B
   C. Both Technician A and Technician B
   D. Neither Technician A nor Technician B
7. Technician A says that when adjusting the air gap on a press-fit clutch plate, the plate should be forced against the feeler gauge. Technician B says that in order to increase the air gap on a splined clutch, shims must be removed. Who is correct?
   A. Technician A
   B. Technician B
   C. Both Technician A and Technician B
   D. Neither Technician A nor Technician B
8. Technician A says that the orifice tube should be installed in no particular direction. Technician B says that the orifice tube is not serviceable in some applications. Who is correct?
   A. Technician A
   B. Technician B
   C. Both Technician A and Technician B
   D. Neither Technician A nor Technician B

Chapter 19 Refrigeration System Retrofit

Introduction

Much confusion and panic surrounded the automotive industry when the ban on R-12 refrigerant was initially announced. It was generally assumed that R-12 would be removed from the market and would no longer be available to anyone. This left consumers and repair facilities alike concerned as to what would become of their R-12 air conditioning units. To add to the panic, early retrofit procedures called for the replacement of nearly the entire refrigeration system. Each of these concerns proved to be false points of worry. R-12 refrigerant is still available to licensed technicians, although it is very expensive, and a basic system retrofit can provide outstanding performance in most converted vehicles.

When speaking about an automotive air conditioning system, the term **retrofit** is used to describe the process of converting an R-12 system to one using an alternative refrigerant. In this text, it will be assumed that the conversion refrigerant is R-134a, because the worldwide automotive industry chose this refrigerant to be the replacement in new as well as in retrofitted automotive air conditioning systems.

System retrofit is usually only required when a leak has severely depleted the refrigerant level or when a major component failure requires that the system be opened for repairs. If an R-12 system is operating properly and the refrigerant level has remained stable, there is no reason to perform a system retrofit.

THE BASIC RETROFIT

The procedures for a basic retrofit are relatively simple and generally do not require major component replacements. The process usually only requires removal of the R-12 refrigerant, installation of new fittings, new label, and the addition of the proper lubricant. In most cases, this retrofit will provide the owner with an air conditioning system performance level comparable to the former R-12 system. Even if the retrofit results in slightly reduced performance, it is usually sufficient for customer satisfaction. According to EPA regulations, any alternate refrigerant used to replace R-12 requires the following:

- Unique service fittings must be used on both the high side as well as the low side of the system. This requirement is intended to reduce the likelihood of cross-contamination of the air conditioning system or the repair facility's refrigeration service equipment.
- Use of the new refrigerant must be noted on a uniquely colored label to distinguish the type refrigerant and lubricant used in the system.
- All R-12 must be removed from the system using dedicated EPA-approved equipment.
- To prevent release of refrigerant to the atmosphere, a high-pressure compressor shutoff switch must be installed on any system equipped with a pressure relief device.
- Separate dedicated EPA-approved equipment must be used for subsequent recovery, recycling, and recharging of R-134a.
- Barrier hoses must be used with alternative refrigerant blends that contain HCFC-22.

R-134A: THE REPLACEMENT REFRIGERANT OF CHOICE

Several refrigerants in addition to R-134a are now listed by the EPA as acceptable for motor vehicle air conditioner (MVAC) use under their Significant New Alternatives Policy

(SNAP) plan. However, except for R-134a, no refrigerant has been endorsed by vehicle manufacturers for use in mobile air conditioning systems. R-134a is considered to be one of the safest refrigerants available based on toxicity data. Extensive tests indicate that R-134a does not pose cancer or birth defect hazards, is not corrosive on steel, aluminum, or copper samples, and is not flammable at ambient temperatures at atmospheric pressure. However, as with any other chemical, R-134a should be handled with respect: work in a well-ventilated area, wear adequate personal protection, avoid open flames, and do not inhale any vapor.

You Should Know

Some mixtures of air and R-134a have been known to be combustible at elevated pressures. For this reason, service equipment and vehicle air conditioning systems should not be pressure- or leak-tested using compressed air.

Interesting Fact

The EPA SNAP program tests and evaluates substitute refrigerants for their effect on human health and the environment. SNAP does not test and evaluate refrigerants for performance or durability.

You Should Know

Although some alternate refrigerants are being marketed as "drop-ins," there is by definition no such thing as a refrigerant that can literally be "dropped in" on top of existing R-12. The use of all alternative refrigerants calls for some level of system modification.

COMPONENT REPLACEMENT GUIDELINES

Because R-12 and R-134a are not identical, a properly converted system requires a different amount of refrigerant and the use of a different lubricant. Once these modifications have been completed, different system operating pressures can be expected. Observing these guidelines will help ensure a successful system retrofit.

- **System Operation:** First and foremost, for any type of system retrofit to be successful, the refrigeration system must be in good operating condition. Performing a retrofit to a poorly operating system can cause rapid failure of worn components. Before performing any retrofit procedures, you should test the system to ensure that it is working properly, and any pending problems must be corrected.

You Should Know

For the purpose of performance-testing the system, if the refrigerant level of a system presented for retrofit is low or empty, every effort should be made to recharge the system with the original refrigerant type. After system condition has been assessed, repairs can be made and the retrofit can be completed.

- **Service Fittings:** New service fittings must be installed in accordance with the type of refrigerant that is being used.
- **System Charge:** The amount of R-134a charged into the system should initially be 80–90 percent of the required charge of R-12. Most manufacturers provide guidelines regarding the amount of R-134a to be used.
- **Lubricant:** The mineral oil used with R-12 cannot be adequately transported through the system by R-134a. Most, but not all, automobile manufacturers chose polyalkylene glycol (PAG) lubricants for use in new and retrofitted air conditioning systems charged with R-134a. PAGs are very hygroscopic; they draw water from the atmosphere when exposed to open air. Because of this, some specialists choose to use polyol ester (POE) lubricants, believing that PAG's hygroscopic nature limits its lubricating ability and causes corrosion in a system. Although it is less hygroscopic than PAG, care must still be taken with POE to ensure that excess moisture does not enter the system. They should be stored in tightly sealed containers to prevent contamination by humidity and to ensure that the vapors do not escape.

You Should Know

Personal protection, such as PVC-coated gloves or barrier creams and OSHA-approved safety goggles, should be used when handling these lubricants. Prolonged skin contact or eye contact can cause irritations such as stinging and burning. One should avoid breathing any vapors produced by these lubricants, and only use them in a well-ventilated area.

- **Flushing:** The amount of mineral oil that can remain in a system after retrofitting without affecting performance is still being debated. The technician should always remove as much of the mineral oil as possible. Removal may require draining components such as the compressor and accumulator, whereas removal of the evaporator and condenser are typically too labor-intensive to make draining the oil practical. Tests have shown that any residual lubricant remaining in the system will not have a significant effect on system performance. If the vehicle manufacturer does not recommend flushing the system during the retrofit procedure, it can be assumed that flushing is not necessary.

- **Hoses and O-Rings:** Tests have shown that lubricants used in an automotive air conditioning system are absorbed into the hose materials to create a natural barrier to R-134a permeation. In most cases, R-12 nonbarrier hoses and O-rings will perform well for R-134a service, provided they are in good condition. Any replacement hose, however, should be of the barrier type. If the fittings were not disturbed during retrofit, replacing O-rings should not be necessary. Most retrofit instructions suggest lubricating replacement green or blue R-134a O-rings with mineral oil rather than PAG oils.
- **Compressors:** Most compressors that function satisfactorily in an R-12 system will continue to function after retrofitting an R-134a system. When a compressor is first operated with R-12, a thin film of metal chloride forms on bearing surfaces to serve as an antiwear agent. This protection continues even after the system has been retrofitted to R-134a. This may explain why new R-12 compressors often fail when installed in an R-134a system without the benefit of a break-in period with R-12. Most new and remanufactured compressors available today are constructed to perform equally well in either refrigerant environment.
- **Desiccants:** R-12 systems often use silica gel or a desiccant designated XH-5, whereas R-134a systems use either XH-7 or XH-9. Some recommend replacement during the retrofit procedure of the accumulator or receiver drier to one having XH-7 or XH-9 desiccant. It is generally agreed, however, that the accumulator or receiver drier should be replaced if the vehicle has over 70,000 miles (112,630 kilometers), is 5 or more years old, or is opened up for major repair.
- **Condensers and Evaporators:** It is generally accepted that if an R-12 system is operating within the manufacturer's specifications, there may be no need to replace the condenser or evaporator. The higher vapor pressures associated with R-134a, however, may result in lost condenser capacity. When planning a retrofit, the technician should consider the airflow and condenser design. The installation of a pusher-type cooling fan mounted in front of the condenser often can improve the performance of a retrofitted air conditioning system. Bent, misshapen, or improperly positioned airflow dams and deflectors also affect performance. Hood seal kits are often recommended for retrofit procedures.

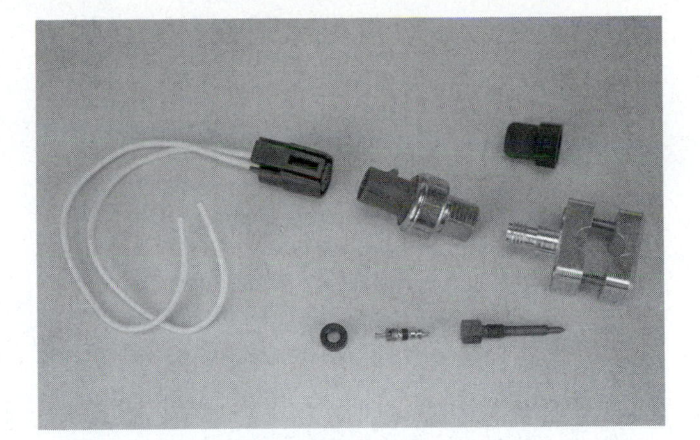

Figure 1. A high-pressure cutout switch intended for retrofit applications.

- **High-Pressure Cutout Switch:** The EPA requires the installation of a high-pressure cutout switch, also called a refrigerant containment switch. Its purpose is to interrupt the clutch coil circuit, thereby stopping the compressor before high-side pressure reaches the point at which it would open the high-pressure relief valve and release refrigerant into the environment. Systems not equipped with a high-pressure cutout switch should have one installed (see **Figure 1**). The switch should be installed in series with the clutch or the clutch relay control circuit.
- **Clutch Cycling Pressure Switch:** The clutch cycling pressure switch may need to be changed for some R-134a retrofits. The difference is that the R-134a switches are calibrated for slightly lower clutch cycling pressures. Lower cycling pressure is sometimes required to achieve maximum system efficiency. The mounting threads of most switches intended for R-134a are metric to prevent the connection of an English-thread R-12 switch in an R-134a system; installation of a replacement switch may require that the accumulator be changed.
- **Metering Devices:** Metering devices should not be changed as a matter of practice when retrofitting a system.
- **Thermostatic Expansion Valve:** The thermostatic expansion valve (TXV) does not have to be replaced when retrofitting a system from R-12 to R-134a. If, however, a TXV is found to be defective, it should be replaced with a model designed for use with the refrigerant currently being used.
- **Orifice Tube:** With the exception of one automobile manufacturer, it is not recommended that the orifice tube be replaced when retrofitting an air conditioning system. Volvo recommends changing the orifice tube to one that has a 0.002 in. (0.0508 mm) smaller orifice.
- **Retrofit Label:** A retrofit label indicating the refrigerant that has been used must be installed to indicate that the

system has been retrofitted (see **Figure 2**). Each different SNAP-approved refrigerant will have its own designated label. The label elements to be filled in include:

- Manufacturer of lubricant and specific part number used
- Quantity of refrigerant added to the system
- Amount of oil added to the system
- Type of oil added to the system
- Name of facility performing retrofit
- Date that the retrofit was performed
- Address of servicing facility
- City
- State
- Zip code

> **You Should Know** *One must be cautioned that not all alternative refrigerants are approved by the EPA for use in motor vehicle air conditioners. Some are considered by the EPA to be dangerous, and heavy penalties are imposed on those who use them. The EPA makes no exceptions and its rules are simple: Use it—get caught—pay the penalty. There are no excuses.*

> **You Should Know** *The Mobile Air Conditioning Society (MACS) has warned on many occasions that several refrigerant products are being offered as substitutes for R-12. Many of these refrigerants contain butane (R-600), ethane (R-170), or propane (R-290). Although they are all refrigerants, they also are very flammable materials.*

CONTAMINATED REFRIGERANT

With the transition to CFC-free air conditioning systems, the likelihood of cross-mixing refrigerants is a growing concern. Mixing two or more different refrigerants in an air conditioning system, recovery equipment, or container will contaminate the refrigerant. The refrigerant is contaminated in that it is no longer "pure" and will not react chemically and physically as intended. Different refrigerants, as well as their lubricants, are not compatible and should not be mixed. When contamination occurs, not only will the system not function properly, but also the contaminated refrigerant can damage expensive equipment, such as a recovery/recycle unit.

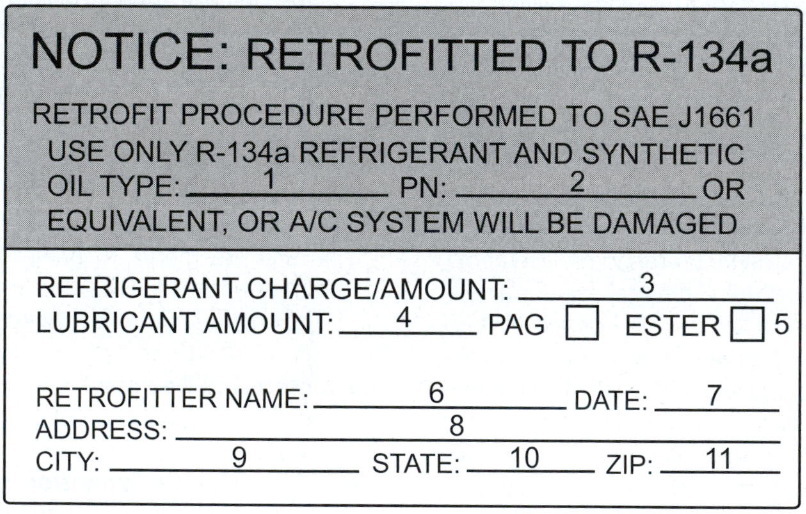

NOTICE: RETROFITTED TO R-134a

RETROFIT PROCEDURE PERFORMED TO SAE J1661
USE ONLY R-134a REFRIGERANT AND SYNTHETIC
OIL TYPE: ___1___ PN: ___2___ OR
EQUIVALENT, OR A/C SYSTEM WILL BE DAMAGED

REFRIGERANT CHARGE/AMOUNT: ___3___
LUBRICANT AMOUNT: ___4___ PAG ☐ ESTER ☐ 5

RETROFITTER NAME: ___6___ DATE: ___7___
ADDRESS: ___8___
CITY: ___9___ STATE: ___10___ ZIP: ___11___

1 Type: Manufacturer of oil (Saturn, GM, Union Carbide, etc.).

2 PN: Part number assigned by manufacturer.

3 Refrigerant charge / amount: Quantity of charge installed.

4 Lubricant amount: Quantity of oil installed (indicate ounces, cc, ml).

5 Kind of oil installed (check either PAG or ESTER).

6 Retrofitter name: Name of facility that performed the retrofit.

7 Date: Date retrofit is performed.

8 Address: Address of facility that performed the retrofit.

9 City: City in which the facility is located.

10 State: State in which the facility is located.

11 Zip: Zip code of the facility.

Figure 2. A R-134a retrofit label.

The purity of refrigerants has been set by SAE purity standards for both R-12 and R-134a. Refrigerant should test at least 98 percent pure when tested with a purity tester. If the refrigerant is less than 98 percent pure, it should be considered contaminated refrigerant and treated as such. The purity standard for recycled R-12 is J1991 and the specified limits are 15 parts per million (ppm) by weight, 4,000 ppm by weight for refrigerant oil, and 330 ppm by weight for noncondensable gases (air). The purity standard for recycled R-134a is J2099 and the specified limits are 15 parts per million (ppm) by weight, 500 ppm by weight for refrigerant oil, and 150 ppm by weight for noncondensable gases (air).

Refrigerant contamination can be determined by observing the pressure-to-temperature relationship of a particular refrigerant. However, a refrigerant identifier is far superior to pressure/temperature comparisons, because at certain temperatures the pressures of R-12 and R-134a are too similar to differentiate with a standard gauge. For example, at 90 degrees F (32.2 degrees C), both 95 percent R-12 and 95 percent R-134a have about the same pressure—111 and 112 psig, respectively. Given that this chart is accurate to ±2 percent, there is really no way of determining which type refrigerant is in the air conditioning system or tank. Also, because other substitute refrigerants and blends may have been introduced into the automotive air conditioning system, they can contaminate a system or tank and may not be detected by the pressure/temperature method. A refrigerant identifier would conclude the refrigerant in our example to be of unknown quantity. If there is any doubt as to the purity of the refrigerant in the vehicle, do not service the air conditioning system unless you are properly equipped.

You Should Know *Refrigerant recovery cylinders must be inspected every 5 years. The test date is stamped on the cylinder's shoulder or collar. Using a recovery cylinder beyond the reinspection date can result in heavy penalties.*

Proper Equipment

In order to properly handle contaminated refrigerant, you must have access to and use "recovery only" equipment. You also must have proper recovery cylinders that meet rigid Department of Transportation (DOT) specifications. These cylinders should be marked "CONTAMINATED REFRIGERANT" for identification. Only contaminated refrigerant should be recovered into this cylinder. Disposable cylinders must not be used for any purpose. Federal law prohibits refilling these cylinders.

Proper recovery of contaminated refrigerant is only one part of your responsibility. Once recovered, the refrigerant must be reclaimed or disposed of. Contaminated refrigerant may be reclaimed to ARI-700-88 (Air Conditioning and Refrigeration Institute) standards, or it may be destroyed by fire. This can be accomplished by sending the refrigerant to an off-site reclamation facility that is equipped to handle such problems. Remember, however, that it is your responsibility to legally dispose of contaminated refrigerant.

THE FUTURE OF REFRIGERANT

For many years now, R-134a has been the refrigerant of choice by manufacturers for use in automotive air conditioning systems. Although R-134a has been very successful in this capacity, it is a greenhouse gas and, when improperly handled, can cause damage to the earth's atmosphere. Because of its success, however, most manufacturers have made a commitment to the continued use of R-134a for the immediate future. However, manufacturers are continuously looking for ways to make systems more efficient and are continually searching for the next generation of environmentally friendly refrigerants.

One alternative that is being considered is the development of a new mobile air conditioning system refrigerant (R-744) that uses carbon dioxide (CO_2), a naturally occurring gas, as the refrigerant. A CO_2 system is similar to today's systems, but the operating pressures are extremely high, 7 to 10 times greater than R-134a systems. Currently, the efficiency of the overall system is much lower than R-134a systems. Less efficient systems require more power to perform the same job, and because we still rely on the internal combustion engine, the overall benefits of the CO_2 system are negated. As research continues, R-744 systems are becoming more efficient.

For many years to come, technicians can expect to see the continued use of R-134a. As research continues and manufacturing techniques continue to evolve, refrigeration system components will become more efficient. As a result, R-134a systems will be able to achieve greater overall efficiency using smaller refrigerant charges than are typical today. This in effect will further reduce the environmental impact of R-134a.

RETROFIT PROCEDURES

General information is given in this chapter regarding the proper and safe practices and procedures for retrofitting an automotive air conditioning system. It is most important, however, to follow the manufacturer's instructions when servicing any particular make and model vehicle.

1. System evaluation: Visually inspect, leak test, and evaluate system performance. Determine if any repairs will be required.
2. Start the engine and adjust its speed to 1,250–1,500 rpm. Set all air conditioning controls to the MAX cold position with the blower on HI speed. Operate for

10–15 minutes to stabilize the system. Return the engine speed to normal idle. Turn off all air conditioning controls. Shut off the engine.

3. Using a refrigerant identifier, determine what type of refrigerant is in the system.
   a. If refrigerant is pure, continue to step 4.
   b. If refrigerant is contaminated, use the recovery method that is appropriate for your service facility.
4. Using dedicated R-12 equipment, recover the refrigerant.
5. Wait 10 minutes. If the system pressures increase, repeat the recovery process. Repeat as often as necessary until the gauge stays at or rises no higher than zero.
6. Disconnect R-12 recovery equipment.
7. Make any needed system repairs at this time.
8. If the accumulator meets the replacement criteria, replace at this time.
9. Select the desired conversion fittings. A number of different fitting styles are available for retrofit applications. Select those fittings that will best fit the vehicle application (see **Figure 3**).
   a. Some fittings are equipped with sealed Schrader valve fittings and may require removal of existing Schrader valves (see **Figure 4**).
   b. Some fittings may be equipped with unsealed Schrader valve fittings and will work in conjunction with existing Schrader valves (see **Figure 5**). If existing Schrader valves are leaking, they should be replaced before conversion valves are installed.

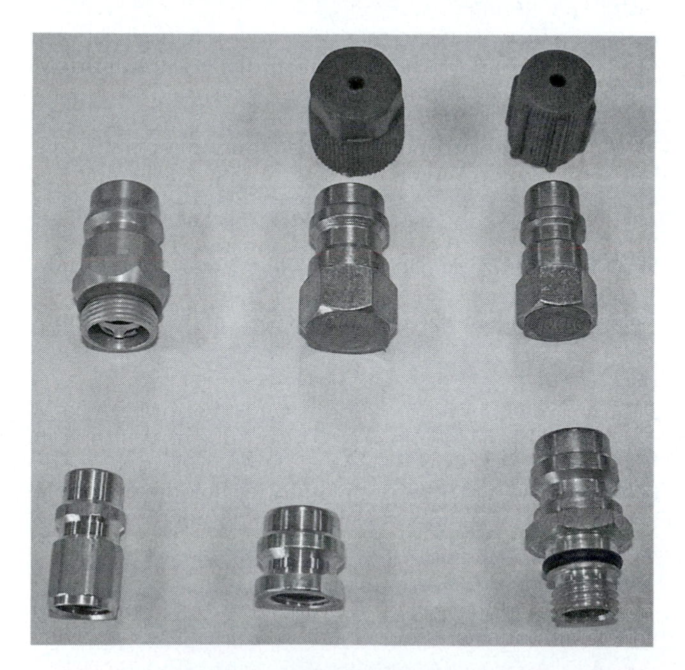

Figure 3. There are various styles of R-134a fittings to fit most all applications.

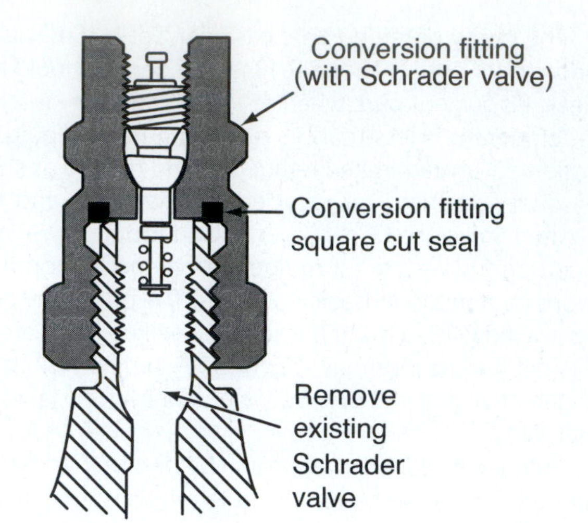

Figure 4. Some replacement fittings require the removal of the Schrader valve.

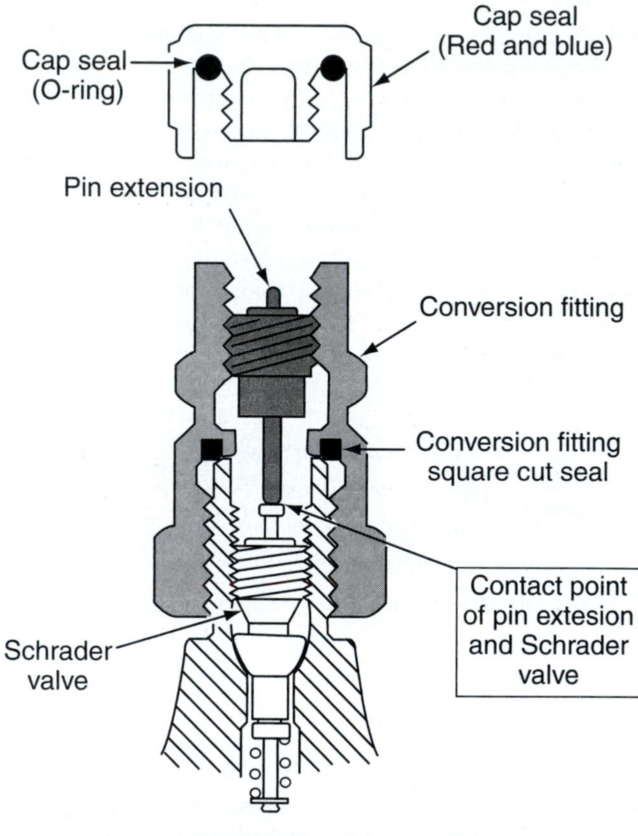

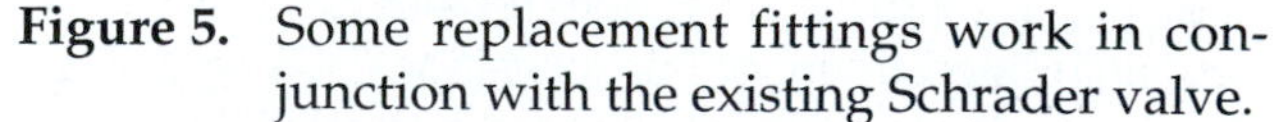

Figure 5. Some replacement fittings work in conjunction with the existing Schrader valve.

   c. A saddle valve clamp can be used in applications in which the original valve cannot be converted or clearance issues require the service fitting to be relocated (see **Figure 6**).

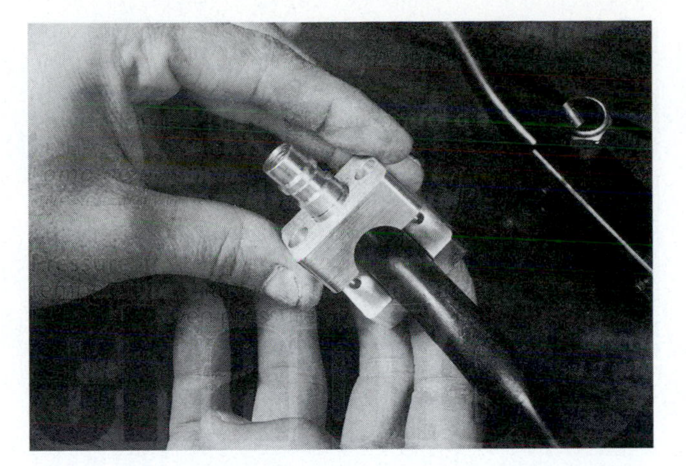

Figure 6. A saddle valve is used to install a fitting where none currently exists.

10. Remove or replace the Schrader valve as needed.
11. Clean the threads of the existing fittings of grease and scale.
12. Place one drop of permanent thread locking compound on the threads. This will prevent subsequent fitting removal and reduce the potential of refrigerant cross-contamination.
13. Install the conversion fittings.
14. If a high-pressure cutout switch is required, install at this time.
15. Connect R-134a service equipment.
16. Open only the high-side access valve. Evacuating from the high side will help to remove any remaining residual R-12 that might remain in the system.
17. Evacuate the system for at least 15 minutes.
18. Install the proper amount and type of refrigeration oil. This is typically about 8 oz (236.59 ml).
19. Recharge the system with the required amount of refrigerant.
20. Complete and attach the retrofit label in a clearly visible position to fixed location (see **Figure 7**).
21. Performance-test the system.
22. Check for leaks.
23. Disconnect all service equipment and install service caps.

> **You Should Know** *If the high-side head pressure is excessive after retrofitting a system from R-12 to R-134a, the installation of an auxiliary cooling fan will help lower this pressure. The auxiliary fan should be wired so that it will run continuously when air conditioning is selected.*

NOTICE: RETROFITTED to R-134 a
Retrofit procedure performed to SAE J1661
CAUTION: System to be serviced by qualified personnel only.
Retrofit Date:
Name of company or individual performing retrofit:
Address:
City / State Zip
Tel # R-134a Charge
Lubricant Type Lubricant Amount
Auxiliary Label Location:

Figure 7. The retrofit label must located under the hood, attached to a rigid body component and must be in plain view.

SADDLE VALVE INSTALLATION

The saddle valve is used for those applications in which it is impossible to install the conventional R-134a conversion fittings or when an existing fitting is damaged. A typical application in which this type of valve is needed is one in which both service valves are installed in very close proximity to one another, preventing the connection of the larger R-134a service hose connectors. This typically occurs when both fittings are installed on the suction and discharge fittings at the back of the compressor.

1. Recover the system refrigerant as needed.
2. Select the proper location for the valve.
   a. Will there be clearance for the hose access adapter?
   b. Will there be adequate clearance to close the hood or replace protective covers?
   c. Will access to other critical components be restricted or blocked?
   d. Is the tubing straight, clean, and sound?
3. Select the proper valve for the application.
   a. For low- or high-side use (the low-side valve is larger).
   b. The size of the tube on which the valve is to be installed.
4. Position both halves of the saddle valve on the tube (see **Figure 8**).
5. Place the screws (usually socket head) and tighten them evenly. Do not overtighten them; 20–30 in.-lb. (2–3 N·m) is usually recommended. A method other than that outlined in steps 5, 6, and 7 might be recommended. Follow the recommendations provided by the manufacturer of the saddle valve when they differ from those given here.
6. Insert the piercing pin in the head of the access port fitting.

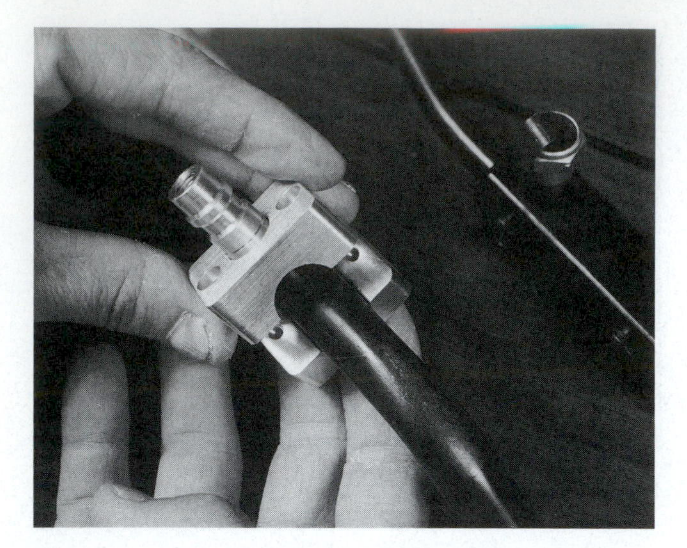

Figure 8. The two halves of the saddle valves are clamped around the line in the selected location.

7. Tighten the pin until the head touches the top of the access port (see **Figure 9**).
8. Remove the piercing pin and replace it with the valve core.
9. Securely tighten the valve core (see **Figure 10**).
10. Install the cap (or pressure switch) on the installed fitting. To ensure compatibility, use only the O-ring included with the saddle valve kit.

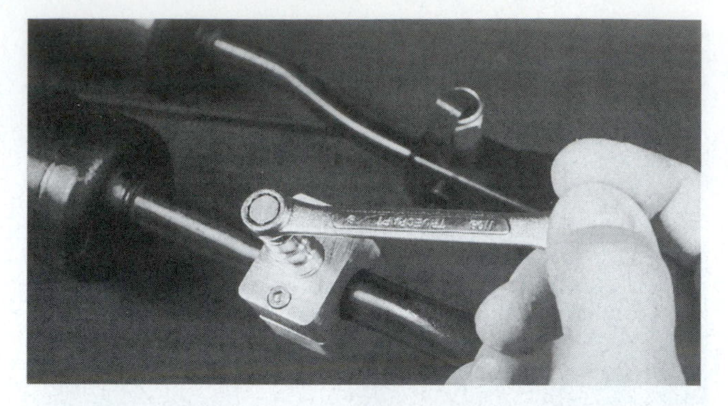

Figure 9. The line is pierced using a special piercing pin.

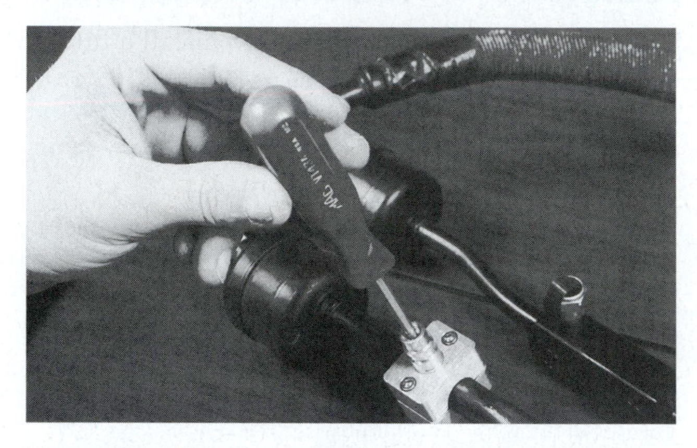

Figure 10. When completed the Schrader valve is installed into the fitting.

Summary

- R-134a is the primary replacement refrigerant for R-12 retrofit applications. The basic retrofits do not require major component replacements and generally provide acceptable performance.
- The basic retrofit usually only requires the removal of existing R-12, installation of new fittings, installation of a new label, and the addition of the proper lubricant.
- R-134a is the only refrigerant that is recommended by vehicle manufacturers as a replacement for R-12. R-134a is considered the safest available alternative to R-12.
- Any existing refrigeration system problems should be corrected before a retrofit is performed. New service fittings must be installed. R-134a capacity is approximately 80–90 percent that of R-12. New lubricant must be installed.
- Most refrigeration system components do not need to be changed as part of a retrofit. SAE standards require that refrigerant should be at least 98 percent pure. The use of a refrigerant identifier is by far the most accurate method of determining refrigerant purity.
- Contaminated refrigerant must be recovered by dedicated equipment or by use of alternative means. Refrigerant—contaminated or otherwise—cannot be stored in disposable containers.
- R-134a has been chosen by most manufacturers as the refrigerant of choice for the immediate future. As systems become more efficient, refrigerant charge amounts will decrease.
- Retrofit fittings may or may not work in conjunction with the existing Schrader valves. Retrofit fittings should be secured with thread-locking compound. Labels should be permanently fixed in a prominent location in the engine compartment.
- Saddle valves can be used when existing fittings are damaged or when the installation of retrofit fittings would make them inaccessible.

Review Questions

1. List minimum requirements for a basic retrofit.
2. Explain why R-134a has been chosen by the majority of manufacturers as the substitute for R-12.
3. Technician A says that all traces of mineral oil must be removed from the refrigeration system before a retrofit can be completed. Technician B says that R-134a and mineral oil will combine to provide adequate system lubrication. Who is correct?
   A. Technician A
   B. Technician B
   C. Both Technician A and Technician B
   D. Neither Technician A nor Technician B
4. Technician A says that all of the system O-rings should be replaced with R-134a compatible materials. Technician B says that O-rings only have to be replaced if they are leaking or have been disturbed. Who is correct?
   A. Technician A
   B. Technician B
   C. Both Technician A and Technician B
   D. Neither Technician A nor Technician B
5. Explain the purpose of the high-pressure cutout switch.
6. Technician A says that refrigerant contamination can be determined by comparing the temperature-to-pressure relationship of the specific refrigerant. Technician B says that use of the refrigerant identifier is the most accurate method of determining refrigerant purity. Who is correct?
   A. Technician A
   B. Technician B
   C. Both Technician A and Technician B
   D. Neither Technician A nor Technician B
7. If the high-side pressure is excessive after a retrofit, what is one possible solution to repair the condition?
8. Technician A says that when retrofitting a refrigeration system, the system should be evacuated through the high side of the system. Technician B says that evacuating the system from the high side will purge the system of any residual mineral oil. Who is correct?
   A. Technician A
   B. Technician B
   C. Both Technician A and Technician B
   D. Neither Technician A nor Technician B
9. Technician A says that contaminated refrigerant can be reclaimed in the shop using most commercial recovery, evacuation, and recharging units. Technician B says that contaminated refrigerant must be reclaimed or destroyed by fire at an off-site facility. Who is correct?
   A. Technician A
   B. Technician B
   C. Both Technician A and Technician B
   D. Neither Technician A nor Technician B

Section 5

Electrical and Electronic Systems

SECTION OBJECTIVES

After you have read, studied, and practiced the contents of this section, you should be able to:

- Explain basic electrical principles such as electrical sources, electron flow, types of circuits, and electron currents in circuits.
- Demonstrate knowledge of electrical measurements and their relationship to each other as defined by Ohm's law.
- Identify series and parallel circuits and describe how electricity flows through each type of circuit to produce work.
- List the components that make a circuit.
- Diagnose and repair circuit faults in circuit protection devices such as fuses, circuit breakers, and fusible links.
- Describe the function of resistance components in circuits.
- Explain how a relay operates in a circuit and diagnose circuit problems that include relays.
- Describe how circuits are used to generate magnetic fields and how HVAC systems utilize the magnetism generated.
- List the components used in a compressor control circuit and explain their function.
- Explain the basic operation of clutch cycling orifice tube, thermostatic expansion valve, and variable displacement compressor control circuits.
- Detail how computer controls are integrated into the compressor control circuit.
- Describe the operation and diagnose faults that occur in the various styles of blower control circuits.
- List the common types of electrical circuit faults and describe the basic techniques used to locate the problems.
- Diagnose problems that occur in common HVAC electrical circuit components.
- Service HVAC electrical circuits and the components that compose them.

Chapter 20 Electrical Theories

Introduction

Understanding electrical concepts is critical for today's automotive technician. In the 1970s, less than 30 percent of vehicle systems were electrically controlled. The modern automobile is closer to 95 percent electrically or electronically controlled. Many technicians are intimidated by the sudden increase in electronics that have been integrated into the vehicles. However, once you are comfortable with the electronics and understand how they work, you will find the modernized systems to be a great ally in diagnosing problems in the systems.

Automotive air conditioning systems have led the industry in the ever-expanding use of electronic controls. Automatic temperature control systems, once available exclusively in luxury vehicles, have become common across all makes and models. Manual control systems have moved from cables and vacuum switches to electronic actuators and solenoids. The HVAC control module is in communication with other modules, such as the powertrain control module, instrument control module, and the body control module. All these elements require electronic controls, and it is up to the modern technician to be able to understand their operation and diagnose problems when they arise.

ELECTRICITY

Electricity is the movement of electrons from an area with an excess of electrons to an area with a lack of electrons. This movement occurs physically in nature when lightning strikes or we rub a balloon on a piece of fabric. This chapter will look at a more controlled application of electricity. The controlled flow of electrons in a fixed circuit, designed to perform a type of work, is used in automotive applications (see **Figure 1**). In order to understand this

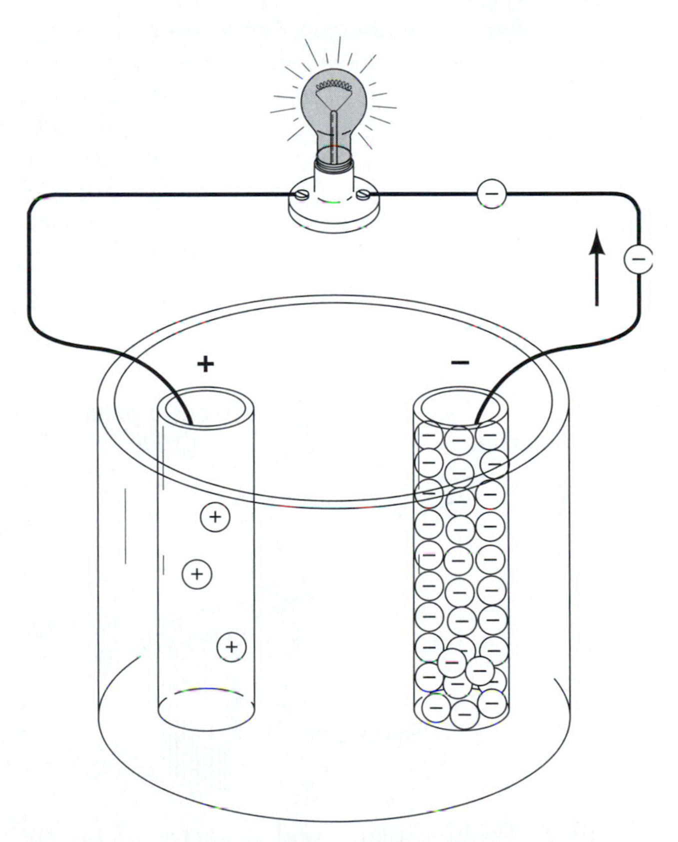

Figure 1. Excess electrons will always try to reach an area missing electrons. If they pass through a load device while moving, work will be done.

form of electricity, a technician must be able to measure the intensity of the force pushing the flow of electrons, how many electrons are flowing, and what is inhibiting the flow of electrons.

SOURCES OF ELECTRICITY

The typical automotive circuit has two sources of electrons to draw from in order to create an electrical flow. In order to have electricity, there must be two areas with an imbalance of electrons. The two methods used to create these areas are chemical and electromagnetic induction.

The chemical method takes place in the battery. A battery starts out with two similar lead plates in an electrolyte solution. When the battery is charged, a chemical reaction takes place that strips electrons from one plate and moves them to the other plate. The plate with the missing electrons becomes the positive plate, whereas the other plate becomes the negative plate with its extra electrons. These plates are connected to the positive and negative posts of the battery (see **Figure 2**). When a conductive pathway called a circuit is connected to the two battery posts, the electrons can flow from the surplus negative plate back to the positive plate.

The electromagnetic induction process occurs in the alternator. A magnet is rotated inside a large coil of wire in the alternator. As the magnet spins, it forces the electrons in the wire away from one end of the wire (see **Figure 3**). This end of the wire leads to the positive (often called the BAT +) post on the alternator. The other end of the wire coil that is gaining electrons is connected onto the alternator case that becomes ground. Circuits can then be connected to the positive alternator post and to the alternator case to flow electrons.

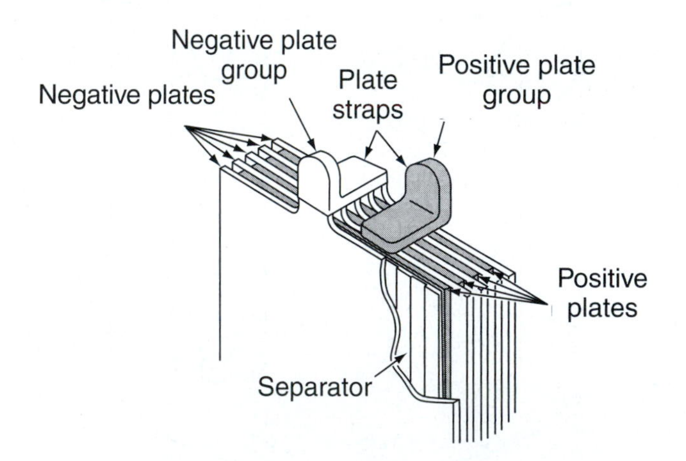

Figure 2. Positive plates and negative plates are respectively strapped together, eventually leading to the battery posts.

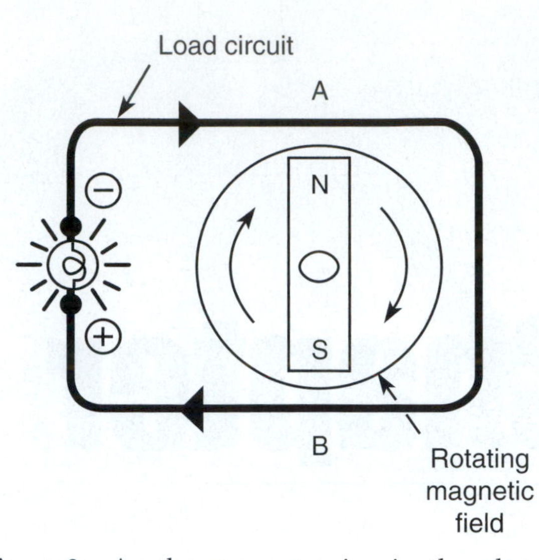

Figure 3. As the magnet spins in the alternator, electrons are forced out the coil windings to the load.

> **You Should Know** *Although physically the electrons move from the negative post to the positive post on the battery, common practice is to consider the flow as traveling from positive to negative. This is referred to as the conventional flow theory. All wiring diagrams and schematics are drawn using this conventional view.*

VOLTAGE

A basic law of physics is that an object at rest (not moving) will stay at rest unless an outside force is applied to it. Electricity is the flow of electrons from one place to another. These electrons, before they start to flow, are attached to an atom. They will stay at that atom unless an outside force is applied to them (see **Figure 4**). This force is known as **electromotive force (EMF)**. Voltage is a measurement of the strength of that force. The higher the voltage applied to a circuit, the more pressure there is to push the electrons through the circuit.

To visualize the effect of voltage on the electrons in a circuit, let's picture a gorilla pushing a large block of wood. The gorilla represents voltage and the block of wood represents an electron. If there is no gorilla, the

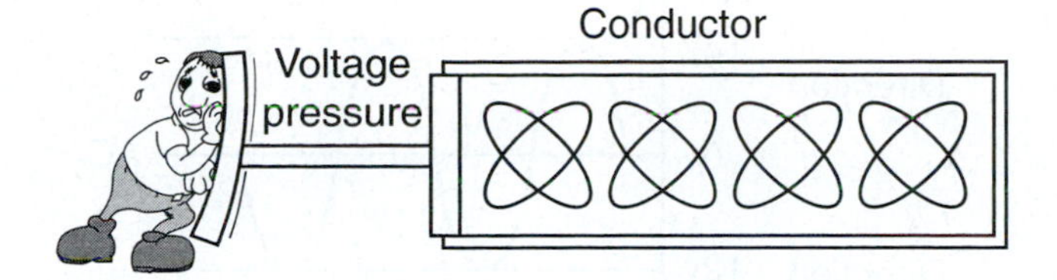

Figure 4. The electrons will not move until an outside force, or push, is applied to them. The intensity of the force is measured as voltage.

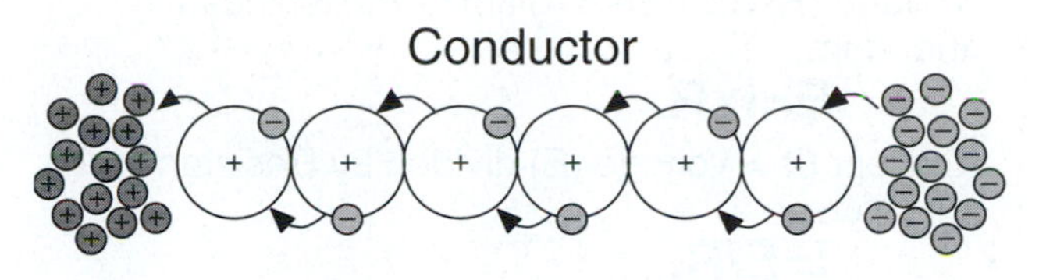

Figure 5. Amperage is an actual measurement of the number of electrons flowing through a point in the wire.

block of wood does not get moved. When we bring in the gorilla, the block gets moved. What would happen if a second gorilla of equal strength began to push on the opposite side of the block? Again, the block would not move. If the voltage on each end of a circuit is equal, then there is no electrical flow. Now let's replace the second gorilla with a chimpanzee. Because the chimp is weaker than the gorilla, the block gets pushed toward the chimp. If the voltage is stronger at one end of the circuit than the other, then the electrons flow toward the weaker force. And if we put both gorillas on one side of the block so they can both push, then we can move twice as many blocks.

> **You Should Know** *When using a voltmeter, the meter does not read the actual voltage where the red probe is placed. The meter tells the difference in voltage between where the red probe is placed and where the black probe is placed. To get a true voltage reading at the red probe, make sure the black probe is placed at a good zero volt ground.*

AMPERAGE

Amperage is a measurement of how many electrons are flowing through a circuit. It is easy to visualize water flowing down a river. It can be seen and even measured in cubic feet per second. Electrical flow is just as real and physical as water flow. Even the term **current** is used to describe both moving water and traveling electrons. If a powerful-enough microscope could be built, we could count the electrons passing through a given point in a wire when a current is flowing through it (see **Figure 5**). The counting would have to be fast, though: if one amp is flowing through a wire, there are 6.28 billion electrons passing by that one point per second.

RESISTANCE

Anything that opposes or restricts the flow of electrons through a circuit is adding **resistance** to the circuit. The unit of measurement for resistance is the **ohm**. Everything has some resistance to the flow of electricity. Electricity does flow easily in some materials, which are called **conductors**. Materials in which electricity does not flow well are called **insulators**. Conductors have a low ohms rating in comparison to the high ohms rating of insulators.

There are five variables that affect the resistance of a wire. The **atomic structure** of the wire, or what it is made of, contributes significantly to how well it conducts electricity. When comparing metals, silver is the best conductor, but it tarnishes too easily to make good wire. Gold is almost as good at conducting and does not tarnish; it would make great wire if it were not so expensive. Aluminum is also a good conductor, but it is not good wire because it tends to get brittle and break. The best all-around choice is copper. It conducts well, is inexpensive, and is sturdy under use. The **length** of a wire is proportional to the resistance of the wire: the longer the wire, the more resistance it has. The **temperature** of the wire changes its resistance. Like everything else, with the exception of some thermistors, as its temperature increases, so does its resistance. The **diameter** of a wire affects its resistance: the larger the wire, the more current it can carry, and its resistance is lower. Finally, the **condition** of the wire can change its resistance. Corrosion, damage, or overheating can change its ability to carry current.

OHM'S LAW

The relationship between volts, amps, and ohms can be expressed as a mathematical formula. **Figure 6** illustrates the calculations for determining any one electrical value if the other two values are known.

Voltage (E) = Current (I) times Resistance (R), therefore

E=IxR.

Current (I) = Voltage (E) divided by Resistance (R), therefore

I=E/R.

Resistance (R) = Voltage (E) divided by Current (I), therefore

R=E/I.

Figure 6. The relationship between volts, amps, and ohms can be expressed as a mathematical formula. The calculations are for determining any one electrical value if the other two values are known.

Although the math formulas are valuable to know, it is more important to understand the relationship between these electrical measurements. Voltage and resistance (ohms) work directly against each other. The relative strength of these two measurements will determine the amount of flow in a circuit. The more voltage there is to push against the resistance, the more flow or amperage there will be. The more resistance in the circuit, the less amperage there will be. To see how the formulas illustrate this concept, start with a circuit that has six volts applied to it and has three ohms of resistance. The calculated amperage would be 6 ÷ 3 = 2 amps. Doubling the volts will double the flow, so 12 ÷ 3 = 6 amps. Doubling the resistance in the original circuit will cut the flow in half, as 6 ÷ 6 = 1 amp. Anything that changes an automotive circuit will result in a change in the flow of electrons through the circuit. Electrical problems are often diagnosed by determining how the flow has been modified, which is why this concept is so important to understand.

DIRECT AND ALTERNATING CURRENT

Current is generated in one of two flow types (see **Figure 7**): alternating current (AC) and direct current (DC). In alternating current, the electrons travel in one direction through the circuit and then reverse direction and travel the opposite way. The alternating current used in household wiring cycles back and forth 60 times a second, or 60 hertz. In automotive applications, some sensors generate an AC signal in which the cycle frequency varies; this is what the associated control module measures. Almost all automotive circuits use direct current. In direct current, the electrons move in only one direction.

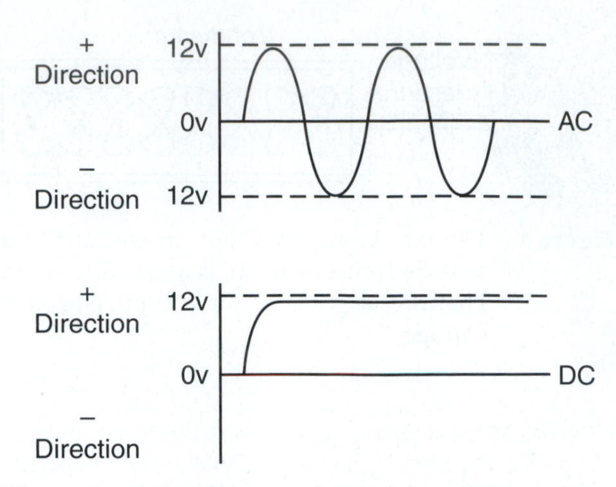

Figure 7. AC voltage cycles from positive to negative while DC voltage stays constant.

Interesting Fact

Although alternating current is easier to generate than direct current, it is nearly impossible to store. DC is easy to store in a battery, which is why car batteries are sometimes referred to as storage batteries. Its ability to be stored is the reason early carmakers chose DC over AC for use in their vehicles.

SERIES CIRCUITS

The distinguishing characteristic of a series circuit is its one path. There is only one possible way for the current to flow through the components. If any part of the circuit fails to allow the electricity to flow through it, then the entire circuit does not function.

Although it is possible to have more than one load in a series circuit, most automotive applications have only one load device. If there is more than one load, then the loads will have to share the available voltage, and the device with the most resistance takes a greater share than the other device. This means that if two similar bulbs are in the same series circuit, both bulbs will burn dimly. There are designed automotive series circuits, such as one using a variable resistor, which changes resistance to modify a gauge reading.

Series circuits have certain laws that pertain to electrical measurements such as voltage, ohms, and amperage. First, voltage is always completely used up in a series circuit, with each part of the circuit consuming voltage based on its resistance. The more resistance, the greater the percentage of the voltage it uses. Second, the total resistance of the circuit is equal to the total of the individual resistances of each part in the circuit. And, third, the amperage is the same everywhere in the circuit.

PARALLEL CIRCUITS

A parallel circuit will have more than one path for the current to follow. As shown in **Figure 8**, there can be several paths, or legs, all of which may operate at the same time. If one leg fails to operate properly, the other legs may continue to function as normal. The operation of one leg does not affect the other legs.

Like series circuits, parallel circuits have certain laws that pertain to electrical measurements such as voltage, ohms, and amperage. The voltage applied to each leg is the same. The total resistance of the circuit is calculated using a mathematical formula, but the total resistance will be less than the leg with the smallest resistance. The total amperage of the circuit is equal to the total of the individual leg amperages. When working with a leg with just one path, the formulas for a series circuit will apply to that leg.

CIRCUIT COMPONENTS

There are certain elements every circuit must contain in order for it to be a functioning part of the vehicle. First, every circuit has to have a **source** of electrons. In automotive applications, this is the function of the battery and alternator. Every circuit also has to have a **load**. A load is the device that is doing the actual work in the circuit, such as a light, a motor, or the coil in a relay. The circuit also has to have a **path** for the electricity to flow through. This includes the wires and ground that connect everything together. It should be noted that the ground path is usually the frame and metal body components, which most circuits use as part of the ground path.

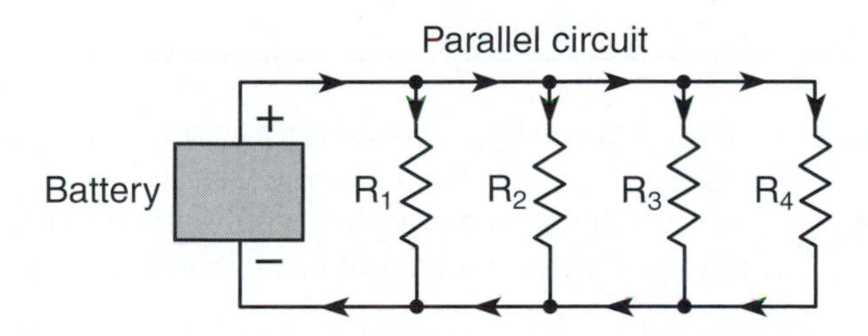

Figure 8. Each path connected directly to power and ground is a separate leg and functions independently of the other legs.

It is possible for the final two parts of a circuit not to be included in certain specialized systems such as the starter cranking circuit. However, almost all other circuits have a **protection** device and a **control** device. Circuit protection includes fuses, circuit breakers, and fusible links. Circuit control devices are the switches that block or redirect the flow of electrons through the circuit. These and other circuit devices will be explored in more detail in the next chapter.

Interesting Fact

Ask many technicians where true ground is on a running vehicle and the answer will be the battery negative post. In reality, when the vehicle is running, the alternator is the source of almost all electron flow, which would make the alternator housing the primary ground.

WIRING DIAGRAMS

As an automotive technician, learning to read a wiring diagram may be the most important skill to develop. The first step to improving this skill is learning the symbols that make up the diagram (see **Figure 9**). Without this knowledge, it would be like trying to read without knowing the alphabet. Once the symbols are learned, following the diagram is like reading a roadmap. When you go on a road trip, you begin at home and follow the roads that take you to your destination. On a wiring diagram, the voltage source is home: from there, you follow the wires and circuit devices to the final destination, which is ground. Do not assume the current is flowing anywhere in the circuit unless you can trace that point in the circuit to both power and ground.

Once you have found the complete path, determine what parts of the circuit should have battery voltage. This should be from the voltage source to a load device. Remember, a load device must do some kind of work. Fuses and switches protect and direct the current flow, but they do not do work. When diagnosing, check this part of the circuit on the vehicle for battery voltage. Next, determine which wires are connected to ground. When checking these wires on the vehicle, there should be little or no voltage on this part of the circuit. If the voltages on the power and ground wires are correct, and the load on the circuit is not functioning, then the load device is malfunctioning.

SYMBOLS USED IN WIRING DIAGRAMS			
	Positive		Temperature switch
	Negative		Diode
	Ground		Zenner diode
	Fuse		Motor
	Circuit breaker	C101	Connector 101
	Condenser		Male connector
	Ohm		Female connector
	Fixed value resistor		Splice
	Variable resistor	S101	Splice number
	Series resistors		Thermal element
	Coil		Multiple connectors
	Open contacts		Digital readout
	Closed contacts		Single filament bulb
	Closed switch		Dual filament bulb
	Open switch		Light emitting diode
	Ganged switch (N.O.)		Thermistor
	Single pole double throw switch		PNP bi-polar transistor
	Momentary contact switch		NPN bi-polar transistor
	Pressure switch		Gauge
	Battery		Wire Crossing

Figure 9. Learning electrical symbols is the first step in learning to read a wiring diagram.

Summary

- Understanding electricity is one of the most important skills needed by the modern technician.
- Electricity is the flow of electrons from one area to another.
- Automotive circuits draw current from the battery, a chemical source, and the alternator, an electromagnetic source.
- Voltage is the push that drives the electrons through the electric circuit.
- Amperage is a measure of the volume of electrons flowing in a circuit.
- Resistance, measured in ohms, is the opposition to flow in a circuit.
- Electricity flows easily in conductors and poorly in insulators.
- Atomic structure, temperature, length, diameter, and condition all affect the flow of electricity through a wire.
- Ohm's law expresses the relationship between electrical values.
- Direct current flows in one direction, alternating current cycles flow in both directions.
- Series circuits have only one path; parallel circuits have more than one path.
- Circuits are made up of a voltage source, protection device, control device, load, and connecting path.

Review Questions

1. Technician A says that circuit loads include fuses and switches. Technician B says that all automotive circuits use direct current. Who is correct?
   A. Technician A only
   B. Technician B only
   C. Both Technician A and Technician B
   D. Neither Technician A nor Technician B
2. Technician A says that an alternator creates electricity using electromagnetic induction. Technician B says that electricity also can be generated using a chemical reaction. Who is correct?
   A. Technician A only
   B. Technician B only
   C. Both Technician A and Technician B
   D. Neither Technician A nor Technician B
3. An ________________ does not conduct electricity well.
4. Which of the following is NOT a required circuit component?
   A. The battery
   B. The blower motor
   C. The connectors
   D. The ground
5. Technician A says that electrons flow from the negative battery post to the positive post. Technician B says that the wiring diagrams show the current flowing from positive to negative. Who is correct?
   A. Technician A only
   B. Technician B only
   C. Both Technician A and Technician B
   D. Neither Technician A nor Technician B
6. Technician A says that amperage is not affected by the length of the wire in a circuit. Technician B says that voltage is a measure of the flow of electrons through a circuit. Who is correct?
   A. Technician A only
   B. Technician B only
   C. Both Technician A and Technician B
   D. Neither Technician A nor Technician B
7. Technician A says that alternators gather electrons from magnetic fields. Technician B says that volts are equal to amps divided by ohms. Who is correct?
   A. Technician A only
   B. Technician B only
   C. Both Technician A and Technician B
   D. Neither Technician A nor Technician B
8. Technician A says that circuits use conductors to create a path for the electrical current to flow. Technician B says that insulators have no use in creating functional circuits. Who is correct?
   A. Technician A only
   B. Technician B only
   C. Both Technician A and Technician B
   D. Neither Technician A nor Technician B
9. Technician A says that longer battery cables could make a starter motor crank slower. Technician B says that everything's resistance increases as it gets hotter. Who is correct?
   A. Technician A only
   B. Technician B only
   C. Both Technician A and Technician B
   D. Neither Technician A nor Technician B
10. A series circuit that has two light bulbs will not work at all if one of the bulbs is removed.
    A. True
    B. False
11. A parallel circuit with two legs that has one bulb in each leg will still give some light if one bulb is removed.
    A. True
    B. False
12. Technician A says that a parallel circuit that has three legs with 3 ohms resistance each will have a total resistance of 9 ohms. Technician B says that a circuit with 6 ohms of resistance and 3 amps flowing through it must have 18 volts applied to it. Who is correct?
    A. Technician A only
    B. Technician B only
    C. Both Technician A and Technician B
    D. Neither Technician A nor Technician B

Chapter 21 Electrical and Electronic Components

Introduction

Understanding the operation of electrical and electronic components can greatly assist the automotive technician in diagnosing failures. Although it is important to know the rules pertaining to circuits discussed in the last chapter, knowing how individual components function in a working circuit will allow a technician to pinpoint exact problems.

CIRCUIT PROTECTION DEVICES

Protection devices are installed in circuits to ensure that too much amperage does not flow through the circuit. In the event of excess current, the protection devices create an open in the circuit that blocks all electrical flow through the circuit. It should be noted that the devices are there to protect the wiring in the circuit, not the load or control components. If the wires are not protected, any amperage above what they are capable of safely carrying can result in the wiring catching fire and igniting materials located close to the wire. The three devices used in automotive circuits are fuses, circuit breakers, and fusible links. (see **Figure 1**) Each of these protection devices relies on the heat generated by the current passing through them to make them work. Circuit designers try to place the protection device as early in the circuit flow as possible in order to provide the most protection.

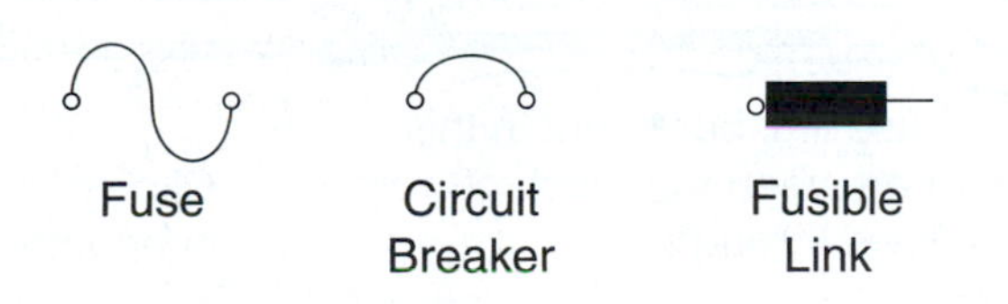

Figure 1. The diagram symbols for fuses, circuit breakers, and fusible links.

Fuses

The most common circuit protection device is the fuse (see **Figure 2**). Although there are several different designs, all fuses function in the same way. They contain a small metal strip through which all the current must pass. If the heat generated by the amperage passing through the resistance in the strip becomes too great, the strip melts, creating an open that blocks all current flow. The size of the strip is dependent on the amperage rating of the fuse: the higher the fuse rating, the thicker the metal strip.

> **You Should Know** *Never replace a fuse with a fuse of a higher amperage rating. Doing so would result in a fire hazard that could destroy the entire vehicle.*

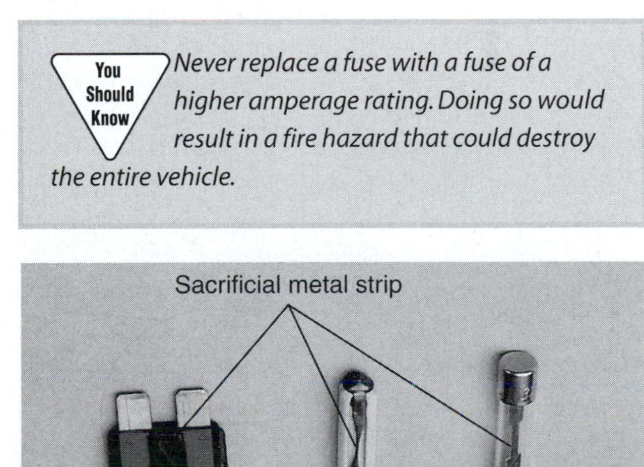

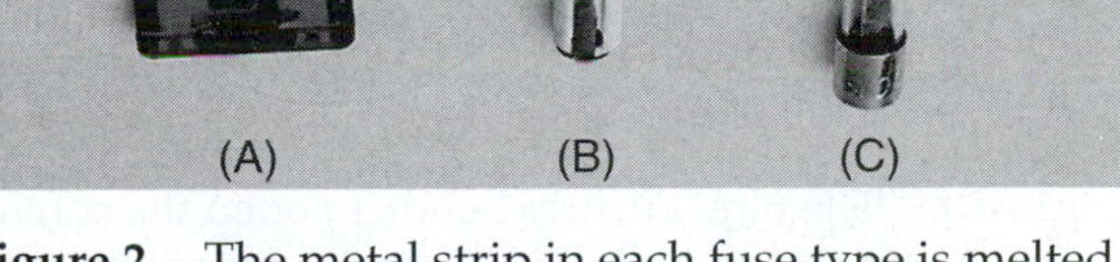

Figure 2. The metal strip in each fuse type is melted, or sacrificed, to save the circuit wiring.

Different manufacturers and varying applications have resulted in several different fuse designs. Among these are three types of blade fuses (referred to as auto fuses, mini fuses, and maxi fuses), ceramic fuses often found in older import vehicles, glass cartridges used on older domestic vehicles, and Pacific element fuses.

The metal strip in blade fuses is housed in a plastic case that is color-coded to identify the amperage rating of the fuse. A small access slot is molded into the plastic on each side of the strip for testing purposes. Despite their similar appearance, blade fuse types are not interchangeable. Always use the correct fuse application.

Pacific element fuses are used in place of fusible links. These cartridge fuses are replaced as a unit and are made in two common styles: pressed-in or bolt-in. The pressed-in version is the more popular design. These fuses are also color-coded to identify their amperage. The element is visible through the clear plastic top of the case for visual testing.

Ceramic and glass cartridge fuses are obsolete designs. The ceramic fuse strip ran along the outside of a ceramic rod, whereas the glass cartridge ran the strip through a glass tube. The metal ends of both types fit into metal clips in the fuse holder where they could be tested for voltage.

Circuit Breakers

Circuit breakers (see **Figure 3**) are nondestructive fuses, that is, when they open to block current flow, they are not damaged. They are used in circuits that, from time to time, may have excess amperage applied to them even when there is no fault in the circuit. An example would be a windshield wiper motor circuit in which the blades are frozen to the windshield during the winter. If the wipers are turned on, the frozen blades will not allow the motor to turn and a high amperage draw will be placed on the circuit. If the circuit were protected by a fuse, the fuse would melt and have to be replaced before the wipers would work again. A circuit breaker in this situation will open but will eventually reset itself so the wipers will work again.

There are two main styles of circuit breakers: mechanical and solid state. Mechanical circuit breakers utilize a bimetal strip through which the current passes. When the strip is in its normal straight position, it has a contact that will mate with a fixed contact in the breaker housing. The two metals in the bimetal strip expand at different rates so that when excess current passes through the strip, it causes the strip to bend. This opens the contacts and stops the flow of electricity. Most mechanical circuit breakers reset themselves as the bimetal strip cools and returns to its straight position. However, some models include a button that must be manually pressed to reset the breaker. Mechanical circuit breakers are usually small black or silver boxes installed in the fuse panel much like a fuse is installed.

Solid-state circuit breakers are Positive Temperature Coefficient (PTC) thermistors. The crystalline material they are made of has many carbon fingers that intertwine to allow electricity to pass through the material when cool. However, when excess current heats the material, the carbon fingers pull away from each other. This restricts the flow of electrons and stops the circuit from operating. Unlike the mechanical version, a tiny amount of current still passes through the PTC breaker, which keeps it from resetting itself. Only after the current has been removed from the circuit will the circuit breaker reset. This style of breaker typically is housed internally in components such as window lift motors, wiper motors, and seat motors.

Fusible Links

Like a fuse, a fusible link (see **Figure 4**) is designed to melt when excess amperage passes through it. It consists of a short piece of wire with a heavy, rubberized insulation. The insulation is designed to catch the melted wire material. The size of the wire is four gauge sizes smaller than the circuit wires it is to protect. The link wire is not special; it melts before the circuit wire simply because it is smaller in diameter. The advantage of fusible links is that they can be placed very near the voltage source because they require no special fuse holders. However, they are cumbersome to repair and are mostly being replaced by Pacific element and maxi fuses.

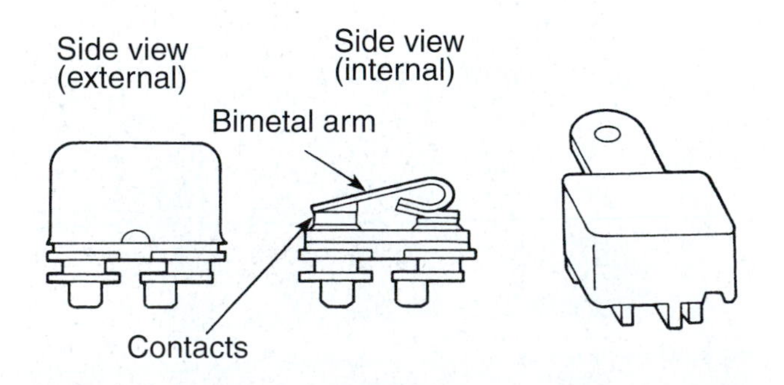

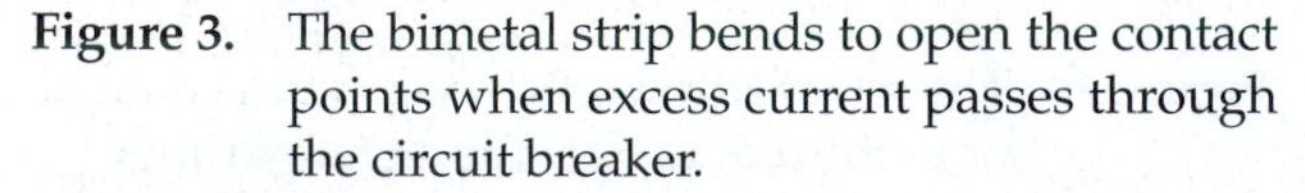

Figure 3. The bimetal strip bends to open the contact points when excess current passes through the circuit breaker.

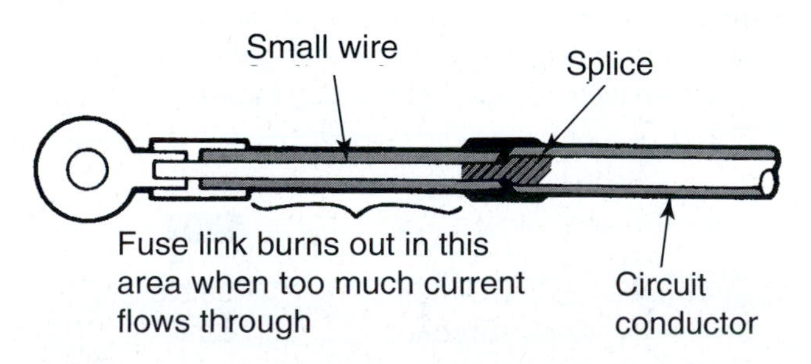

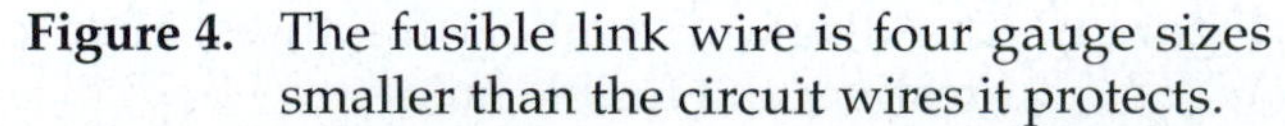

Figure 4. The fusible link wire is four gauge sizes smaller than the circuit wires it protects.

WIRES

Automotive wiring is generally copper-strand wiring. Copper is used instead of other metals because it offers the best compromise of conductivity, cost, availability, and durability. The wire is stranded, which means it is made up of many little strands, because it is more flexible and durable than solid core wire. The metal wires are covered with a plastic insulator layer, which is color-coded to assist in tracing the wires in the vehicle.

RESISTORS

Resistors are load devices designed to reduce the amount of current flow or drop the voltage in a circuit. As a load device, they must do some sort of work. Resistors change electrical energy into heat. However, with the exception of cigarette lighters and heated windshields, resistors are seldom used primarily to produce heat. In automotive applications, there are two main types of resistors: fixed resistors and variable resistors.

Fixed Resistors

Fixed resistors (see **Figure 5**) are usually made of carbon-graphite with a wire extending from each end. Although there are thousands of these resistors in the control modules throughout the vehicle, they are seldom serviceable separately. A stepped resistor is a series of fixed resistors with connection points between each resistor. A control switch directs the current through the different connections on the stepped resistor, varying the number of resistors the current must pass through. The more resistors the current passes through, the less amperage that is available to the circuit primary load device.

Variable Resistors

Variable resistors have the ability to change their resistance in response to a changing condition, such as movement, temperature, or pressure. Two common variable resistors that respond to movement are the rheostat and the potentiometer. They are similar in construction in that they both have a moveable sweep arm with a contact that moves along the length of a resistor surface. The two-wire rheostat (see **Figure 6**) has one wire connected to the end of the resistor and one wire connected to the moveable contact. As the contact moves closer to the end of the resistor with the wire, the current passes through less of the resistor. This reduces the total resistance of the rheostat and is used to control the current flow through the circuit. A rheostat is commonly used in fuel tank level gauge circuits. The fuel gauge is wired in series with a rheostat that has its moveable contact attached to the sending unit float. As the float moves up and down with the fuel level, it moves the contact. The changing current results in the fuel gauge changing its reading.

Figure 5. This symbol is for a fixed resistor.

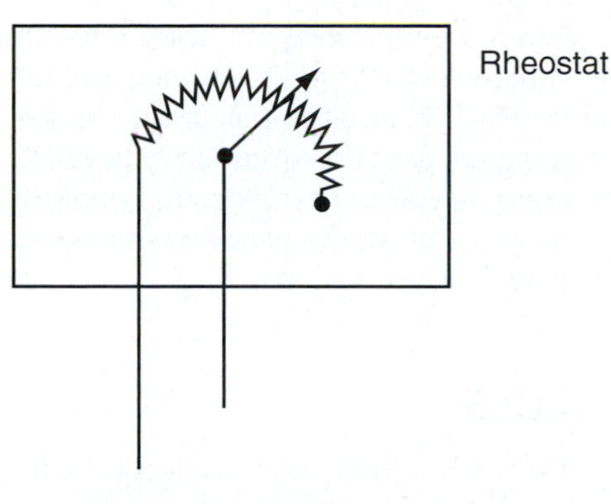

Figure 6. The rheostat is used in circuits to control the current to another load device.

A potentiometer (see **Figure 7**) is constructed exactly like the rheostat, except both ends of the resistor have a wire attached to it giving it a total of three wires. This third wire is attached to a ground, whereas one resistor wire is attached to a constant voltage supply. This keeps the current flow through the resistor constant. Potentiometers are used to generate a voltage signal. The moveable contact arm wire is connected to a high-resistance circuit in a control module. Because the current through the resistor is constant, the voltage along the resistor surface is proportionally constant. The closer the moveable contact is to the constant source voltage wire, the higher the voltage signal to the module. This system is used in many applications to measure movement, such as throttle position, blend door, and seat position.

Other variable resistors change resistance in response to conditions other than movement. Thermistors are devices

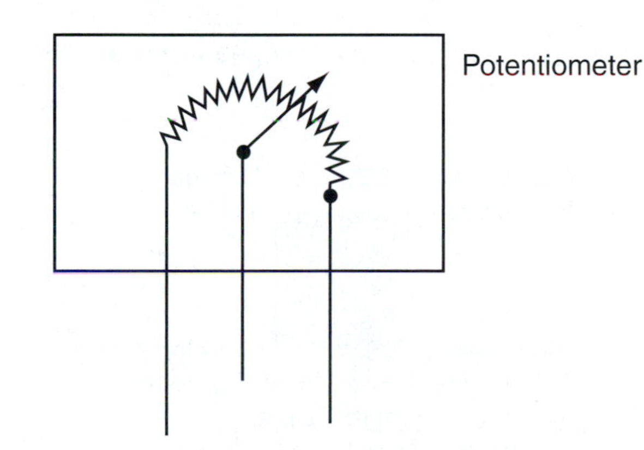

Figure 7. The potentiometer is a variable resistor used to send a voltage signal to a control module.

that change their resistance in response to temperature. They are unique in that they have a negative temperature coefficient (NTC). That means their resistance goes down as their temperature goes up. Almost everything else responds just the opposite way to heat; resistance goes up. There are other variable resistors that change in response to pressure, light, vibration, and acceleration.

SWITCHES

Switches are used to control or block the flow of electricity through a circuit. One method of describing switches indicates if they are normally open or normally closed. If no outside force acts on the switch, a normally open switch will block the flow of electricity. A normally closed switch will allow the current to flow. In wiring diagrams, switches are usually drawn in their normal position unless otherwise indicated.

Switches are also described by the number of poles and throws they have. Poles are the number of inputs and throws are the number of outputs. So a single-pole single-throw (SPST) switch has one input and one output (see **Figure 8**). A single-pole double-throw (SPDT) switch has one input and two possible outputs (see **Figure 9**). Anything above two outputs, such as a fan speed selector switch, would be a single-pole multi-throw (SPMT) switch (see **Figure 10**).

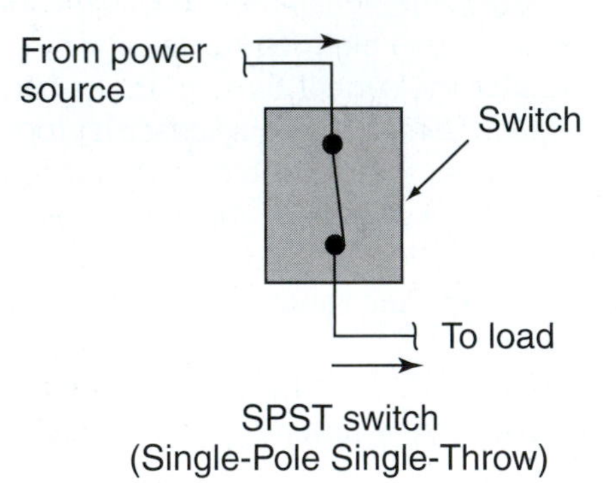

Figure 8. The SPST switch has one input and one output circuit.

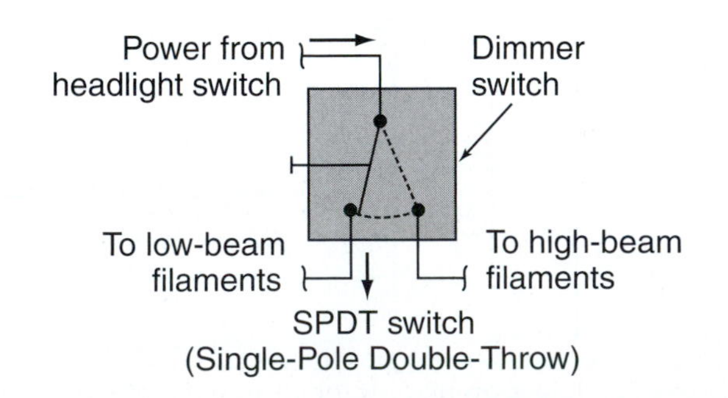

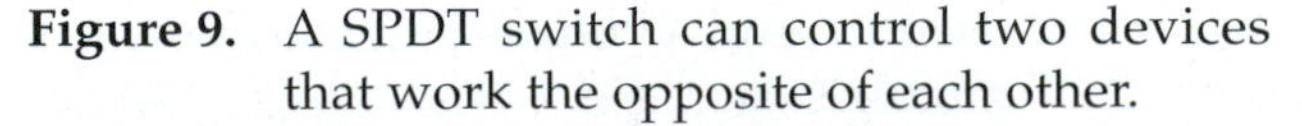

Figure 9. A SPDT switch can control two devices that work the opposite of each other.

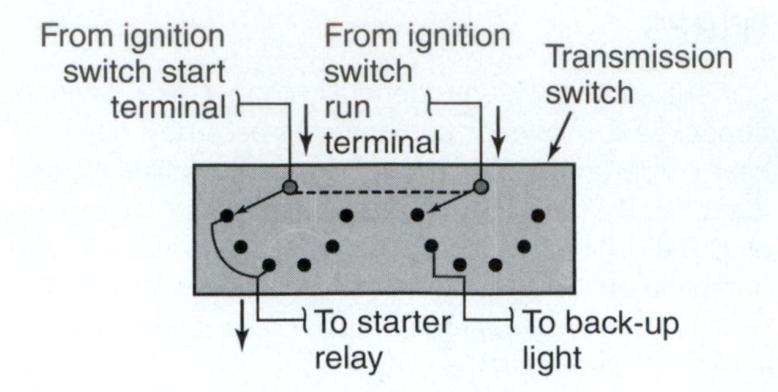

Figure 10. These ganged switches move together through multiple output positions.

At times, there are two or more switches that are controlling separate circuits that must move in unison. When one moves, they all move. This is often found in manual HVAC control heads in the mode selector switch. Separate switches control the blower enable circuit and the AC compressor circuit. When the operator moves the mode switch, both switches are moved at the same time to the selected mode. Switches that move together are called ganged switches and are represented on wiring diagrams by a dashed line connecting the switch arrows together.

RELAYS

A relay is an electromagnetic device in which a small current energizes a magnetic coil. The magnetism from the coil then moves an electric switch that can activate a circuit with a larger current. Thus, a small amount of current can control a large amount of current. Relays are used extensively in circuits where a computer is controlling an electric device. A good example of this is a modern electric radiator cooling fan circuit. The cooling fan motor can often draw as many as 20 amps. If a computer module tried to control this circuit directly, then all 20 amps would have to flow through the computer. This level of amperage is well above what most modules can handle and would burn up the module's internal circuits. **Figure 11** illustrates how a relay would be incorporated into the fan circuit. Now the only current flowing through the module is from the magnetic coil, which is usually less than half an amp.

The switch inside the relay is usually a single-pole single-throw type switch. This type of relay usually has four legs or terminals, the two coil legs in opposite corners and the two switch legs in the other two corners (see **Figure 12**). This allows the relay to be installed in a mounting socket in either direction and still function properly. Other relays have single-pole double-throw switches that use five legs and can only be installed one way. Most relays today have a mini-diagram molded into the side of the case, which identifies relay operation and terminal identification.

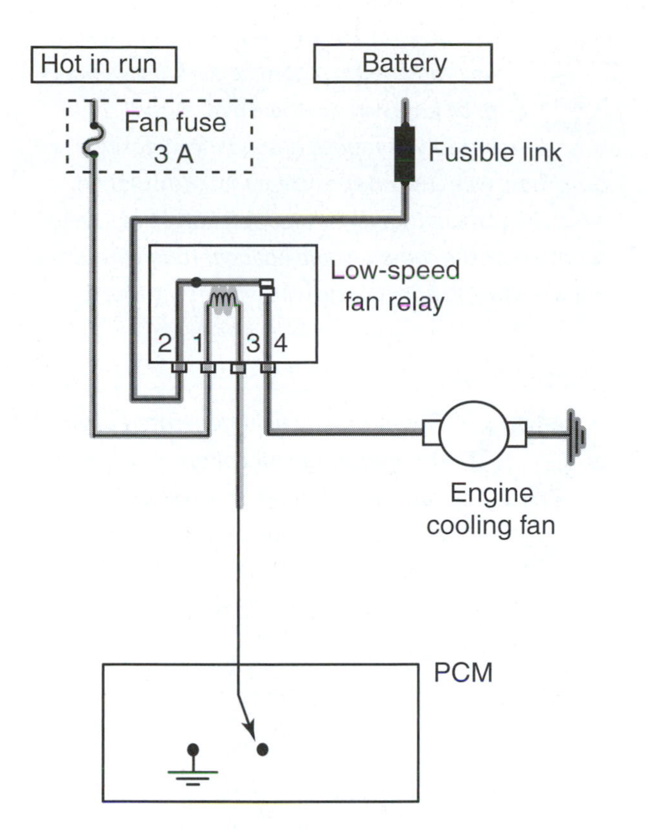

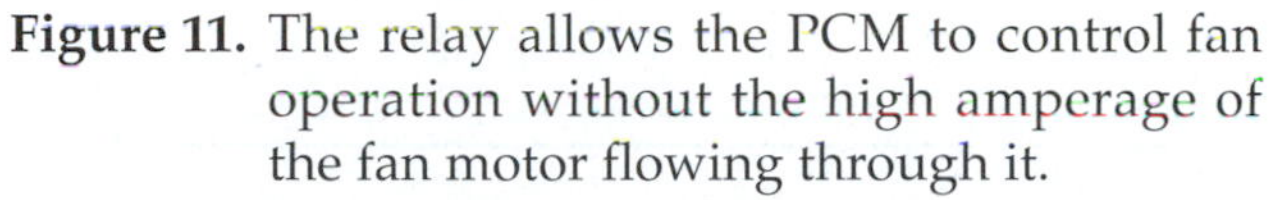

Figure 11. The relay allows the PCM to control fan operation without the high amperage of the fan motor flowing through it.

Figure 12. These four terminal relays can be installed in either direction without damaging the circuit operation.

SOLENOIDS

Solenoids (see **Figure 13**) are electromagnetic devices that when energized result in a movement. Solenoids consist of a coil with a rod through the center of it. The rod typically has a spring attached to it to keep it at one end of the rod's travel. When a current is passed through the coil, the magnetic field generated around the coil moves the rod to the other end of its travel. What makes the solenoid useful is the many devices that can be actuated by the moving rod. Such items as vacuum control valves, fuel injectors, and transmission shift valves are all examples of a working solenoid.

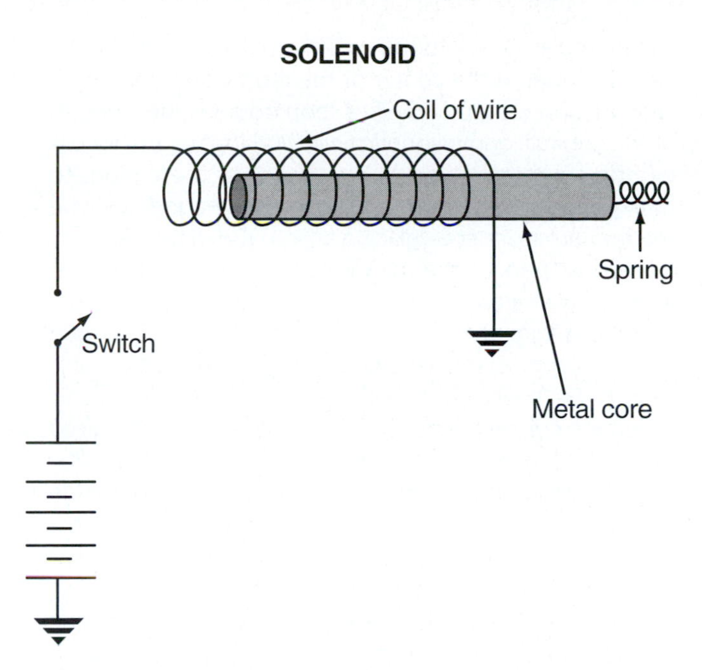

Figure 13. The magnetic field generated by the coil overcomes the spring tension to move the metal rod.

MAGNETIC COILS

Any time an electric current passes through a wire, a magnetic field is created around the wire. If this magnetic field is strong enough, it can be made useful. One way to strengthen the magnetic field is to apply more voltage to the circuit. This increases current flow that makes the magnetic field stronger. This has limitation in that too much current could burn up the wire.

Think about the magnetic field as making little circles of magnetism around the wire. If the wire is formed into a loop (see **Figure 14**), then the inside edges of the magnet

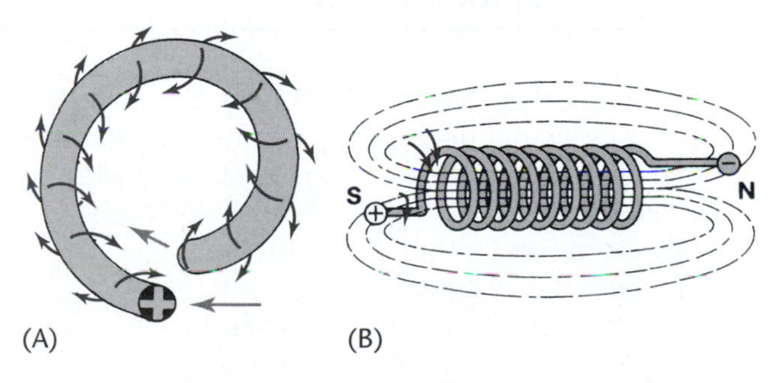

Figure 14. Forming a coil with the wire intensifies the magnetic field.

circles move closer together. This makes the magnetic field stronger in the center of the loop circle. Forming a second loop on top of the first loop doubles the strength of the original loop magnetic field. Working magnetic coils combine hundreds or even thousands of these loops to make extremely strong magnets. The magnet can be made even stronger by placing an iron rod in the center of the loops to focus the magnetic loops, called flux lines, into a smaller area.

The magnetic coil has a multitude of uses. It is used in relays, solenoids, sensors, and even ignition coils. A common HVAC application is on the compressor clutch coil. When the coil on the front of the compressor is energized inside the pulley assembly, it creates a strong magnetic field, transforming the entire hub assembly into a large magnet. The magnet pulls the clutch drive plate against the pulley assembly. The drive plate then turns the compressor shaft activating the compressor.

A coil is a load device. Despite the fact that it is just one long coil of wire, the length of the wire creates enough resistance that a current passing through the straight wire does not burn the wire up. Even the coil in a small relay may consist of several hundred feet of wire.

You Should Know *The wire used in magnetic coil loops appears to be bare wires as there is no plastic insulator wrapped around the metal wire core. These wires use an epoxy material as insulation, which gives them that slightly reddish tint. If there were no insulator, the current would not go through the loops but would jump from wire to wire where they touch.*

Interesting Fact *The magnetic field generated around an operating coil collapses when the current is shut off. This collapsing field generates its own voltage in the coil that can exceed 200 volts. That is why most coils have a clamping diode or resistor to help dissipate the voltage spike.*

Summary

- Understanding component operation is key to successful diagnostic work.
- Circuit protection devices such as fuses, circuit breakers, and fusible links block current flow when excess amperage is detected in the circuit they are protecting.
- Automotive wiring is stranded copper material to be flexible, durable, and economical.
- Resistors reduce current flow and drop voltage by changing electrical energy into heat.
- Fixed resistors are used to reduce current and voltage in a circuit.
- Variable resistors can be used to control a device or send an input signal to a control module.
- Switches are used to control or block the flow of electricity through a circuit.
- A relay is an electromagnetic device that allows a small current to control a large current.
- A solenoid uses a coil's magnetic field to move a rod that activates some device.
- Coils use the natural magnetic field surrounding a wire to do some sort of work.

Review Questions

1. Technician A says that fusible links are becoming less common on our modern vehicles. Technician B says that fuses are circuit loads that protect the circuit. Who is correct?
   A. Technician A only
   B. Technician B only
   C. Both Technician A and Technician B
   D. Neither Technician A nor Technician B

2. Technician A says that excess voltage always blows fuses. Technician B says that excess amperage always blows fuses. Who is correct?
   A. Technician A only
   B. Technician B only
   C. Both Technician A and Technician B
   D. Neither Technician A nor Technician B

3. Circuit breakers need not be replaced when exposed to excess current.
   A. True
   B. False
4. Solid state circuit breakers block current flow as long as there is a current applied to them.
   A. True
   B. False
5. Two technicians are discussing whether silver would make good automotive wiring. Technician A says that silver wires would have very good conductivity. Technician B says that there would be problems at the connectors. Who is correct?
   A. Technician A only
   B. Technician B only
   C. Both Technician A and Technician B
   D. Neither Technician A nor Technician B
6. Technician A says that resistors always produce heat when a current passes through them. Technician B says that there are few fixed resistor automotive applications. Who is correct?
   A. Technician A only
   B. Technician B only
   C. Both Technician A and Technician B
   D. Neither Technician A nor Technician B
7. Technician A says that some variable resistors change in response to movement. Technician B says that some variable resistors change in response to temperature. Who is correct?
   A. Technician A only
   B. Technician B only
   C. Both Technician A and Technician B
   D. Neither Technician A nor Technician B
8. Technician A says that rheostats have three wires. Technician B says that potentiometers have two wires. Who is correct?
   A. Technician A only
   B. Technician B only
   C. Both Technician A and Technician B
   D. Neither Technician A nor Technician B
9. Mechanical circuit breakers use a ______________ ______________ that bends when heated.
10. Technician A says that reversing the polarity on a solenoid causes the rod to move. Technician B says that an unprotected coil could damage sensitive electronics when it is turned off. Who is correct?
    A. Technician A only
    B. Technician B only
    C. Both Technician A and Technician B
    D. Neither Technician A nor Technician B
11. Technician A says that magnetic flux lines run parallel to the wire. Technician B says that coils are load devices. Who is correct?
    A. Technician A only
    B. Technician B only
    C. Both Technician A and Technician B
    D. Neither Technician A nor Technician B
12. Technician A says that a poor ground wire will not affect a potentiometer because it can ground through the moveable contact wire. Technician B says that a rheostat could be used in the dash illumination dimmer circuit. Who is correct?
    A. Technician A only
    B. Technician B only
    C. Both Technician A and Technician B
    D. Neither Technician A nor Technician B

Chapter 22 Compressor Control Circuits

Introduction

Compressor control circuits are designed with several concerns in mind. Primarily, they must activate the air conditioning compressor when the correct HVAC mode is selected on the control panel. Depending on the system design, the circuit may also control compressor activity to regulate the temperature of the evaporator core, protect the refrigeration system from excessively high pressures, or keep the system from operating when it is low on refrigerant, which would starve the compressor of lubricant. On modern cars, it also must be able to shut down the compressor under heavy engine load or overheating conditions.

COMPRESSOR CLUTCH COILS

The compressor clutch coil (see **Figure 1**) is a large electromagnet located within the compressor clutch assembly. It is made from one long continuous insulated wire wrapped around a center mounting hub. Each end of the wire has a terminal in the coil connector to attach to the compressor control circuit wiring. When a current is passed through the coil, the magnetic field it creates pulls the compressor drive plate up against the clutch pulley. The drive force from the engine is transferred to the drive plate that then turns the compressor. When the current to the coil is shut off, electromagnetic induction creates a voltage spike with a reverse polarity to the control circuit. If left unchecked, it could damage electronic controllers and processors throughout the vehicle. A clamping diode (see **Figure 2**) is installed in the wires leading to the coil in order to dissipate the spike.

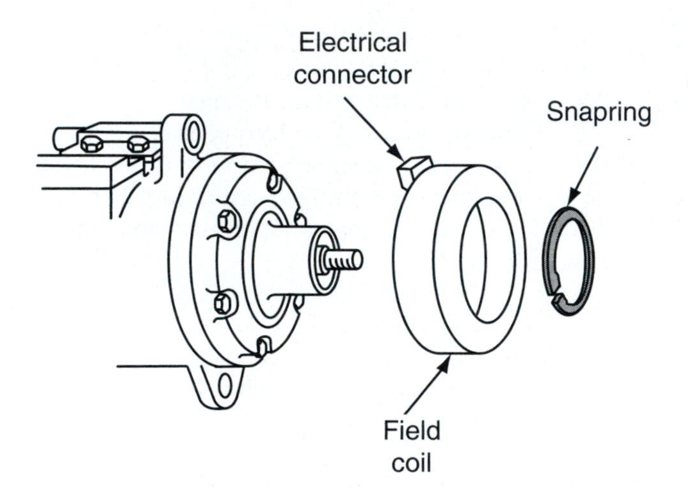

Figure 1. The clutch of field coil creates a strong magnetic field that bonds the clutch components together to activate the compressor.

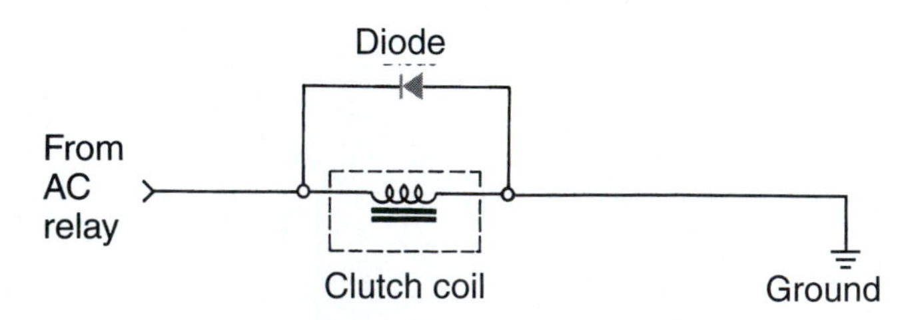

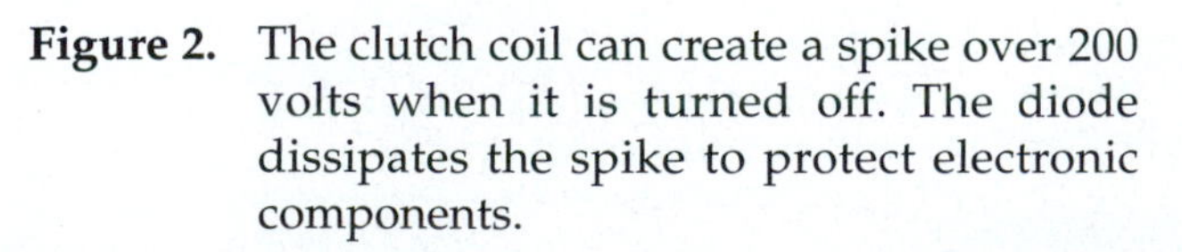

Figure 2. The clutch coil can create a spike over 200 volts when it is turned off. The diode dissipates the spike to protect electronic components.

In the early days of automotive air conditioning, the clutch coil spun with the pulley. The current passed through a brush set to complete the circuit.

SWITCHES

Switches, unlike sensors, have only two possible positions. They are either open, which blocks the flow of electricity, or closed, which allows electricity to flow. The switches used in compressor control circuits have some outside force acting on them that makes them operate. This force may be temperature, pressure, a magnetic field, or even a human finger. Whatever the force, the activity of the switch usually results in the compressor clutch being activated or deactivated.

Pressure Cycling Switch

A pressure cycling switch (see **Figure 3**) is found on many clutch cycling orifice tube (CCOT) systems. It is designed to keep the evaporator core at its correct temperature. The switch accomplishes this by turning the compressor off when the low-side pressure drops below a certain level. The compressor then remains off until the pressure rises above a preset level, at which point it reactivates. This keeps the pressure in the evaporator within an average controlled range. Chapter 6 explains the direct relationship between pressure and temperature in a saturated system. So, by controlling the low-side pressure, the pressure cycling switch controls the evaporator temperature.

An additional benefit of this switch is low refrigerant protection. If the refrigerant level drops too low, then the low-side pressure also will drop. The switch will then shut down the compressor, protecting it from lubricant starvation.

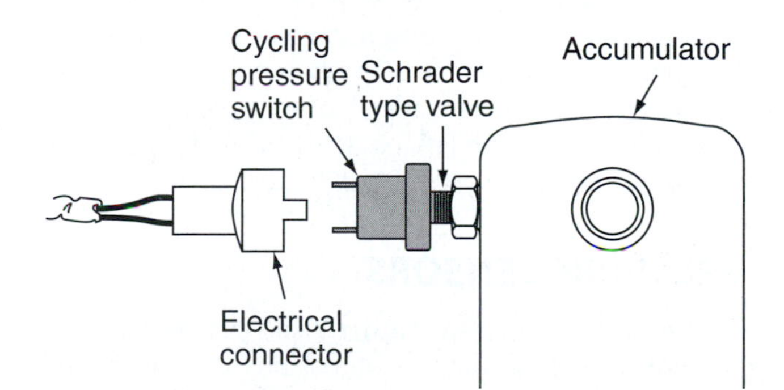

Figure 3. The pressure cycling switch maintains an average temperature in the evaporator to keep it just above freezing.

The pressure cycling switch is usually mounted on the accumulator tank, screwing onto a service port and sealed with an O-ring. The opening and closing pressures vary with the system and refrigerant used. Typical switch opening pressures are in the 22 psi to 26 psi range, whereas closing pressures tend to run from 42 psi to 48 psi. Some switches have an adjustment screw between the electrical connector blades that can be used to make small changes in the settings. Moving the screw raises or lowers both the opening and closing switch pressures.

Thermal Cycling Switch

The thermal cycling switch (see **Figure 4**) functions similarly to the pressure cycling switch, but it relies on temperature to operate instead of pressure. It utilizes a capillary tube filled with gas to sense temperature. The capillary tube is inserted into the fins of the evaporator core, or in some cases into an access port in the discharge tube of the evaporator. Gas within the tube expands and contracts with the temperature of whatever the tube is inserted into. The gas moves a flexible diaphragm that opens and closes a switch. The switch and diaphragm are calibrated so that when the gas reaches 32 degrees F, it opens the switch. This then shuts down the compressor, keeping the evaporator from freezing up. As the gas in the tube warms up, the switch closes, reactivating the compressor. One disadvantage of the thermal switch is that it cannot detect low refrigerant levels. Some other device must be used to protect the system from this problem.

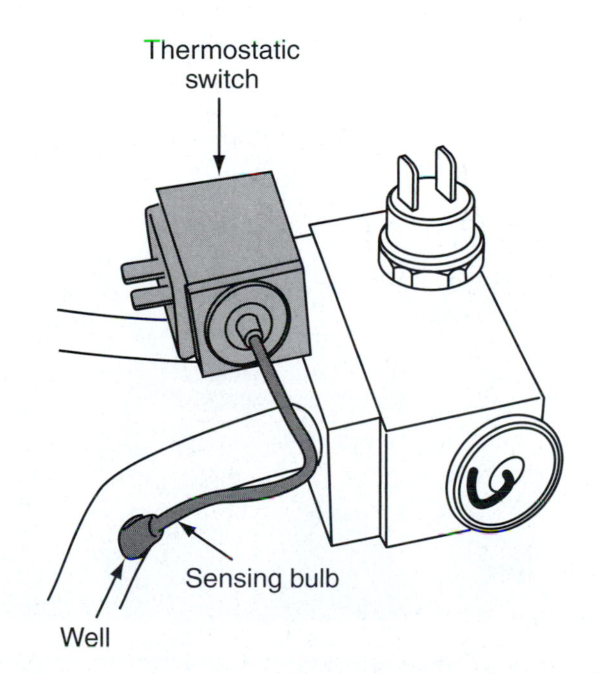

Figure 4. The thermal cycling switch directly measures the refrigerant temperature to cycle the compressor clutch.

High-Pressure Cutoff Switches

High-pressure cutoff switches (see **Figure 5**) are used to protect the system from dangerously high pressures. Certain conditions, such as an overheated condenser or a restriction in the refrigerant flow, could result in pressures exceeding 600 psi. This could result in a hose rupturing or a component exploding. To prevent this, the high-pressure switch opens somewhere in the 400 psi to 450 psi range.

Most systems have a high-pressure relief valve in the refrigerant system as well. In the event of dangerously high pressure, the relief valve physically opens and allows refrigerant to vent into the atmosphere until the pressure is reduced. The venting of refrigerant is an environmental concern. High-pressure switches are designed to shut down the compressor at a pressure less than the pressure relief valve discharge pressure. This prevents undesirable refrigerant releases. Many older vehicles are not equipped with high-pressure switches. Federal law requires the installation of a high-pressure switch if one of these older vehicles is being retrofitted to another refrigerant.

High-pressure switches may be mounted in several locations. They are found on the liquid line, the discharge line, and screwed into the condenser itself. If located in one of these places, the switch usually has two wires running to it, one for each side of the switch. Some compressors have a cavity that accepts a high-pressure switch. Some of these applications also have two wires running to the switch. However, some switches have only one wire and the switch itself is case grounded to the compressor.

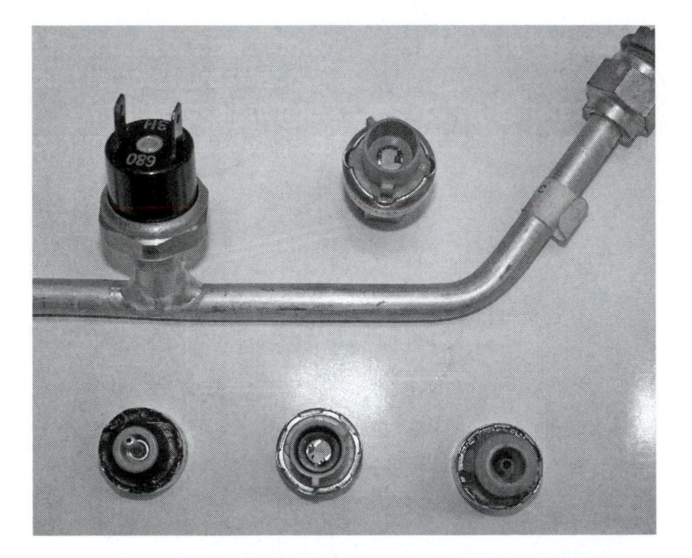

Figure 5. These high-pressure switches are responsible for protecting the system from dangerously high pressure and activating the condenser cooling fan.

Figure 6. This compressor-mounted switch protects the system from low refrigerant charges and cold ambient temperatures.

High-Side Low-Pressure Switch

Some designs use a compressor control switch that opens under low-pressure conditions on the high side of the refrigerant system. This switch is referred to as a high-side low-pressure switch (see **Figure 6**). It can have two purposes. First, it is used as a low refrigerant indicator. The high side of the refrigerant system is at its lowest pressure before the compressor engages. If the system is low on refrigerant, then the pressure will be low and the switch will open, stopping the compressor from ever engaging. If the compressor does start, the high-side pressure will go up, keeping the compressor on. Second, it can keep the compressor from activating if the ambient temperature is too low. It is undesirable to activate the compressor if it is too cold outside, typically around 45 degrees F. If the ambient temperature is cool enough, it will lower the high-side pressure below the switch cutoff point, keeping the compressor from activating.

High-Pressure Fan Switch

For many years, vehicles with electric cooling fans automatically activated the cooling fan whenever the AC compressor was activated. Although simple in design, the fan was not always needed, which wasted fuel and increased emissions. A high-pressure fan switch activates the condenser cooling fan only when needed. The switch contacts close when refrigerant high-side pressure reaches 300 psi to 320 psi. The fan remains in operation until the pressure drops to the 230 psi to 250 psi range.

PRESSURE SENSORS

Pressure sensors (see **Figure 7**) are used primarily on the high side of the system but may be found on the low side as well. These sensors are three wire potentiometers that vary their resistance in response to pressure. They are always used in systems that use a control module to activate compressor control. The module uses the sensor to

Figure 7. The three-wire pressure sensor gives the control module the actual operating pressure.

measure the actual operating conditions in the system and control devices such as the compressor and condenser cooling fan.

COMPRESSOR CIRCUITS

Like everything else on the vehicle, the compressor control circuit has evolved significantly over the years. What began as a simple on/off switch circuit has become a microprocessor-based system capable of adjusting performance for a variety of variables. The following paragraphs will examine circuits from the simplest to the more complex. Each circuit will have diagnostic hints to help in working with the circuit.

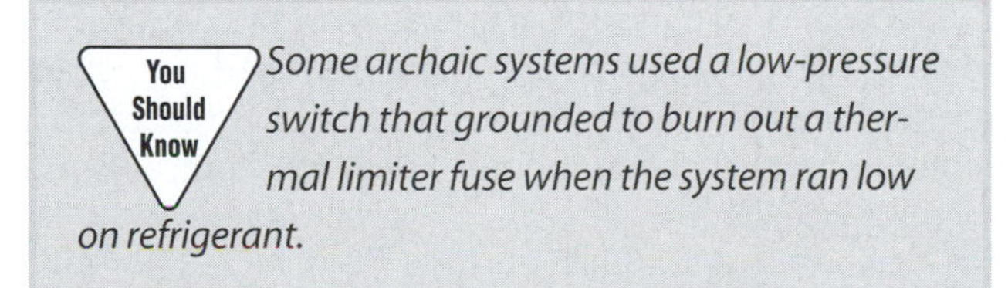

Some archaic systems used a low-pressure switch that grounded to burn out a thermal limiter fuse when the system ran low on refrigerant.

Basic clutch cycling and thermostatic expansion valve circuits were used primarily up to 1981, when various versions of powertrain control modules (PCM) became prevalent on many makes of automobiles. Computers were incorporated into engine controls in order to improve fuel economy and decrease emissions. If the AC compressor could be deactivated under heavy engine load circumstances, both fuel economy and emissions would be improved. Therefore, control of the compressor became a function of a computer module: either the PCM or one that is in communication with it.

Basic Clutch Cycling Orifice Tube Circuit

The basic clutch cycling orifice tube or CCOT circuit (see **Figure 8**) relies on a cycling switch to control compressor operation. The cycling switch may be either pressure or thermal in design. Operation of the circuit is identical in either case. The circuit activation switch is usually incorporated in the mode selector switch. Once the compressor switch is activated in the control head, the cycling switch energizes the compressor clutch coil in response to the refrigeration system pressure or temperature, cycling the clutch to maintain the proper evaporator temperature.

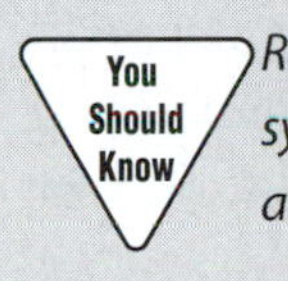

Rapid cycling of the compressor clutch on systems using a pressure cycling switch is an indication of low refrigerant level.

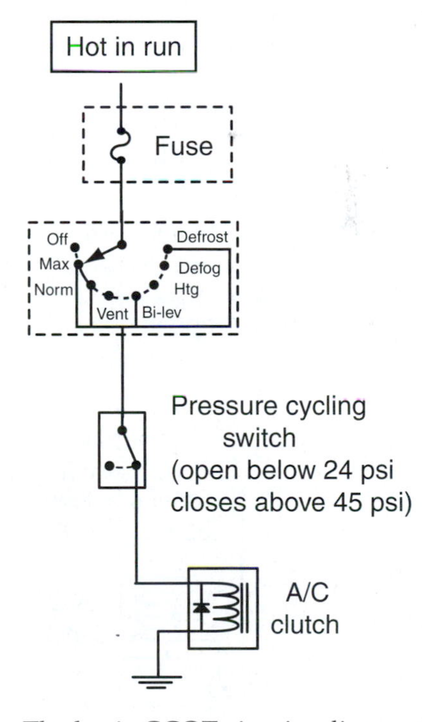

Figure 8. The basic CCOT circuit relies on the pressure cycling switch for compressor control.

Basic Thermostatic Expansion Valve Circuit

The basic thermostatic expansion valve or TXV circuit (see **Figure 9**) does not need to directly control the temperature of the evaporator. The TXV controls refrigerant flow. The circuit does need to protect the compressor from low refrigerant levels and from the evaporator freezing up. A thermal cutoff switch, which is almost identical to a thermal cycling switch, is placed in the evaporator fins to shut off the compressor if the evaporator temperature reaches 32 degrees F. This switch is different from the cycling switch in that it is a backup to the TXV and not the primary control device. The circuit also has a low-pressure switch to deactivate the clutch when refrigerant levels reach too low a level.

CCOT with Computer Control

One new component added to computer-controlled compressor circuits is a relay that is controlled by the computer. An early version of the circuit (see **Figure 10**) left the circuit much as it was before, except for how the power is applied to the circuit. Instead of the control head supplying power to the circuit itself, it sends an AC request signal to the PCM. Programming inside the PCM

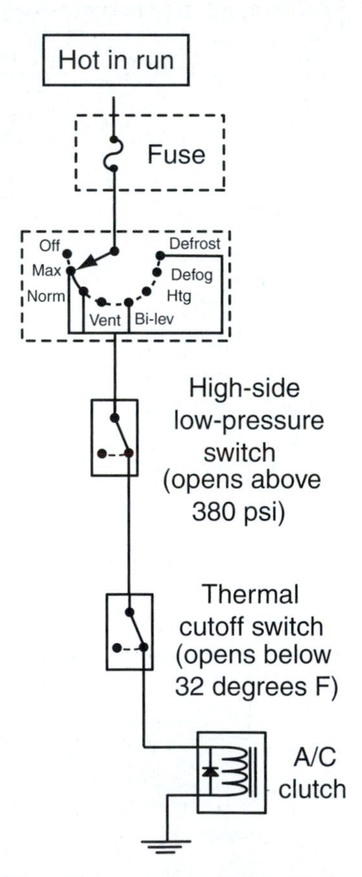

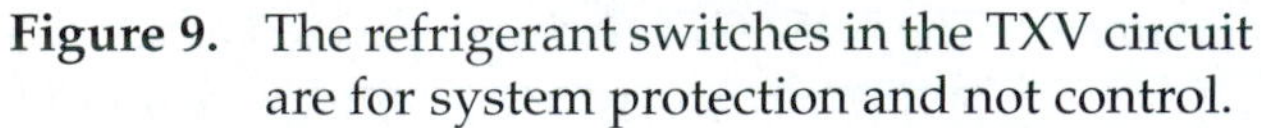

Figure 9. The refrigerant switches in the TXV circuit are for system protection and not control.

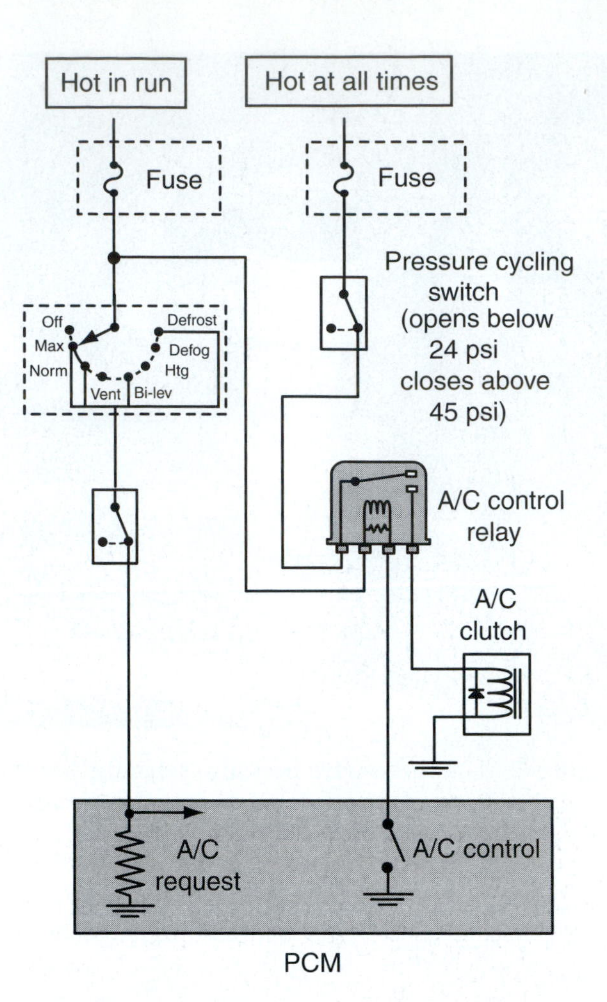

Figure 10. The relay allows for computer control of the compressor based on driving conditions.

determines if engine conditions are right for activating the AC compressor. If everything is correct, the PCM closes an internal switch that supplies a ground path for the coil side of the relay. The relay switch then closes and supplies power to the compressor clutch circuit. Notice that this circuit now incorporates a high-pressure cutoff switch to protect the system from high-pressure damage.

Later versions of this circuit (see **Figure 11**) moved the switches from the actual compressor clutch circuit to the AC request circuit. The actual operation of this design is no different from the version previously described. A feedback circuit is included from the compressor coil circuit to the computer to let the computer know the compressor is actually operating.

Variable Displace with Pressure Switch

The variable displacement compressor refrigeration system is a constant clutch design, which means there is no need to cycle the clutch. The circuit (see **Figure 12**) sends an AC request from the control head to the PCM. The PCM also receives a signal from the high-pressure

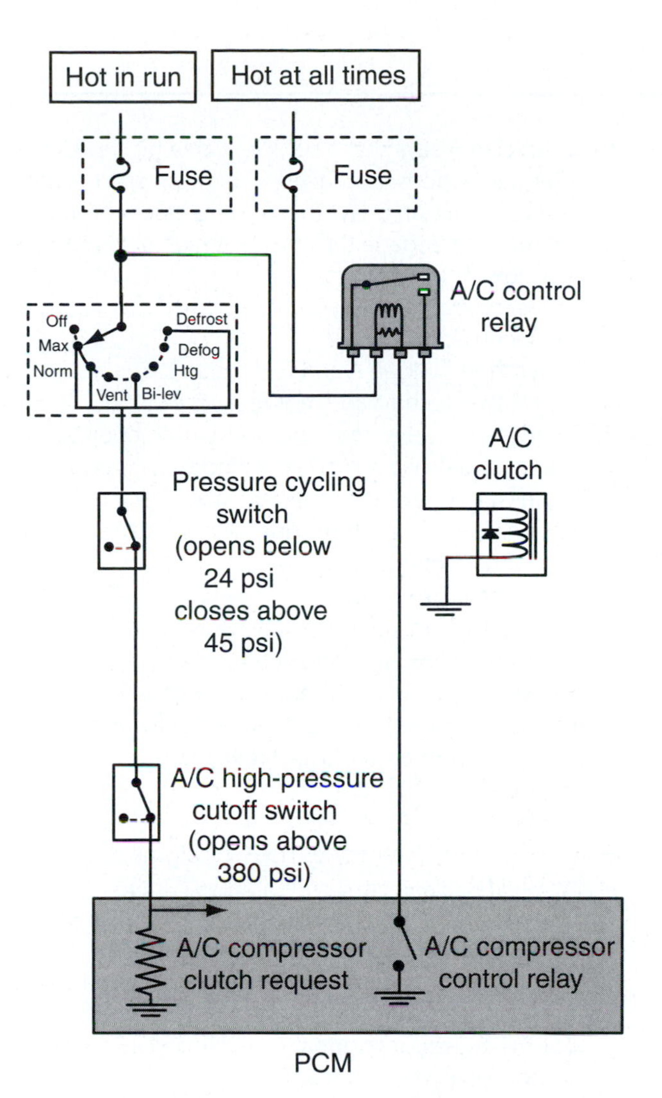

Figure 11. The pressure sensors on the AC request circuit allows for more precise system controls.

sensor indicating the pressure on the high side of the refrigeration system. If the PCM determines the refrigerant

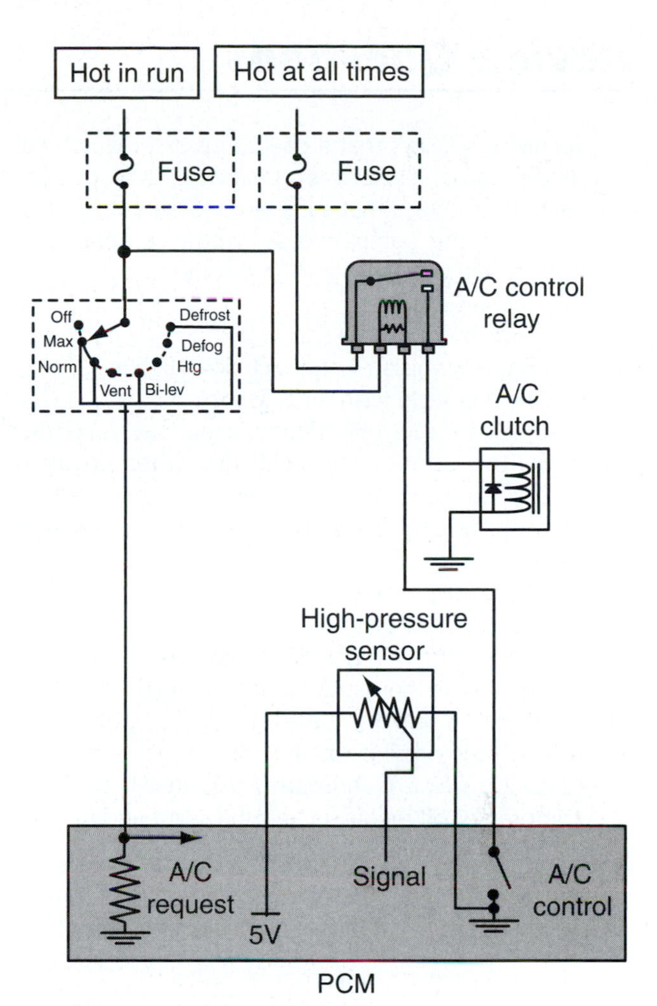

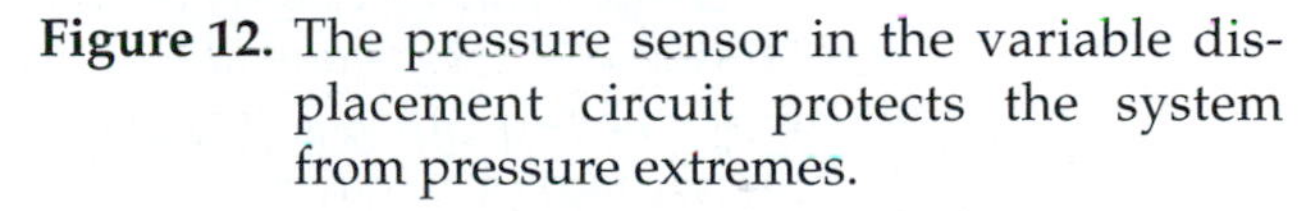

Figure 12. The pressure sensor in the variable displacement circuit protects the system from pressure extremes.

is not low and the ambient temperature is warm enough, the PCM then activates the AC relay, activating the compressor.

Summary

- Compressor control circuits must ensure operating conditions are acceptable before activation of the compressor.
- Compressor clutch coils generate large magnetic fields to engage the clutch components.
- Switches are on or off; sensors can create a range of readings.
- Pressure and thermal cycling switches are the primary refrigerant control devices on orifice tube systems.
- High-pressure cutoff switches protect the components by disengaging the compressor when dangerously high pressures are present.
- Low-pressure switches protect the compressor from lubricant starvation if the refrigerant level is low.
- High-pressure fan switches operate the condenser cooling fan only when it is needed to lower the high-side pressure, saving fuel and reducing emissions.
- Compressor circuits have evolved from simple on/off controls to sophisticated electronic circuits.
- CCOT circuits rely on a pressure or thermal cycling switch for their primary refrigerant flow control.
- Computers now control compressor operation to improve fuel economy and emissions.

Review Questions

1. Technician A says that a bad compressor clutch coil diode could cause the engine to not operate. Technician B says that sufficient refrigerant must be present in the system in order for the compressor to operate without damage. Who is correct?
   A. Technician A only
   B. Technician B only
   C. Both Technician A and Technician B
   D. Neither Technician A nor Technician B
2. Technician A says that friction is the main force that drives the compressor clutch disc when engaged. Technician B says that switches are used to tell the compressor control computer the exact pressure in the system. Who is correct?
   A. Technician A only
   B. Technician B only
   C. Both Technician A and Technician B
   D. Neither Technician A nor Technician B
3. Technician A says that a pressure cycling switch turns the AC compressor on and off at the evaporator freeze-up point. Technician B says that it turns the compressor off below its freezing point and back on well above the freeze-up point. Who is correct?
   A. Technician A only
   B. Technician B only
   C. Both Technician A and Technician B
   D. Neither Technician A nor Technician B
4. Technician A says that a pressure cycling switch can protect the compressor from low refrigerant levels. Technician B says that a thermal cycling switch cannot protect the compressor from low refrigerant levels. Who is correct?
   A. Technician A only
   B. Technician B only
   C. Both Technician A and Technician B
   D. Neither Technician A nor Technician B
5. Technician A says that high-pressure cutoff switches deactivate the compressor when pressure rises over 600 psi. Technician B says that these switches are always located in the wire leading to the clutch coil. Who is correct?
   A. Technician A only
   B. Technician B only
   C. Both Technician A and Technician B
   D. Neither Technician A nor Technician B
6. Technician A says that a vehicle with a high-pressure relief valve does not need a high-pressure cutoff switch. Technician B says that a compressor control circuit may not engage the coil if the high-side pressure is too low. Who is correct?
   A. Technician A only
   B. Technician B only
   C. Both Technician A and Technician B
   D. Neither Technician A nor Technician B
7. Technician A says that the condenser cooling fan should always be on if the AC compressor is engaged. Technician B says that the two wires attached to a pressure sensor are the signal wire and the ground wire. Who is correct?
   A. Technician A only
   B. Technician B only
   C. Both Technician A and Technician B
   D. Neither Technician A nor Technician B
8. A thermal limiter fuse is a new device for protecting the compressor from a lack of refrigerant.
   A. True
   B. False
9. A TXV design system does not typically need to cycle the clutch coil.
   A. True
   B. False
10. Technician A says that the PCM supplies power to the AC relay to activate the compressor. Technician B says that the AC request signal is generated in the HVAC control head. Who is correct?
    A. Technician A only
    B. Technician B only
    C. Both Technician A and Technician B
    D. Neither Technician A nor Technician B
11. Technician A says that a variable displacement system will not engage if the ambient temperature is 42 degrees F. Technician B says that the system will engage, but the compressor will cycle frequently. Who is correct?
    A. Technician A only
    B. Technician B only
    C. Both Technician A and Technician B
    D. Neither Technician A nor Technician B

Chapter 23 Blower Control Circuits

Introduction

The blower control circuit is designed to provide adequate airflow through the air handling system, ensuring proper heat exchange and passenger comfort. Under heavy heat load conditions, the air requirement might be quite large, requiring the blower motor to move large volumes of air. Electrically, this equates to high amperage draws. When blower speed is set to high, some models may run as high as 20 amps.

Although manual blower motor circuits are designed with three, four, and five available fan speeds, this chapter will use a common four-speed example. Other combinations are similarly designed but use a different number of resisters. Some automatic systems use pulse-width modulated (PWM) control to provide an infinite number of blower speeds.

BLOWER MOTORS

A direct current blower motor drives the HVAC fan. The motor itself has no provision for speed control. Speed is controlled solely by the amperage that reaches the motor through the resistors or pulsed through the power module. Some motor designs have two terminals, which are for power and ground wires. Others have one power wire and are case grounded. Because many HVAC housings are plastic, a ground terminal is screwed to the mounting flange for a ground wire.

SWITCHES

If the power to the blower circuit is activated by the mode selector control, a switch in the mode assembly transfers power to the circuit in all mode positions but OFF. The blower speed switch (see **Figure 1**) has one input, but it may have as many as five outputs. The number of outputs depends on the number of fan speeds, one for each speed.

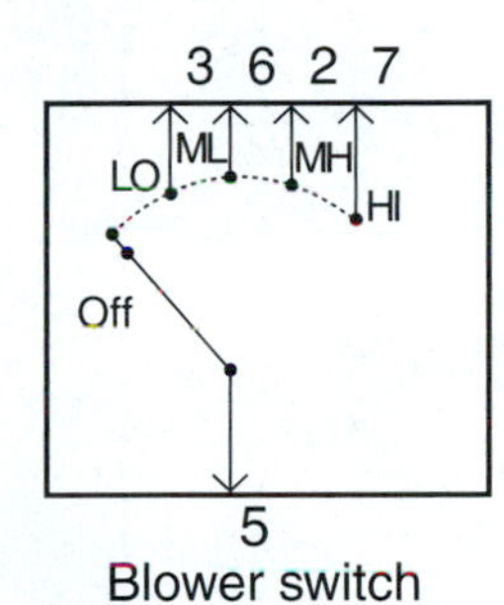

Figure 1. The blower speed selector is a single-pole multi-throw switch.

RESISTORS

Almost all manual blower circuits use a stepped resister. **Figure 2** illustrates two common types of stepped resistors: the wire coil and the printed circuit. Although visibly quite different, functionally they are identical. Both must be mounted so that the resistor elements protrude into the air distribution plenum. The heat generated by the resistors is removed by the airflow in the plenum. Some designs incorporate a thermal fuse on the output terminal of the resistor assembly (see **Figure 3**). Excessive heat will melt the fuse, which will stop all blower speeds that go through the resistor assembly.

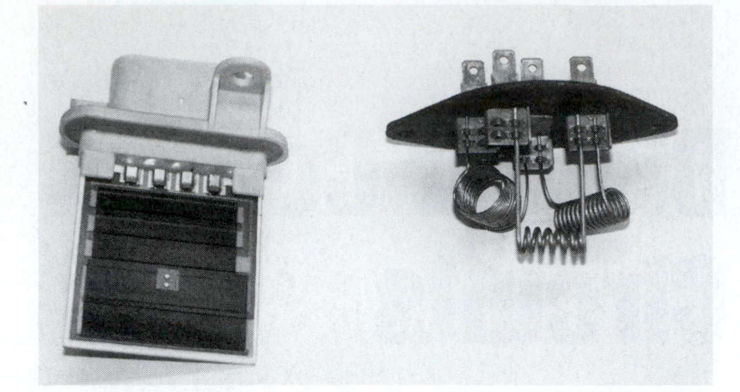

Figure 2. These stepped resistors have multiple entry contacts to vary resistance for the blower motor circuit.

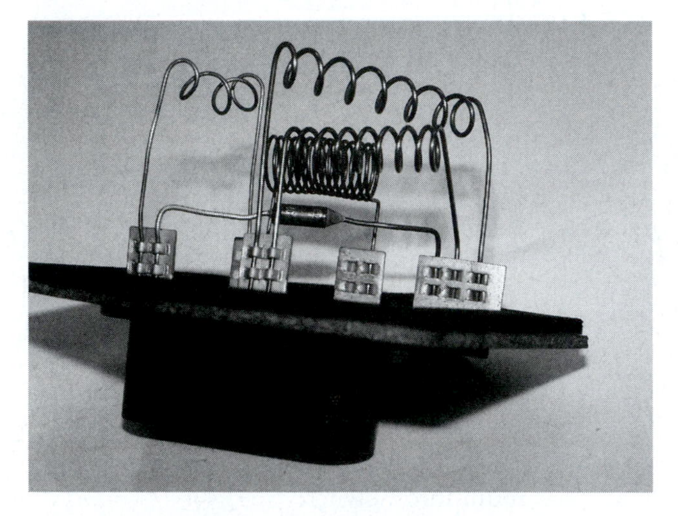

Figure 3. The thermal fuse in the center of the picture will melt if excess heat is generated by the resistors.

BLOWER CIRCUITS

Blower circuits fall into two main design groups: resistor control and pulse-width modulated. Resistor circuits are either manual switch design or electronically controlled relays. Both designs rely on directing the current through a series of resistors, the slower the blower speed desired, the more resistors the current is routed through. Pulse-width modulated systems use a power module to cycle on and off full battery voltage to the blower motor. The following circuit descriptions are typical of their design. Variations from system to system will be found.

Basic Blower Motor Circuit

The basic blower motor control circuit (see **Figure 4**) is first switched at the mode control switch where power

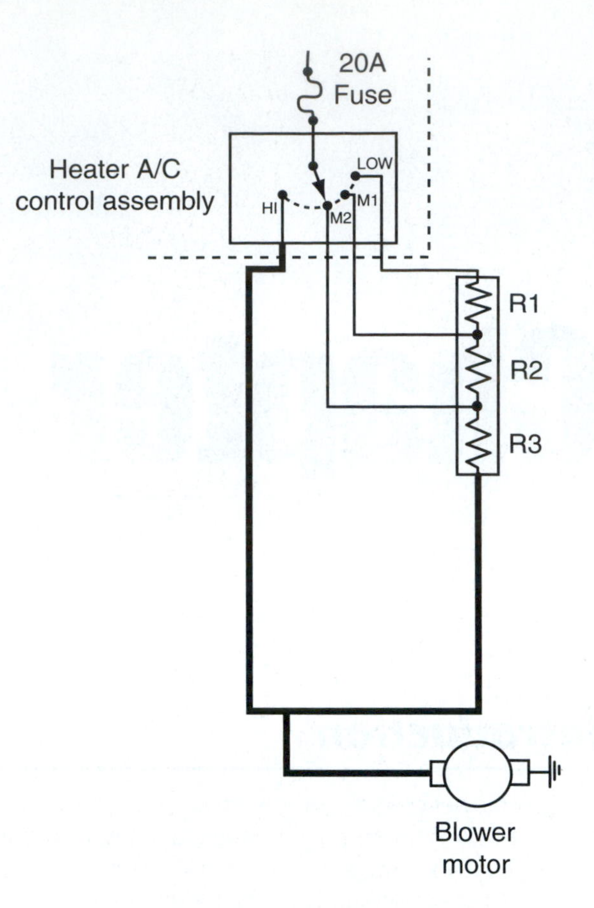

Figure 4. This basic blower motor control circuit was the industry standard for years.

to the remainder of the circuit is passed in all modes except OFF. From the mode switch, the circuit then proceeds to the fan speed selector switch. Depending on operator input, the current can then travel through one of three wires to the blower resistor assembly. On the LOW-speed setting, the current must pass through three resistors before reaching the blower motor. On medium-1, it passes through two resistors and one resistor on medium-2. On high, the circuit is a direct path to the blower motor. The circuit then proceeds to ground after the blower motor.

Ground Side Switching Circuit

One problem with the basic blower motor circuit is the high amperage that passes through the circuit when the motor is on HIGH. The amperage passing through the contacts of the speed selector switch produces heat that can be damaging to the contacts. In addition to the heat, the full charging system voltage at the switch contacts can produce arching when the selector switch is turned from high speed to a lower setting. This arching is like a tiny torch on the switch contact surface, cutting away at the metal.

Interesting Fact

Air is a poor conductor of electricity. In order for electricity to jump across the air gap of a spark plug, large amounts of voltage are required. Many vehicles use 40,000 or more volts to create the spark, but the current flow at that voltage is .001 amps or less. High voltage is required to jump an air gap, not amperage.

The heat and arching have resulted in rapid switch failure in many vehicles. Placing the blower motor directly after the fuse in the circuit (see **Figure 5**) greatly improves switch reliability. Now the switches are located on the ground side of the blower motor. When high blower speed is selected, there is no change in the amperage passing through the switch compared to the basic blower design described earlier. However, because it is on the ground side of the motor, there is practically no voltage at the switch. This eliminates the arching when the switch is moved, thus improving the switch reliability.

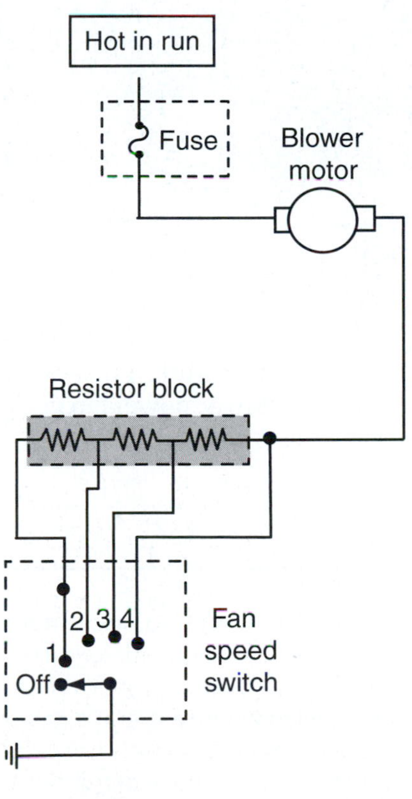

Figure 5. Putting the blower load first in the circuit eliminates most of the voltage at the switch when in HIGH.

You Should Know

When a switch fails from excess heat, the electrical connector that plugs into it is often damaged as well. Inspect and replace the connector and terminals when there is any sign of damage.

High-Speed Blower Relay Circuit

Another method of improving blower selector switch reliability is not to pass the large current through the switch when high blower speed is requested. This is accomplished through the use of a high blower speed relay circuit (see **Figure 6**). The single-pole double-throw switch in the relay allows two different power sources to be attached to the power wire leading to the blower motor. When the speed selector switch is in any position besides HIGH, the relay is not activated. This allows the normal current flow from the stepped resister to travel through the normally closed switch contacts in the relay to the blower motor. The blower motor still operates as before.

When the fan speed selector switch is placed in HIGH, the current does not go to the blower motor. Instead, a small amount of amperage travels to the high blower relay

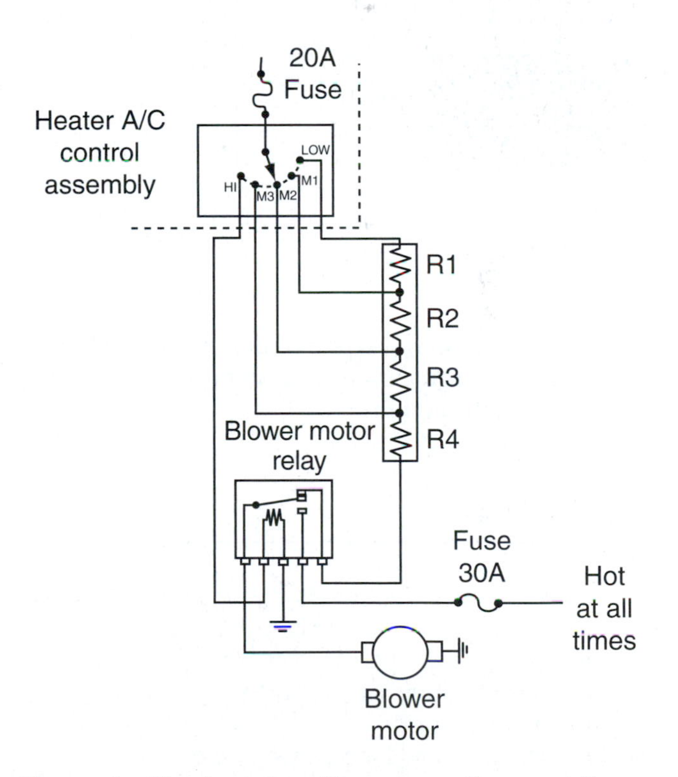

Figure 6. The fan relay allows a small current flowing through the switch to control the large current to the fan motor when in HIGH.

coil. The relay switch then moves to allow battery voltage to flow through a fuse directly to the blower motor. Because the high amperage does not travel through the switch, its life span is increased.

Relay Based Circuit

Manual systems that have electronic control heads and automatic systems cannot utilize manual switches for blower speed control. These systems often utilize multiple relays to control blower motor speed. The electronic control head can easily provide a ground path to activate the proper relay to achieve the desired fan speed. The circuit (see **Figure 7**) still utilizes a number of resistors to achieve the required resistance to regulate blower speed.

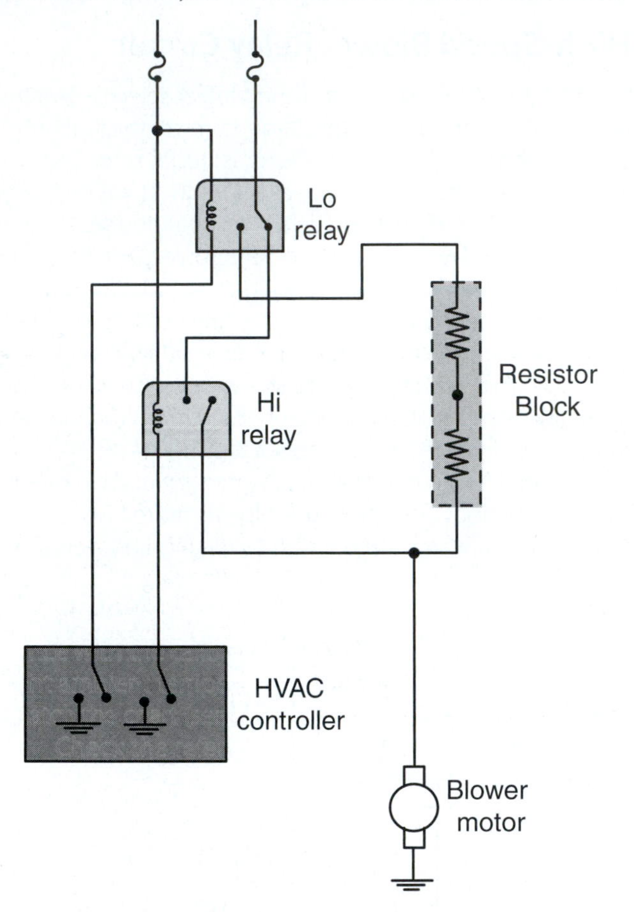

Figure 7. This relay based circuit may be used in an electronic manual system or a fully automatic system.

Power Module Circuit

Most automatic temperature control systems use a power module (see **Figure 8**) to control blower speed. The HVAC controller sends a digital signal to the power module indicating the desired blower speed. The power module then pulse-width modulates current directly from battery power to the blower motor. It varies motor speed by cycling the current from full on to off in fixed time durations. A full description of PWM is included in Chapter 28.

Figure 8. The power module pulse-width modulates the blower power circuit for infinite blower speeds.

Summary

- Blower control circuits use either resistors or pulse-width modulation to vary the current to the blower motor.
- In order to move the large volume of air required to maintain passenger comfort, blower motors may draw upward of 20 amps.
- Blower motors are direct current motors.
- Stepped resistors are used in many manual blower circuits.
- Ground side switching and high blower relays are two methods used to improve fan speed switch durability.
- The blower motor is installed before the switch in a ground side switching blower motor circuit.
- Blower motor drive current passes through the relay and not the switch when blower speed is set to HIGH in high blower relay circuits.
- Relay based circuits pass none of the motor drive current through the control head.
- The power module cycles current to the blower motor on most automatic temperature control systems.

Review Questions

1. Technician A says that high amperage is harmful for blower speed switch contacts. Technician B says that the amperage causes pitting on the contact surfaces. Who is correct?
   A. Technician A only
   B. Technician B only
   C. Both Technician A and Technician B
   D. Neither Technician A nor Technician B
2. Technician A says that blower motors are grounded through the plenum case. Technician B says that variable resistors are used to control blower speed. Who is correct?
   A. Technician A only
   B. Technician B only
   C. Both Technician A and Technician B
   D. Neither Technician A nor Technician B
3. Technician A says that a three-speed blower circuit will use two resistors. Technician B says that some resistors are built on printed circuits. Who is correct?
   A. Technician A only
   B. Technician B only
   C. Both Technician A and Technician B
   D. Neither Technician A nor Technician B
4. Technician A says that the thermal fuse protects the circuit from excess amperage flow in high speed. Technician B says that the high blower relay has five terminals. Who is correct?
   A. Technician A only
   B. Technician B only
   C. Both Technician A and Technician B
   D. Neither Technician A nor Technician B
5. Two technicians are discussing a high-speed blower relay with a bad coil. Technician A says that the blower will only work on high speed. Technician B says that the blower will work on all speeds except high. Who is correct?
   A. Technician A only
   B. Technician B only
   C. Both Technician A and Technician B
   D. Neither Technician A nor Technician B
6. Technician A says that relay based systems use one relay for each speed. Technician B says that the power module directs battery power to the blower motor. Who is correct?
   A. Technician A only
   B. Technician B only
   C. Both Technician A and Technician B
   D. Neither Technician A nor Technician B
7. Blower motors run faster when more amperage is applied to them.
   A. True
   B. False
8. The mode switch can turn the blower motor off on some vehicles.
   A. True
   B. False
9. Technician A says that most automatic temperature control systems use relay based blower speed control circuits. Technician B says that enough voltage can cause electricity to jump across an air gap. Who is correct?
   A. Technician A only
   B. Technician B only
   C. Both Technician A and Technician B
   D. Neither Technician A nor Technician B
10. ________________ switching allows electrical controls devices to be more reliable.

Chapter 24 Electrical and Electronic System Diagnosis

Introduction

Diagnosing electrical and electronic faults requires a structured technique. The use of published diagnostic routines can be extremely helpful in organizing a technician's diagnostic and repair procedures. Diagnosis of electrical circuits pertains to systems that do not contain semiconductors or processors. Problems with these circuits usually involve getting voltage, current, and ground to the right places at the right times. Using a DVOM to trace the flow of electricity from its source, through the wires and components of the circuit to ground, is a critical skill that must be mastered. In order to do this, the technician must understand the principles of electricity (see Chapter 20).

Applying the skills used for electrical diagnosis is equally needed for doing electronic diagnosis on processor-based systems. The technician also must be able to understand what the processor is "thinking." Thinking means knowing how the processor will control the system it is overseeing based on the inputs the processor is receiving. The use of a scan tool greatly aids in communicating this information to the technician.

CIRCUIT FAULTS

All circuit faults fall into one of three categories: opens, shorts, and unwanted resistance. Of the three, opens are probably the most common. An open (see **Figure 1**) is a break in the circuit that does not allow current to flow. A switch that is not closed could be considered an open in the circuit. However, when the term "open" is used in reference to a problem, it is more likely to be a broken wire or a component like a light or fuse that has melted the electrical path. When a circuit has an open, the load is not working.

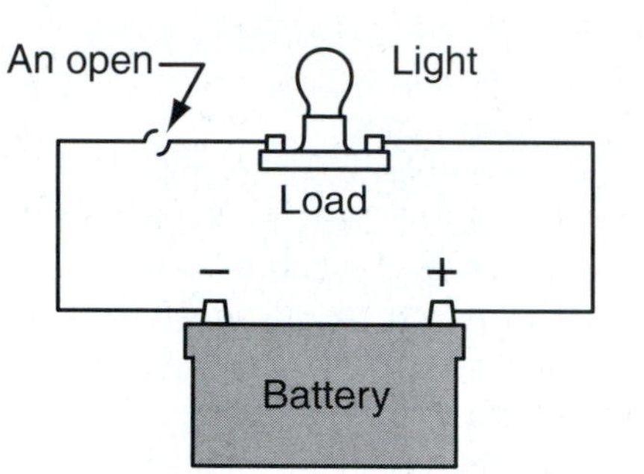

Figure 1. An open in a circuit blocks all current flow.

A short occurs when the electrical current is flowing somewhere it is not supposed to go. The electricity is taking a shortcut on its path. There are two types of shorts. One type is a short to power. **Figure 2** shows two separate circuits with a short to power between them. When either switch is closed, both load devices are activated. Shorts to power are often caused by two wires melting or rubbing together and are indicated by loads working when they are not supposed to. The other type is a short to ground. These shorts occur when a wire that normally has voltage on it touches a ground (see **Figure 3**). A pinched wire or a wire that has rubbed against a body panel or engine for too long often causes this. Shorts to ground are often noted by a blown fuse.

The third type of circuit fault is unwanted resistance (see **Figure 4**). Unwanted resistance can be caused by a dirty connection or corrosion on a contact. A dirty battery cable is one of the most common examples of this

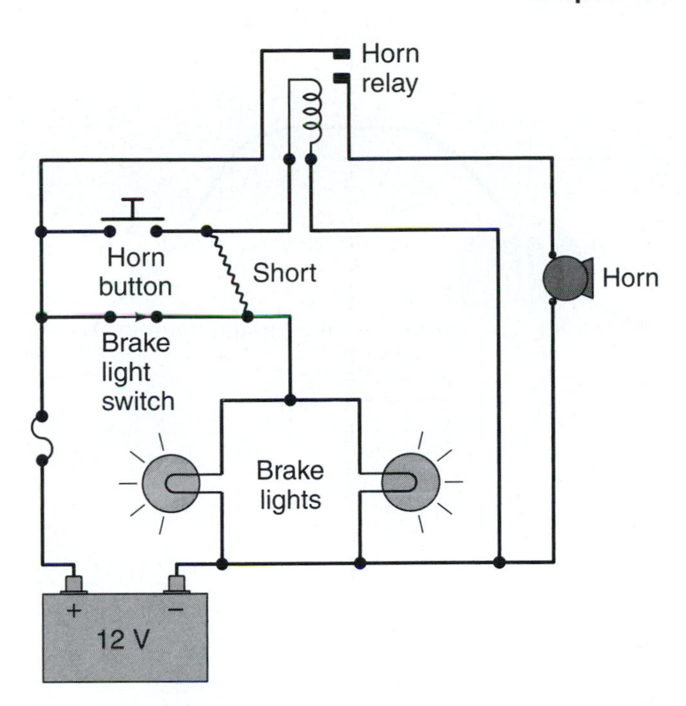

Figure 2. This short to power will operate the brake lights and the horn at the same time, regardless of which button is pressed.

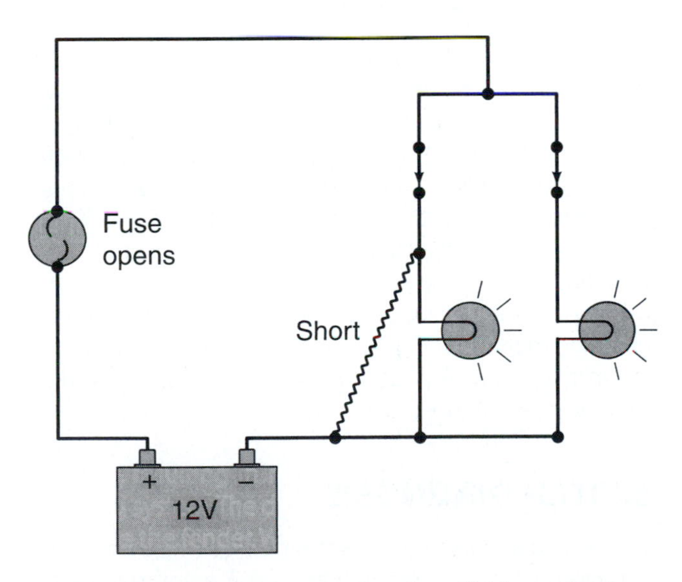

Figure 3. This short to ground will blow the fuse whenever the left switch is closed.

problem. Resistance problems often result in the circuit load not working or working poorly, such as a motor turning slower than it should or a light bulb burning dimly.

CIRCUIT PROTECTION DIAGNOSIS

Diagnosis of fuses, circuit breakers, and fusible links is primarily about ensuring that there is voltage on both sides of the device. Many technicians will pull a fuse to

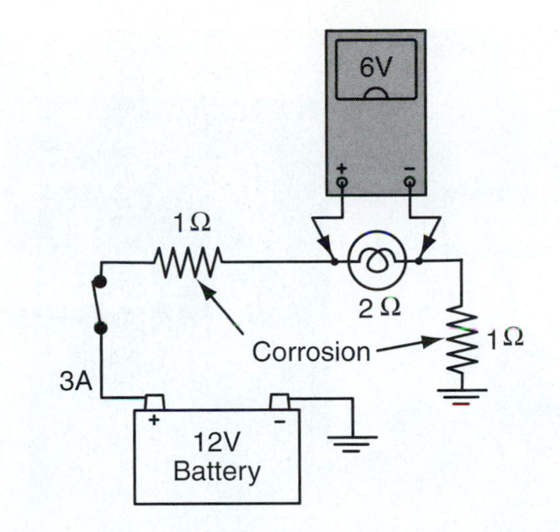

Figure 4. Corroded connections in this circuit have dropped the available voltage to the bulb from 12 volts to 6 volts.

Figure 5. Testing fuses with a test light is more reliable than a visual inspection.

inspect it visually. Although this is often effective, it can lead to problems as well. Fuses can fail in ways that are not visible to the eye. A test light (see **Figure 5**) or DVOM is the preferred check. With power to the system activated, check each side of the fuse for voltage. If it is present, then the fuse is good. This has the added benefit of assuring the circuit has power to the fuse when it should. Circuits can fail between the power source and the fuse.

Circuit breakers can be checked using the same voltage test as a fuse. However, accessing the circuit breaker terminals can be difficult. If this is the case, for fuse panel–mounted breakers, remove the unit and use an ohmmeter across the two legs (see **Figure 6**). If the resistance is less than .5 ohms, it is good. If you are checking a PTC circuit breaker, remember that they do not reset until all voltage is removed from the circuit.

A fusible link has a heavy rubberized insulation around the fuse wire. It is designed not to burn when the fuse wire

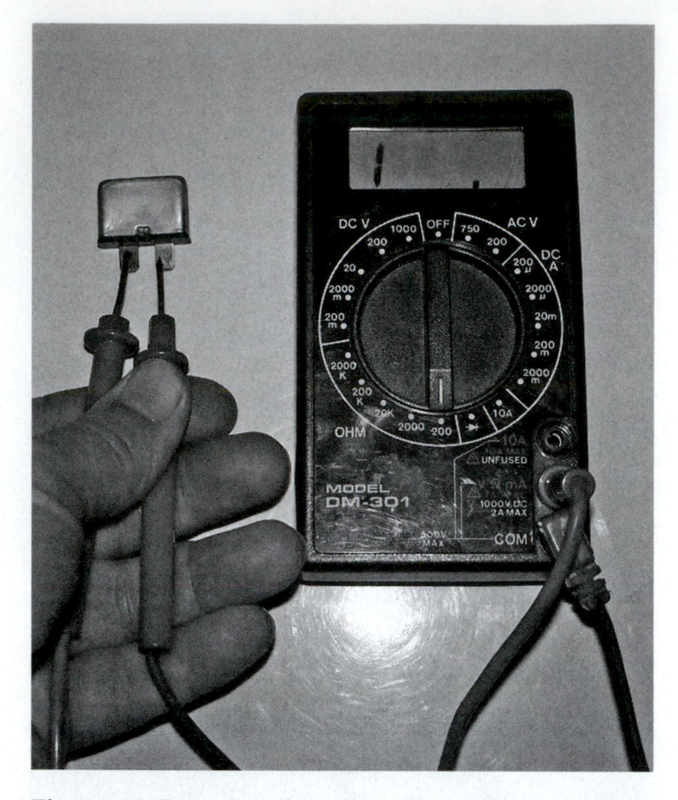

Figure 6. Remove circuit breaker and test with an ohmmeter.

melts. However, burning can still occur. If the insulation is burned, replace the link. If the insulation looks intact, that does not mean the link is good. Using a test light or DVOM probe to pierce the wire looking for voltage is not recommended, as this can damage the insulation and allow moisture to get inside it. To test the fusible link, merely pull on the ends of the link. If it stretches, the wire inside it is melted and the link should be replaced.

> **You Should Know** *When replacing a fusible link, never use more than 9 inches of new link wire, as the smaller diameter of the link wire will add unwanted resistance to the circuit.*

RESISTOR DIAGNOSIS

Diagnosis of resistors typically consists of checking the unit with an ohmmeter. Although simple, there are a couple of concerns that should be remembered. Whenever checking a component with an ohmmeter, it is a good practice to remove it from the circuit if possible. Because the ohmmeter puts out a small current to check the resistance, there could be parallel paths the current could

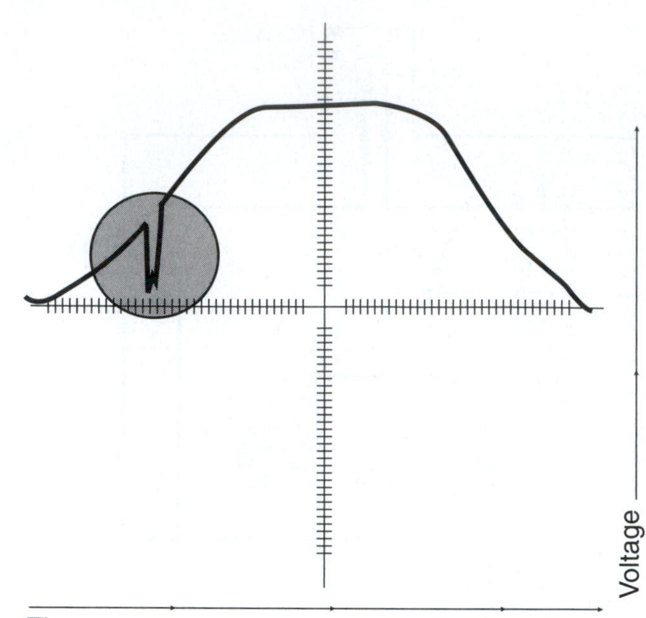

Figure 7. A variable resistor with a defective segment may drop the voltage to zero when the bad spot is reached.

follow if the resistor is still installed in the circuit. Also remember that the temperature of the resistor could alter its resistance. If comparing the reading to a specification, make sure the specification is within the actual temperature range. If testing a variable resistor, remember it could have a malfunction at only one point through its entire range. Unless the resistor is checked at that one point, the problem will not show on a regular ohmmeter. A better method would be to check the variable resistor with a graphing meter (see **Figure 7**) or an ohmmeter with a minimum/maximum function while the resistor is worked through its entire operating range.

SWITCH DIAGNOSIS

Testing a switch is best done installed in the circuit it is controlling. Open the switch and hook a voltmeter from the switch input wire to a known good ground. The volt reading should be near battery voltage. If not, determine why voltage is not reaching the switch. Next, move the voltmeter ground wire to the output switch wire and close the switch. This will perform a voltage drop test on the switch (see **Figure 8**). With the circuit activated, read the voltage on the meter. Anything near battery voltage indicates the switch has an open and should be replaced. The voltage reading should be near zero. Any reading above .5 volt may indicate excessive resistance in the switch contacts. If the circuit is still not functioning, check the other components in the circuit.

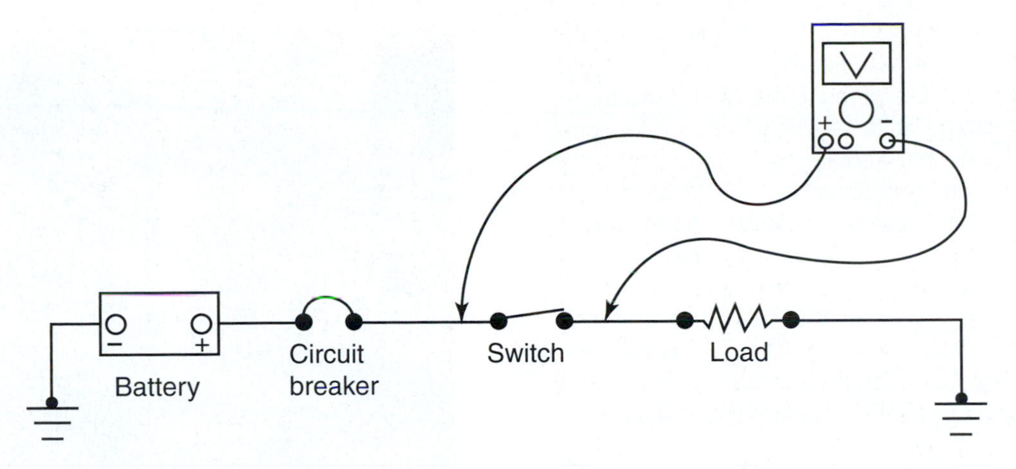

Figure 8. A voltage drop in excess of .5 volts when the switch is closed may indicate a problem.

RELAY DIAGNOSIS

Testing a relay out of the vehicle involves testing the two separate circuits within the relay, the coil circuit and the switch circuit. To test the coil circuit, connect an ohmmeter across the coil terminals. These are usually labeled 85 and 86. Compare the resistance reading with the specification and replace the relay if bad. If the relay has a normally closed switch, connect the ohmmeter to the switch terminals labeled 87a and 30. The resistance should be less than .5 ohms. To test a normally open switch, the relay must be activated. Use a fused jumper wire to connect 12 volts to terminal 85 of the coil. Use another jumper wire to connect coil terminal 86 to ground. A click should typically be heard when this circuit is complete. Next, connect the ohmmeter to terminals 87 and 30 on the relay switch and check the resistance. Again the reading should be less than .5 ohms.

The transistors used in semiconductor circuits often serve the same purpose as a relay, allowing a small current to control another, often larger, current.

MAGNETIC COIL AND SOLENOID DIAGNOSIS

Checking any coil is identical to checking a relay coil. Use an ohmmeter across the two terminals and compare the reading to the specification. The physical operation of a solenoid may be checked by using fused jumper wires to connect the solenoid to a good 12-volt battery. When the circuit is complete, a clicking should be heard and the solenoid rod, if visible, should move.

DIAGNOSING BASIC CCOT COMPRESSOR CIRCUIT

To diagnose an inoperative compressor clutch, start by disconnecting the connector at the cycling switch. Use a jumper wire (see **Figure 9**) to bypass the switch at the connector terminals. If the compressor activates, the refrigerant level is too low to activate the switch or the switch is bad. Use a pressure gauge to check the refrigerant pressure. If the pressure reading is above 45 psi, replace the switch. If the pressure is below specification, determine the cause of the refrigerant loss and recharge the system.

If the jumper wire does not activate the compressor, use a DVOM to check for voltage at the switch connector. If there is no voltage, check the HVAC fuses. Replace any

Figure 9. If the compressor activates when the pressure cycling switch is bypassed, the system is low on refrigerant or the switch is bad.

blown fuses and determine why they are bad. If there are no bad fuses, the problem is probably in the control head switches or an open in the wiring to the cycling switch.

If there is voltage at the cycling switch connector, leave the jumper wire in the connector and disconnect the compressor clutch coil connector. Use the DVOM to check for voltage at the coil connector to a known good ground. If there is no voltage, there is an open between the cycling switch and the coil. If there is voltage, connect the DVOM between the compressor power wire and the compressor ground wire. No voltage indicates a bad compressor ground. If there is voltage, the compressor coil is likely bad. Check the coil resistance to confirm the diagnosis.

DIAGNOSING BASIC THERMOSTATIC EXPANSION VALVE CIRCUITS

Diagnosing the basic thermostatic expansion valve circuit is very similar to diagnosing the CCOT circuit. The operation of thermal and high-pressure switches may be checked by bypassing them with a jumper wire. If the compressor clutch activates when a device is bypassed, determine why that component failed to complete the circuit. If none of the control switches are at fault, then the voltage on the circuit must be traced until the open circuit or unwanted resistance is found.

DIAGNOSING COMPRESSOR CIRCUITS WITH COMPUTER CONTROLS

The inclusion of a relay in the compressor control circuit can be an aid in diagnosing problems with the circuit. Remove the compressor relay and install a jumper wire in the relay socket for relay terminals 30 and 87 (see **Figure 10**). If the compressor clutch activates, then there is no problem in the clutch circuit itself. The problem is either a bad relay or the relay is not being activated. To check the relay, use the diagnostic routine described earlier or swap the relay with an identical relay that is known to be good.

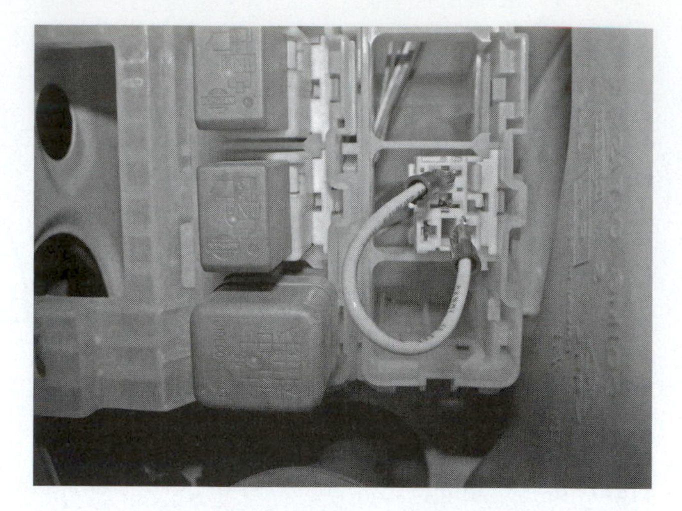

Figure 10. Jumping terminals 30 and 87 will activate the compressor if the clutch coil circuit is good. If it does not activate, start diagnosis with coil circuit.

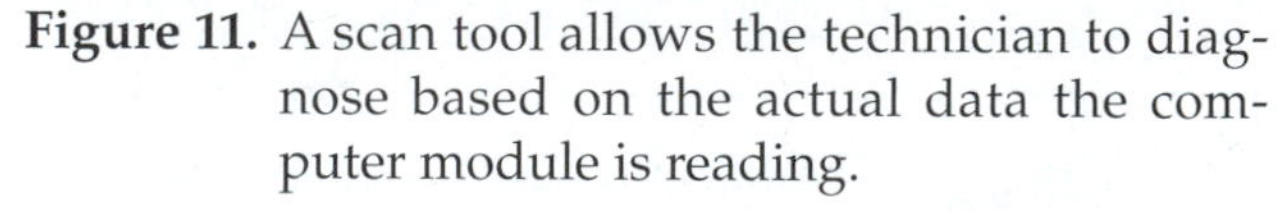

Figure 11. A scan tool allows the technician to diagnose based on the actual data the computer module is reading.

If the relay is good, then the problem is in the relay coil circuit or the computer is not activating the circuit. First, make sure there is power for the coil in the relay socket. This power usually comes directly from a fuse. Test with a DVOM or test light. If there is no voltage, check for a blown fuse or an open circuit.

Next, attach a scan tool to the diagnostic link on the vehicle (see **Figure 11**). Check for any trouble codes that might effect HVAC operation. Repair as necessary. Examine the data stream information. If there is an "AC REQUEST" field, watch to see if it cycles from YES to NO when the AC button is turned on and off on the control head. If it does not change, there is a problem in the circuit. This could be in the control head or the wiring between the two modules. Some scan tools are capable of manually activating the AC relay. If so equipped, activate the relay, making sure everything that was unplugged in previous testing is reconnected. If the compressor still does not activate, there is a problem in the circuit between the module and the relay, or the module itself is bad.

If the AC compressor does come on when the scan tool manually activates the relay, then the control module has a logic reason for not turning the compressor on during normal operation. Examine the data stream information for a parameter that is out of specification. Pay close attention to any data items that are AC-related, such as a high-pressure reading or pressure switch. Also look for unusual engine conditions such as overheating or wide-open throttle.

BLOWER MOTOR DIAGNOSIS

Diagnosing an inoperative blower motor is similar to checking any load device. If the component has power and ground applied to it, but it does not work, then the component is bad. Testing for power and ground is best done with a DVOM. To test the blower motor, turn the vehicle ignition on and set the HVAC settings so the blower motor is on high. With the blower motor connected to its control circuit, back probe the input power wire to the blower motor with the meter. There should be sufficient voltage at that point to power the blower motor. If there is not enough voltage, check the circuit leading to the blower motor for an open or excessive resistance. If the voltage on the power wire is satisfactory, then move the DVOM lead to back probe the ground wire on the blower motor (or the housing if it is case-grounded). If the voltage on the ground wire is near zero and the motor is not running, then the blower motor is bad. If there is voltage on the ground wire, then the circuit may have an open or have excessive resistance.

Summary

- Open circuit faults block the flow of electricity from its intended destination.
- A circuit has a short when the current flow takes a shortcut from its correct path.
- Unwanted resistance faults keep load devices from operating at their maximum output.
- Circuit protection devices are best checked by testing for voltage on each side of the fuse element.
- Fixed resistors are best checked using an ohmmeter with the resistor removed from its circuit.
- A graphing or memory equipped ohmmeter is useful in diagnosing variable resistors.
- Relays are tested by actuating the electric coil and checking switch operation.
- CCOT circuits are checked by bypassing the cycling switch and checking the circuit for correct voltage.
- Relays make excellent points to begin diagnosis of electrical circuits.
- Any load device that does not work when it has power and ground is defective.

Review Questions

1. Technician A says that pulling on a fusible link is a valid test. Technician B says that blade fuses should be visually inspected. Who is correct?
   A. Technician A only
   B. Technician B only
   C. Both Technician A and Technician B
   D. Neither Technician A nor Technician B
2. Technician A says that high resistance in a circuit can cause a fuse to melt. Technician B says that an electrical open fault is more likely to cause a fuse to fail. Who is correct?
   A. Technician A only
   B. Technician B only
   C. Both Technician A and Technician B
   D. Neither Technician A nor Technician B
3. Technician A says that a short to power may cause components to act strangely. Technician B says that all circuit breakers reset themselves after they cool down. Who is correct?
   A. Technician A only
   B. Technician B only
   C. Both Technician A and Technician B
   D. Neither Technician A nor Technician B
4. Technician A says that resistor ohm measurements may need to be temperature-compensated. Technician B says that resistance in a component should be checked when it is operating in the circuit. Who is correct?
   A. Technician A only
   B. Technician B only
   C. Both Technician A and Technician B
   D. Neither Technician A nor Technician B
5. Two technicians are discussing a voltage drop test performed on a closed switch in a circuit. The voltage reading from the test indicated 13 volts. Technician A says that the reading indicates a good switch. Technician B says that the switch is defective. Who is correct?
   A. Technician A only
   B. Technician B only
   C. Both Technician A and Technician B
   D. Neither Technician A nor Technician B
6. The resistance of a relay coil is being tested. The ohmmeter shows the coil has 42 ohms of resistance. Technician A says that the relay coil is good. Technician B says that he does not have enough information to evaluate the coil yet. Who is correct?
   A. Technician A only
   B. Technician B only
   C. Both Technician A and Technician B
   D. Neither Technician A nor Technician B

7. An inoperative CCOT compressor circuit is being discussed. Technician A says that bypassing the cycling switch will activate the AC relay. Technician B says that the compressor will engage when the cycling switch is disconnected. Who is correct?
   A. Technician A only
   B. Technician B only
   C. Both Technician A and Technician B
   D. Neither Technician A nor Technician B
8. Technician A says that a scan tool can activate the AC compressor on some vehicles even if it is low on refrigerant. Technician B says that bypassing the relay switch may indicate if the compressor clutch coil is good. Who is correct?
   A. Technician A only
   B. Technician B only
   C. Both Technician A and Technician B
   D. Neither Technician A nor Technician B
9. Technician A says that swapping identical relays is a common diagnostic procedure. Technician B says that a compressor may not engage on some vehicles if the throttle position sensor is bad. Who is correct?
   A. Technician A only
   B. Technician B only
   C. Both Technician A and Technician B
   D. Neither Technician A nor Technician B
10. Technician A says that a blower circuit with ground-side switching may have voltage on the blower motor ground terminal. Technician B says that a blower motor that is running slow may not have enough resistance in the control circuit. Who is correct?
    A. Technician A only
    B. Technician B only
    C. Both Technician A and Technician B
    D. Neither Technician A nor Technician B
11. A short to ____________ often causes unexpected devices to activate when current is applied to a circuit.

Chapter 25 Electrical and Electronic System Service

Introduction

After an electrical problem is diagnosed, the repair must be made to return the vehicle to its original operating condition. A professional technician knows a job was well done when the system works correctly and the customer cannot tell the vehicle was ever worked on. This can be extremely important when the electrical problem is buried somewhere deep inside the dash. Returning the vehicle without any new cracks or rattles makes for a happy customer. For most operations, it is highly recommended that good service information such as an electronic database or the manufacturer's service manual be consulted.

CIRCUIT PROTECTION SERVICE

Service circuit protection devices such as fuses and circuit breakers by replacement. Fusible links must be repaired by splicing a new piece of fusible link wire in place of the melted one (see **Figure 1**). Take care to use the same size and rating of wire as that used originally on the vehicle. The replacement wire should not be more than 9 inches long.

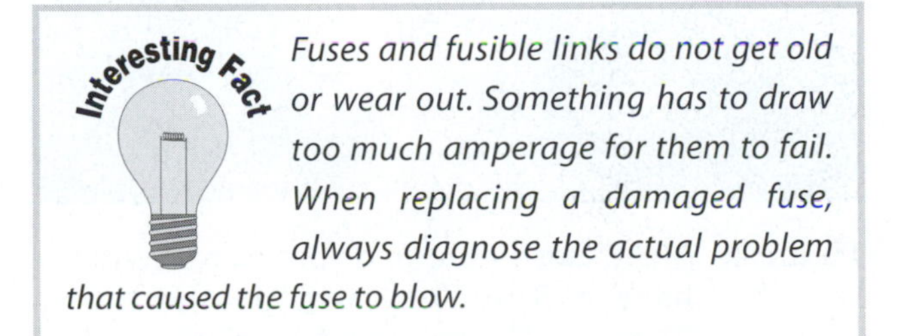

Fuses and fusible links do not get old or wear out. Something has to draw too much amperage for them to fail. When replacing a damaged fuse, always diagnose the actual problem that caused the fuse to blow.

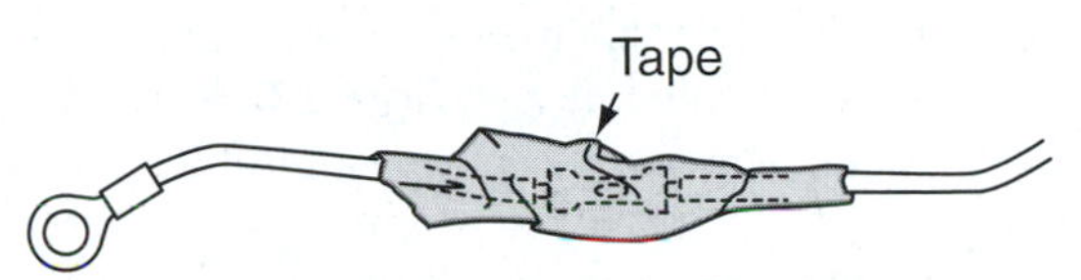

Typical repair using the eyelet terminal fuse link of the specified gauge for attachment to a circuit wire end.

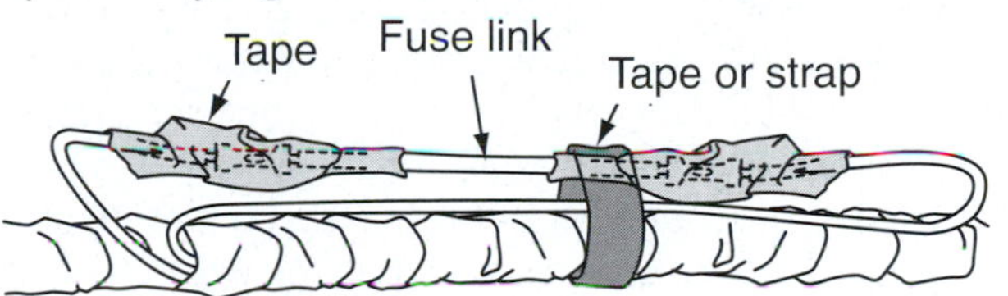

Typical repair for any in-line fuse using the specified gauge fuse link for the specified circuit.

Figure 1. A new section of fusible link wire the same diameter as the original must be spliced in place of the damaged link.

CONTROL HEAD SWITCH SERVICE

Replacement of switches in the control head will be required when the switch fails to perform as expected. This procedure entails removal of the control head for most models (see **Figure 2**). Care should be taken when removing trim panels covering the control head. These panels typically have several screws holding them in place, whereas others are held in place by small clips. Look for hidden screws in vents or adjacent trim panels that must be removed first. Prying on a panel when screws are still holding it in place can do considerable damage. Some vehicles may require removal of the center console.

After the trim panel is removed, remove the control head mounting screws. Gently slide the control head

Figure 2. Once the control head is removed, switch replacement entails removal of the switch retaining screws.

forward until access to the electrical connectors is possible. Removing one connector may give additional access to harder-to-reach connectors (see **Figure 3**). Bowden cables or vacuum lines also may need to be disconnected. When adequate access is available to the switch mounting screws, remove them and install the new switch. Some applications may require the removal of the switch handle. A clip or small set screw may hold it in place. After replacement, reverse the installation procedure. When guiding the head back into place, make sure that all the wiring returns to its original location. Failure to do so may result in a pinched or grounded wire. Also, make sure that no other connectors were disconnected when the work was done in the dash. The radio wires are often nearby and are easily loosened.

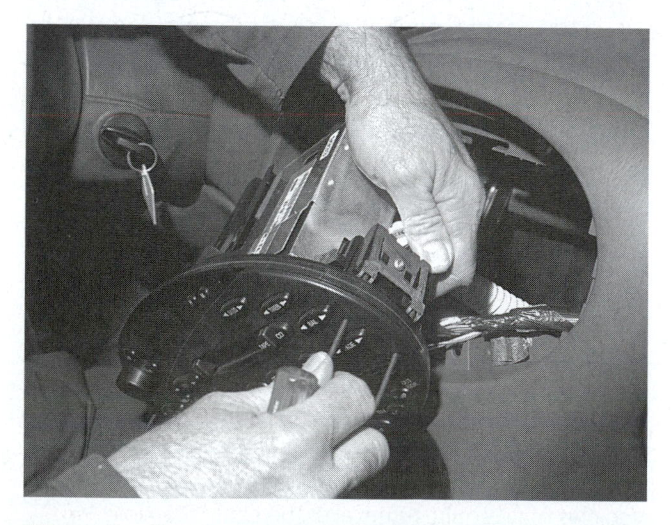

Figure 3. Removing one electrical connector may allow greater access to harder to reach connectors.

You Should Know *When replacing a switch, always inspect the attaching wires for damage. If any signs of heat or corrosion are present, they must be repaired.*

REFRIGERANT SWITCH SERVICE

Replacement of refrigerant pressure switches that are mounted in cavities in the compressor will require refrigerant recovery, system evacuation, and recharging. These switches are usually held in place by a snapring or small bolt and retainer. If there is adequate room to remove the snapring or retainer, the procedure may be done on the vehicle after all pressure is removed from the system. Otherwise, removal or repositioning of the compressor will be required. Most of the switches are sealed using an O-ring. Carefully remove the old O-ring without scratching the compressor housing. Lubricate and install a new O-ring and install new switch.

You Should Know *All O-rings should be lubricated with mineral oil before installation. PAG and ester oils are not recommended because they are hydroscopic (absorb moisture from the air) and can cause corrosion at the seal point.*

Most hose-mounted pressure switches have a Schrader valve on the hose fitting (see **Figure 4**). If there is a valve,

Figure 4. Most hose-mounted pressure switches have a Schrader valve on the fitting. Replace O-ring and lubricate before installation.

remove the electrical connector and unscrew the switch, making sure not to twist the metal hose. These valves are sealed with an O-ring. Lubricate and install a new O-ring before installing the new switch. Reconnect electrical connector.

> **You Should Know** *A few high-pressure switches do not have Schrader valves. Removing them without recovering the refrigerant will release all of the refrigerant in the system. If you are unsure, consult a service manual.*

BLOWER MOTOR REPLACEMENT

The location of a blower motor will determine the procedure used to replace it. Regardless of where it is mounted, some basic concerns apply to any blower motor replacement. The blower motor mounting flanges are usually attached onto a plastic case with self-tapping screws. Overtightening these screws can easily strip the plastic material (see **Figure 5**). If a larger diameter screw is used to repair the stripped hole, do not use a longer screw. It may hit the fan blades and lock operation of the motor. Motors that have sound insulators or rubber bushings must have these transferred to the new motor when installed. Make sure that the polarity of the wires is correct when installing the new motor. If hooked up backward, the motor will spin in the wrong direction.

When planning how to remove the blower motor, remember that the fan has to come out with it. That means an additional 4 to 6 inches of clearance will be required to remove the entire motor and fan. Do not force the unit out of the housing. This could damage the fan. Once it's removed, inspect the fan for damage. A problem with the fan could damage the new motor if not replaced. Also, look inside the housing. Remove any debris and repair any other problems that might be spotted.

Many models place the motor in the engine compartment. They are typically easy repairs, except for on a few vehicles that may require fender or fender liner removal. Some components may have to be removed in order to gain access to the motor. Once access is cleared, remove the cooling hose (see **Figure 6**), if it is equipped with one, and install it on the new motor during installation. Then disconnect the motor wires and remove the four to six screws around the motor flange. Some models may have a sealant around the flange that will try to hold the motor in place. A gentle pry with a screwdriver should loosen the motor. Once it's loose, slide the motor and fan straight out of the opening. If the fan is to be reused, transfer it onto the new motor and reverse the process to install.

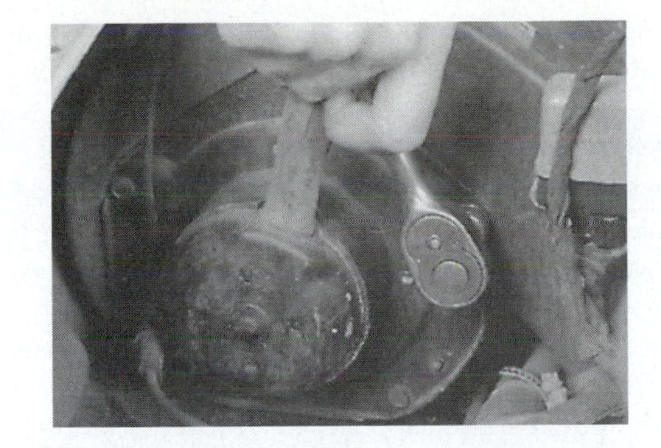

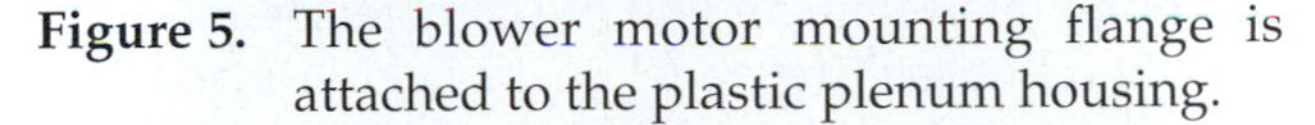

Figure 5. The blower motor mounting flange is attached to the plastic plenum housing.

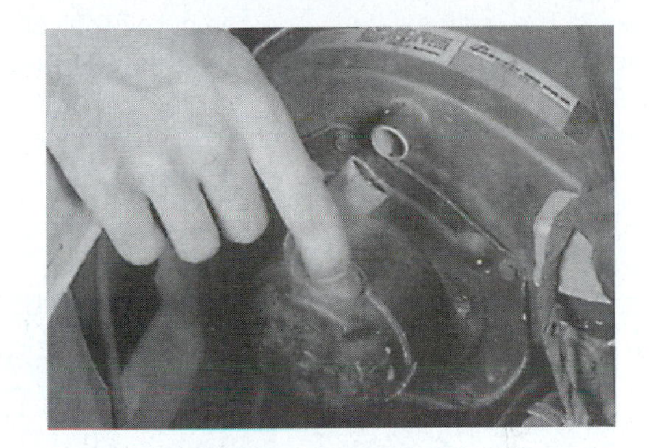

Figure 6. Always transfer the motor cooling hose to the new motor.

> **You Should Know** *If the blower motor is case-grounded, one of the screws is not for mounting purposes but for attaching a ground wire terminal. When transferring the terminal to the new motor, it must be installed in the same flange hole. Make sure a good electrical path is made for the ground circuit.*

Blower motors that mount under the dash may be more of a challenge. Once access to the motor is made, the actual replacement of the unit is the same as under the hood repairs. Depending on the model, access may be as simple as removing a plastic cover panel. Other designs may entail glove box removal, carpet relocation, partial removal of the dash, or even removing sections of the plenum assembly.

COMPRESSOR CLUTCH COIL REPLACEMENT

Some clutch coils are pressed onto the front compressor case. Replacing these pressed-on coils typically is not recommended. In the event of coil failure, the compressor and clutch assembly is replaced as a unit. On compressors with removable clutches, the procedure for replacing the clutch coil is very similar for most compressors. There may be special tools required to remove and reinstall some of the clutch components. If there is enough clearance to work and install the pullers, the compressor does not need to be removed from the vehicle. The compressor drive belt will need to be removed. The refrigerant does not need to be drained if the compressor is not removed from the vehicle.

The drive disk is first removed. Some disks are pressed on and require a special puller to remove. Others are held on using a retainer nut or bolt. The compressor pulley can then be removed. They are usually held in place by a snapring. A puller aids in the removal of the pulley (see **Figure 7**). The coil is located under the pulley. The coil can be held in place by a snapring or bolts (see **Figure 8**). Make note of the position of the coil terminals. The new coil must be installed in the same position. Reinstall the other clutch components, making sure that the clutch air gap is correct (see **Figure 9**).

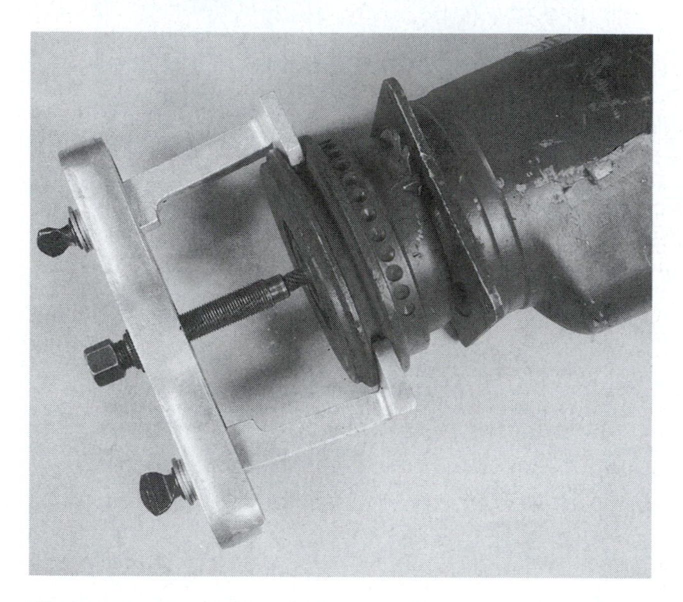

Figure 7. Special tools are a necessity for replacing clutch components.

Figure 8. Removal of the snapring will allow the removal of the clutch coil.

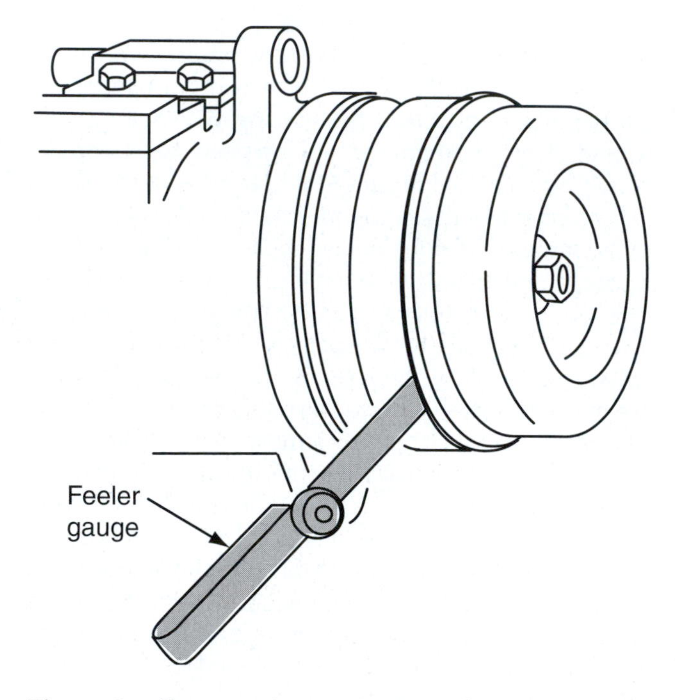

Figure 9. Correct air gap is crucial to the correct operation of the clutch.

Summary

- Fuses and circuit breakers are replaced as units when defective.
- Fusible link wire must be spliced in place of a melted fusible link.
- Fuses and fusible links do not wear out or get old.
- Correct repair procedures must be used when removing dash trim panels to avoid damage.
- When replacing any electrical component, always inspect connecting wires and terminals for damage.
- Always replace and lubricate O-rings during service.
- Compressor-mounted pressure switches are held in place by a snapring.
- Hose-mounted switches usually screw on a Schrader fitting.
- Gaining access to a blower motor is usually more difficult than the actual motor replacement.
- Special tools are required for compressor clutch coil replacement.

Review Questions

1. Two technicians are discussing a vehicle with a blown fusible link. Technician A says to use a replacement wire that is the same gauge and type as the wire in the circuit. Technician B says that additional diagnosis must be done in addition to replacing the damaged link. Who is correct?
   A. Technician A only
   B. Technician B only
   C. Both Technician A and Technician B
   D. Neither Technician A nor Technician B
2. Technician A says that all dash trim panels are held in place with screws. Technician B says that dash trim panels just pull off with snap clips. Who is correct?
   A. Technician A only
   B. Technician B only
   C. Both Technician A and Technician B
   D. Neither Technician A nor Technician B
3. Technician A says to always make sure the wires have clearance when installing a control head. Technician B says that a gentle tug on the control head will disconnect the hidden electrical connector behind it. Who is correct?
   A. Technician A only
   B. Technician B only
   C. Both Technician A and Technician B
   D. Neither Technician A nor Technician B
4. Technician A says that an air ratchet is a handy tool for installing the blower motor flange attaching screws. Technician B says to lubricate O-rings with any refrigerant oil when replacing. Who is correct?
   A. Technician A only
   B. Technician B only
   C. Both Technician A and Technician B
   D. Neither Technician A nor Technician B
5. Technician A says that refrigerant recovery is required when replacing compressor-mounted pressure switches. Technician B says that refrigerant recovery is required when replacing hose-mounted pressure switches. Who is correct?
   A. Technician A only
   B. Technician B only
   C. Both Technician A and Technician B
   D. Neither Technician A nor Technician B
6. Some compressors must be replaced if their clutch coil is defective.
   A. True
   B. False
7. Technician A says that wire polarity should be checked when installing a new blower motor. Technician B says that failure to inspect the wires near a defective switch could result in a customer comeback complaint. Who is correct?
   A. Technician A only
   B. Technician B only
   C. Both Technician A and Technician B
   D. Neither Technician A nor Technician B
8. Technician A says that refrigerant has to be recovered before the clutch coil is replaced. Technician B says that the drive belt must be removed. Who is correct?
   A. Technician A only
   B. Technician B only
   C. Both Technician A and Technician B
   D. Neither Technician A nor Technician B
9. The compressor must be removed from the vehicle to replace the clutch coil.
   A. True
   B. False
10. Replacement fusible links must not be more than _______________ long.

Section 6

Air Distribution Systems

SECTION OBJECTIVES

After you have read, studied, and practiced the contents of this section, you should be able to:

- Describe the function and operation of HVAC air distribution systems.
- Discuss the air input and discharge options available in air distributions and the benefits of each option.
- List the styles of air distribution cases and describe how they are serviced differently.
- Discuss the different operating modes available in air distribution systems.
- Describe the airflow doors used in air distribution systems and how they operate.
- Explain how dual zone climate controls work.
- Discuss the purpose of cabin air filters and describe how they are serviced.
- Describe the operation of manual air distribution controls.
- Explain how Bowden cables and vacuum motors are used to position airflow doors in manual control systems.
- Discuss how electronic actuators are used in manual and automatic control systems to control door position.
- Describe how computers are used in automatic temperature control systems to control all operating aspects of the air distribution system.
- Explain how sensors operate and how the sensor circuits are diagnosed when a fault occurs.
- Diagnose air distribution problems such as insufficient airflow, inadequate cooling or heating, and incorrect discharge air location.
- Describe basic air distribution service procedures.

Chapter 26 Air Distribution

Introduction

In order for the HVAC system to provide for passenger comfort properly, heat must be removed from or added to the air in the passenger compartment. The transportation of heat from one location to another is the function of the refrigeration and heating systems. It is the **air distribution** system that provides a sealed pathway for the air delivered to the passenger compartment, while controlling its temperature, volume, quality, and discharge location.

Control of the air distribution system may be manual, automatic, or semi-automatic. In a manual system, the operator controls discharge air temperature, blower speed, and where it blows out. An automatic control system only requires the operator to set the desired interior temperature. An electronic controller will then adjust the operation of the system to maintain that temperature within the vehicle. A semi-automatic system will require the operator to manually adjust some of the controls, such as the discharge temperature, whereas other adjustments such as blower speed may be automatic.

Regardless of the type of control system used, many components of air distribution systems are common to all vehicles and systems. This chapter explores the basic function of the air distribution system and the common elements found in them. It features vehicles with air conditioning and heating. Although they are not covered in the chapter, heater-only units are similar but without the evaporator core.

AIR INLET OPTIONS

There are two options for the air entering the air distribution system, which are commonly called outside fresh air and recirculate air (see **Figure 1**). The source used is determined by the design of the system and operator input. On many vehicles, the air intake source is a function of which operating mode the operator selects. The system will draw in outside fresh air in all modes except for MAX AC, which switches to recirculate air. On other vehicles, the air inlet source is manually selectable and there is a recirculate air button on the control panel (see **Figure 2**).

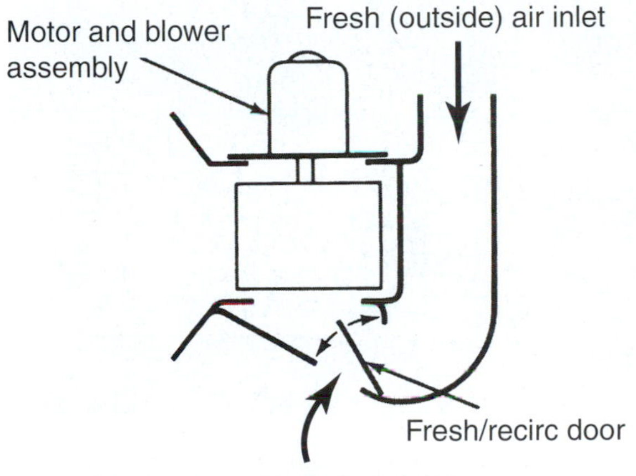

Figure 1. The two-position air inlet door selects between fresh outside air and inside recirculated air.

Outside Fresh Air Mode

In outside air mode, the air enters the air distribution system from outside the vehicle. The air typically enters through the cowl in front of the windshield (see **Figure 3**). Because this area is exposed to rain, the inlet ducts are routed so that the moisture is trapped at a low spot in the

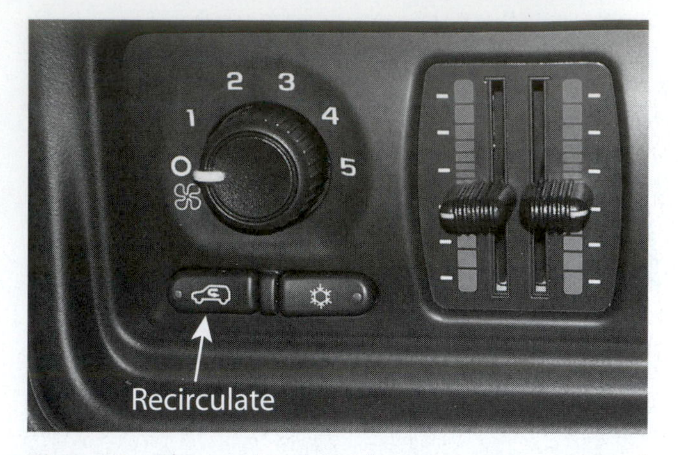

Figure 2. The recirculate button gives manual control over the air inlet control door.

Figure 3. The wire mesh over this fresh air intake keeps the leaves and debris from entering the air distribution system.

duct and allowed to drain off. There is also a grill over the entrance to keep out leaves and other debris. The air is then drawn into the fan chamber, where it is pushed toward the rest of the system.

With today's well-sealed passenger compartments, forcing air into it from the outside creates a higher pressure in the passenger area than in the surrounding atmosphere. This pressure helps to keep engine exhaust, with its deadly carbon monoxide (CO) gas, out of the passenger compartment.

Outside air mode is useful in AC mode when the temperature inside the vehicle is hotter than the ambient temperature. This will produce a cooler vent temperature and cool the vehicle off quicker.

Outside air is the default setting for the air intake door. The door is moved to this position when the vehicle is turned off. This is particularly noticeable when the door is controlled by an electric motor and the vehicle engine is shut off. The passengers can hear the motor running as it moves the door to the fresh air position. Keeping the evaporator core exposed to outside air helps it to dry better when wet from condensation. This prevents the formation of mold and mildew that can cause odors in the passenger area.

You Should Know *It is important to keep leaves out of the AC system because they get trapped at the evaporator core and decompose. The rotting leaves mix with the moisture off of the evaporator and become corrosive, and are consequently a leading cause of evaporator failure.*

Recirculate Mode

In **recirculate** mode, the air enters the air distribution system from inside the vehicle. The air typically enters from under the dash on the front passenger side (see **Figure 4**). The air is then drawn into the fan chamber where it is pushed toward the rest of the system. Some systems will completely seal the outside air intake when in recirculate mode. However, most vehicles will continue to draw in up to 20 percent outside air, which continues to slightly pressurize the passenger compartment. This is important in keeping exhaust fumes from entering the vehicle.

Recirculate air is useful in MAX AC mode when it is hotter outside the vehicle than inside the vehicle. This mode will produce the coolest possible vent temperature. Recirculate may produce some problems on cool humid days because it tends to fog up the windshield. There are some vehicles, primarily light trucks and SUVs, that will force the air intake to recirculate mode if the

Figure 4. The recirculate door opens to allow air from the passenger compartment to enter the HVAC system.

refrigeration system is placed under a heavy load. If the high-side refrigerant pressure exceeds a certain limit, usually between 320 to 380 psi (2,206 to 2,620 kPa), the AC controller moves to recirculate mode. Switching to this mode will eventually make the evaporator colder, reducing the load on the refrigeration system and lowering the pressure.

Interesting Fact

Some people believe the blower motor runs faster in recirculate mode. Actually, there is no change in the fan operation. The sound of the blower motor operation is louder because the occupants of the vehicle hear not only the air coming out of the vents, but also the inside air being sucked into the system.

BLOWER ASSEMBLIES

All HVAC systems need a method of forcing air through the evaporator and heater cores and then out to the passenger compartment. This is accomplished with the use of an electric motor and squirrel cage fan (see **Figure 5**). The fan and motor assembly are almost always mounted just after the air intake and before the evaporator and heater core (see **Figure 6**). A few cases will place it farther downstream in the airflow.

Figure 5. The squirrel cage fan and blower motor drive the air through the air distribution system.

The direct current motor is a sealed, nonserviceable unit. Many are case-grounded, with one input power wire. Because the motors are usually mounted onto a nonconductive plastic case, a ground terminal is attached to the motor housing with a sheet metal screw. If the motor is not case-grounded, it will have two terminals in the electrical connector: one power and one ground. To reduce the amount of noise from the blower motor, some motors are mounted in rubber or have an insulating cover placed over them. Many motors have a hose or passageway that runs from the motor to the airflow passages. This connection allows a small amount of air to be drawn across the brushes of the motor, keeping the motor cooler. Removing the heat helps to prolong the life of the motor.

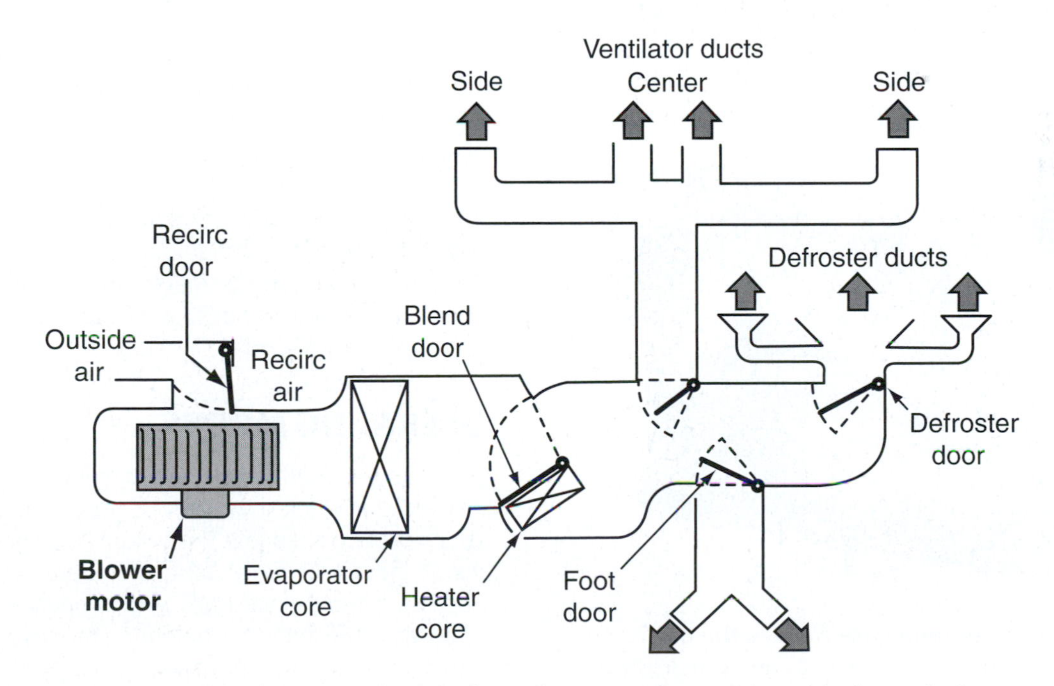

Figure 6. The blower assembly draws air directly from the air intake and forces it through the evaporator core to the rest of the system.

The fan blade resembles the wire wheel hamsters run on in their cage. There are several ways the fan may be mounted on the output shaft of the motor: press-fit, nut, clip, or set screw. If it is a press-fit, the assembly is almost always replaced as a unit. When spun, the centrifugal force of the fan blades draws air from the center of the fan and forces it outward. Because the fan is nondirectional, blowing air outward in all directions, the case must direct the air in the desired direction.

PLENUM ASSEMBLY

The main housing for the air distribution system is called the plenum assembly. Even though the evaporator core is a part of the refrigeration system and the heater core is a part of the cooling system, they are both housed in the plenum assembly. The unit also contains the blend door and all the mode doors. There are two main styles of plenum assemblies: the **combined case** and the **split case**. The combined case (see **Figure 7**) houses all the components already listed in addition to providing mounting for the fan and motor assembly. This type of case is typically used on larger vehicles and may be mounted under the dash, under the hood on the right firewall, or, in some cases, may extend from under the dash through the firewall. The split case style (see **Figure 8**) is used in smaller vehicles and is almost always mounted entirely under the right dash. This type of assembly is actually made up of a series of smaller cases. Typically, there is a case for the input air selector and blower assembly, an evaporator core case, and a case housing the heater core and all the remainder of the control doors.

Figure 7. This combined case houses the entire air distribution system for the vehicle. Note the four hose connections for the heater and evaporator cores.

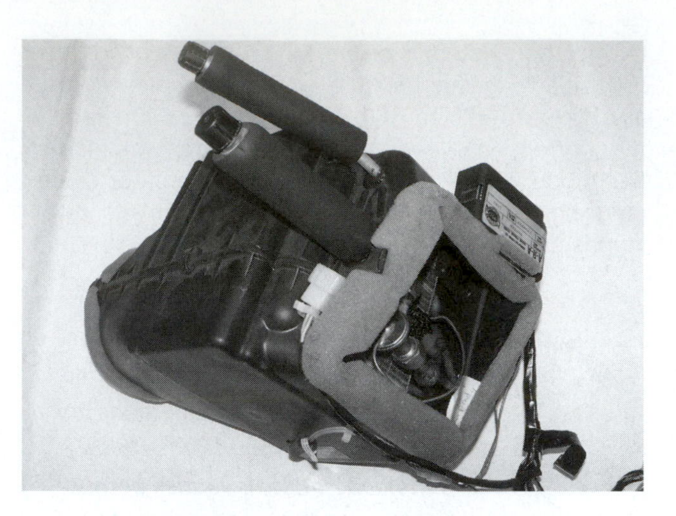

Figure 8. This section of a split case only contains the evaporator core and thermal expansion valve.

DISCHARGE AIR VENT OPTIONS

Up to this point in the system, the air has been drawn in by the fan, cooled or heated by the evaporator and heater cores, and is ready to be discharged into the passenger compartment. Depending on the operating mode selected, there are three discharge options: **defrost**, **panel**, and **floor**. Some vehicles also may have a small vent tube to the side windows to help keep them defogged.

The defrost vent is located in the top of the dash near the base of the windshield. Air flowing from this vent will blow on the bottom of the windshield and then travel upward along the rest of its surface. Typically, a small amount of air will still flow to the floor vents in order to maintain passenger comfort.

The panel vents are located across the front of the instrument panel, which allows the air to blow directly on the occupants. This setting produces the greatest cooling effect for the passengers.

The floor discharge ports direct the airflow at the feet of the occupants. This option is used in heat mode. Because hot air rises, the warm air will then travel upward to their bodies. A small amount of air is diverted to the defrost vent to keep the windshield clear of fog.

OPERATING MODES

A key element of any air distribution system is to deliver the air where it will provide the greatest comfort and safety for the passengers. The system is designed to control the air input, blower operation, temperature control, and discharge air so that the HVAC system operates as efficiently as possible. The following operating modes are commonly found on many makes of automobiles. Naturally, there are variations by different manufacturers, but they are primarily differences in nomenclature, not design. Two main differences include the ways in which the AC compressor and recirculate mode are activated. The following modes handle these automatically, whereas the controls on many newer vehicles allow the operator to control one or both of these independently.

> **You Should Know** *The mode selected does not directly control the actual discharge temperature of the air. The air can be hot even on MAX AC if the temperature control is turned to full warm.*

MAX AC Mode

MAX AC mode (see **Figure 9**) is designed to provide the greatest cooling capacity possible. The system is put into recirculate, the AC compressor is activated, and, on some vehicles, the blower motor is forced to HIGH. The system is forced into recirculation because the cooler the air is going into the inlet, the colder it will be coming out. A general rule of thumb is that if an AC system can produce a 20-degree F (11-degree C) drop in the outlet temperature as compared to the inlet temperature, it is working correctly. This means that if it is 85 degrees F (29 degrees C) inside the vehicle, then you should expect the discharge temperature to be less than 65 degrees F (18 degrees C). This cool air will then drop the cabin temperature even further, resulting in even cooler discharge air. Without the recirculate activated, a 95-degree F (35-degree C) day would result in only a 75-degree F (24-degree C) discharge temperature and will not get cooler as the interior temperature drops.

The discharge air flows from the panel vents, where it will strike the passengers. This allows greater evaporative and conductive cooling on the skin. Once it reaches its cooling potential, MAX AC mode also puts the least load on the AC system, making it the most economical of the AC settings.

AC Mode

In **AC mode** (see **Figure 10**), the AC compressor is activated and the air inlet is set to outside air. The discharge air is set to the panel vents and the blower motor is under normal control. This mode is effective for cooling off a hot vehicle that has been parked in the sun or when it is desirable to draw in fresh air from outside the vehicle.

BILEVEL Mode

As the name implies, **BILEVEL** (see **Figure 11**) allows the discharge air to exit at two levels instead of the normal one vent level. On most vehicles, BILEVEL flows from the panel and the floor outlets. The ductwork in the case is designed to allow most of the air to flow to the panel vents but slightly warmer air to flow out of the floor vents. The AC compressor is usually activated and the air inlet is set to outside air. Some models may not activate the AC compressor.

VENT Mode

VENT mode (see **Figure 12**) is the most fuel efficient for stop-and-go city driving. The AC compressor is turned off, the air inlet is on outside air, and the air is discharged through the panel vents. The air can be no cooler than the outside ambient temperature. It may be made warmer by adjusting the temperature control on the control panel warmer.

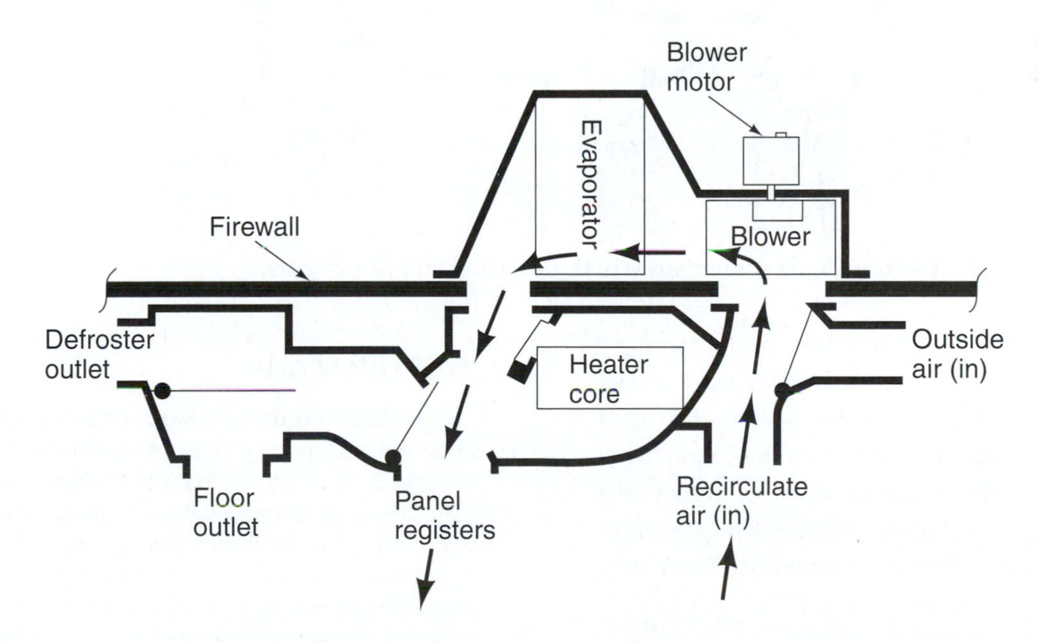

Figure 9. MAX AC is the coldest HVAC setting since it is more efficient to chill the cooler inside air on a hot day.

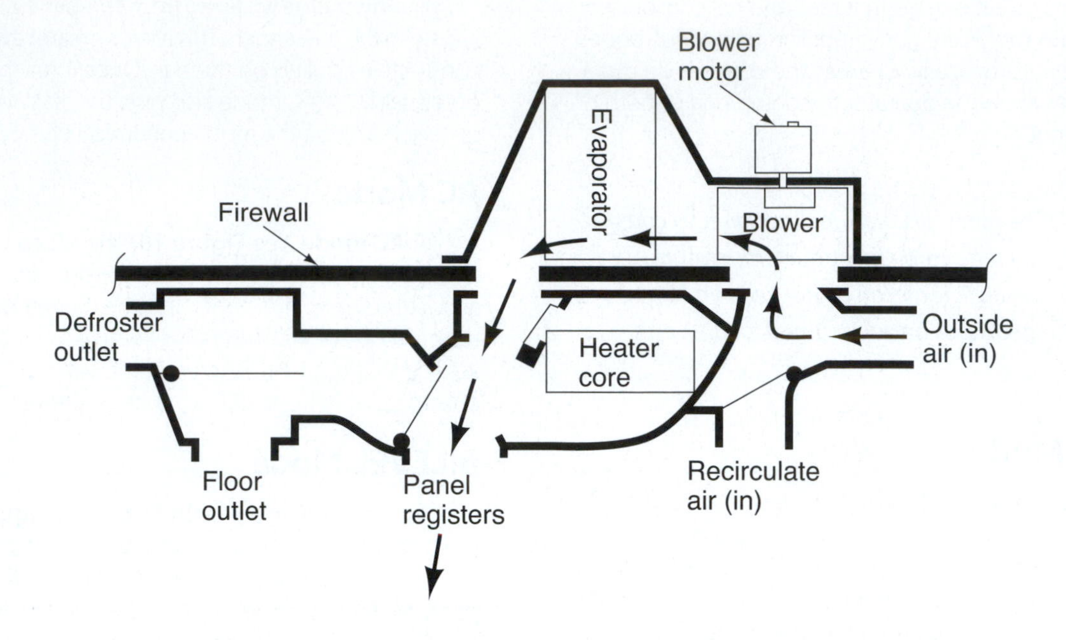

Figure 10. Regular AC mode draws outside fresh air in to be cooled by the evaporator core.

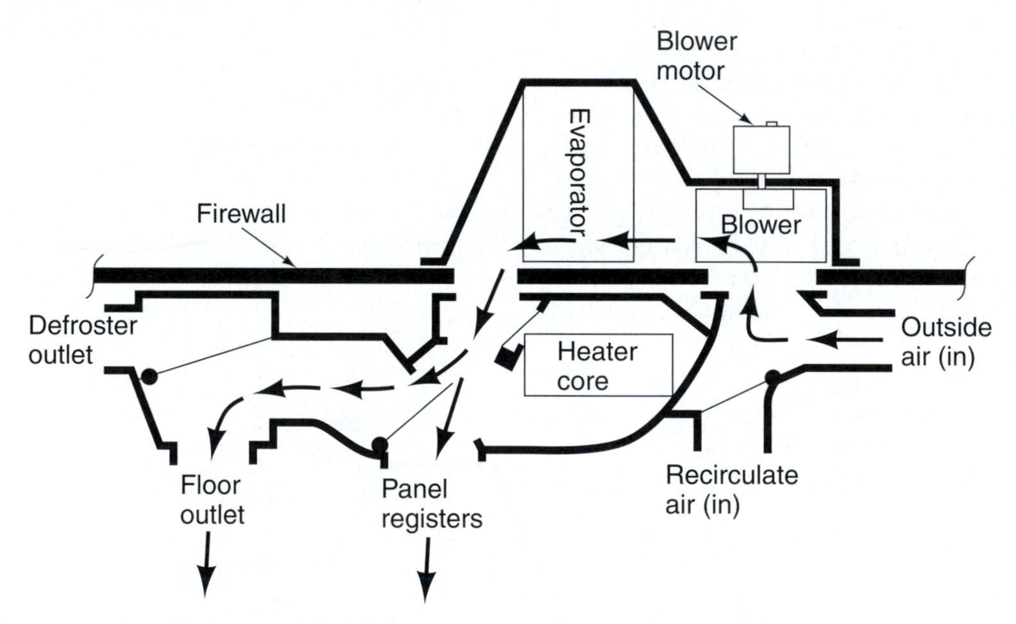

Figure 11. BILEVEL discharges the cooled air to both the panel and floor vents.

Interesting Fact

Fuel economy studies have shown that, at highway speeds, it is more economical to run the AC than to roll the windows down. Naturally, the most economical combination is with the windows up and the AC compressor off, such as in VENT or HEATER mode.

HEATER Mode

HEATER mode (see **Figure 13**) is designed to maximize the warming capacity of the air distribution system. The air is discharged from the floor vent and the AC compressor is turned off. The air inlet is set to outside air. Some air is routed to the defroster to reduce windshield fogging. Operating the heater when the engine is cold, as in a winter morning startup, is not recommended. The air coming out of the ducts will be cold and it will prolong

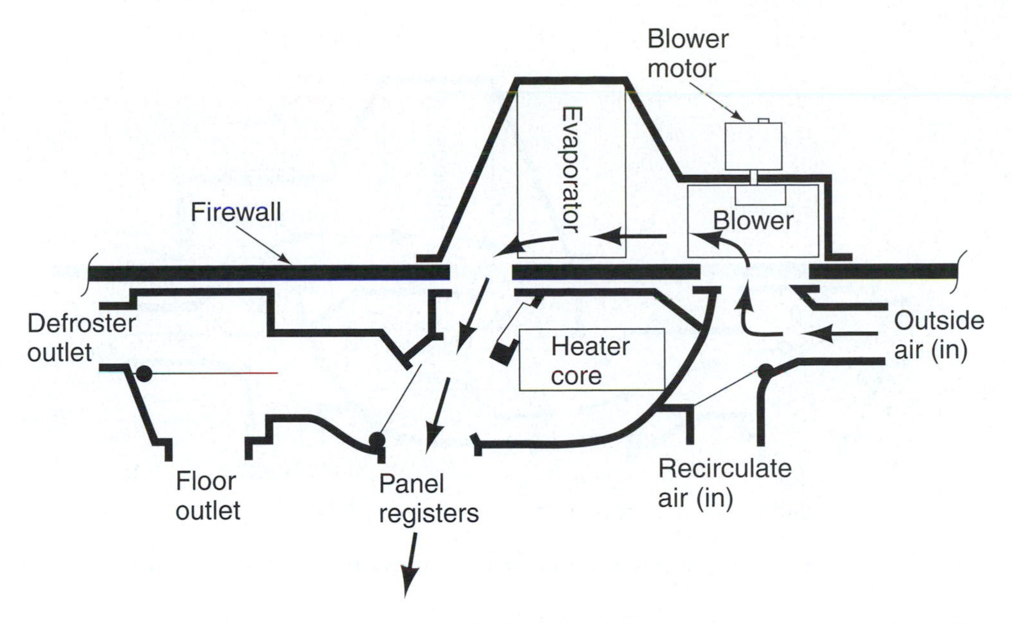

Figure 12. VENT mode air flow is identical to AC mode, but the AC compressor is not activated.

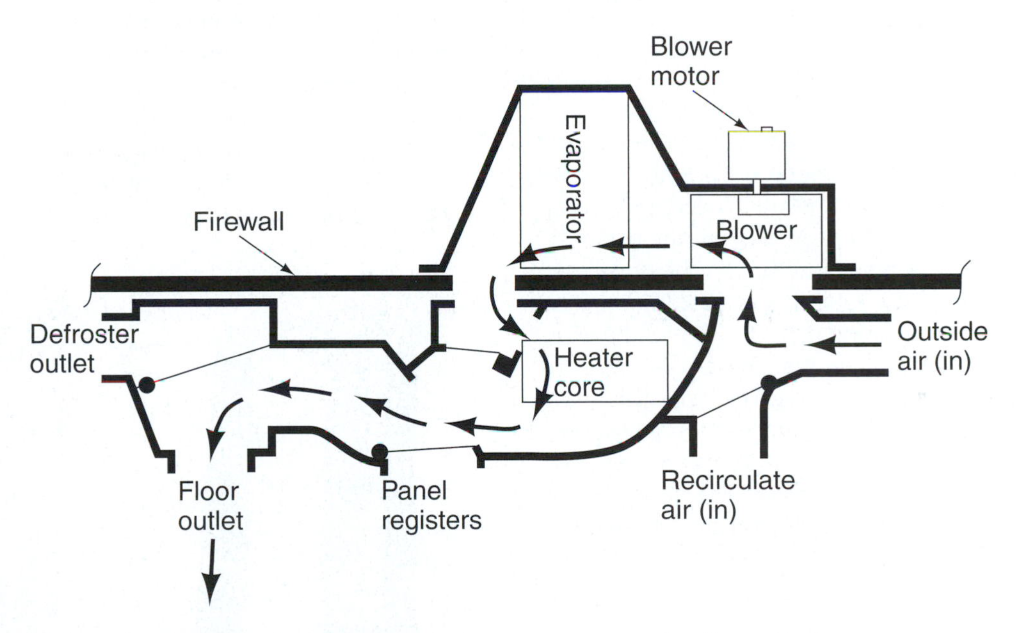

Figure 13. HEATER mode channels the air through the heater core to the floor discharge vents.

engine warm-up time because running the heater reduces engine temperature. If the heater is used at all, fan speed should be kept to a minimum until the engine is warm.

DEFROST Mode

DEFROST mode (see **Figure 14**) is almost identical to AC mode, with the exception that the discharge air is directed to the defrost vents on top of the dash so that the conditioned air is blown on the windshield. By activating the AC compressor, the fan drives the air through a chilled evaporator core, condensing the moisture in the air onto the evaporator. This allows dry air to be blown on the windshield, which does a better job of removing moisture from the glass. Most compressor control systems have a shutoff mechanism that turns off the compressor if ambient temperature is cooler than approximately 45 degrees F. Warming the air after it is dehumidified also aids in defogging the windshield.

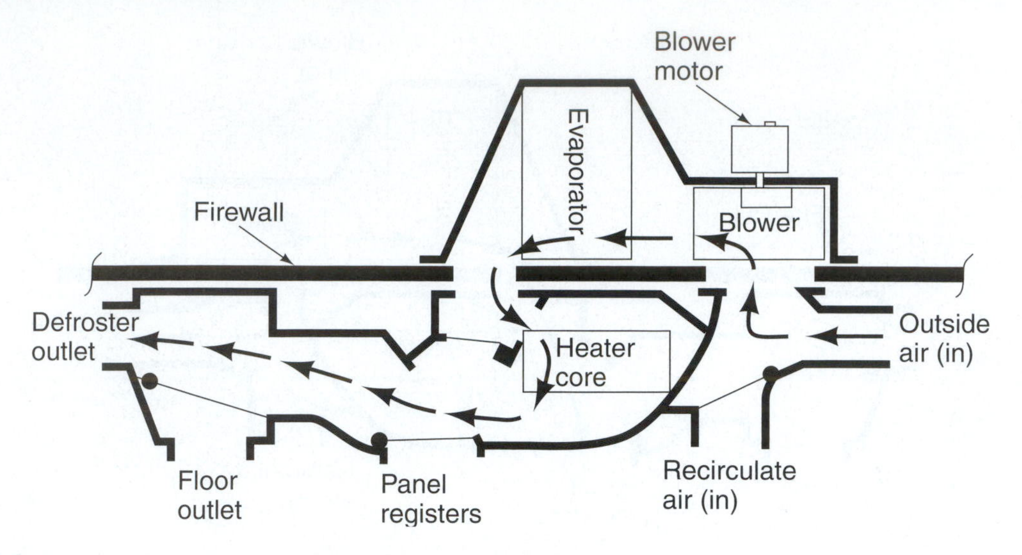

Figure 14. DEFROST mode is most efficient when the air is dried by the cold evaporator core and warmed by the heater core.

CONTROL DOORS

Using an airflow control door, also called an air valve, activates the majority of the controllable functions in the air distribution system. The only exception is the blower speed. These doors are usually rectangular with foam or rubber edges to seal against the duct walls and door stops. The doors will pivot on an axis either through the center of the door or along one side of the door (see **Figure 15**). Those pivoting along the middle act as a double door, opening or closing on both sides of the axis. Regardless of the design, all the doors have some type of actuator that causes them to move. The actuators may be an operator-controlled Bowden cable, a vacuum motor, or an electric motor. These actuators will be covered in more detail in later chapters. The doors themselves may be two-position, three-position, or variable-position type.

Blend Door

The **blend door** (see **Figure 16**) is the primary device for controlling the temperature of the discharge air. It is a variable position door that is functional in all HVAC operating modes. It is also called the temperature door. It controls the air temperature by regulating the percentage of discharge air that flows through the heater core. When the temperature control on the control head is in the full cold position, the blend door allows none of the discharge air to flow through the heater core. When the temperature control on the control head is in the full hot position, the blend door allows all of the discharge air to flow through the heater core. For all other temperature control settings, the blend door allows a proportional percentage of air through the heater core. On vehicles that use manual controls, the operator has direct control over the blend door and must move it when comfort or environmental levels change. An automated controller will adjust the blend door to maximize the HVAC performance to whatever temperature the operator has input. A system with dual zone climate

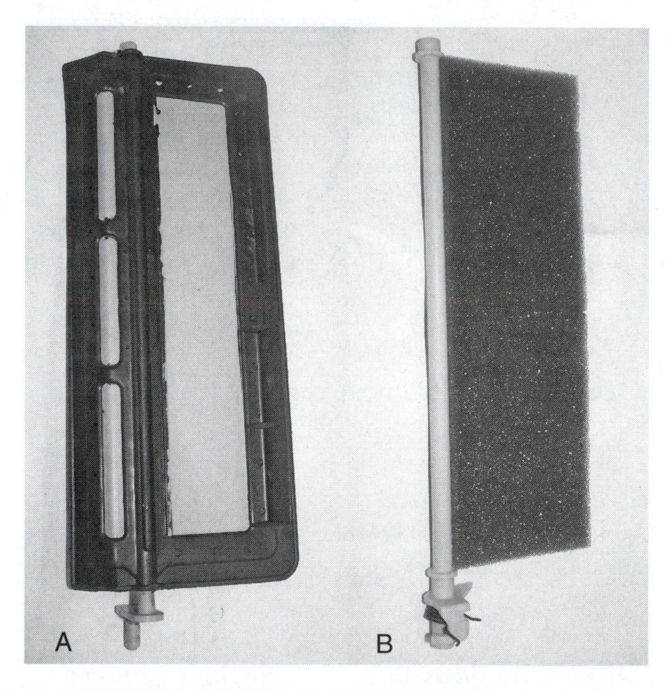

Figure 15. Door A has a center axis and controls two openings simultaneously. Door B is a more traditional side pivot.

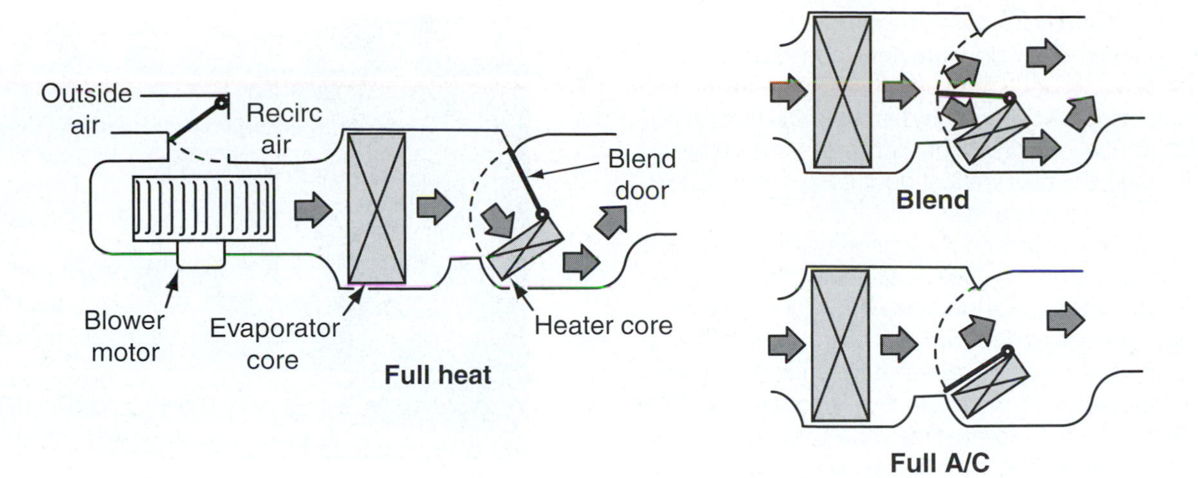

Figure 16. The blend door is the primary air temperature control in ALL operational modes.

control will have two blend doors, one for each side of the passenger compartment. A few vehicles have a restrictor door after the heater core that further blocks airflow to the core in MAX AC mode.

The blend door is situated just after the evaporator core. If the AC compressor is activated, then the air reaching the blend door will be chilled. It should be pointed out that the refrigeration system is always working at maximum potential. It attempts to keep the evaporator core as cold as possible without it freezing up. The evaporator core temperature does not change in response to operator temperature requests. The cold air must be warmed by opening the blend door enough to allow all or part of the air to pass through the heater core if the operator wants the discharge air warmer.

Mode Doors

After the air passes through the evaporator core and, if required, the heater core, where the air is discharged is determined by the mode doors. **Figure 17** illustrates a combined case that uses two mode doors. The first door that the conditioned air reaches is the heater-defrost door. This door is capable of being placed in three positions, labeled A, B, and C. When the door is in position A, the air is diverted toward the AC-defrost door that has two positions, labeled A and B. When the AC-defrost door is in position A, the air is discharged through the panel vents. This is how the air travels in MAX AC, AC, and VENT modes of operation. When the heat-defrost door is in position B, the AC-defrost door will be in position A. This setting is used in

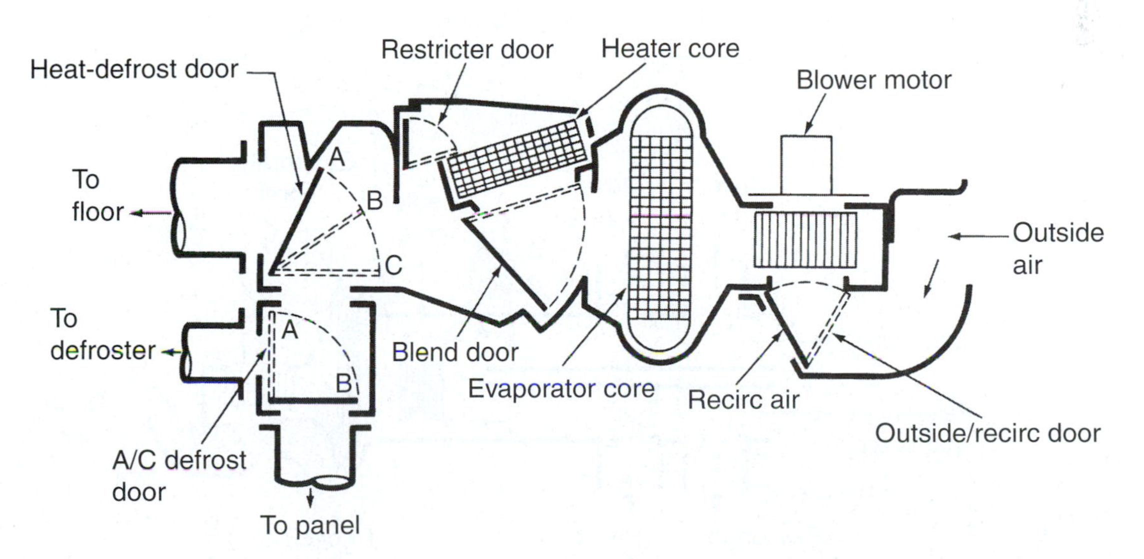

Figure 17. This plenum has two mode doors, one two-position and one three-position door.

BILEVEL mode and will allow some of the air to flow to the floor vents and the remainder to flow to the panel vents. If the heat-defrost door is in position A and the AC-defrost door is in position B, the HVAC system is in DEFROST mode and the air flows through the defroster vents. When in HEATER mode, the heat-defrost door is in position C, forcing the air to the floor vents.

There are many other door arrangements in use. Some systems may use only one door, whereas others may use three or four doors to control the airflow. One system has no doors in the actual ductwork, but each discharge outlet has an individual door that opens only when that desired mode is selected. A new system similar to the individual door system uses a device that resembles a windup curtain on a roller. There are slots in the curtain that allow air to flow through it. When a particular vent mode is desired, an electric motor winds the curtain on its roller until the slot openings line up with that desired vent outlet.

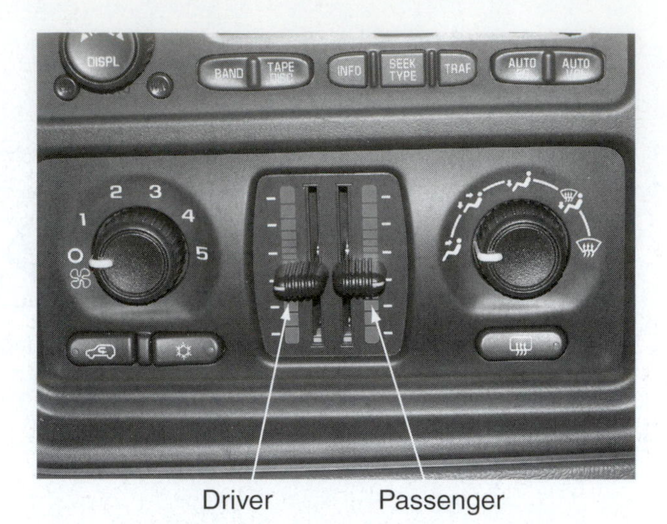

Figure 19. Dual temperature controls allow for greater comfort for all occupants.

DUAL ZONE

Dual zone climate control is designed to allow the driver and front passenger to adjust the air outlet temperature on their side of the vehicle to their personal preference. Both sides of the system must operate in the same mode. The key to the dual zone system is a redundant duct after the evaporator core. The system splits into two separate systems at that point, with each half having its own blend door (see **Figure 18**). The blend door controls the amount of air flowing through the heater core for that side of the vehicle. The cabin occupants move the blend door by adjusting their temperature control (see **Figure 19**). It should be noted that if one control is set to full cold or full warm, the other control cannot produce colder or warmer air from the other side of the system. Both sides will be working at their maximum efficiency.

CABIN AIR FILTERS

Air conditioning systems have always cleaned the air in addition to cooling it. The condensation that forms on the evaporator core traps dust particles as the air moves through the core. The dust is then carried out of the system as the water droplets drain from the system. In order to further improve the quality of the air, many modern vehicles are now equipped with a **cabin air filter** (see **Figure 20**), which cleans the air as it is circulated through the air distribution system. Some models place the filters just before the evaporator core, which forces all of the air going through the system through the filter in all operating modes. Other designs only filter the air that enters the vehicle through the outside air inlet. Most of these filters are paper element type filters that are similar to the flat square

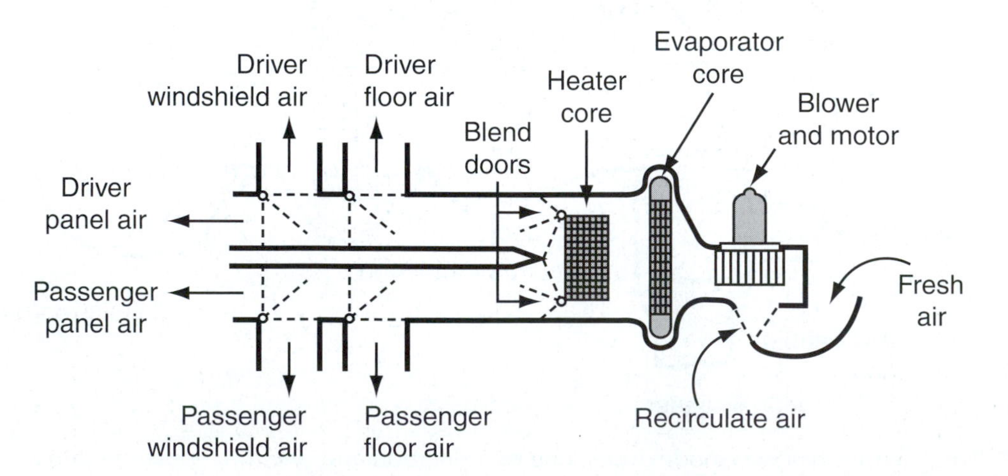

Figure 18. Both sides of the dual zone system have their own blend door and discharge ducts.

Figure 20. The cabin filter access door has been removed to see the filter located in the cowling. It cleans all fresh air entering the HVAC system.

air filters used on many cars today. Some of the filters also include a charcoal element to absorb odors and gasses as they pass through the system. The filters are a service item and must be replaced in order to maintain the efficient operation of the system.

CONDENSATION DRAINS

Cold air cannot support as much water as hot air. That is why a cold glass of water attracts so much moisture on a hot, humid day. The air around the glass cools and drops its moisture onto the glass as condensation. The exact same thing happens at the evaporator core as warm air passes over its fins. The air drops its humidity and coats the evaporator core with water. All evaporator core cases are equipped with a drain to channel this water outside the vehicle.

Under certain humid weather conditions, the droplets on the evaporator fins can turn to ice and a visible frost vapor can be blown out the discharge vents. Some people have mistaken this vapor for refrigerant leaking from the system. In order to prevent this from happening, manufacturers have placed a porous sheet of foam on the discharge side of the evaporator core to capture the blown frost.

CONTROL HEADS

The HVAC control head is the operator interface with the system. There are as many designs for these dashboard-mounted units as there are models of vehicles. However, they all provide the same basic functions for controlling the system. The head must be able to control the blower motor speed, air discharge temperature, operating mode, compressor activation, and system air input source. If the vehicle is equipped with a rear HVAC system, the front control head will be capable of controlling both systems. There will usually be a second control head in the rear of the vehicle that will allow the rear passengers to control the second unit. The front control head is the primary controller and can override the rear controller. Later chapters will explore the main two divisions of control heads: manual and electronic. Although both accomplish the same task, they go about it in very different ways.

REAR AIR

The available area under the dash for an evaporator core in most vehicles is very limited. This size restriction means the evaporator core can handle only a certain amount of heat. As vehicles get larger, one evaporator cannot handle the heat load generated within the cabin to keep the passengers comfortable. Therefore, most vans and SUVs are now equipped with dual or rear HVAC units. The second units are roof mounted above the rear doors or along the side panels next to the rear passenger seats.

These second plenums have almost all of the features of the front units. They are equipped with evaporator and heater cores, blower assemblies, blend doors, and mode doors. The one area in which they do differ from the front unit is their air intake. These second units operate in the recycle mode at all times. They lack the ability to draw in outside air. The evaporator core has its own expansion device, but, other than that, it shares the rest of the refrigeration system with the front evaporator core. Long hoses bring refrigerant back to the rear evaporator core and hot water to the rear heater core.

The rear HVAC units usually have an auxiliary control head that is accessible to the rear passengers. This allows them to control output temperature, blower speed, and operating mode on the rear unit only. The front control head has the ability to override the auxiliary control head.

Summary

- The air distribution system controls the temperature, volume, quality, and location of the air discharged into the passenger compartment.
- Air intake options are outside fresh air and inside recirculate air.
- Recirculate air produces the coldest air conditioned discharge temperature
- A squirrel cage fan and blower motor forces air through the air distribution system.

- The plenum assembly contains the evaporator core on a split system. In a combined system, it also houses the blower assembly, heater core, blend door, and mode doors.
- The HVAC system discharges air through three vent options: defrost, panel, and floor.
- Operating modes include MAX AC, AC, BILEVEL, VENT, HEATER, and DEFROST. On many vehicles, these modes will automatically control the input air control and AC compressor operation. Newer vehicles usually allow the operator to control these functions directly.
- Dual zone climate control may allow the passenger to vary the output air temperature by as much as 30 degrees F.
- Cabin air filters improve the cleanliness and aroma of the conditioned air.

Review Questions

1. Two technicians are discussing a vehicle with dual zone climate control. Technician A says that if the driver has the main control set to MAX AC with the temperature control on full cold, the passenger can set their control for 30 degrees colder. Technician B says that if the driver has the main control set to MAX AC, the passenger can set their control for DEFROST. Who is correct?
   A. Technician A only
   B. Technician B only
   C. Both Technician A and Technician B
   D. Neither Technician A nor Technician B
2. The air intake is set to outside air in all operating modes *except* for:
   A. MAX AC
   B. BILEVEL
   C. VENT
   D. DEFROST
3. Technician A says that one section of a split plenum assembly may contain only the evaporator core. Technician B says that a combined case also can house the heater core and the operating doors. Who is correct?
   A. Technician A only
   B. Technician B only
   C. Both Technician A and Technician B
   D. Neither Technician A nor Technician B
4. Technician A says that the AC fan blows harder in RECIRCULATE mode. Technician B says that BILEVEL delivers most of the air to the panel vents. Who is correct?
   A. Technician A only
   B. Technician B only
   C. Both Technician A and Technician B
   D. Neither Technician A nor Technician B
5. Technician A says that the AC compressor always works in BILEVEL mode. Technician B says that the compressor is never on in VENT mode. Who is correct?
   A. Technician A only
   B. Technician B only
   C. Both Technician A and Technician B
   D. Neither Technician A nor Technician B
6. The ______________________ is the main control device for the air discharge temperature.
7. Technician A says that the discharge air is always cold in AC mode if the system is working correctly. Technician B says that the discharge air is always hot in HEATER mode. Who is correct?
   A. Technician A only
   B. Technician B only
   C. Both Technician A and Technician B
   D. Neither Technician A nor Technician B
8. Technician A says that the cabin temperature will be coldest in MAX AC mode. Technician B says that regular AC mode will be just as cold if the temperature lever is set to its coldest position. Who is correct?
   A. Technician A only
   B. Technician B only
   C. Both Technician A and Technician B
   D. Neither Technician A nor Technician B
9. Technician A says that the AC compressor may be activated by mode selection. Technician B says that the operator pressing a button may activate the AC compressor. Who is correct?
   A. Technician A only
   B. Technician B only
   C. Both Technician A and Technician B
   D. Neither Technician A nor Technician B
10. Technician A says that BILEVEL flows air to the panel and floor vents on most vehicles. Technician B says that BILEVEL flows air to the defrost and floor vents on most vehicles. Who is correct?
    A. Technician A only
    B. Technician B only
    C. Both Technician A and Technician B
    D. Neither Technician A nor Technician B
11. ____________________ mode may be the most efficient AC setting when the interior air is warmer than the ambient temperature.
12. Technician A says that rear AC units work in recirculate mode all the time. Technician B says that rear HVAC controls can override the front dash controls. Who is correct?
    A. Technician A only
    B. Technician B only
    C. Both Technician A and Technician B
    D. Neither Technician A nor Technician B

Chapter 27 Manual Air Distribution Systems

Introduction

Air distribution controls must allow the operator to adjust air output mode, blower speed, air temperature, compressor activity, and intake air source. **Manual air distribution** control has been the industry standard since the first heater was installed in an automobile. It has only been since the advent of the automotive computer that automated temperature controls, which will be studied in Chapter 28, have become commonplace. The operator has direct control over the operation of each adjustment in a manual control system. If the operator wants the temperature warmer or cooler, the fan to blow harder or softer, or the air to blow out of a different vent, then the operator must physically change the setting using the **control head**.

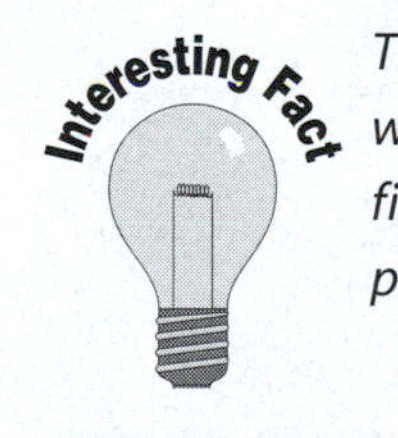

The earliest climate control device was a small firebox that could be filled with coals to warm the feet of passengers.

Traditionally, manual control has indicated the use of strictly mechanical devices. Modern vehicles, even those with manual HVAC controls, use more electronic devices to control the various outputs of the system. This chapter will explore all manual systems, both mechanical and electronic.

MANUAL CONTROL HEADS

The manual control head (see **Figure 1**) is the interface between the operator and the HVAC system. The head consists of a set of levers, dials, switches, and buttons, each

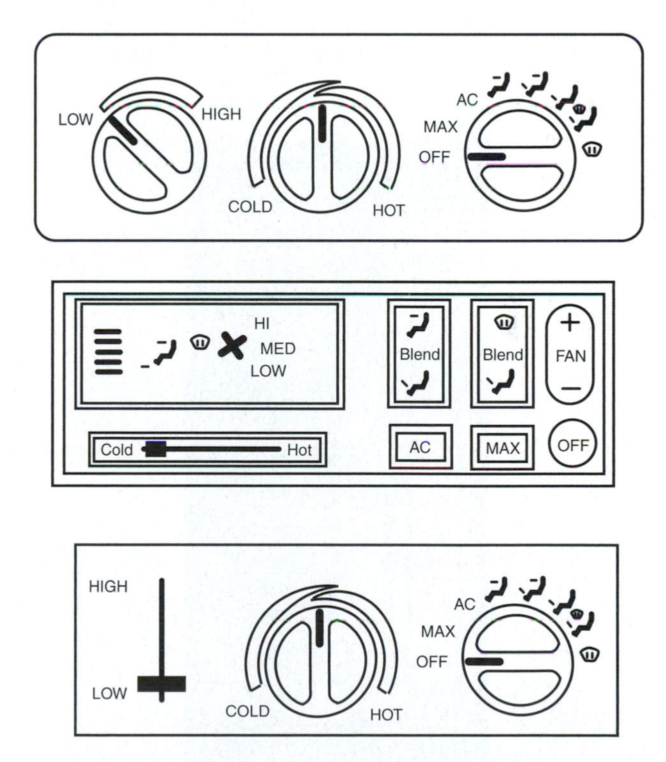

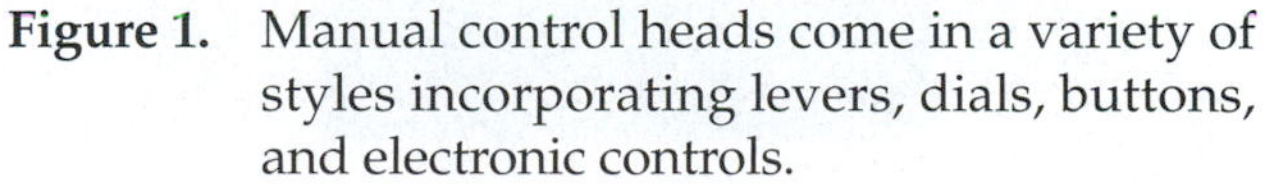

Figure 1. Manual control heads come in a variety of styles incorporating levers, dials, buttons, and electronic controls.

designed to control a certain function of the system. All control heads have certain functions in common. They all have a mode controller. The mode control directs where the air distribution system is to discharge the conditioned air. (See Chapter 26 for a detailed description of each operating mode.) On some vehicles, the mode control also activates the AC compressor and the air intake door. Another standard control head device is the temperature controller. The temperature control has direct control over the blend door. Labels on the control head bezel indicate which direction the control is to be moved to increase or decrease the discharge air temperature. The warmer this control is placed, the more air the blend door allows to flow through the heater core. On vehicles with dual zone climate control system, there will be two temperature controls, each adjusting their own blend door. The third standard control head function is blower speed control. This switch regulates the amount of current flowing to the blower motor by diverting the circuit through a stepped resistor. In addition to these standard controls, vehicles that do not control the AC compressor and the intake air door with the mode controller will have a separate button to directly control these items.

Levers

Levers (see **Figure 2**) are one of the simplest forms of panel controls. They are also called slide controls. The end of the lever protrudes through a slot in the control head. The pivot for the lever is mounted behind the bezel in the control head mechanism. When the operator moves one end of the lever, the lever works another device within the head, such as a Bowden cable or a rotary vacuum switch. Both of these devices are described later in this chapter. Levers also can be used to control a variable resistor in an electronic system or to activate a switch in a compressor control circuit.

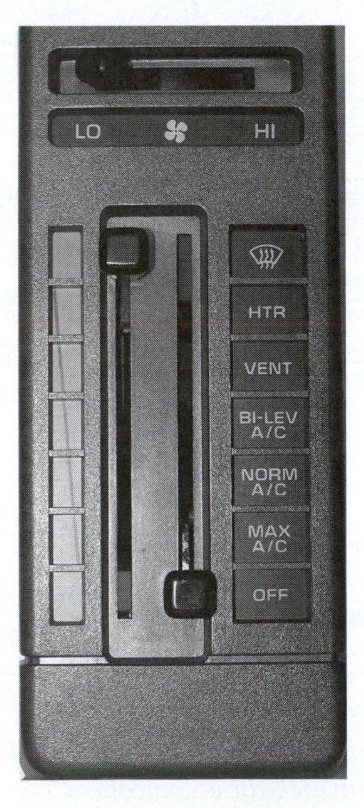

Figure 2. Levers are used to control mode, temperature, and blower speed.

Dials

On mechanical control heads, dials (see **Figure 3**) usually function as rotary levers. They are capable of attaching to any device that a lever can attach to, such as a Bowden cable, rotary vacuum switch, variable resistor, or an electrical switch. Dials have become more popular in recent years because they take up less room on the control panel than levers. Dials are used for mode control, temperature control, and blower speed control.

Push Button

Some models use a push button assembly (see **Figure 4**) for mode control. Each operating mode has a separate button on the control head. When the operator presses the button for a particular mode, that button stays depressed and any other button previously pressed is pushed back up. Internally, the button control assembly must switch vacuum routing to the correct actuators and activate any required electrical switches to function in that mode.

Blower Speed Switches

Blower speed switches may be dials, levers, or buttons. On vehicles with mechanical heads, the switch routes the electrical current through a series of resistors. The more resistance it sends the current through, the slower the

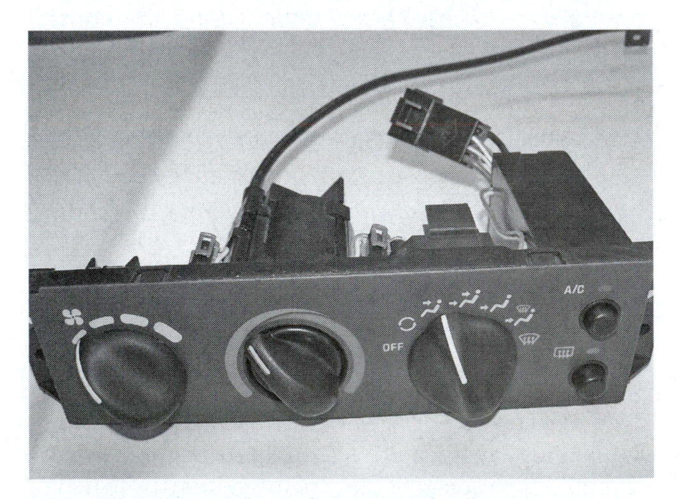

Figure 3. Dials are used as rotary levers, capable of attaching to any device that a lever can, such as a Bowden cable, rotary vacuum switch, variable resistor, or an electrical switch.

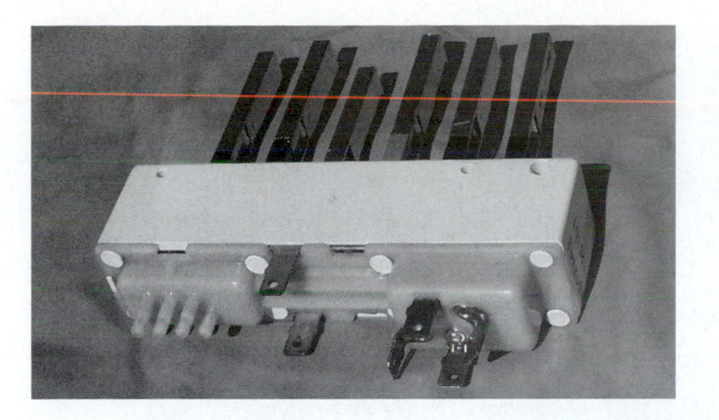

Figure 4. Each mode function has its own button on push button control heads.

blower motor runs. On electronic manual control heads, the blower speed switch activates a series of relays. The relays in turn direct the blower current through the blower resistors.

Single Function Buttons

The control panel also can have buttons that control a single HVAC function (see **Figure 5**). The most common are the AC and recirculate air buttons. The AC button activates the AC compressor. This button is usually available in all operating modes but usually requires the fan to be in an active (not off) position. The recirculate air button switches the air intake to draw inside air into the HVAC system. It may not be available on some models when defrost or heater mode is selected. The recirculate button also may activate the AC compressor as well. Although not an actual part of the HVAC system, some vehicles also may have a rear window defogger button on the control panel as well.

Figure 5. Single buttons can be used to select the intake air mode and activate the AC compressor.

BOWDEN CABLES

A Bowden cable (see **Figure 6**) is a simple device for transferring mechanical movement from one location to another. It consists of a flexible tubular housing with a sliding steel cable running through the middle of it. When used as a control device, both ends of the housing are held in place. One housing end will be attached to the control head, where the sliding cable is connected to a moveable device, such as a lever. As the lever is moved back and forth, the cable is moved in and out of the housing. This forces the other end of the cable in and out of the other end of the housing. This end can then move a device such as a blend door. There is often an adjustment at one end or the other to accommodate wear and manufacturing variations.

> **Interesting Fact**
> *The Bowden cable was invented by Frank Bowden (1848–1921), who was the founder of the Raleigh Bicycle Company.*

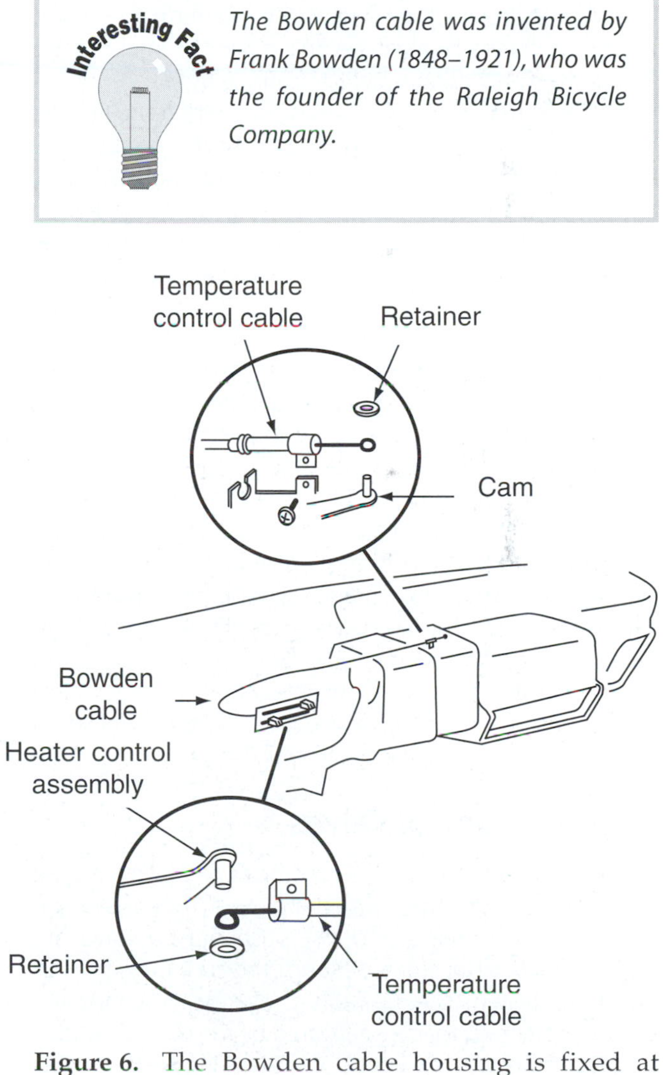

Figure 6. The Bowden cable housing is fixed at both ends allowing the cable to transfer movement.

VACUUM CONTROLS

Vacuum controls were the primary means of controlling the mode control doors for many years. They proved to be safe and reliable as well as easy to manufacture. Unfortunately, they are also bulky under today's streamlined dashes. They also can steal a little vacuum from the engine, which can be detrimental to its performance. Although still in use, they are quickly being replaced by electric controls.

Figure 7. Vacuum reservoirs store engine vacuum for use by the air handling system when engine vacuum is low.

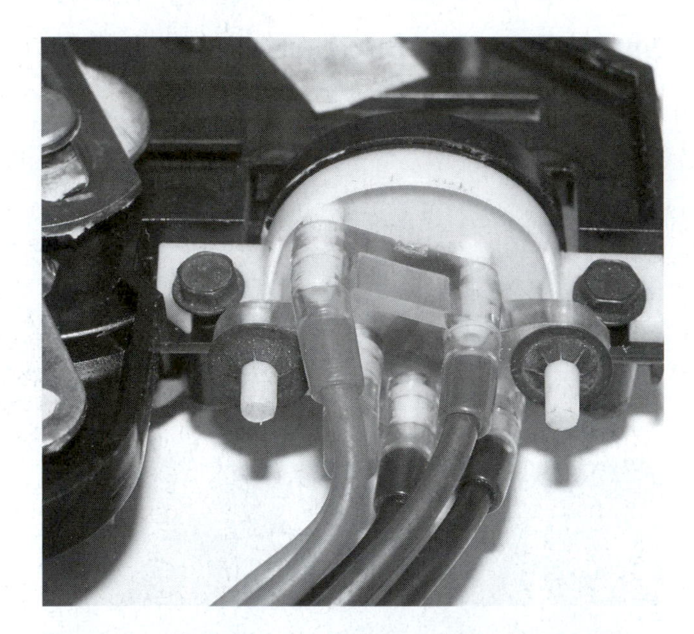

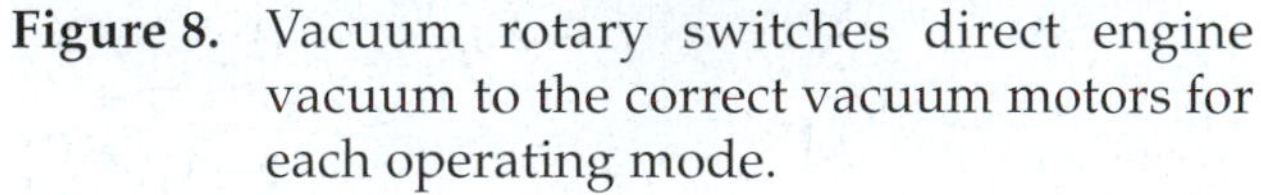

Figure 8. Vacuum rotary switches direct engine vacuum to the correct vacuum motors for each operating mode.

Vacuum Sources

Vacuum is any pressure that is less than atmospheric. The vacuum used by the HVAC system can come from two sources: a gasoline engine or a vacuum pump. Gas engines create vacuum when the piston is moving downward in the cylinder and the intake valve is open. This generates a low-pressure area in the cylinder that draws air in from the intake manifold. If air is restricted from flowing into the intake manifold to replace this air, such as when the throttle is closed, then it creates a low-pressure (vacuum) area in the intake manifold. A port can be drilled into the intake manifold so that other control systems can use the vacuum.

An external vacuum pump is used when the engine cannot generate sufficient vacuum by itself to satisfy the needs of the auxiliary systems using the vacuum. This is always the case in diesel motors, as they do not naturally generate a vacuum because they do not have a throttle. Gasoline motors also may use a vacuum pump if the demands on the vacuum supply are so heavy that it could interfere with the operation of the engine. Regardless of the source, the vacuum is carried through rubber and plastic hoses to where it can be utilized by the system.

There is one problem with using vacuum to operate actuators. When a gasoline engine is operated under a heavy load or at wide-open throttle, the vacuum supply drops to zero. Unchecked, this could cause all of the vacuum actuators to move to their default position. To prevent this, vehicles have a vacuum storage reservoir (see **Figure 7**) to supply the system with vacuum under these conditions. In order to keep the vacuum in the reservoir from being drawn back into the engine, there is a one-way check valve that keeps the vacuum in the storage tank. On some models, the valve is built into the tank itself.

Vacuum Control Valves

There are three common methods of controlling the flow of vacuum to the correct actuators. Two of the systems, the rotary valve and the push button valve, are mechanical devices that function at the control head itself. The third system uses a bank of electrically controlled vacuum solenoids that are activated by the control head.

The rotary vacuum control valve (see **Figure 8**) consists of two plastic disks clamped on each side of a rubber spacer. The rear disc has six to eight vacuum hose ports. One of the ports is for the source vacuum input from the engine. The remaining ports go to different actuators. The front disc and the rubber spacer rotate together when the operator moves the mode selector. There are small channels formed in the rubber spacer that, when rotated, will connect the vacuum input port with the desired actuator port(s) for each mode position. The valve has one position for each mode. For example, if the mode controller had seven possible modes, then the rotary switch will have seven positions.

Push button vacuum switches, as described earlier, have an internal vacuum switching device that directs the vacuum to the correct actuator when each mode is pressed on the control panel.

Figure 9. Vacuum solenoid packs allow electronic control heads to control manual vacuum motors.

Vehicles with electronic control heads have no mechanical device for routing vacuum. These heads must rely on a vacuum solenoid pack (see **Figure 9**) to control the vacuum to the actuators. A vacuum solenoid pack has one solenoid for each actuator in the system. Source vacuum and battery voltage is supplied to each solenoid in the pack. When the control head wants to activate an actuator, the head provides an electrical ground for the related solenoid. The solenoid valve then opens and allows vacuum to the actuator to move the door.

Vacuum Motors

A vacuum motor (see **Figure 10**) or actuator uses the differential force between atmospheric pressure and vacuum to create movement. The motor consists of a chamber with a flexible diaphragm across its center, dividing the chamber into two sections. One side of the chamber is exposed to atmospheric pressure, whereas the other side has a port that connects it to vacuum. There is a spring in the vacuum side of the chamber that pushes the diaphragm toward the atmosphere side. When vacuum is applied to the motor, the atmospheric pressure becomes greater than the spring pressure and pushes the diaphragm in the opposite direction. When a rod is connected to the center of the diaphragm and it extends through the atmosphere side of the chamber, it can move a door to two positions. Another style of vacuum motor that has two different vacuum sections can move the door to three positions.

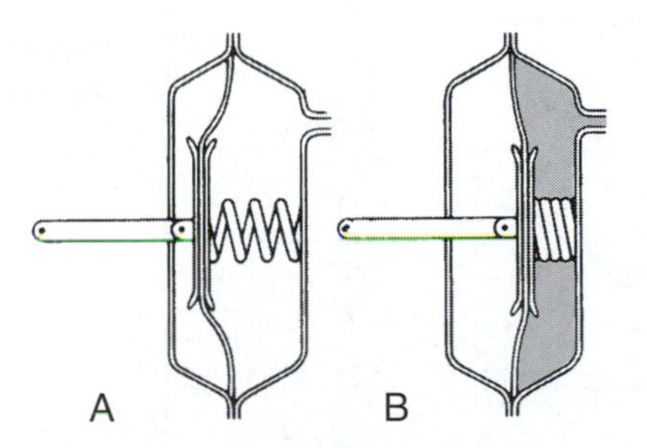

Figure 10. The vacuum motor without vacuum (A) is positioned by the internal spring. When vacuum is applied (B), the diaphragm retracts, moving the door.

ELECTRIC MOTOR ACTUATORS

Modern automobiles are more likely to use electronic controls that use electric motors to move the air handling doors in the air distribution system. The electric motor is housed in an actuator assembly (see **Figure 11**) that mounts on the side of the case and attaches to the door shaft. There are several different styles of actuators. How the door needs to be controlled can indicate the style of actuator that is used.

The simplest design is that of a door that has only two positions, such as an air inlet door. **Figure 12** shows the wiring for this type of actuator. When the control head needs to switch the door position, it supplies voltage to one wire, while the other one remains grounded. There is no feedback to the control head indicating the actual position of the door. The program in the control head moves the door for a predetermined amount of time and then terminates the voltage supply. To position the door back to the original location, the controller reverses the polarity of the wires.

Figure 11. Several different types of electric actuators use this same housing.

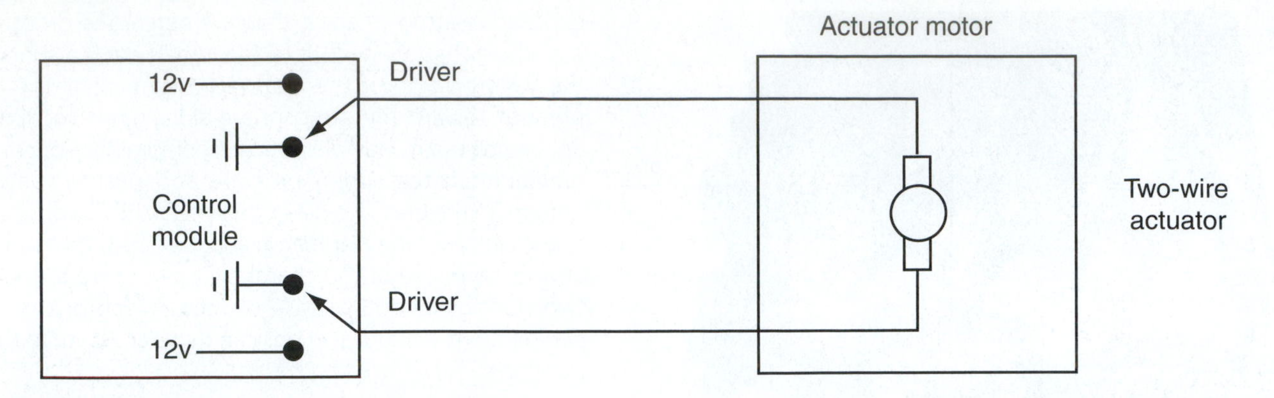

Figure 12. These two-wire actuators can only position the door in two positions, one at each end of its travel.

Figure 13 shows a second type of actuator. The control head moves the door exactly like the first actuator described above. However, there is a variable resistor in the actuator that changes resistance as the motor moves the door. This change in resistance produces a feedback voltage to the control head, indicating the position of the door. This allows the control head to adjust the door position over a wide range of variable positions.

This type of system calibrates itself from time to time. To accomplish this, the control head moves the door all the way in one direction and takes a reading from the feedback circuit. It then moves the door to its opposite stop and takes another feedback reading. The control head can then calculate the full range of movement for the door and allows it to position the door anywhere between the two stops.

The feedback design works well for controlling a blend door. To change the output temperature, the operator moves the temperature control knob. This sends a voltage signal to the electronic controller in the control head, similar to the feedback signal from the actuator. The controller then determines the percentage of heat the operator wants and moves the blend door until the feedback from the door actuator matches the signal from the control knob.

The most advanced style of actuator has a logic module built directly into the actuator itself. The logic module has the actual control over the door motor. **Figure 14** shows how the control head sends a variable voltage signal to the actuator that is based on the operator's adjustment of the temperature control knob. The module then calculates the correct position for the blend door and adjusts it accordingly.

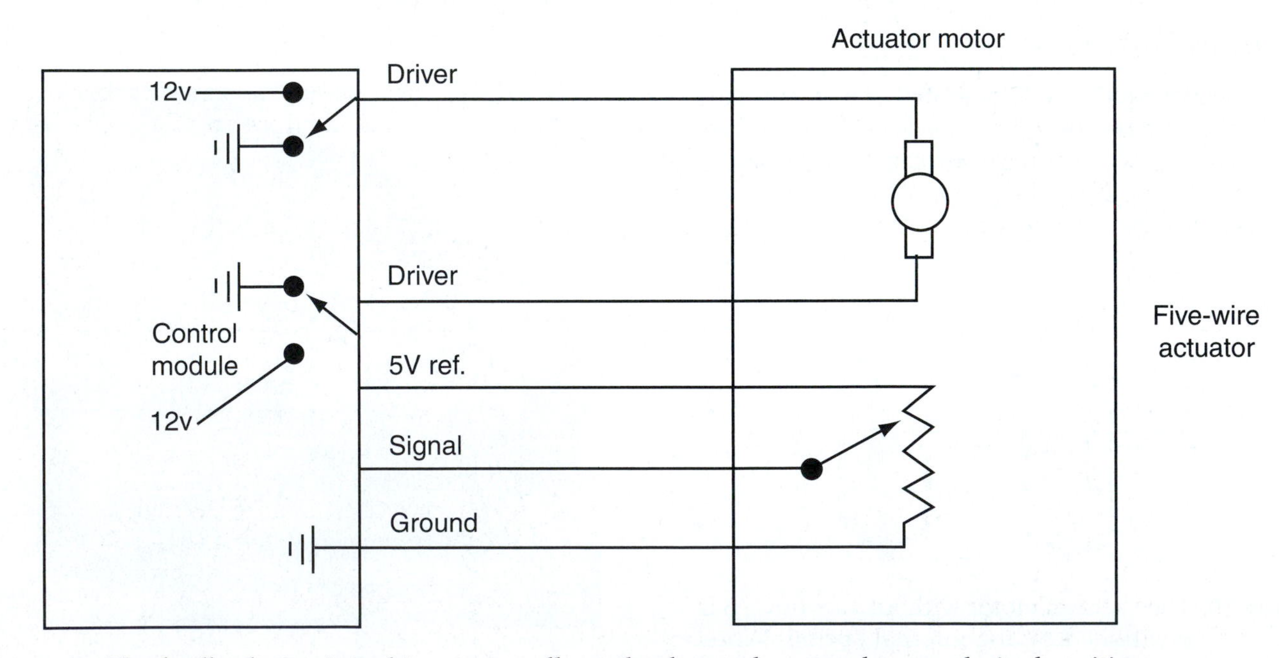

Figure 13. The feedback circuit in this actuator allows the door to be moved to any desired position.

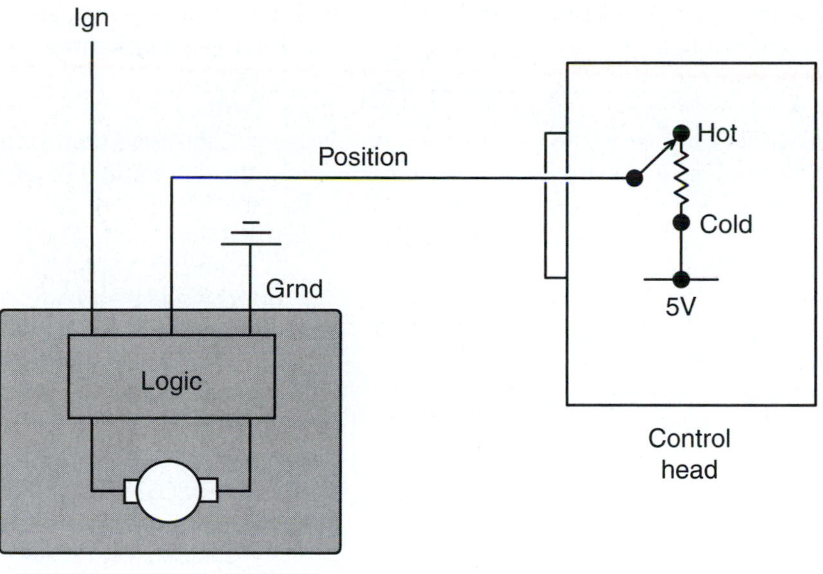

Figure 14. The logic module in the actuator decodes the signal from the control head and positions the door.

Summary

- Manual controls traditionally have meant a mechanical device. Today they are more likely to be electrical controls.
- The operator must make all operational adjustments to the HVAC system for manual control systems.
- The control head is the interface between the HVAC system and the operator, allowing mode, temperature, and blower speed adjustment.
- Levers and dials are used to control Bowden cables, vacuum switches, variable resistors, and electric switches on the control head.
- Single function buttons on many control heads control AC compressor and recirculate door operation.
- Vacuum is any pressure less than atmospheric pressure and can be used with atmospheric pressure to create movement.
- A rotary valve, push button switch, or a solenoid pack may control vacuum systems.
- Depending on design, vacuum motors may be used to position mode doors in two or three positions.
- Electric actuators can control a door to a varying range of positions.
- Mode doors are controlled by vacuum motors or electric actuators. Blend doors are controlled by an electric actuator or Bowden cable.

Review Questions

1. Technician A says that vacuum is always present in a gasoline engine. Technician B says that an auxiliary pump is needed for a diesel engine to generate vacuum. Who is correct?
   A. Technician A only
   B. Technician B only
   C. Both Technician A and Technician B
   D. Neither Technician A nor Technician B
2. Technician A says that a Bowden cable can be used to control the blend door. Technician B says that the mode control switch regulates the temperature of the discharge air. Who is correct?
   A. Technician A only
   B. Technician B only
   C. Both Technician A and Technician B
   D. Neither Technician A nor Technician B
3. The ______________________ is the interface between the operator and the HVAC system.
4. Technician A says that the AC compressor may be activated by the mode control switch. Technician B says

that the AC compressor may be controlled by a button on the control panel. Who is correct?

A. Technician A only
B. Technician B only
C. Both Technician A and Technician B
D. Neither Technician A nor Technician B

5. Technician A says that mode levers are more popular on late model vehicles. Technician B says that electronic controls have become more popular. Who is correct?
   A. Technician A only
   B. Technician B only
   C. Both Technician A and Technician B
   D. Neither Technician A nor Technician B
6. Technician A says that a relay is used to control the current to individual vacuum solenoids. Technician B says that vacuum pushes on the diaphragm to move the doors. Who is correct?
   A. Technician A only
   B. Technician B only
   C. Both Technician A and Technician B
   D. Neither Technician A nor Technician B
7. Electrical actuators with two wires do not include a feedback circuit.
   A. True
   B. False
8. Some of the newer electrical actuators have an internal chip that controls the door.
   A. True
   B. False
9. Technician A says that some blower speed switches rout the current to a resistor assembly. Technician B says that some of the blower switches must use relays to direct the current. Who is correct?
   A. Technician A only
   B. Technician B only
   C. Both Technician A and Technician B
   D. Neither Technician A nor Technician B
10. Technician A says that recirculate air is available in all modes if it has a single function control button. Technician B says that recirculate air may not be available in DEFROST mode. Who is correct?
    A. Technician A only
    B. Technician B only
    C. Both Technician A and Technician B
    D. Neither Technician A nor Technician B

Chapter 28 Automatic Temperature Control

Introduction

Automatic Temperature Control (ATC) provides the greatest level of comfort and convenience for the automotive passenger. The operator only needs to input the desired temperature of the passenger compartment. The electronic controls will then adjust all of the system parameters to bring the cabin to that temperature as quickly and comfortably as possible. The ATC system accomplishes this by monitoring key sensors that gather information about the operating environment. These sensors can monitor the temperature in the passenger compartment, the outside ambient temperature, the air temperature coming out of the various vents, the intensity of the sun shining on the vehicle, and the position of the air management doors.

ATC CONTROL HEADS

An ATC control head (see **Figure 1**) closely resembles the control heads used in electronic manual systems. The operator will still have access to mode, temperature, blower speed, and other controls (covered in Chapter 27). Two main items set ATC heads apart from their manual counterparts. One is the addition of an AUTO button, the other is how the temperature control is labeled on the face of the control head. Instead of the generic warmer to colder or red to blue descriptors, an ATC system will have actual numeric values representing desired cabin temperature. It will usually be in the form of a digital readout, capable of displaying values ranging from 60 to 90 degrees. The ATC system has two operational modes: full automatic control and semi-automatic control.

Figure 1. Automatic control heads can closely resemble manual controls. Note the AUTO positions on the dials.

Full Automatic Operation

To run the HVAC system in automatic mode, the operator need only press the AUTO button and select the desired interior temperature. Once this is done, any HVAC buttons that were previously selected are ignored. The HVAC controller now has total control over the operation of the system. The controller begins the control cycle by gathering information from its sensors. These sensors, which will be detailed later in this chapter, include ambient temperature, interior temperature, duct temperatures, sunload, engine coolant temperature, and engine RPM. The controller then uses this information to determine the HVAC setting that will best bring the cabin temperature to match the desired temperature input by the operator. It is capable of adjusting the input air door, blend door, all the mode doors, blower speed, and AC compressor operation. After all adjustments are made, the controller again surveys the input sensors to see how the adjustments have changed the passenger environment. It continues this cycle of gathering information and making adjustment as long as the vehicle is running or the operator makes a manual change on the control head.

Examining the blower operation gives a good example of how this control cycle works. If the ATC system is in AUTO mode and the operator sets the desired temperature to 68 degrees, the HVAC controller will examine its input values. If the actual cabin temperature is 93 degrees, the controller will recognize a large difference in the two temperatures. Because large differences mean passenger discomfort, improvements need to be made quickly. The controller will put the blower motor to its highest speed. However, it is undesirable to leave the blower motor on high until the two temperatures are equal, as it would leave the passengers feeling chilled. Calculations in the controller slow the blower speed down as the desired and actual temperatures approach each other. Of course, the controller also will be making similar changes in blend door position and other settings to bring the temperatures in line.

The typical adjustable range for the desired temperature is 65 degrees to 85 degrees. However, the control head is capable of being adjusted to 60 degrees or 90 degrees as well. At the extreme settings, the controller does not make continuous adjustments to the controls. The system is placed in maximum cold or maximum heat mode respectively. The HVAC controller will continue this peak output until the operator changes the system settings.

Semi-Automatic Operation

Despite the natural design of the system, manual operation of many of the control functions is possible. The operator can manually set blower speed, air intake, AC compressor operation, and air discharge mode. To enter manual operation, simply press the related button (or buttons) on the control panel (see **Figure 2**). The designated device will be moved to the selected position, as long as it does not conflict with basic operational design, such as selecting recirculate air when the system is in defrost mode. An important thing to remember is that any setting not selected for manual operation will remain under automatic control. Temperature control always remains in automatic mode. The controller will attempt to maintain the set temperature to the best of its ability under the manual settings selected.

Figure 2. Operator input to HVAC operation is still available for all functions.

Interesting Fact

Automatic temperature control is even functional when the control head is set to OFF. The HVAC controller will continue to adjust the blend door based on the last desired temperature setting and the sensor readings. Any airflow through the ducts will be conditioned by the amount of air flowing through the heater core.

The first time a manual button is pressed, it does not change the setting for that device. When in automatic mode, the device is switched to manual mode, but at the same setting that it was operating. For example, if the fan is operating at medium speed, and the operator pushes the fan speed increase button, the fan stays at medium until the FAN or AUTO button is pushed.

HVAC CONTROLLER

The HVAC controller is a microprocessor capable of evaluating information and formulating decisions based on that information (see **Figure 3**). Depending on the design of the vehicle, it can be located in a variety of places. In many of our modern vehicles, it is located in the HVAC control head. In other designs, the control head is only an input device for the operator. If this is the case, the controller may be a separate module dedicated strictly to the HVAC system or it may be incorporated into a larger device, such as a Body Control Module (BCM) or Instrument Panel Module (IPM). Some older designs had an additional device called a programmer that contained some electronics as well as the actuators that controlled the system. Regardless of where it is located, the controller keeps the system operating efficiently to maximize the comfort level of the occupants.

SENSORS

The sensors provide the interface between the real world and the electronic control module. They allow the computer to feel what the occupants are feeling: the cabin temperature, the discharge duct temperature, and the sun shining on the vehicle. The sensors also provide information about how the system is operating, such as door positions and how fast the vehicle is running. This is the critical information that the ATC system needs in order to adjust the system properly to maintain optimum performance.

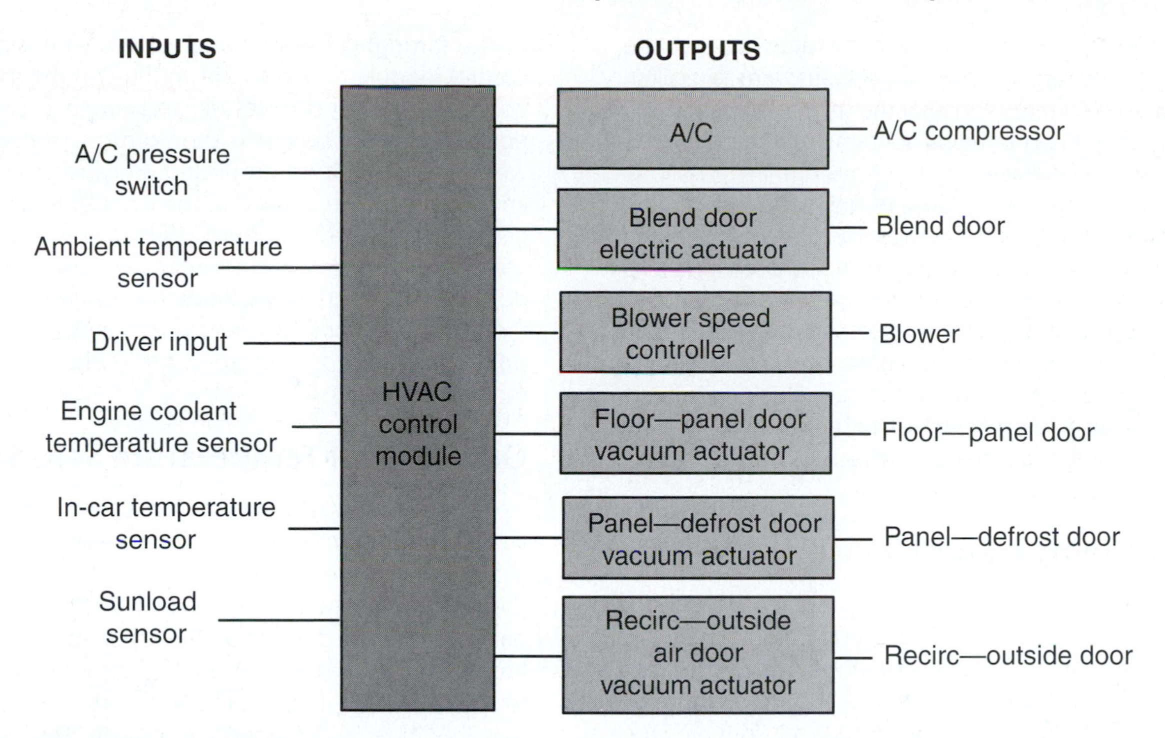

Figure 3. The HVAC control module evaluates sensor inputs to control operation of the system actuators.

Ambient Temperature Sensors

The ambient or outside temperature sensor (see **Figure 4**) lets the HVAC controller know the air temperature surrounding the vehicle. It is usually mounted on the support brackets in front of the condenser; however, a few older models used to mount it in the driver's door jam. The sensor is a negative temperature coefficient thermistor. This means that the sensor changes its resistance with temperature changes and that its resistance goes up as it gets colder and goes down as it gets warmer. The sensor is calibrated to have a certain resistance at a specific temperature. This information is used to display the outside temperature on the dash and to help control the output levels of the HVAC system.

The ambient air temperature circuit (see **Figure 5**) originates in the control module as a 5-volt reference signal. It then passes through a fixed internal resistor. It is at this point, after the internal resistor, that the control module monitors the voltage on the circuit. The circuit then leaves the module and connects to the ambient temperature sensor. After that, it returns to the module where the circuit is grounded. Because the module knows the voltage applied to the circuit, the ohms specification of the internal

Figure 4. The ambient temperature sensor allows the HVAC controller to compensate for outside conditions.

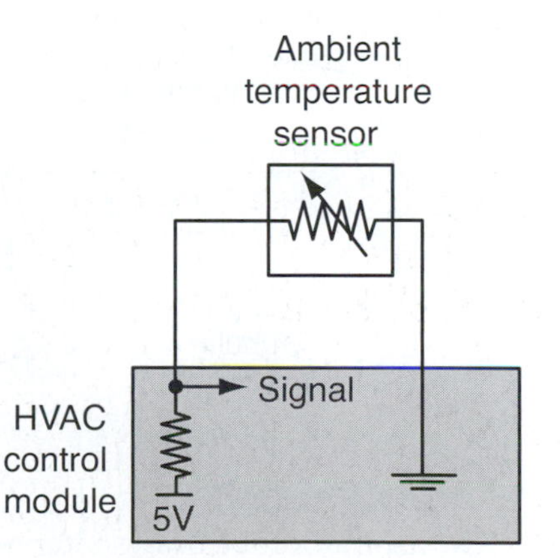

Figure 5. The ambient temperature sensor circuit includes the sensor, which is an NTC thermistor.

fixed resistor, and can read the voltage value between the resistor and the sensor, it can do a simple Ohm's law calculation to derive the resistance of the sensor. This resistance can be used to look up the sensor temperature in an internal table in the module.

Problems can arise because of the mounting location of the ambient temperature sensor. On a hot engine restart or slow stop-and-go driving, the sensor might pick up heat from the engine cooling system and the condenser. In order to prevent false readings, software in the control module can adjust this reading for better accuracy. The software uses information such as vehicle speed, engine temperature, ambient temperature when the vehicle was shut off, as well as timers to help correct for a more accurate reading.

Interior Temperature Sensor

The interior or inside temperature sensor lets the HVAC controller know the air temperature of the passenger compartment. It is usually mounted inside the dash within a small tube, called an aspirator tube (see **Figure 6**), which runs from a port in the dash panel to a low-pressure area in the air management plenum. The low pressure draws air into the tube from the passenger compartment and across the sensor. Some vehicles also use a small fan to force cabin air across the sensor. The sensor is a negative temperature coefficient thermistor that changes its resistance with temperature variations, its resistance going up as heat is removed and going down as heat is added. The sensor has a specific resistance at a specific temperature. The data from this sensor is the primary input for adjusting the automatic temperature control system. Some makes only use the interior sensor for the first 10 to 15 minutes of operation and then switch to the output duct temperature sensors.

The interior air temperature circuit originates in the control module as a 5-volt reference signal and then passes through a fixed internal resistor. At this point, the control module monitors the voltage on the circuit. The circuit then leaves the module and connects to the interior temperature sensor. It then returns to the module where the circuit is grounded. Because the module knows the voltage applied to the circuit, the ohms specification of the internal fixed resistor, and can read the voltage value between the resistor and the sensor, it can do a simple Ohm's law calculation to derive the resistance of the sensor. This resistance can be used to look up the sensor temperature in an internal table in the module.

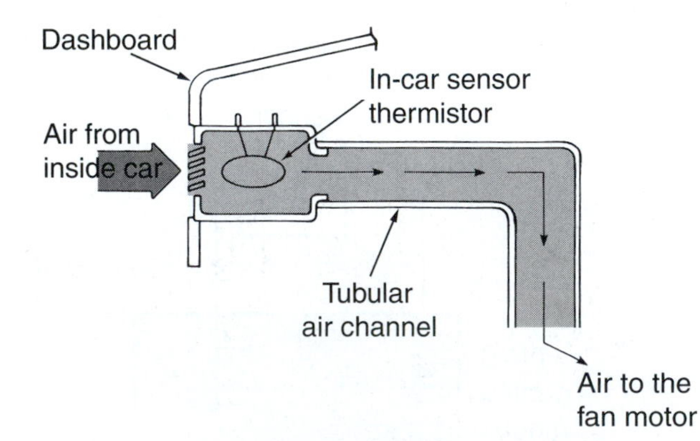

Figure 6. The aspirator tube draws interior air across the sensor to a low-pressure area in the plenum.

Output Duct Temperature Sensors

The output duct temperature sensors (see **Figure 7**) let the HVAC controller know the temperature of the air as it comes out of the specific duct. On single HVAC systems, there are usually two sensors: one mounted in the center panel discharge duct and the other mounted in the heater discharge duct. On dual zone systems, there are four sensors split across each side of the vehicle to provide for driver and passenger controls. The sensor is a negative temperature coefficient thermistor that changes its resistance with temperature variations, its resistance going up as heat is removed and going down as heat is added. The sensor has a specific resistance at a specific temperature. The data from this sensor is used for the primary input for adjusting the automatic temperature control system on some makes after the first 10 to 15 minutes of operation.

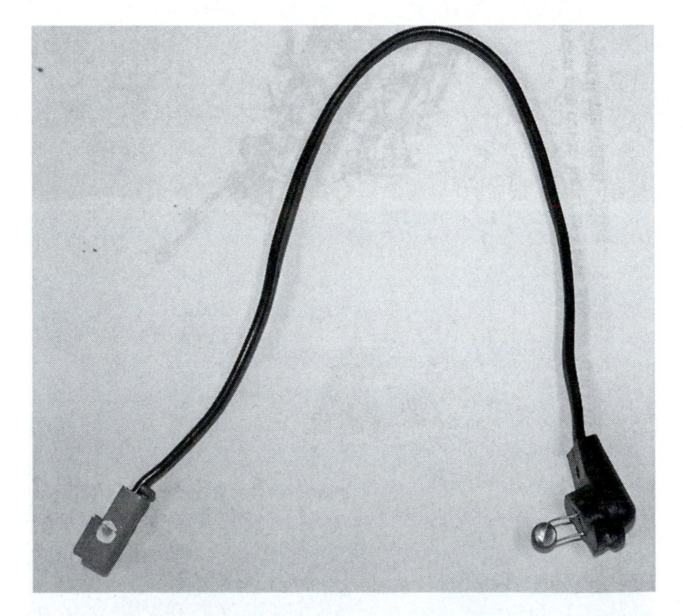

Figure 7. The tiny head of this duct temperature sensor extends into the duct to monitor the air entering the passenger compartment.

The output duct temperature sensor circuits operate identically to the ambient and interior air temperature circuits. Refer to those sections of the chapter for a description of the circuit operation.

Sunload Sensors

Sunload sensors (see **Figure 8**) are photo-diode devices. In darkness, they block current flow through the circuit. When light shines on it, it allows current flow. As the light intensifies, it allows more current to flow through it. The sunload sensor circuit originates in the control module as a 5-volt reference signal. It then passes through a fixed internal resistor. The control module monitors the voltage on the circuit at this point. The circuit then leaves the module and connects to the sunload sensor. After that, it returns to the module where the circuit is grounded. If the module sees 5 volts on the sensor circuit, it assumes there is no sunlight shining on the vehicle. As the voltage drops, the module knows the sunlight is intensifying.

Sun heat load is an important factor in keeping the passenger compartment comfortable. A sunny day requires much more heat to be removed from the cabin air than a cloudy day. High light readings from the sunload sensor can cause the fan to blow at a higher speed and the blend door to keep the air cooler in order to remove the additional heat. Single systems will have one sunload sensor in the middle of the top of the dash. Because it can be sunnier on one side of the vehicle than the other, dual zone systems have one on each side of the dash: one for the driver control and one for the passenger control. The more intense the sun is on that side of the vehicle, the cooler the discharge air temperature on that side.

You Should Know

Sunload sensors do not respond to fluorescent light. The system will interpret exposure to fluorescent light as darkness.

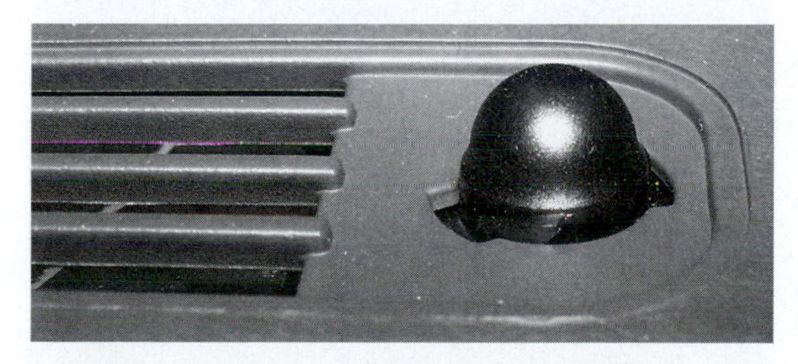

Figure 8. The sunload sensor adjusts HVAC controls to improve passenger comfort on sunny days.

ACTUATORS

The actuators used to control door operation with ATC are very similar to those used with electronic manual control systems. A few systems still use vacuum control motors controlled by electric solenoids, as did many of the older system designs. Electric motor actuators are by far the most common in use today. The automatic systems all use the actuators described in Chapter 27. However, there is an additional two-wire design in use in automatic systems. Although older two-wire actuators were only capable of positioning the door in one of two positions, the newer style can stop the door in a variety of positions. To accomplish this, the HVAC controller must first calibrate itself to the actuator and door assembly. First, it moves the door from its full open position to its full closed position. When it does this, the controller counts the number of voltage ripples (see **Figure 9**) caused by the motor brushes skipping from contact to contact on the motor commutator. Each ripple represents a certain percentage of door movement. The controller can move the door to one of its stops and then move the door a certain number of ripples until the desired position is obtained. For example, if there are 930 ripples from stop to stop, and the controller wants the door to be 30 percent open, then it can move the door to its full closed position and move the door 310 ripples to the 30 percent open position. This type of procedure is why some ATC systems have a relearn time if battery power is lost. The HVAC controller must recalculate the information it stores in its memory for the different doors.

BLOWER CONTROL

Control of the blower motor in ATC systems can be accomplished using several relays, much like in a relay-based manual control system. However, it is much more common to use a blower control module (see **Figure 10**) as an interface between the HVAC controller and the blower motor. In this system, the HVAC controller determines the desired blower speed and sends an electronic

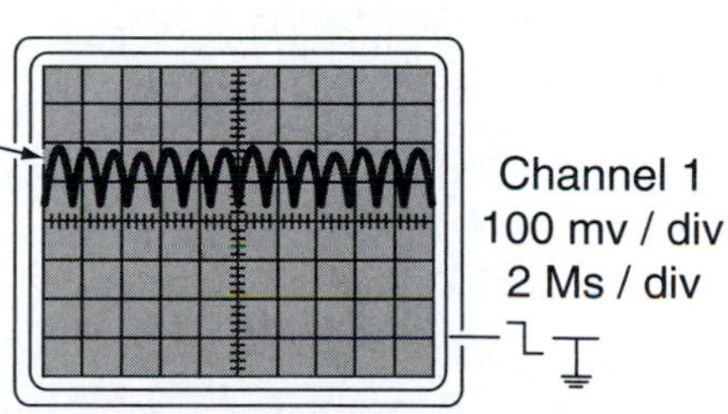

Figure 9. The controller counts the ripples as the door moves from stop to stop. It can then move the door to any position by moving a calculated number of ripples.

Figure 10. The power module pulse-width modulates the blower power circuit for infinite blower speeds

signal indicating the desired speed. The blower control module then pulse-width modulates the blower motor. This is a process in which the motor is turned fully on at predetermined intervals. The controller then modifies how long the motor stays on during these intervals. For example, the turn-on interval may be every tenth of a second. This means that the fan motor is turned on 10 times a second. If a slow fan speed is desired, then the on time may be only one-hundredth of a second. That means that during a 1-second period, the motor will be on for 10/100 of the second (1/100 times 10 on cycles per second) and off for 90/100 of a second, resulting in a slow or 10 percent fan speed. As faster fan speeds are desired, the motor will still only be turned on 10 times a second, but the duration of on time will be increased. For maximum speed, the on time for each cycle in this example would be 10/100 of a second, resulting in 100 percent on time. One advantage of this design is that there are no preset fan speeds such as high, medium, or low. The speed is infinitely variable to meet the needs of the passenger's comfort level.

DIAGNOSING AUTOMATIC TEMPERATURE CONTROL PROBLEMS

Diagnosing problems with automatic temperature control systems will depend significantly on the type of problem and the design of the electronic system used. Most modern automatic temperature control systems have self-diagnostics that monitor many of the operating parameters of the HVAC system. When a monitored parameter moves out of range, the system sets a trouble code. Depending on the vehicle, the controller also may turn on an indicator light or flash one of the HVAC display icons. Other vehicles require the operator to notice a decrease in system performance and seek service.

The service technician must determine if a poorly performing ATC system is experiencing a mechanical refrigeration problem or if the system electronic controls are malfunctioning. During the initial visual inspection, the instrument panel and HVAC control head should be checked for malfunction indicators. If any are found, a scan tool is used to retrieve information from the HVAC controller.

Diagnostic Code Malfunctions

ATC problems that generate diagnostic codes usually require accessing the manufacturer's troubleshooting information. This material covers how the system must be operated before the computer will test for the problem, what conditions must be present for the problem to be detected, and what the controller does to compensate for the problem.

Figure 11 illustrates a scan tool display with a problem with the ambient temperature sensor circuit, code B0159. The service manual for the vehicle indicates that the HVAC control module monitors the ambient air temperature circuit for shorts to ground, shorts to voltage, and open circuits. The circuit is tested whenever the ignition switch is turned on. The trouble code is set whenever the controller detects that the circuit signal is less than .09 volts or greater than 4.90 volts. The controller uses a default value for its control calculations whenever the code is set.

The diagnostic procedure in the service information takes the technician through multiple steps to verify the integrity of the circuit. The first step in the procedure is to verify that the problem is still present in the system. The process then checks the condition of the power side of the

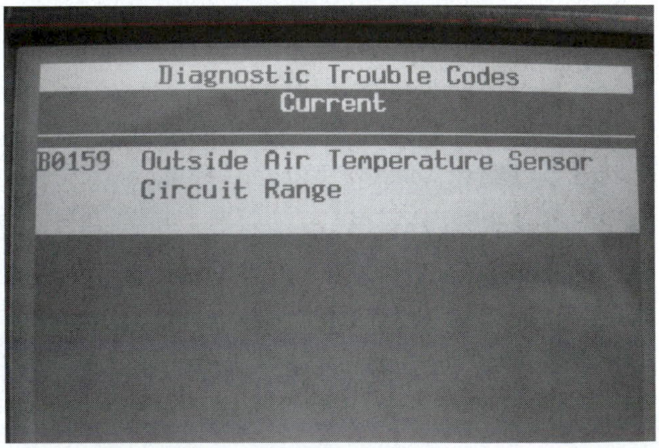

Figure 11. Diagnostic trouble codes pinpoint problems encountered by the HVAC controller.

circuit and then the ground side of the circuit. The condition of the circuit terminals is also checked. If the entire circuit proves to be in good condition, the procedure then recommends replacing the actual sensor.

> **You Should Know** *Some service technicians with poor diagnostic skills may replace the component associated with the trouble code as their first repair step. Although this may fix the vehicle in some cases, it usually results in wasted time, unnecessary part purchases, and unhappy customers.*

Diagnosing No-Trouble Code Problems

If an ATC system is not performing to expectations, but there is no diagnostic code generated, the service technician must isolate the problem to the part of the system that is malfunctioning. If the customer complaint is poor cooling, the mechanical refrigeration system should be checked. The refrigeration functions the same on an ATC system as it does in a manual control system. Verify that the compressor is operating, the refrigerant pressures are correct, and that the evaporator is getting cold. If a problem is detected, perform repairs as found in the refrigerant service chapter.

> **You Should Know** *ATC systems that monitor system operation often block compressor engagement when a low refrigerant condition is detected. The compressor will remain off until the associated code is cleared from system memory, even if additional refrigerant is added to the system.*

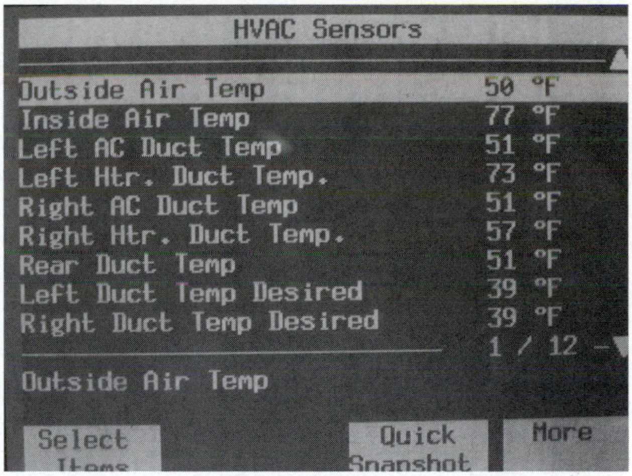

HVAC Sensors

Outside Air Temp	50 °F
Inside Air Temp	77 °F
Left AC Duct Temp	51 °F
Left Htr. Duct Temp.	73 °F
Right AC Duct Temp	51 °F
Right Htr. Duct Temp.	57 °F
Rear Duct Temp	51 °F
Left Duct Temp Desired	39 °F
Right Duct Temp Desired	39 °F
	1 / 12

Outside Air Temp

Select Items | Quick Snapshot | More

Figure 12. Scan tool data can be a valuable aid in performance evaluations.

If the refrigeration system is working properly, the technician needs to evaluate the automatic system in much the same manner that the controller would. The controller relies on input information from its sensors. **Figure 12** shows many of the input parameters available to the controller. The technician must evaluate these values, comparing them to each other and the actual vehicle conditions to determine if any of them are out of range. If any sensor value is questionable, diagnose the circuit as described in Chapter 29.

If all the inputs seem to be within an acceptable range, then the operation of the actuators should be evaluated. This evaluation can be made by manually adjusting the system controls to force the actuator to change its position. The technician should observe the actuator's movement, both physically and its scan tool reading, to make sure it is moving through its full range of movement.

Summary

- Automatic temperature control provides the greatest level of comfort and convenience for the automotive passenger.
- In AUTO mode, the HVAC controller monitors and adjusts all parameters of the air handling system.
- In semi-automatic operation, the operator may manually control air discharge mode, intake air source, AC compressor activation, and blower speed.
- ATC attempts to control the air discharge temperature in all modes of operation, even OFF.
- The ambient temperature sensor monitors the air surrounding the vehicle.
- The interior temperature sensor is the primary control reference during initial operation of the system.
- The output duct sensors monitor the temperature of the air coming out of the ducts and can be used to control the discharge air temperature.
- The sunload sensor can modify output air controls to compensate for additional sun heat load.
- The blower motor is pulse-width modulated to adjust fan speed.

Review Questions

1. Technician A says that the operator may control the fan speed even after the system is put into AUTO mode. Technician B says that the ATC system will increase cooling capacity on a sunny day. Who is correct?
   A. Technician A only
   B. Technician B only
   C. Both Technician A and Technician B
   D. Neither Technician A nor Technician B
2. Technician A says that the ATC control head has all the buttons and controls that the manual head has. Technician B says that the ATC system does closed-loop control from 60 to 90 degrees. Who is correct?
   A. Technician A only
   B. Technician B only
   C. Both Technician A and Technician B
   D. Neither Technician A nor Technician B
3. Technician A says that the ATC system will select desired air discharge mode automatically based on the operator input desired temperature. Technician B says that ambient temperature sensor resistance goes up when heated. Who is correct?
   A. Technician A only
   B. Technician B only
   C. Both Technician A and Technician B
   D. Neither Technician A nor Technician B
4. Technician A says that pressing the fan speed increase button once will increase the blower motor speed. Technician B says that, depending on cabin temperature, pressing the AUTO button may slow the blower speed. Who is correct?
   A. Technician A only
   B. Technician B only
   C. Both Technician A and Technician B
   D. Neither Technician A nor Technician B
5. Technician A says that most of the sensors use a 5-volt reference voltage. Technician B says that the HVAC controller monitors the sensor circuits between the two resistors. Who is correct?
   A. Technician A only
   B. Technician B only
   C. Both Technician A and Technician B
   D. Neither Technician A nor Technician B
6. Technician A says that the ambient sensor is mounted above the dash. Technician B says that interior temperature sensors are mounted above the dash. Who is correct?
   A. Technician A only
   B. Technician B only
   C. Both Technician A and Technician B
   D. Neither Technician A nor Technician B
7. Technician A says that the sunload sensor is an NTC thermistor. Technician B says that dual zone systems use two sunload sensors. Who is correct?
   A. Technician A only
   B. Technician B only
   C. Both Technician A and Technician B
   D. Neither Technician A nor Technician B
8. Technician A says that ATC systems can use relays to control the blower motor. Technician B says that pulse-width modulated motors are turned on more frequently to increase their speed. Who is correct?
   A. Technician A only
   B. Technician B only
   C. Both Technician A and Technician B
   D. Neither Technician A nor Technician B
9. Technician A says that the HVAC controller is always located in the ATC control head. Technician B says that dual zone systems have two ambient air temperature sensors. Who is correct?
   A. Technician A only
   B. Technician B only
   C. Both Technician A and Technician B
   D. Neither Technician A nor Technician B
10. Ambient sensor readings can be affected by engine temperature.
    A. True
    B. False
11. Controller logic tries to compensate for engine temperature.
    A. True
    B. False

Chapter 29 Air Distribution System Diagnosis

Introduction

The air distribution system, whether manual or automatic, has proven to be very reliable through the years. However, that does not mean it cannot fail. Diagnosing problems with the system begins by having a good basic understanding of how the different components work.

The material in this chapter is meant to be a supplement to good service information, not a replacement for it. When working with a vehicle, always consult the manufacturer's service recommendations. This information is available in service manuals and aftermarket service information data systems.

INSUFFICIENT AIRFLOW

Customers may complain that the volume of air coming from the ducts is inadequate. This is often a result of poor blower motor operation. Please refer to Chapter 24 for diagnosis of the blower motor. Non–blower motor problems are usually the result of an obstruction in the airflow. Vehicles equipped with cabin air filters may need to have the filters serviced. A dirty filter can certainly restrict airflow through the system.

Air distribution systems are designed so that all of the airflow must pass through the evaporator core. These cores are designed to maximize heat transfer as the air travels through them; thus, the webbing between the evaporator tubes makes for fairly small openings. If enough material gets trapped in the passages, airflow will be blocked. One common cause of the foreign material in the evaporator is a missing screen over the fresh air inlet that sucks in leaves and other debris (see **Figure 1**). Some customers take their pets with them in the car wherever they go. Recirculate mode draws hair into the system very easily, clogging passages. Small plastic bags also can be sucked into the system. Even a gradual buildup of dirt and grime over years of service can limit airflow.

Figure 1. Anything that obstructs airflow through the system, such as these leaves, can lead to customer complaints.

VACUUM SYSTEM DIAGNOSIS

The most common problem encountered with vacuum controls is a lack of vacuum. The vacuum source on most vehicles is the engine, which means that the vacuum must be transported to the HVAC system. This is accomplished using either small rubber or plastic hoses to carry the vacuum where it is needed. Because these hoses are exposed to the heat of the engine compartment, they become brittle and are easily broken by engine vibrations or by technicians while servicing the vehicle (see **Figure 2**).

Figure 2. Any loss of vacuum in the system, such as this split hose, could cause the HVAC controls to not function properly.

When vacuum is lost to the HVAC system, the most common customer complaint is that the discharge air all flows out the defroster vents. This is a safety issue. Defrost is the default output when the system fails. Any other discharge location could leave the windshield fogged up, obscuring the driver's vision.

To test the source vacuum, find the hose that supplies the vacuum to the control device. This could be at the vacuum solenoid pack under the dash or at the rotary control valve behind the control head. Disconnect the hose and connect a vacuum gauge to it (see **Figure 3**). Start the vehicle and allow it to idle while you check the reading on the gauge. Anything less than 17 in-Hg indicates that the HVAC system is not getting enough vacuum. If there is a problem, check the vacuum port at the engine that supplies the vacuum to the HVAC system or the vacuum pump, if it is so equipped. If there is sufficient vacuum at the port, then there must be an obstruction or a break in the line. Don't forget that there is also a vacuum reservoir and a one-way check valve in the system, which could be a potential trouble source.

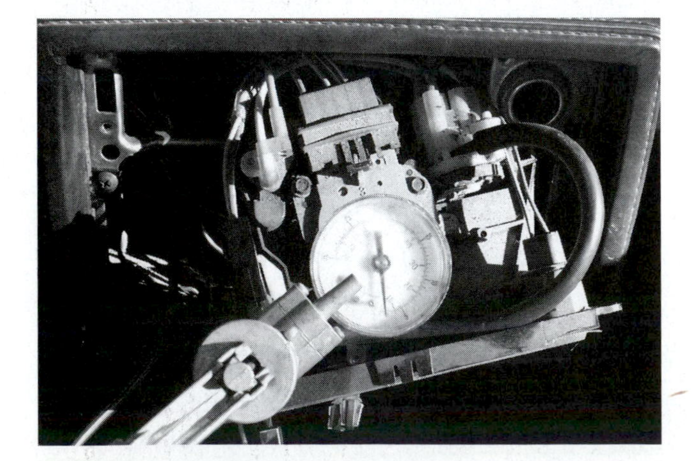

Figure 3. Insufficient source vacuum to the control valve could cause control problems.

You Should Know

Do not forget that a vacuum leak in systems other than the HVAC system could rob enough vacuum from the system to affect its performance.

To find a vacuum leak, a visual inspection should first be performed. Look for cracked, broken, or puckered lines and connectors. An audio check also can be useful. Start the vehicle and, being careful not to injure yourself on the rotating or hot parts, listen for the hissing noise that is caused by a vacuum leak. A mechanic's stethoscope (see **Figure 4**) can be extremely helpful in pinpointing the exact location of the leak. Remove the metal rod holder on the stethoscope, leaving the open hose at the end. Then slowly move the hose around the engine area until the hissing is loudest, indicating the leak source. There are also electronic vacuum leak detectors available that can be used to find the leak.

A "smoke machine" also can be effective in finding a vacuum leak. This machine generates a nontoxic smoke. Connect the output smoke hose of the machine to the HVAC vacuum system. If there is a leak, the smoke will start to escape from the system at that point. Do not forget to look under the dash for leaking smoke as well.

Another common customer complaint is a hissing noise coming from the dash. This problem may only occur in certain operating modes. There also may be a complaint of improper operation of the discharge air locations or that the system is slow to change from one mode to another. The most likely cause of this complaint is the rotary vacuum switch on the control panel. Slide the control head out away from the dash and squeeze the two halves of the rotary switch together. If the hissing changes, replace the rotary switch. If the hissing persists, use the vacuum leak diagnostics described earlier to find the leak source.

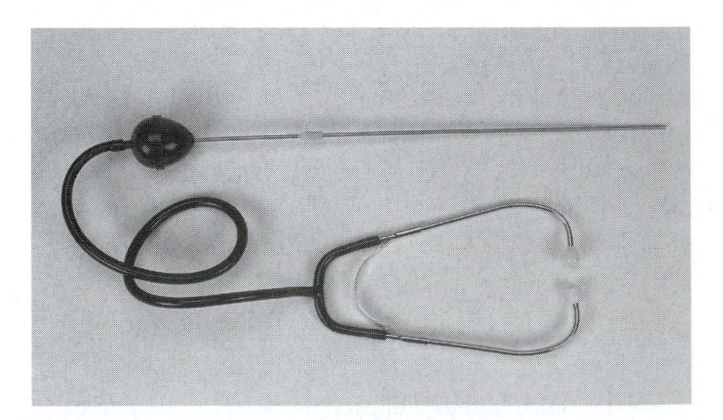

Figure 4. A mechanic's stethoscope with the tube end open is an invaluable tool for finding vacuum leaks.

The vacuum motors themselves are typically very reliable. To test a vacuum motor, remove the vacuum hose on the motor and connect a hand-operated vacuum pump to the port. Apply 18 in-Hg to it and observe operation of the motor. If the motor fails to hold the vacuum, it is bad and needs to be replaced. If it holds a vacuum but the motor does not move, there could be one of two problems. Either the motor is frozen internally, in which case it needs to be replaced, or the door that the motor controls is stuck. To determine the exact cause, disconnect the motor from the door and repeat the test. If the door is stuck, disassembling the plenum may be required.

> **You Should Know** *It is not uncommon for pennies, pencils, and other objects to fall down the defroster vents and cause one of the doors to become stuck.*

BOWDEN CABLE DIAGNOSIS

Bowden cables, although very simple in design, can be a source of problems in the HVAC system. They may become out of adjustment, either as a result of normal wear or from technician error. Because these cables commonly are used on blend doors, adjustment problems could cause unwanted air to flow through the heater core or not enough air to flow through it. This could result in insufficient heating or cooling. There are provisions on many vehicles to adjust the Bowden cable. However, if no adjustment is provided, part replacement may be required if there is a problem.

One quick test of a Bowden-controlled blend door is to move the temperature control lever or knob quickly back and forth through its full range of operation (see **Figure 5**). A small thud should be heard at both ends of its travel, which indicates the door is traveling all the way to its stop.

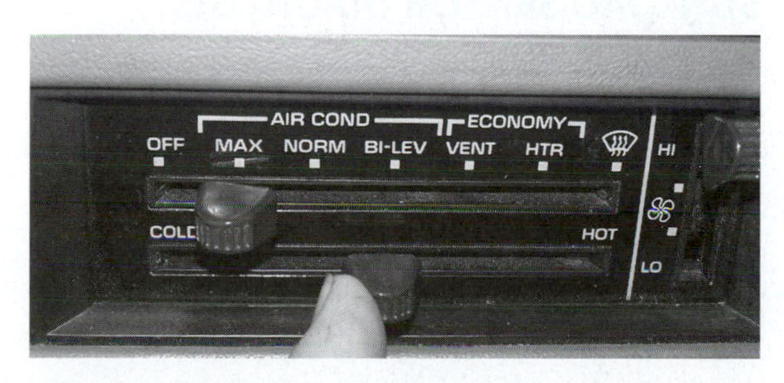

Figure 5. Move the temperature control lever through its full range of operation. A small thud should be heard at both ends of its travel.

There is another test to ensure maximum cooling is taking place. Operate the vehicle until the engine is warm and put the vehicle HVAC system to maximum cooling. Allow the system to stabilize and take and record a temperature reading from the center panel vent. Restrict coolant flow through one of the heater hoses and observe the temperature at the center vent. If the discharge temperature drops more than 5 degrees, you probably have a blend door that is not fully closing.

A Bowden cable may occasionally start to stick or bind as it moves in and out of its housing. Removing it and holding it vertically when lubricating the housing ends and working the cable back and forth may help the problem. However, replacing the assembly is the preferred repair. If the cable or housing become kinked or bent, it must be replaced.

ELECTRONIC CONTROL HEAD DIAGNOSIS

The electronic control head is a microprocessor-based control device. If it fails to work as expected, diagnosis can be limited to evaluating its inputs and outputs. Always check the power and ground circuits first in any electronic component evaluation. Remember, this type of component can have more multiple power circuits and ground circuits. If power and ground are good, check the operator displays. If these fail to function, or if it is normally accessible from a scan tool and it will not communicate, the head is bad.

Never condemn a control unit for failing to control a system properly without first evaluating the inputs to the processor. The old computer expression, "Garbage In, Garbage Out," applies equally well to control processors. If the data from the sensors is incorrect, the controller will not be able to properly adjust the operation of the system.

TEMPERATURE SENSOR DIAGNOSIS

All temperature sensor circuits operate the same way. This makes it simple to develop some basic techniques for testing these devices. Many vehicles will have data stream information available from the HVAC control module, which can be read by using a scan tool. The internal diagnostics in the module can monitor the sensor readings and set a trouble code if they are out of range. If the information is scan tool–accessible, check for trouble codes and also check what the scan tool temperature reading for that sensor is. If there are no trouble codes and the reading is correct for the air surrounding the sensor, then there is probably no problem with the sensor. If the readings are incorrect, the following checks can help to pinpoint the problem.

Check the sensor resistance. Unplug the wiring to the sensor and use a DVOM set on the Ohm's position and measure across the terminals of the sensor (see **Figure 6**).

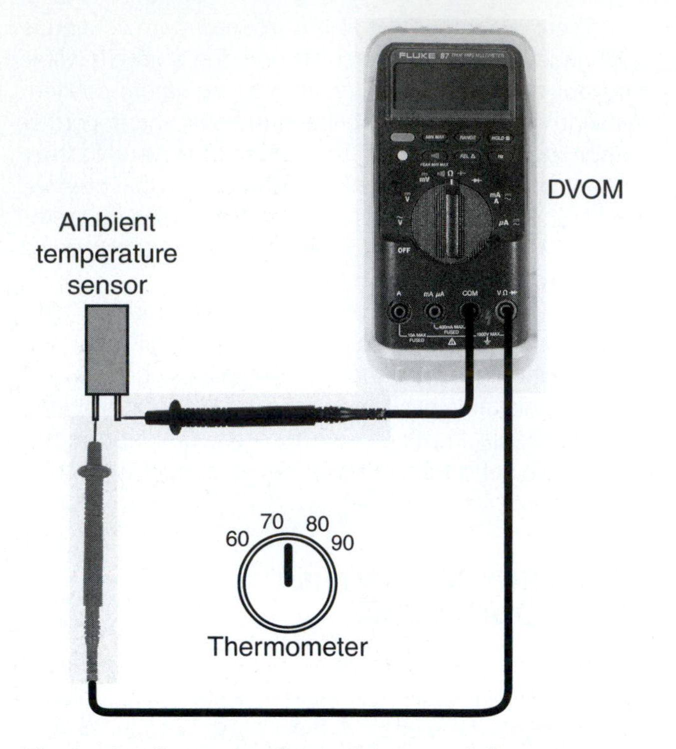

Figure 6. Compare the resistance of the sensor to the current temperature on the chart in the repair manual.

Use the service information to compare the sensor resistance with the calibration chart for that resistor at the current ambient temperature. If the sensor reading does not match the chart (see **Figure 7**), replace the resistor.

Check the reference voltage side of the circuit. With the sensor unplugged, use the DVOM to measure the voltage on the sensor signal wire at the connector to a good ground. It should be at or just above 5 volts. If the voltage is low, check the wiring for an open or a bad connection. Repair the circuit if a problem is found. If the wire is good, check the power and ground circuits to the HVAC controller. The controller is bad if it has power and ground but cannot output the 5-volt reference. The controller is also bad if the 5-volt reference is too high.

Temperature vs Resistance

°F	°C	Ohms
−40	−40	100,700
−04	−20	28,660
32	0	9,400
50	10	5,670
68	20	3,515
86	30	2,235
104	40	1,460
140	60	667
176	80	332

Figure 7. Locating the relative resistance of the sensor should cross over to ambient temperature.

You Should Know *Never push the DVOM probes into the female terminals of the connector. This will distort their shape and cause a bad connection when the connector is plugged back in. Find an adaptor of the correct size to probe the terminal.*

Check the ground side of the circuit. With the sensor unplugged, use the DVOM to measure the voltage from the sensor signal wire on the connector to the ground wire on the connector. It should read 5 volts as well. If it does not, there is an open in the ground side of the circuit. Check the circuit wires and all the grounds on the controller. Also check the connections at the controller. Repair as needed. If no problem is found in the wiring, the ground path through the controller might be damaged. The controller will need to be replaced if this is the case.

If no problem is found during these procedures, but the scan tool data for the sensor indicates a lower than actual reading, then there is probably unwanted resistance in the circuit. The controller interprets the extra resistance as coming from the sensor. Because the sensor resistance is supposed to go up as the temperature drops, the controller thinks the sensor is cooler than it actually is.

A scan tool also can be used to do a quick check on the temperature sensor circuit. Unplug the temperature sensor, turn the vehicle on, and look at the data for that sensor on the scan tool. Most circuits will read −40 degrees F when the sensor is disconnected. Next, jumper the two wires on the connector. This should produce the maximum reading for the circuit, usually in the 350 degrees F to 525 degrees F range.

SUNLOAD SENSOR DIAGNOSIS

The photo-diode sunload sensor circuit functions almost identically to the temperature sensor circuit. They both rely on a 5-volt reference with an internal resistor and a sensor ground. Diagnosis of this part of the circuit can be accomplished in the same way as the temperature circuit.

ELECTRONIC ACTUATORS

Diagnosis of electronic actuators will vary depending on the style of actuator involved. Two-wire actuators should have an equal voltage on each wire, either high or ground, until they are commanded to move. When activated, one wire will have high voltage and the other will be grounded. Back-probe the actuator wires with a DVOM and command the door to move. If the voltage is correct and the door is not moving, the actuator is bad. The motor wires on a five-wire actuator without the logic module can be checked out in the same manner. Actuators with a logic module are diagnosed in much the same way that a control module is checked. Make sure that the actuator has power and ground, then check for an input signal from the HVAC controller.

Most ATC systems will have diagnostic codes that are set if the HVAC control module cannot properly adjust one of its doors using an electronic actuator. Use the code diagnostic chart whenever it is available.

Summary

- Foreign material can obstruct airflow through the evaporator core, resulting in poor passenger cooling.
- Sufficient engine vacuum is essential to the proper operation of the HVAC vacuum control system.
- Vacuum controls default to the defrost system when inoperative.
- Bowden cables must be properly adjusted to ensure maximum HVAC performance.
- Electronic control heads with power and ground should be replaced if inoperative.
- Diagnosis of temperature sensor operation is accomplished using basic serial circuit and variable resistor diagnostic procedures.
- Diagnosis of electronic actuators will vary with the design style of the component.

Review Questions

1. Technician A says that an engine performance problem could affect HVAC door operation. Technician B says to check the vacuum hoses if all modes direct airflow to the windshield. Who is correct?
   A. Technician A only
   B. Technician B only
   C. Both Technician A and Technician B
   D. Neither Technician A nor Technician B
2. Technician A says that insufficient AC cooling is always a problem with the refrigeration system. Technician B says to replace the HVAC control head if the electronic display does not function. Who is correct?
   A. Technician A only
   B. Technician B only
   C. Both Technician A and Technician B
   D. Neither Technician A nor Technician B
3. Technician A says that a defective sunload sensor could result in cooler than normal interior temperatures. Technician B says that two-wire actuators should have equal voltage on both wires when it is moving. Who is correct?
   A. Technician A only
   B. Technician B only
   C. Both Technician A and Technician B
   D. Neither Technician A nor Technician B
4. Technician A says that a scan tool reading from an ambient temperature sensor of 35 degrees F on a 90-degree F day may be caused by unwanted resistance in the sensor circuit. Technician B says that if no problem is found in an improperly reading sensor circuit, then the module may be bad. Who is correct?
   A. Technician A only
   B. Technician B only
   C. Both Technician A and Technician B
   D. Neither Technician A nor Technician B
5. Technician A says that poor airflow may be caused by a dirty filter. Technician B says that poor airflow may be caused by a dirty evaporator. Who is correct?
   A. Technician A only
   B. Technician B only
   C. Both Technician A and Technician B
   D. Neither Technician A nor Technician B
6. A ______________________ can be a valuable visual aid in finding small vacuum leaks.
7. Technician A says that 21 in-Hg may cause a problem in the vacuum control system. Technician B says that a mechanic's stethoscope may be useful on some vehicles to diagnose mode door operation problems. Who is correct?
   A. Technician A only
   B. Technician B only
   C. Both Technician A and Technician B
   D. Neither Technician A nor Technician B

8. Technician A says to replace a vacuum motor if the associated door does not move when vacuum is applied to it. Technician B says to replace the sensor if the reference voltage at the sensor is too high. Who is correct?
   A. Technician A only
   B. Technician B only
   C. Both Technician A and Technician B
   D. Neither Technician A nor Technician B
9. A cable-controlled blend door should produce an audible ____________ when it is moved to the end of its travel.
10. Technician A says that temperature sensors are variable resistors. Technician B says that all temperature sensors have the same resistance at a specific temperature. Who is correct?
    A. Technician A only
    B. Technician B only
    C. Both Technician A and Technician B
    D. Neither Technician A nor Technician B
11. Two technicians are discussing a temperature sensor diagnostic check. When voltage is measured between the two wires, the DVOM reads zero volts. Technician A says that there may be a ground problem. Technician B says that there may be a problem with the module. Who is correct?
    A. Technician A only
    B. Technician B only
    C. Both Technician A and Technician B
    D. Neither Technician A nor Technician B
12. Technician A says that poor diagnostic techniques can actually damage the system. Technician B says that all electric actuator motors get their power from the HVAC controller. Who is correct?
    A. Technician A only
    B. Technician B only
    C. Both Technician A and Technician B
    D. Neither Technician A nor Technician B

Chapter 30 Air Distribution System Service

Introduction

Air distribution service can be as simple as removing a few screws or as complicated as removing the entire dash assembly from the vehicle. For most operations, it is highly recommended that good service information, such as an electronic database or the manufacturer's service manual, be consulted.

INTAKE AIR

The fresh air intake may need service from time to time. The screen over the intake must be kept clean of leaves and other debris. A shop vacuum cleaner does an excellent job of removing trash from this area (see **Figure 1**). Some vehicles route the intake air through a water trap to remove rain. The trap has a drain that can get clogged, resulting in moisture being drawn into the air handling system. The drain is on the right side of the vehicle inside the fender. When servicing, check the intake for obstructions such as plastic bags, paper, or leaves.

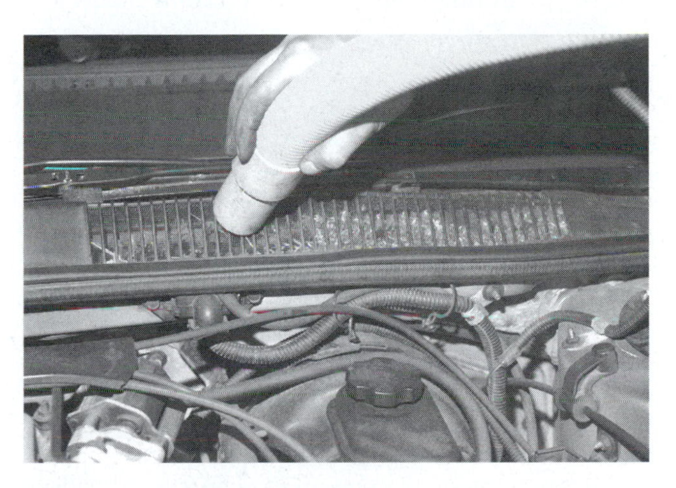

Figure 1. A shop vacuum does an excellent job of removing intake air obstructions.

CABIN AIR FILTERS

Cabin air filters are a relatively new idea in automotive air management systems. They need to be replaced in order to assure efficient operation of the system. Filter replacement should be incorporated into the vehicle's regular maintenance schedule. There are two common styles of cabin air filters: fresh air intake mounted and plenum mounted. Fresh air intake models only filter air drawn in from outside the vehicle. They are usually serviced outside the vehicle on the right cowling, near the passenger windshield wiper. Raise the hood and look for a removable panel in this area. Remove the panel and replace the filter located underneath it. Most of the panels are simple latch-and-lift designs. However, a few are more complicated, requiring the removal of clips or screws. Be sure to remove all of the retainers required in order to prevent having to pry open a panel. This could result in damaged components.

Plenum-mounted filters are serviced from inside the vehicle. These filters clean all the air, fresh and recirculate, that flows through the air handling system. Some models have a filter service door that is easily accessible. Remove the door and replace the filter. If the filter door is not accessible, various dash panels or the glove box may have to be removed. Some models have a trap door in the rear of the glove box to gain access to the filter (see **Figure 2**).

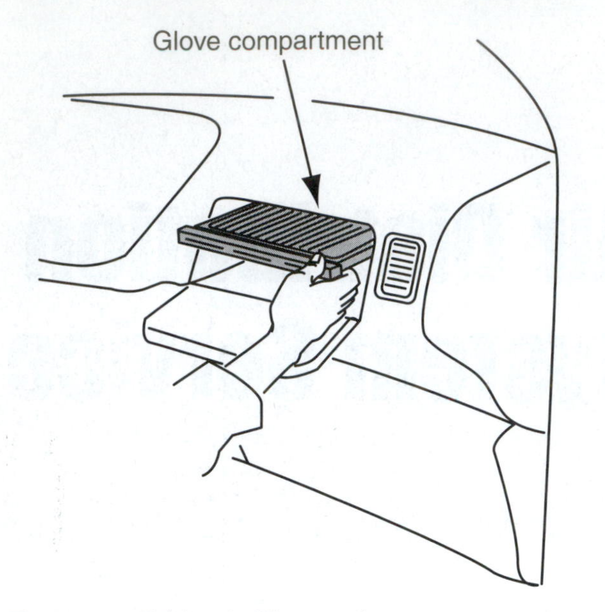

Figure 2. Cabin air filter replacement is performed through the glove box on some models.

> **You Should Know** *Some models use a two-part filter. In order to have a filter large enough to clean all of the air but still be serviced through a small trap door, the filter is split into halves. Make sure that both filters are replaced during servicing.*

BLOWER ASSEMBLIES

Blower motor replacement is a common air handling service. See Chapter 25 for details on blower motor service.

PLENUM ASSEMBLIES

Removal of the plenum assembly is the most challenging of air handling service procedures. Worst-case procedures may call for removal of the instrument cluster, dropping the steering column, removing the center console, the dash cover, and the entire dash assembly. Only then may the plenum assembly be removed. Although not all vehicles require such drastic measures, it is not uncommon to have to remove sections of the dash assembly or to take part of it loose and place it out of the way when removing the plenum. Regardless of the severity of the procedure, it is not acceptable to have screws or brackets left over after the repair is completed. This will lead to rattles or leaks that the customer is sure to complain about.

One procedure common to almost all plenum assembly removals is disconnecting the heater hoses and air conditioner refrigerant lines. Care must be taken not to damage these components if they are going to be reused. The plastic and aluminum heater core outlets are very easy to damage when the hose is removed (see **Figure 3**). It is cheaper to make a small cut in the hose and replace the hose than to ruin the heater core. Always use two wrenches when removing the refrigerant lines (see **Figure 4**). Hold the stationary fitting with one wrench and turn the fitting on the hose. Using one wrench could twist the evaporator core line and cause the core to need replacing.

When removing the plenum from the vehicle, it often is necessary to rotate the case when working it out from under the dash. This could cause the residual coolant in the heater core and the oil in the evaporator core to spill out on the vehicle carpeting. To avoid this problem, purge as much coolant as possible from the heater core by

Figure 3. The plastic outlets on newer heater cores can be damaged if proper procedures are not followed.

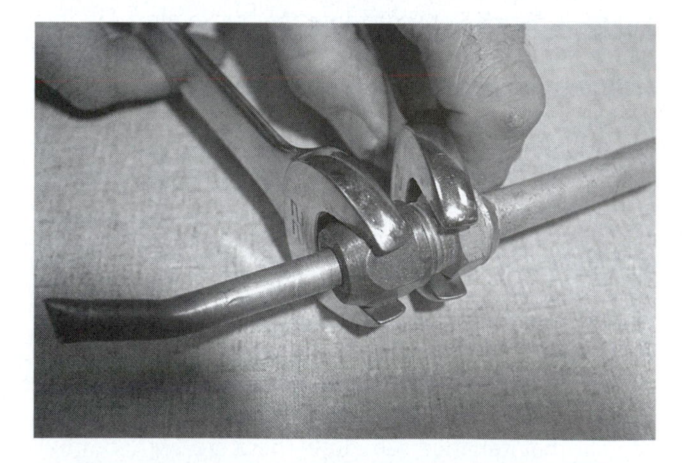

Figure 4. Always use two wrenches when disconnecting fittings to avoid damaging the lines.

applying light shop air pressure to one of the heater hose fittings while simultaneously directing the purged coolant into a drain pan with a short section of heater hose from the other fitting. Shop air is not recommended for the evaporator core because it introduces moisture into the residual oil in the core. Use plugs to seal all four core fittings so that any remaining fluids do not leak onto carpet. Plastic also can be placed over the carpet to further prevent any damage.

When reinstalling the plenum, an assistant is useful in guiding the core fittings through the holes in the firewall. Never let the weight of the plenum rest on the core fittings, as this could damage the cores. Have the assistant start the plenum, attaching bolts when the assembly is held in place. Also, make sure all the wires and vacuum hoses are routed correctly when installing the plenum. Pinched wires and hoses that do not reach their connectors could otherwise result.

EVAPORATOR AND HEATER CORE REPLACEMENT

Most evaporator and heater core replacement procedures involve removing the plenum assembly from the vehicle. Once the plenum is removed from the vehicle, the core can be opened and the core replaced. This usually involves removing a series of screws and splitting the halves of the case open (see **Figure 5**). However, some cases are permanently sealed from the factory. Replacing the core for these models requires cutting open a section of the case to create a removal door (see **Figure 6**). This must be done carefully because the removed section must be reused to close the opening. A silicone sealant is used to reseal the housing.

There are a few designs that may not require removal of the entire plenum. Some heater cores only require removal of an access door or cover. Vehicles with split cases

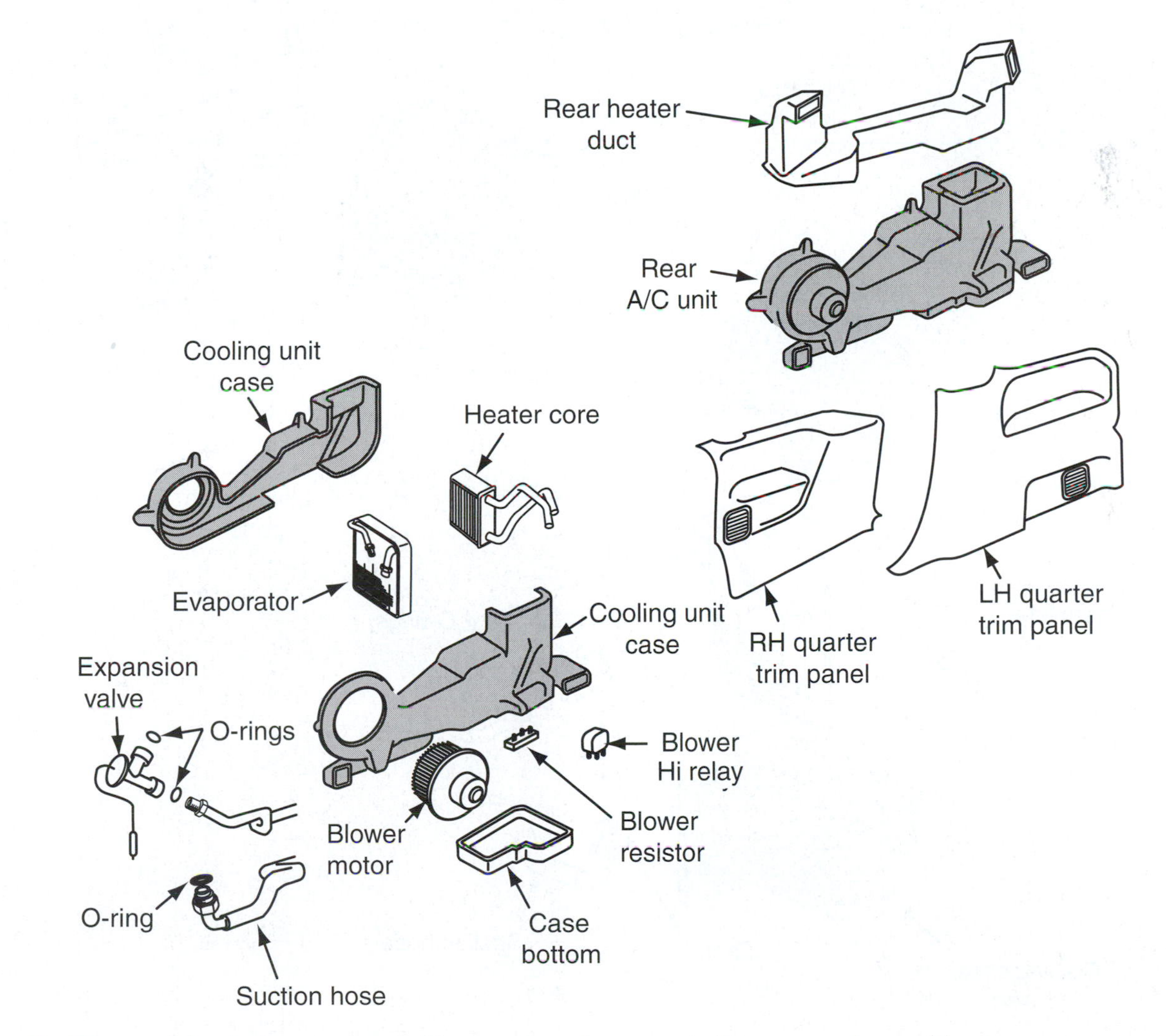

Figure 5. The combined plenum case splits open to gain access to the evaporator and heater cores.

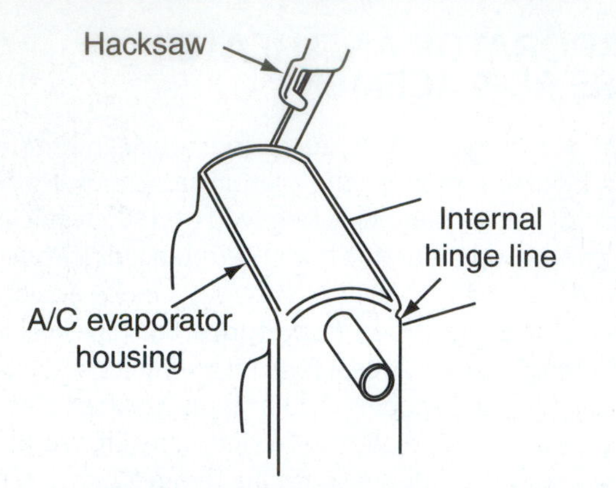

Figure 6. The top of this case must be cut on three sides to create an access door for the evaporator.

will only require removal of that portion of the case that houses the defective core (see **Figure 7**). Hose removal and carpet damage are still concerns on these applications as well.

BOWDEN CABLE SERVICE

The replacement of a Bowden cable requires removal of the control head to access one end of the assembly. The location of the other end of the cable will dictate what other components must be removed for access. The cable and housing must be unhooked at each end. The cable is attached using a metal clip or a ringlet at the end of the cable is slipped over a lever end. The housing may attach using a clip or a small bolt. When installing the new cable,

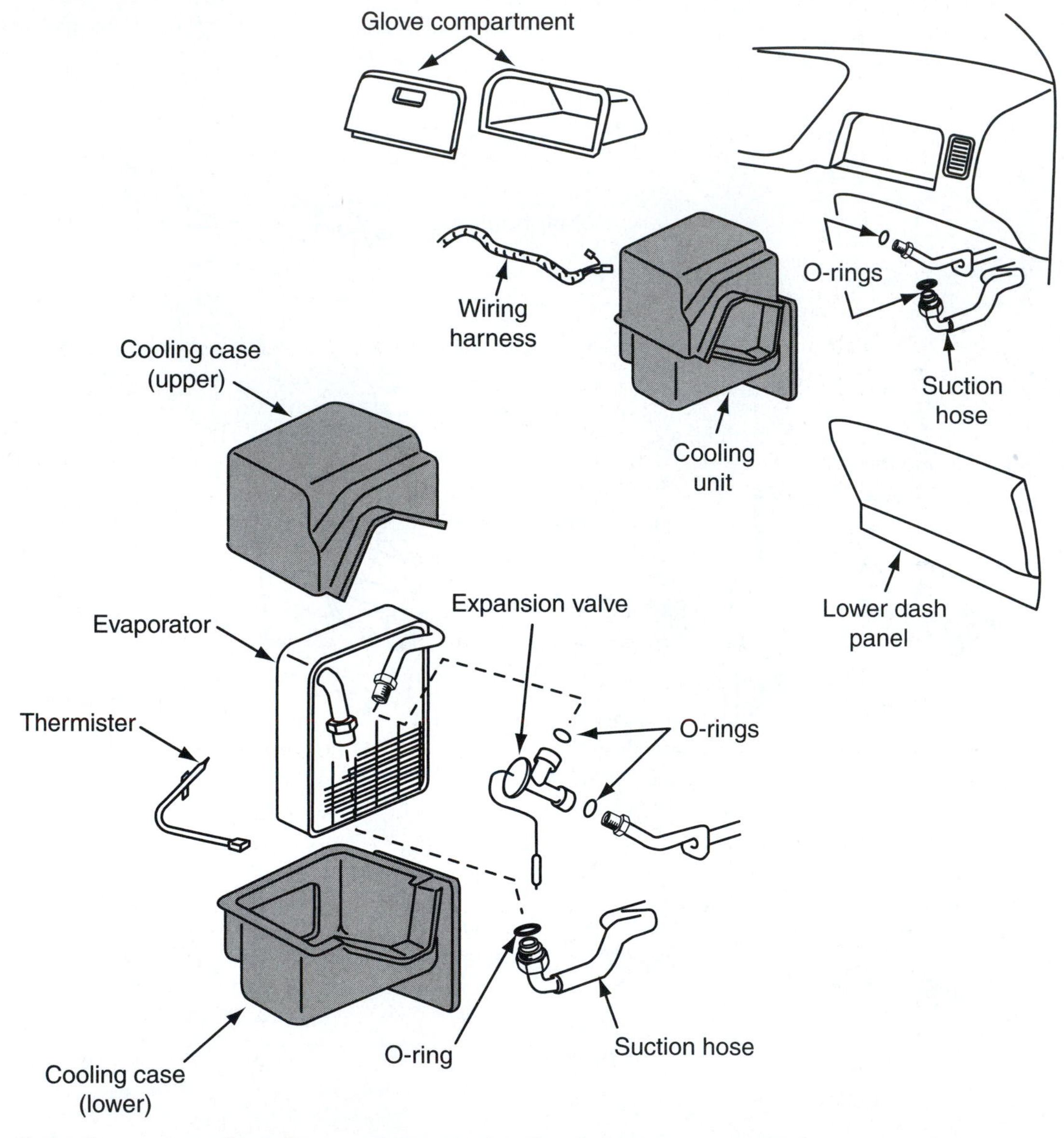

Figure 7. Only the section of a split case that contains the defective core must be removed. However, other case sections may have to be removed for access.

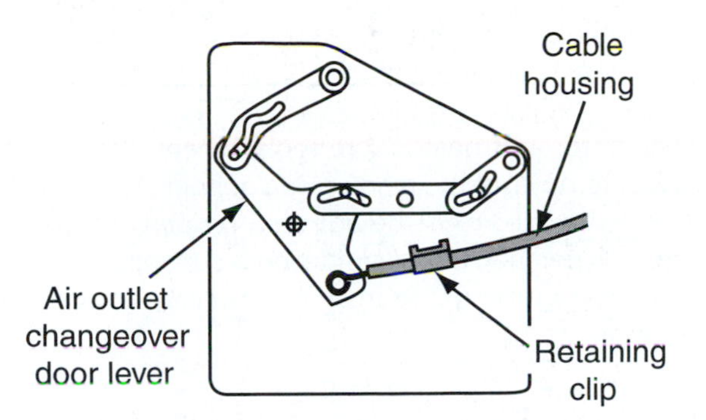

Figure 8. The cable housing can be mounted anywhere in the clip to adjust the cable travel for proper door operation.

be careful not to kink it, as this will damage the ability of the cable to slide in the housing.

Most Bowden cables have some means of adjusting the cable travel. The adjustment may be at the cable attachment point or at the point where the cable housing attaches (see **Figure 8**). The adjustment is used to make sure that the component it is controlling operates in its correct range of movement.

> **You Should Know** *Bowden cables are often used on blend door applications. The cable is correctly adjusted when a slight thud is heard from the door at each end of its travel when the temperature control is moved from full cold to full hot and back.*

VACUUM COMPONENT SERVICE

Most vacuum motors are held in place by screws or clips. A few are held in place with metal brads that must be drilled in order to replace the motor. After the motor housing has been detached, be careful not to break the door arm lever where it attaches to the motor. Many of these are plastic and are easily damaged. When the motor is removed, move the door it actuates through its full range of motion to ensure it is not binding or obstructed.

Vacuum storage reservoirs are installed in many places. For many years, they resembled coffee cans or large plastic balls (see **Figure 9**) mounted in the engine compartment where they were easily replaced. Today, they are just as likely to be mounted on the plenum assembly, where partial dash removal may be required.

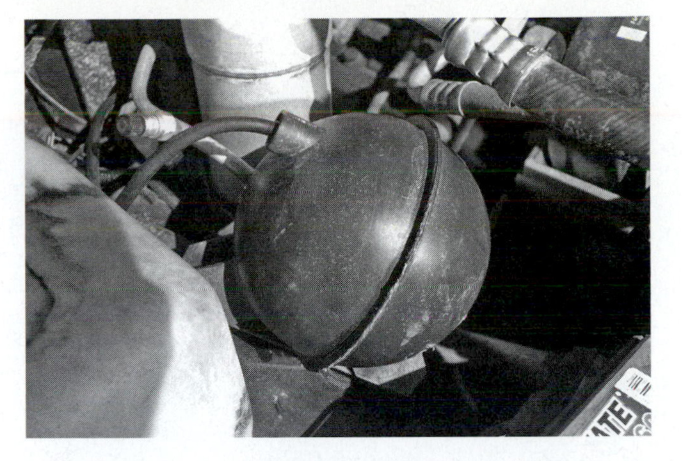

Figure 9. The vacuum storage ball can be damaged if not mounted properly.

Vacuum control switches are installed on the control heads. Removal of the head should allow easy replacement of the switch. Make sure that the tab that actuates the rotary valve is lubricated and is inserted in the slot on the lever that actuates it.

ELECTRIC DOOR ACTUATORS

Replacement access may be difficult for some electric door actuators. Partial dash removal may be required. Most are held in place by three small screws. Once they are removed, the actuator slides off its mount. There are several styles of actuators that closely resemble each other. Be sure that the replacement part is correct for the application. Once installed, some styles will require calibration. Removing power from the HVAC control module will clear its memory and cause it to recalibrate on some designs. Other applications will require the use of a scan tool.

EVAPORATOR CASE DRAINS

A plugged evaporator case drain can cause condensation to overflow onto the carpeting. Find the drain and use a small blast of shop air to remove the obstruction. If standing under the vehicle, watch out for the sudden downpour of water. If shop air is not practical, a small wire may be used to unplug the passage.

Summary

- Intake air systems must be kept clean and free of obstructions to work properly.
- Cabin air filters are located either in the fresh air intake or in the plenum assembly. Replacement should be part of the regular maintenance schedule.
- Plenum removal may require extensive interior repairs. Professional service skills are required to ensure complete restoration of the vehicle.
- Evaporator and heater core replacement typically involves removal of the plenum and opening the case.
- Bowden cables must be adjusted properly to ensure correct operation of the controlled component.
- Electric door actuators may need recalibration after installation.

Review Questions

1. Technician A says that low HVAC airflow may be a result of improper maintenance of the vehicle. Technician B says that the low airflow could be from parking under a tree. Who is correct?
   A. Technician A only
   B. Technician B only
   C. Both Technician A and Technician B
   D. Neither Technician A nor Technician B
2. Technician A says that the plenum must be removed to replace the heater core. Technician B says that having a few extra screws after plenum replacement is not a concern. Who is correct?
   A. Technician A only
   B. Technician B only
   C. Both Technician A and Technician B
   D. Neither Technician A nor Technician B
3. Technician A says that heater core replacement on a vehicle with a split plenum may not require recovery of the refrigerant. Technician B says that twisting a heater hose is the best way to remove it from the heater core. Who is correct?
   A. Technician A only
   B. Technician B only
   C. Both Technician A and Technician B
   D. Neither Technician A nor Technician B
4. Technician A says that it is good procedure to plug the heater and evaporator core lines before removal of the plenum. Technician B says to blow the heater and evaporator cores out with shop air. Who is correct?
   A. Technician A only
   B. Technician B only
   C. Both Technician A and Technician B
   D. Neither Technician A nor Technician B
5. Technician A says that Bowden cables are adjusted to remove all slack from the cable. Technician B says that Bowden cables are not used on blend doors. Who is correct?
   A. Technician A only
   B. Technician B only
   C. Both Technician A and Technician B
   D. Neither Technician A nor Technician B
6. Technician A says that vacuum motors must be calibrated after installation. Technician B says that electric door motors must always be calibrated after installation. Who is correct?
   A. Technician A only
   B. Technician B only
   C. Both Technician A and Technician B
   D. Neither Technician A nor Technician B
7. Evaporator case drains remove rainwater that may enter the system.
   A. True
   B. False
8. There is no need to verify electric motor actuator part numbers before installation because the different styles are not interchangeable.
   A. True
   B. False
9. Technician A says that some cabin air filters clean all of the air in the air management system. Technician B says that some cabin air filters only clean air that enters through the recirculate door. Who is correct?
   A. Technician A only
   B. Technician B only
   C. Both Technician A and Technician B
   D. Neither Technician A nor Technician B

Appendix

ASE PRACTICE EXAM FOR HEATING & AIR CONDITIONING

1. Technician A says that a properly written repair order will contain information about the customer, specific information about his or her vehicle, and a detailed description of the customer's concern.
   Technician B says that the repair order is the communication link between the customer and the technician. Who is correct?
   A. Technician A
   B. Technician B
   C. Both Technician A and Technician B
   D. Neither Technician A nor Technician B
2. Technician A says that service bulletins are very general in nature and cover a wide variety of concerns on several different vehicles.
   Technician B says that service bulletins may be issued by many different sources. Who is correct?
   A. Technician A
   B. Technician B
   C. Both Technician A and Technician B
   D. Neither Technician A nor Technician B
3. Technician A says that at the very least the vacuum pump oil should be serviced after 10 hours of operation.
   Technician B says that the vacuum pump oil should never have to be changed unless it is connected to a contaminated system. Who is correct?
   A. Technician A
   B. Technician B
   C. Both Technician A and Technician B
   D. Neither Technician A nor Technician B
4. Technician A says that latent heat can be measured with an infrared thermometer.
   Technician B says that sensible heat can be considered "hidden" heat. Who is correct?
   A. Technician A
   B. Technician B
   C. Both Technician A and Technician B
   D. Neither Technician A nor Technician B
5. Technician A says that heat transfer occurs when heat energy moves from one body of matter to another.
   Technician B says that conduction, convection, and evaporation are forms of heat transfer. Who is correct?
   A. Technician A
   B. Technician B
   C. Both Technician A and Technician B
   D. Neither Technician A nor Technician B
6. Technician A says that when pressure is applied to a liquid in a sealed container, the temperature of the liquid will increase.
   Technician B says that when heat is applied to a liquid in a sealed container, the pressure will increase. Who is correct?
   A. Technician A
   B. Technician B
   C. Both Technician A and Technician B
   D. Neither Technician A nor Technician B
7. Technician A says that the use of a heat exchanger in an air conditioning system allows heat energy to be transferred from the chemical refrigerant to the atmosphere.
   Technician B says that the use of a heat exchanger in an air conditioning system allows heat energy to be

transferred from the atmosphere to a chemical refrigerant. Who is correct?

A. Technician A
B. Technician B
C. Both Technician A and Technician B
D. Neither Technician A nor Technician B

8. Technician A says that as refrigerant releases heat within the evaporator, it changes from a vapor to a liquid.
Technician B says that as refrigerant absorbs heat within the condenser, it changes to a vapor. Who is correct?
A. Technician A
B. Technician B
C. Both Technician A and Technician B
D. Neither Technician A nor Technician B

9. Technician A says that since R-12 is no longer produced, it is acceptable to top off an existing R-12 system with R-134a.
Technician B says that, once mixed, R-12 and R-134a cannot be separated using on-site recycling equipment. Who is correct?
A. Technician A
B. Technician B
C. Both Technician A and Technician B
D. Neither Technician A nor Technician B

10. Technician A says that mineral oil can readily combine with R-12.
Technician B says that synthetic oils are used with R-134a. Who is correct?
A. Technician A
B. Technician B
C. Both Technician A and Technician B
D. Neither Technician A nor Technician B

11. Technician A says that the ASE specialty area certification A-7 Heating and Air Conditioning meets or exceeds the requirements set forth by the EPA for the refrigerant recovery program.
Technician B says that any technician that performs vehicle service that could result in refrigerant being vented into the atmosphere must receive EPA certification. Who is correct?
A. Technician A
B. Technician B
C. Both Technician A and Technician B
D. Neither Technician A nor Technician B

12. Technician A says that it is legal under federal guidelines to add refrigerant to a system that has a preexisting leak.
Technician B says that if a vehicle presented for service is known to have a refrigerant leak, all refrigerant must be removed from the system to prevent additional leakage. Who is correct?
A. Technician A
B. Technician B
C. Both Technician A and Technician B
D. Neither Technician A nor Technician B

13. Technician A says that the vacuum valve within the radiator cap allows coolant to escape the radiator when internal pressure exceeds the opening point of the valve.
Technician B says that the operation of the vacuum valve helps to maintain a consistent level of coolant in the radiator. Who is correct?
A. Technician A
B. Technician B
C. Both Technician A and Technician B
D. Neither Technician A nor Technician B

14. Technician A says that the engine cooling system thermostat can be located on the inlet side of the cooling system.
Technician B says that engine cooling system thermostat can be located on the outlet side of the cooling system. Who is correct?
A. Technician A
B. Technician B
C. Both Technician A and Technician B
D. Neither Technician A nor Technician B

15. Technician A says that the heater core temperature is a direct reflection of engine cooling system temperature.
Technician B says that a thermostat that is stuck in the open position will cause poor heater system output. Who is correct?
A. Technician A
B. Technician B
C. Both Technician A and Technician B
D. Neither Technician A nor Technician B

16. Technician A says that if a vehicle equipped with an electric cooling fan overheats only when sitting still, the problem is most likely related to a malfunctioning electric cooling fan.
Technician B says that the problem is most likely related to a thermostat that is not opening. Who is correct?
A. Technician A
B. Technician B
C. Both Technician A and Technician B
D. Neither Technician A nor Technician B

17. Technician A says that oily fog on the surface of the windshield is an indication of a leaking heater core.
Technician B says that wet passenger compartment carpet can be caused by a leaking windshield. Who is correct?
A. Technician A
B. Technician B
C. Both Technician A and Technician B
D. Neither Technician A nor Technician B

18. Technician A says that the thermostatic expansion valve provides a fixed restriction between the condenser and the evaporator.

Technician B says that the thermostatic expansion valve can be located at the condenser outlet. Who is correct?

A. Technician A
B. Technician B
C. Both Technician A and Technician B
D. Neither Technician A nor Technician B

19. Technician A says that evaporator temperature in most FOT systems is controlled by cycling the compressor on and off.
Technician B says that a restriction of the FOT inlet screen can lead to eventual compressor damage. Who is correct?
A. Technician A
B. Technician B
C. Both Technician A and Technician B
D. Neither Technician A nor Technician B

20. Technician A says that a clogged oil bleed hole in the accumulator can cause compressor damage.
Technician B says that a low refrigerant level can be a cause of compressor damage. Who is correct?
A. Technician A
B. Technician B
C. Both Technician A and Technician B
D. Neither Technician A nor Technician B

21. Technician A says that if a remote sensing bulb and capillary tube were to lose their charge, the expansion valve would remain in the fully opened position.
Technician B says that a mispositioned remote sensing bulb could cause the expansion valve to remain in the fully opened position. Who is correct?
A. Technician A
B. Technician B
C. Both Technician A and Technician B
D. Neither Technician A nor Technician B

22. Technician A says that only the exact amount of oil removed from the refrigeration system should be replaced.
Technician B says that when an unknown amount of oil has been lost from the system, 4 oz (118.30 ml) should be added. Who is correct?
A. Technician A
B. Technician B
C. Both Technician A and Technician B
D. Neither Technician A nor Technician B

23. Technician A says that equal high and low pressures can result from a faulty compressor.
Technician B says that equal high and low pressures can result from an open clutch coil. Who is correct?
A. Technician A
B. Technician B
C. Both Technician A and Technician B
D. Neither Technician A nor Technician B

24. Technician A says that evaporator inlet and outlet temperatures should be approximately the same.
Technician B says that a rapidly cycling compressor clutch is an indication of a refrigerant overcharge condition. Who is correct?
A. Technician A
B. Technician B
C. Both Technician A and Technician B
D. Neither Technician A nor Technician B

25. Technician A says that higher than normal suction and discharge pressures can be caused by an inoperative cooling fan.
Technician B says that higher than normal suction and discharge pressures can be caused by an external restriction in the condenser. Who is correct?
A. Technician A
B. Technician B
C. Both Technician A and Technician B
D. Neither Technician A nor Technician B

26. Technician A says that the refrigeration system must be fully charged before leak testing is performed.
Technician B says that the suspected leak area should be flushed with compressed air before leak testing is performed. Who is correct?
A. Technician A
B. Technician B
C. Both Technician A and Technician B
D. Neither Technician A nor Technician B

27. Technician A says that rapid fluctuation of the pressure needle may indicate that a compressor has a weak cylinder.
Technician B says that compressor noise can be caused by loose mounting brackets. Who is correct?
A. Technician A
B. Technician B
C. Both Technician A and Technician B
D. Neither Technician A nor Technician B

28. Technician A says that refrigeration system O-rings should be lubricated with the same type of lubricant that is used in the system.
Technician B says that O-rings should only be replaced when a leak is suspected. Who is correct?
A. Technician A
B. Technician B
C. Both Technician A and Technician B
D. Neither Technician A nor Technician B

29. Technician A says that insulating tape is used to isolate the capillary tube of a thermostatic expansion valve from the inlet tube.
Technician B says that failure to insulate a tube-mounted capillary tube will result in improper thermostatic expansion valve operation. Who is correct?
A. Technician A
B. Technician B
C. Both Technician A and Technician B
D. Neither Technician A nor Technician B

30. Technician A says that during a retrofit, if a compressor is found to have an external pressure relief valve, a high-pressure cutout switch must be installed in the compressor control circuit.
Technician B says that fittings installed during a retrofit procedure should be secured with a thread-locking compound. Who is correct?
    A. Technician A
    B. Technician B
    C. Both Technician A and Technician B
    D. Neither Technician A nor Technician B
31. Technician A says that automatic blower controls vary the voltage applied to the motor.
Technician B says that blower power modules adjust motor speed based on a variable signal from the HVAC controller. Who is correct?
    A. Technician A
    B. Technician B
    C. Both Technician A and Technician B
    D. Neither Technician A nor Technician B
32. Technician A says that dual zone climate control systems use two blend doors and two heater cores.
Technician B says that automatic temperature control systems will continue to adjust blend door position when blower control is in manual mode. Who is correct?
    A. Technician A
    B. Technician B
    C. Both Technician A and Technician B
    D. Neither Technician A nor Technician B
33. Technician A says that ground-side switching reduces the amount of current the fan switch must handle.
Technician B that says relays are used to reduce the amperage passing through the fan switch. Who is correct?
    A. Technician A
    B. Technician B
    C. Both Technician A and Technician B
    D. Neither Technician A nor Technician B
34. Technician A says that pressure cycling switches may be located in the compressor power circuit.
Technician B says that pressure cycling switches may be an input sensor to the HVAC controller. Who is correct?
    A. Technician A
    B. Technician B
    C. Both Technician A and Technician B
    D. Neither Technician A nor Technician B
35. Two technicians are discussing a vehicle with a failed cooling fan. Technician A says that the pressure cycling switch should turn off the compressor. Technician B says that the pressure sensor and HVAC controller will shut down the compressor. Who is correct?
    A. Technician A
    B. Technician B
    C. Both Technician A and Technician B
    D. Neither Technician A nor Technician B
36. Technician A says that all circuits include a circuit protection device.
Technician B says that voltage on a circuit will increase as resistance decreases. Who is correct?
    A. Technician A
    B. Technician B
    C. Both Technician A and Technician B
    D. Neither Technician A nor Technician B
37. Technician A says that the coldest A/C duct temperatures are usually produced when the air inlet door is set to recirculate.
Technician B says that certain conditions could produce colder air with the door in the fresh air position. Who is correct?
    A. Technician A
    B. Technician B
    C. Both Technician A and Technician B
    D. Neither Technician A nor Technician B
38. Technician A says that an improperly adjusted blend door may cause inadequate A/C cooling.
Technician B says that two-wire actuator systems can have door position capacity. Who is correct?
    A. Technician A
    B. Technician B
    C. Both Technician A and Technician B
    D. Neither Technician A nor Technician B
39. Technician A says that a temperature sensor reading of -40°F may indicate the sensor wires are shorted together.
Technician B says temperature sensor resistance increases with increased temperature. Who is correct?
    A. Technician A
    B. Technician B
    C. Both Technician A and Technician B
    D. Neither Technician A nor Technician B
40. Technician A says that modern air distribution systems do not require any periodic maintenance.
Technician B says shop air should not be used to clean evaporator cores. Who is correct?
    A. Technician A
    B. Technician B
    C. Both Technician A and Technician B
    D. Neither Technician A nor Technician B

Bilingual Glossary

Absolute zero The temperature at which no energy is present in a material.
Cero absoluto *Temperatura en la cual no hay energía en un material.*

AC Mode HVAC operating mode where air is drawn in from outside the vehicle, cooled, and discharged through the dash vents.
Modo de corriente alterna *Modo de operación HVAC dónde el aire se obtiene de afuera del vehículo, se enfría y se desecha por las salidas de aire en el tablero.*

Actuators A device that changes electrical signals provided by the computer into mechanical actions.
Actuadores *Dispositivo que cambia las señales eléctricas que proporciona la computadora en acciones mecánicas.*

Aerated A condition that exists when air becomes mixed with a liquid, creating small bubbles within the liquid.
Aireado *Condición que existe cuando el aire se combina con un líquido, creando así pequeñas burbujas en el líquido.*

Air Distribution Ductwork and associated controls that directs the air through the HVAC system.
Distribución del aire *Ductos? y controles correspondientes que dirigen el aire a través del sistema HVAC.*

Alloy A material that is manufactured from two or more materials that provides specific desirable results.
Aleación *Material que se fabrica de dos o más materiales que proporcionan resultados deseados específicos.*

ASE See National Institute for Automotive Service Excellence

ASE *Vea Instituto Nacional para la Excelencia en el Servicio Automovilístico*

Automatic Temperature Control (ATC) HVAC operating mode that self regulates its parameters to maintain a cabin temperature set by the occupants.
Control automático de la temperatura (CAT) *Modo de operación HVAC que autorregula sus parámetros para mantener la temperatura en cabina que seleccionan sus ocupantes.*

Axial Plate The axial plate is an offset concentric plate attached to the driveshaft at an angle. As the plate rotates it forces the pistons to and fro.
Placa separadora *La placa separadora es una placa concéntrica desviada que está sujeta en ángulo al eje motor. Al dar vuelta la placa, fuerza los pistones hacia enfrente y hacia atrás.*

Azeotrope A new compound that assumes chemical properties and characteristics that are different from either of the parent chemicals.
Mezcla de temperatura de ebullición constante *Un nuevo material para pulimentar que asume propiedades químicas y características que son diferentes de cualquiera de las sustancias químicas originales.*

Balance pressure A pressure equal to that found in the evaporator, which assists the TXV in making smooth valve adjustments.
Presión de balance *Presión que es igual a la que se encuentra en el evaporador, la cual ayuda al TXV a hacer los ajustes de la válvula más suaves.*

Ballast resistor A large capacity resistor that is able to absorb a large amount of voltage, reducing the amount

of current that is allowed to reach a cooling fan motor, effectively reducing its speed.
Resistencia de estabilización *Un resistor de gran capacidad que puede absorber una gran cantidad de corriente, al reducir la cantidad de corriente que es permitida para alcanzar un motor de abanico enfriador, y que reduce su velocidad efectivamente.*

Bidirectional controls A scan tool mode in which the technician can control various computer actuators.
Control bidireccional *Modo de una herramienta de análisis en la cual el técnico puede controlar varios actuadores de la computadora.*

Bilevel HVAC operating mode where air is drawn in from outside the vehicle, cooled, and discharged through the dash and floor vents.
Nivel doble *Modo de operación HVAC en donde el aire de obtiene de afuera del vehículo, se enfría y se desecha por conductos de ventilación en el tablero.*

Blend Door An air flow control device which regulates the volume of air flowing through the heater core.
Puerta de unión? *Dispositivo que controla el flujo de aire el cuál regula el volumen del aire que fluye a través del núcleo del calentón.*

Boiling point The temperature at which molecules begin to vaporize.
Punto de ebullición *Temperatura en la cuál se empiezan a evaporar las moléculas.*

Cabin Air Filter Device designed to remove contaminates from the passenger compartment air.
Filtro de aire de la cabina *Dispositivo diseñado para quitar los contaminantes del aire del compartimiento de los pasajeros.*

Capillary tube A copper tube filled with heat-sensitive gas used to help the TXV adjust.
Tubos capilares *Tubos de cobre que están llenos de un gas sensible al calor que se usa para ayudar a adaptar el MRT.*

Cavitation A condition in which the coolant pump is starved for coolant, which results in poor coolant pump efficiency and the introduction of air into the system.
Cavitación *Condición en la que la bomba enfriadora se queda sin enfriador, lo cuál resulta en una baja eficiencia de la bomba enfriadora y la introducción del aire al sistema.*

Clean Air Act The Clean Air Act put forth a set of polices and procedures that the United States would adopt to deal with the problem of ozone depletion. Most of the rules and regulations within the act were a direct result of the recommendations made by the Montreal Protocol. The CAA was signed into law on November 15, 1990.
Acta del aire limpio *El acta del aire limpio estableció un conjunto de políticas y procedimientos que adoptarían los Estados Unidos para enfrentar el problema del agotamiento del ozono. La mayoría de las reglas y regulaciones en el acta fueron el resultado directo de las recomendaciones que hizo el protocolo de Montreal. El AAL se legislo como ley el 15 de noviembre de 1990.*

Cohesion The mutual attraction that molecules have for one another.
Cohesión *Atracción mutual entre las moléculas.*

Combined A plenum assembly which houses all of the air handing components in one complete case.
Combinado *Cámara impelente ensamblada que guarda todos los componentes que tratan con el aire en un alojamiento completo.*

Comfort factor The effects that external factors have on the customer's perception of the operation of the air conditioning system.
Factor de comodidad *Efectos que tienen los factores externos en la percepción del cliente sobre la operación del sistema de aire acondicionado.*

Comebacks A comeback is a situation in which a customer has to make a return trip to the service facility either because the concern was not corrected or because another problem may have been created during correction of the original concern.
Recurrencia *La recurrencia es una situación en la que el cliente tiene que regresar al taller ya sea porque tiene la preocupación no se corrigió o porque pudo haberse creado otro problema durante la corrección del problema original.*

Compound Gauge A gauge that has the ability to measure either vacuum or pressure.
Manovacuómetro *Medidor que tiene la habilidad de medir tanto el vacío como la presión.*

Condenser A heat exchanger located at the front of the vehicle in front of the radiator that is used to transfer heat from the refrigerant to the atmosphere.
Condensor *Intercambiador de calor localizado al frente del vehículo enfrente del radiador y que se usa para transferir calor del refrigerante a la atmósfera.*

Control Head Dash mounted interface which allows the occupants to adjust the HVAC operation.
Cabeza de control *Interfase montada en el tablero que permite a los ocupantes que ajusten la operación del HVAC.*

Control point The point at which the variable displacement control valve activates or deactivates to make a change in displacement.
Punto de control *Punto en el que la válvula de control de cilindrada variable se activa o desactiva para hacer un cambio en la cilindrada.*

Crankshaft The part of a reciprocating compressor on which connecting rods and pistons are attached.
Cigüeñal *Parte de un compresor recíproco en el que están sujetos bielas y pistones.*

Defrost To remove condensed water vapor from the inside of the windshield.

Deshielar *Quitar el vapor de agua condensado dentro del parabrisas.*

Defrost Mode HVAC operating mode where air is drawn in from outside the vehicle, cooled and then heated, and discharged through the vent at the base of the windshield.
Modo de deshielar *Modo operador HVAC en donde el aire se obtiene de afuera del vehículo, se enfría y luego se calienta, y se desecha a través de una salida de aire en la base del parabrisas.*

Depressor A device found in an air conditioning service hose that is used to lift a service fitting Schrader valve from its seat.
Depresor *Dispositivo que se encuentra en la manguera de servicio del aire acondicionado que se usa para levantar de su asiento una válvula Schrader de empalme de servicio.*

Desiccant A chemical drying agent.
Desecante *Agente químico secante.*

Distilled water Water that has had minerals removed.
Agua destilada *Agua a la cuál se le han quitado los minerales.*

Environmental Protection Agency: The agency of the U.S. Federal Government that is charged with establishment and enforcement of environmental protection standards.
Agencia de Protección al Ambiente *Agencia del gobierno federal de los Estados Unidos que está encargada de establecer y reforzar los estándares de protección del ambiente.*

EPA See Environmental Protection Agency
EPA *Vea Agencia de Protección del Ambiente*

Evaporator A heat exchanger installed in a position in which it is exposed to the air within the vehicle's interior.
Evaporizador *Intercambiador de calor instalado en una posición en la que se expone al aire en el interior del vehículo.*

Externally balanced An external tube is used to supply balance pressure to the diaphragm.
Externamente balanceado *Tubo externo que se usa para proporcionar presión de balance al diafragma.*

Factory Fill The coolant or any other fluid that was installed in the vehicle when it was built. "SAE 10W30 motor oil was the factory fill in this vehicle."
Relleno de fábrica *Enfriador o cualquier otro líquido que se instaló en el vehículo cuando se fabricó. "El aceite de motor SAE 10W30 fue el relleno de fábrica de este vehículo".*

Ferrule A sleeve that is placed around a fitting that, when compressed, forms a permanent clamp around the fitting.
Guía flexible de descarga *Manga que se coloca alrededor del relleno que, cuando se le comprime, forma una brida de fijación alrededor del relleno.*

Fixed displacement The amount of refrigerant displaced and expelled into the air conditioning system during each revolution of the compressor shaft remains the same regardless of system operating conditions.
Desplazamiento fijo *La cantidad de refrigerante desplazado y echado al sistema de aire acondicionado durante cada revolución del eje del compresor permanece igual sin importar las condiciones de operación del sistema.*

Floor Area of the passenger compartment near the occupants feet.
Piso *Área del compartimiento del pasajero cerca de los pies de los ocupantes.*

Global warming The gradual warming of the earth's atmosphere, which can result in the gradual melting of ice caps and the alteration of weather patterns.
Recalentamiento del globo terráqueo *El recalentamiento gradual de la atmósfera de la tierra, el cual podría resultar en un descongelamiento gradual de las capas de hielo y la alteración de los patrones climatológicos.*

Hall Effect Switch A type of solid state switch that is sensitive to a magnetic field and is used to turn pull a line voltage to ground.
Interruptor de efecto Hall *Un tipo de interruptor de estado sólido que es sensible al campo magnético y que se usa para cambiar y jalar una línea de voltaje a tierra.*

Hard water Water that contains large amounts of minerals.
Agua dura *Agua que contiene grandes cantidades de minerales.*

Heat exchanger A component that is similar in construction to a radiator that facilitates the transfer of heat between ambient air and the chemical contained within the exchanger.
Intercambiador de calor *Componente similar en construcción al radiador y que facilita la transferencia de calor entre el aire ambiental y el químico que contiene el intercambiador.*

Heater Mode HVAC operating mode where air is drawn in from outside the vehicle, warmed, and discharged through the floor vents.
Modo del calentador *Modo de operación HVAC en donde el aire de obtiene de afuera del vehículo, se tibia, y se desecha a través de salidas de aire en el piso.*

HOAT Hybrid Organic Additive Technology; this refers to a specific additive package that is used in antifreeze.
TAOH *Tecnología de Aditivo Orgánico Híbrido; esto se refiere a un paquete de aditivo específico que se usa en el anticongelante.*

Hydrolysis A chemical reaction that takes place when two chemicals combine to form one or more other substances.
Hidrólisis *Una reacción química que sucede cuando dos químicos se combinan para formar una o más sustancias.*

Hygroscopic Chemicals that have a tendency to attract and absorb water.
Higroscópico *Químicos que tienen la tendencia de atraer y absorber el agua.*

IAT Inorganic Additive Technology; this refers to a specific additive package that is used in antifreeze.
TAI *Tecnología de Aditivo Inorgánico; esto se refiere a un paquete de aditivo específico que se usa en el anticongelante.*

Intermittent Problem A vehicle problem that occurs at random intervals for a random period of time and then returns to normal operation.
Problema intermitente *El problema en un vehículo que sucede en intervalos irregulares por un período de tiempo variable y luego regresa a su operación normal.*

Internally balanced Balance pressure is supplied through a passage drilled inside the expansion valve.
Balanceado internamente *Presión de balance que se recibe a través de un pasaje taladrado dentro de la válvula de expansión.*

Latent heat of vaporization The amount of energy that is stored in a molecule that is required to turn them to vapor.
Calor de vaporización *Cantidad de energía que se almacena en una molécula que se requiere para convertirlos en vapor.*

Manifold A tube or block that provides multiple common passages and component mounting areas that direct gases or liquids.
Colector *Tubo o bloque que proporciona múltiples pasajes comunes y áreas de montaje componentes que dirigen los gases o líquidos.*

Manual Air Distribution HVAC operating mode in which the occupants must regulate the operating parameters of the HVAC system.
Distribución manual de aire *Modo de operación HVAC en el que los ocupantes deben regular los parámetros de operación del sistema HVAC.*

Matter Matter is anything that occupies space and has mass.
Materia *La material es cualquier cosa que ocupa espacio y tiene masa.*

Max AC Mode HVAC operating mode where air is drawn in from the inside the vehicle, cooled, and discharged through the dash vents.
Modo máximo de corriente alterna *Modo de operación de HVAC en donde el aire se obtiene de afuera del vehículo, se enfría y se desecha a través de las salidas de aire del tablero.*

Metering The active process by which the expansion valve senses evaporator load conditions and adjusts refrigerant flow accordingly.
Medición *Proceso activo en el que las válvulas de expansión detectan las condiciones de carga del evaporador y ajustan el flujo del refrigerante según sea necesario.*

Metering Device The metering device controls the temperature of the evaporator by limiting the amount of refrigerant that is allowed to enter the evaporator.
Aparato de medición *El aparato de medición controla la temperatura del evaporador al limitar la cantidad de refrigerante que se permite entrar al evaporador.*

Miscibility The ability of a liquid to completely mix with another liquid.
Miscibilidad *Habilidad de un líquido para mezclarse completamente con otro líquido.*

Modulating The active process in which the expansion valve moves in a range from fully closed to fully open to regulate the amount of refrigerant allowed to enter the evaporator.
Modular *Proceso activo en el que la válvula de expansión se mueve en una gama de cierre completo a abertura completa para regular la cantidad de refrigerante que se permite entrar al evaporador.*

Montreal Protocol See Protocol on Substances that Deplete the Ozone Layer
Protocolo de Montreal *Vea el Protocolo sobre las sustancias que destruyen la capa de ozono.*

National Institute for Automotive Service Excellence A nonprofit organization that certifies automotive service professionals through voluntary testing.
Instituto Nacional para la Excelencia en el Servicio Automovilístico *Organización no lucrativa que certifica a los profesionales de servicio automovilístico a través de evaluaciones voluntarias.*

OAT Organic Additive Technology; this refers to a specific additive package that is used in antifreeze.
TAO *Tecnología de Aditivo Orgánico; esto se refiere a un paquete de aditivo específico que se usa en el anticongelante.*

Oil charge An oil charge is a 2-ounce (.060L) quantity of oil packaged in a sealed container pressurized with a 2-ounce (.057Kg) charge of refrigerant; the refrigerant is used as a propellant to force the oil into the refrigeration system.
Cambio de aceite *Un cambio de aceite es la cantidad de .06L (2 oz.) de aceite envasado en un contenedor hermético a presión con .057 Kg (2 oz.) de carga de refrigerante; el refrigerante se usa como un propulsor para forzar al aceite hacia el sistema de refrigeración.*

Panel Area of the dash which faces the occupants.
Panel *Área de un tablero frontal a los ocupantes.*

PCM Powertrain Control Module
MCM *Módulo de Control del Motor*

PH Level The pH level is a measurement of how acidic a material is. PH is measured on a scale of zero to 14 with zero indicating a pure acid solution while 14 indicates a pure alkaline solution.
Nivel de PH *El nivel de PH es una medida de la acidez de un material. El PH se mide en una escala de cero a 14, en*

donde el cero indica una solución de ácido puro, mientras que el 14 indica una solución alcalina pura.

Power dome The gas-filled control diaphragm found on H-valve style TXVs.
Lomo de potencia *Diafragma de control lleno de gas que se encuentra en la válvula H estilo MRT.*

Pressure to Temperature relationship The specific temperature that a chemical maintains when maintained at a given pressure or vice versa.
Relación de temperatura a presión *Temperatura específica que mantiene un químico cuando se conserva a una presión dada o viceversa.*

Protocol on Substances that Deplete the Ozone Layer Officially laid-out guidelines for the worldwide phase out of CFC materials as well as other ozone depleting materials. The protocol was ratified by 25 nations on September 16, 1987.
Protocolo sobre las sustancias que destruyen la capa de ozono *Reglamentos delineados oficialmente para la eliminación por etapas a nivel mundial de los materiales CFC (clorofluorocarbonos) como también otros materiales que destruyen el ozono. El protocolo lo ratificaron 25 naciones el 16 de septiembre de 1987.*

Pulse width modulation (PWM) Pulse width modulation is a method of controlling electronic devices by rapidly turning the voltage on and off.
Módulo de control del motor (MCM) *El módulo de control del motor es un método para controlar los dispositivos electrónicos al rápidamente encender y apagar el voltaje.*

Recirculate HVAC operating condition that draws air into the air distribution system from inside the passenger compartment.
Recirculación *Condición de operación HVAC en donde el aire se obtiene del sistema de distribución del aire de adentro del compartimiento de los pasajeros.*

Recovery A process in which A/C system refrigerant is captured and stored in a container.
Recuperación *Proceso en el que el refrigerante del sistema de corriente alterna se captura y se guarda en un contenedor.*

Recycle A process in which air, moisture, and other impurities are removed from the refrigerant.
Reciclar *Proceso en el que el aire, la humedad y otras impurezas se sacan del refrigerante.*

Reed valve A thin strip of steel that covers either a suction (inlet) port or a discharge (outlet) port and responds to cylinder suction or pressure.
Válvula de laminas *Una tira delgada de acero que cubre un puerto (entrada) de succión o un puerto (salida) de descarga y responde a la succión o presión del cilindro.*

Relative humidity Relative humidity is expressed as a percentage. The percentage represents how much moisture is in the air relative to how much moisture the air can hold.
Humedad relativa *La humedad relativa se expresa en un porcentaje. El porcentaje representa cuánta humedad hay en el aire relativo a cuánta humedad puede tener el aire.*

Remote sensing bulb A bulb attached to the end of a capillary tube used to measure evaporator temperature.
Bulbo termostático sensitivo *Bulbo sujeto al final de un tubo capilar que se usa para medir la temperatura del evaporador.*

Retrofit The process of converting an R-12 system to one using an alternative refrigerant.
Reajuste *Proceso de convertir un sistema R-12 a uno que use un refrigerante alternativo.*

Root Cause The primary factor that is causing the customer's concern.
Causa de raíz *Factor principal que causa la preocupación del cliente.*

Scotch Yoke A method of moving compressor pistons by turning a crankshaft driven eccentric within a common yoke in which the pistons are attached.
Horquilla de articulación Scotch *Método de mover los pistones del compresor al dar vueltas al cigüeñal manejado en excéntrico dentro de una horquilla de articulación común a la que están sujetos los pistones.*

Slugging A condition in which the compressor pumps excessive amounts of liquid.
Llegada de líquido *Condición en que el compresor bombea cantidades excesivas de un líquido.*

Solenoid A device that changes electrical energy into mechanical movement.
Solenoide *Dispositivo que cambia la energía eléctrica en movimiento mecánico.*

Staking A process in which the metal surrounding the bearing is disturbed creating a lip that holds the bearing in place.
Escalonamiento vertical *Proceso en el que el metal que rodea a un cojinete se perturba creando un labio que mantiene en su lugar al cojinete.*

Subcooling The number of degrees that a liquid is below its vaporization point.
Subenfriamiento *El número de grados que está un líquido debajo de su punto de evaporación.*

Split Case A plenum assembly that can be divided into smaller components, each of which contains different air handing devices.
Alojamiento hendido *Un ensamblado a presión que puede dividirse en componentes más pequeños, en el que cada uno contiene diferentes dispositivos de manejar el aire.*

Throttling The act of providing a pressure drop across the system between the high side and low side of the system by providing a restriction.
Estrangulación de gases *Acto de proporcionar una baja de presión a través del sistema entre el lado alto y el lado bajo del sistema al proporcionar una restricción.*

Throws Offset areas of the crankshaft in which the connecting rods are attached.
Alcances *Áreas desviadas del cigüeñal en donde están sujetas las bielas.*

Troposphere The lowest part of the atmosphere, up to about 7 miles (11 kilometers).
Troposfera *La parte más baja de la atmósfera, hasta aproximadamente 11 kilómetros (7 millas).*

TXV Thermostatic expansion valve
MRT *Manorreductor termostático*

Variable displacement Variable displacement compressors have the ability to vary the amount of refrigerant that is displaced and expelled into the system in response to system conditions.
Cilindrada variable *Los compresores de cilindrada variable tienen la habilidad de variar la cantidad de refrigerante que se desplaza y se expele dentro del sistema como respuesta a las condiciones del sistema.*

Vent Mode HVAC operating mode where air is drawn in from outside the vehicle and discharged through the dash vents without being heated or cooled.
Modo de ventila o salida de aire *El modo de operación HVAC en donde el aire se obtiene de afuera del vehículo y se desecha a través de las salidas de aire del tablero sin que se caliente o se enfríe.*

Water Jackets Passages within the cylinder block and cylinder head that are used to circulate coolant.
Cámaras de agua *Pasajes dentro del monobloque y la culata que se usan para circular el fluido refrigerador.*

Index